Petrochemical Machinery Insights

Petrochemical Machinery Insights

Heinz P. Bloch

Butterworth-Heinemann is an imprint of Elsevier
The Boulevard, Langford Lane, Kidlington, Oxford OX5 1GB, United Kingdom
50 Hampshire Street, 5th Floor, Cambridge, MA 02139, United States

Library of Congress Cataloging-in-Publication Data
A catalog record for this book is available from the Library of Congress

British Library Cataloguing-in-Publication Data
A catalogue record for this book is available from the British Library

ISBN: 978-0-12-809272-9

For information on all Butterworth-Heinemann publications
visit our website at https://www.elsevier.com/

Publisher: Joe Hayton
Acquisition Editor: Brian Guerin
Editorial Project Manager: Edward Payne
Production Project Manager: Punithavathy Govindaradjane
Cover Designer: Greg Harris

Typeset by SPi Global, India

Dedication

To the memory of Les Kane, who suggested
the *HP in Reliability* columns and started
to publish them in 1990.
As my editor, he valued integrity in both
engineering and publishing
above all else.

Contents

List of Figures

List of Figure Credits

AEGIS-SGR—15.1.1

AESSEAL, Inc—4.5.2; 4.6.2; 9.1.1; 9.1.2; 9.1.3; 9.1.4; 9.1.5; 14.1.1; 14.1.2; 14.1.3; 17.1.5; 18.7.1; 18.7.2; 21.1.1; 21.1.2; 21.1.3; 21.1.4; 21.1.5; 21.1.6; 21.1.7; 21.1.8; 21.2.1; 21.2.2; 21.2.3; 21.2.4; 21.3.1; 21.3.2; 21.3.3; 21.3.6; 21.3.7; 21.4.1; 21.4.2; 21.7.1; 21.7.1T; 23.1.2; 23.3.2; 23.3.3; 27.1.1; 30.2.4; 30.2.5; 32.6.1; 34.3.2; 35.3.1; 39.3.3; 42.8.1

Aerzen USA—37.1.2

Aronen/Boulden/Chemours—17.4.1; 17.4.1T; 17.4.2; 20.2.1; 20.2.2; 20.2.3

Autenrieth/Phillips—17.1.1; 17.1.2; 17.1.3; 17.1.4

Baldor Electric—7.1.2

Barringer/Taylor—30.1.1; 30.5.2

Bloch/Bloch—4.1.5; 4.4.3T; 16.2.1; 19.5.1; 26.6.1; 30.2.1; 32.5.1; 34.4.2; 39.1.1; 42.3.1T

Bloch/Budris—2.1.2; 2.1.5; 2.2.1; 2.2.2; 2.2.3; 2.2.4; 3.2.2; 4.5.3; 4.6.1; 8.1.3; 11.3.1; 11.3.1T; 11.3.2; 11.4.1; 11.4.2; 14.2.1; 16.1.1; 17.1.1T; 17.1.2T; 17.5.1; 17.5.2; 17.5.3; 17.5.4; 21.3.4; 21.3.5; 24.2.1; 25.2.1; 25.2.2; 30.8.1; 30.11.1; 30.11.2; 33.1.2T; 33.1.3T; 40.1.1; 42.3.1; 42.3.2; 45.2.1T; 45.2.2; 45.3.1T

Bloch/Geitner—9.1.1T; 30.3.1; 39.3.2; 39.4.1; 39.4.2

Bloch/Alkho—7.1.1; 33.2.2; 39.1.2

Bloch/Piotrowski—3.1.1

Bradshaw/Hawa—24.1.3

Colfax/Don Ehlert/LSC-Houston, TX—4.5.4; 4.5.6; 5.2.1; 5.2.2; 12.3.1; 23.1.1; 23.1.3; 23.1.4; 23.1.5; 23.1.6; 23.1.7; 23.1.8; 23.1.9; 23.1.10; 23.1.11; 23.1.12; 23.1.13; 23.2.1;23.2.2; 23.2.3; 23.3.1; 27.1.2

Dengyosha Machine Works, Japan—3.1.4; 5.1.1; 5.1.2; 5.1.3; 5.1.4; 30.1.2; 32.1.1; 32.2.1

Dresser-Rand—18.8.1; 45.6.1

Elliott Group, Jeannette, PA—43.1.1; 43.1.2

Emile Egger, Cressier, Switzerland—3.1.3; 30.9.1

FlexElement Texas, Houston, TX—8.1.2

Flexitallic, Deer Park, TX—13.1.1; 13.1.2

General Electric—6.4.1; 42.8.3

Hasbargen/Weigand—4.5.1T

Hydro Inc., Chicago, IL—1.1.1; 1.1.2; 3.1.2; 3.2.1; 4.2.5; 16.1.2; 16.1.3; 21.8.1; 22.1.1; 21.1.2; 22.1.3; 22.1.4; 22.1.5; 22.1.6; 22.1.7; 22.1.8; 22.1.9; 22.1.10; 22.1.11; 22.1.12; 22.1.13; 22.1.14; 22.2.1; 22.3.1; 22.3.2; 22.3.3; 22.3.4; 22.3.5; 22.3.6; 22.4.1; 22.4.2; 22.4.3; 22.4.4; 22.5.1; 22.5.2; 22.6.1; 30.2.2; 30.2.3; 30.5.1; 30.12.1; 30.12.2; 30.12.3; 33.1.1; 33.2.1

L.A. Turbine, Ventura, CA—43.1.3; 43.1.4

Ludeca, Miami, FL—2.1.3; 2.1.4; 25.1.1; 32.7.1; 34.1.1; 44.1.1

MAN Turbo, Germany—6.2.1; 6.2.2; 6.2.3; 10.1.1; 10.1.2; 46.1.1

Matthews, Bob—4.5.1; 4.5.5; 24.2.2; 34.4.1; 42.8.2

Monroe, Perry—12.1.1; 12.3.2; 34.3.1

Murray, Malcolm—2.1.1

MRC/SKF—4.3.1

Neumann & Esser—31.2.1; 31.2.2; 31.3.1; 31.4.1; 31.4.2; 31.5.1; 31.5.2; 31.5.3

NSK—4.4.1

R&W, Germany—8.1.1

Rienstra/UTE—45.5.1

SAFWAY Suspended Scaffolding—41.1.1; 41.1 2; 41.1.3

SKF AMERICA, Kulpsville, PA—4.1.1; 4.1.2; 4.1.3; 4.1.4; 4.1.1T; 4.2.1; 4.2.2; 4.2.3; 4.2.4; 4.3.2; 4.4.2; 4.5.9; 24.1.1; 24.1.2; 29.1.1; 30.5.4; 45.1.1; 45.2.1

TLV, Japan—39.5.1; 39.5.2; 39.5.3

Trico Corporation, Pewaukee, WI—4.5.7; 17.5.5; 17.8.1; 24.2.1

Torrington Co.—17.6.1

Turbocam, Barrington, NH—5.1.5; 16.3.1; 16.3.2; 16.3.3; 35.2.1

UNKNOWN/UNTRACEABLE—4.5.8; 6.4.2; 25.3.1; 30.4.1; 30.5.3; 30.8.2; 37.1.1

United States Government Chemical Safety Board (CSB)—21.6.1T

Westinghouse Electric Company—39.3.1

Worthington Pump/Igor Karrassik—28.1.1; 28.1.2; 30.5.5

Foreword

As a reliability professional who knows the needs of the industry, I tackled this writing task with the primary intent of updating machinery reliability-related topics. Suffice it to say that more than ever, modern industry is affected by equipment reliability. But it would be a serious mistake to assume that a particular facility or corporate entity can achieve its reliability goals without the support and cooperation of many interacting job functions. Moreover, equipment reliability is greatly influenced by the support and continuity given by management. I therefore included subject categories on organization, management, and training. Equipment reliability is also affected by the implementation skills of the people in the trenches, so to speak, and by the perceptions of everyone between them and higher management. Every one of us fits in somewhere and every one of us influences asset reliability. By way of an automobile analogy, the driver and maintenance technician and design engineer carry equal weight. If one of them slacks off, reliability becomes illusory. We need to apply the same logic in our plants. We must accept that our respective responsibilities overlap, that everybody matters and fulfills a role.

For accessibility, the material had to be neatly indexed and cross-referenced without undue complexity. Its relevance had to be reconfirmed by engineers and marketers who—like the author—were no longer compelled to submit their findings for approval to higher authorities whose agenda would inevitably conflict with that of the pure engineer. Such approval is often withheld in order to discourage publishing anything that could even remotely be viewed as a competitive advantage, or viewed as an endorsement, or a critically important fact. But ignorance is as dangerous as information overload; I therefore tried to come down somewhere in the middle between silence and overload.

Conveying just the right amount of useful information to the intended target audience is one of the great challenges in reliability engineering. Failure analysis, remedial steps, or desirable substitute approaches must be explained, perhaps more than just once. Good work processes encompass sound explanations. Even if impeded by an often unproductive up-and-down chain-of-command approach, good work processes must be adopted to be successful. Unproductive up-and-down wavering is occasionally found in stifling Pareto-confirming work environments. In places where Pareto's law prevails, 80% of the people are doing all the talking and 20% are doing all the work. I waited until the depressing possibility of being trapped in such a merry-go-round was no longer a concern of mine. I can now be critical whenever criticism is appropriate. When appropriate, I can submit constructive criticism and layout and explain which reliability strategies and concepts have stood the test of time. I have found these strategies embraced by well-managed best-of-class companies; the strategies are interwoven and their implementation is shared among many job functions. In compiling this book, I could finally take whatever time was needed to give readers pertinent facts. My hope was to explain these facts in a manner that gets to the point quickly and accurately, nonjudgmental—but uncompromisingly truthful—nevertheless.

Preface

This text was compiled because of many requests to update, revise, and reissue some of the hundreds of *HP in Reliability* columns I had written in the years from 1990 to 2016. Because my mechanical engineering career actually began with graduating from the New Jersey Institute of Technology (Newark, New Jersey) in 1962, I was at first inclined to call this book "50 Years of Machinery Notes." Then again, the mere thought (or even the remote possibility) of such a title scaring away readers made me rethink. Nobody would want to sift through 50 years of notes. So, I decided to give it the title "Petrochem Machinery Insights," which sounds better and, hopefully, not too pompous. The intent and scope of the book is easily explained: You should not have to read volumes of textbooks to find the one important nugget of information that you have tried to locate, or that's really the only item of interest to you.

My target readership includes operating technicians, maintenance professionals, reliability engineers, and midlevel and senior managers. Some readers will be working in refinery and process plant shops; others could be field mechanics, millwrights, project engineers, midlevel managers, and project executives. All of them are people whose actions influence asset reliability. For many years, I have seen merit in giving these widely divergent job functions easy access to relevant material. I was always pleased when subject matter experts (SMEs) agreed with me, and many SMEs, whether active or retired, are probably aware of the significance of this material. Yet, while this group of peers and colleagues are equally knowledgeable, they may not have had similar opportunities to transfer their experience into written words. Time, motivation, and opportunity have to come together as we publish relevant highlights of our work-related insights.

For my part, and also because of the encouragement passed on by many readers, I assumed that they (and you) would benefit from a no-nonsense text. And so, the text comprises many updates of the material I had written from the early 1990s until possibly only 3 or 4 months before this book went into print. I also assumed that readers wanted me to select and craft and condense the pages of this book while leaving out consultant-conceived generalities. They've probably heard enough nonoffending, no value-adding generalities. Chances are that their equipment life expectancies, or the profitability of their facilities, fell short of reasonable projections. Could failure to meet expectations be rooted in issues that are not popular to pursue? Many SMEs agree with our contention that basing one's actions on vague generalities will be ineffective, wasteful, or even dangerous. Could lack of success be rooted in a mere anecdote being passed down and being applied stripped of its original context? Could it be because, in the past, action was often initiated by listening to opinions instead of facts? Is it possible that the core material, the monthly Reliability Columns previously published in a premier engineering journal, was either not being read or had been forgotten by the intended audience? It became clear to me that this writing project had to be tackled by re-reviewing the many folders and files containing background material to my monthly *HP in Reliability* columns. These and another potential pool of source material would be some of the hundreds of articles I had compiled and issued in the years since 1989. In that year, I had finally obtained a word processor with adequate storage capacity to file away thousands of pages. Wading through these accumulated pages and the associated stack of personal material, I tried to be mindful of its relevance for an audience made up of many job functions, skills, age groups, backgrounds, and talents. I was also thinking of the task of future translators who surely would not want to waste time struggling with, or deciphering, fuzzy or hidden messages.

Here, then, is the final product. I hope that you will find it useful.

Acknowledgments

I consider myself fortunate to have met, over the past six decades, hundreds of companies and individuals who cheerfully responded to my requests for literature. Whenever necessary, I obtained their permission to embed excerpts from their prior writings or illustrative materials for use in my columns, books, tutorials, conference papers, and articles. As I later either restored or resurrected the contact, I have tried to communicate with many of their successor companies. My attempts were not always successful. At times, there was no response to my request. That's easy to understand: People are busy and times have changed. Information sharing is not always encouraged. Still, many companies and individuals in the United States (some in Germany, Saudi Arabia, Switzerland, and Japan) responded favorably and even enthusiastically to my requests—I am greatly indebted to them. They are acknowledged in the image credits you will find elsewhere in the front matter of this book.

Introduction

This is not just another asset reliability text. The book represents an unusual collection of relevant material. It's a guideline text that explains how asset management, to be useful, must be further separated into application and implementation details. It's a deliberately crafted synopsis or anthology of reliability improvement matters. The reader might consider it a book of essays on reliability themes. These themes are both stand-alone and interwoven; either singly or collectively, they give insight into certain maintenance details and reliability concepts. For a certainty, these concepts and details reach far beyond the endless reams of consultant-conceived generalities that one finds in some publications—publications that often try very hard to please their advertisers.

While it's not my intention to say that true reliability professionals must shun or disregard consultants' input, it is also true that great benefits are traceable to reliable homegrown SMEs with the ability to spell out and pursue actionable detail. As this text goes to print, we can observe training needs that often go unacknowledged. The unintended consequence is lingering issues with budgets for grassroots and existing plants. For lack of training, we accept cost estimates that cover only the outlay for the cheapest, potentially maintenance-intensive equipment. As these cost estimates then become rigidly frozen allocations and appropriations, all hoped-for future reliability achievements are doomed. By mistakenly accepting the notion that assets can be improved once the plant is running, the uninformed decision-makers frequently lock asset owners into a cycle of unprofitability and downtime risk. I liken this pathway to the foolishness of attempting to turn a low-cost two-seater sports coupe into a safe and solid school bus. No amount of wishful thinking will let you do that in a profitable manner.

A simple example of an actionable detail alludes to lubricant delivery flaws in thousands of process pumps. Process pumps in the (primarily) 10–200 kW size range usually incorporate rolling element bearings, and many of these bearings fail long before they have reached their published design lives. Planning and carrying out machinery quality assessment (MQA) is important in this regard; MQA should provide the answer. We review component design details early in the project and visualize probable installation and maintenance practices that will likely prevail once the equipment is delivered or after the plant is commissioned. During MQA, we should identify and find risky elements or issues that usually culminate in repeat failures. So why do so many bearings fail early? Why do some process pumps fail repeatedly or randomly? We should care about the matter.

User questions on why pump bearing lubrication providers do not offer answers need to be explored. By being observant and applying the right failure analysis approach, we may learn more about why the industry often struggles with elusive failure causes. Answers can be found if we work with manufacturers that have, and will share, application engineering know-how. This text will provide ample details on how we can benefit from working with them.

Another example of an actionable detail is found on the pages in this text that deal with little-known pump types or a number of pages on well-represented process pumps that deserve to be upgraded. Again, the details on how and why certain parts should be upgraded must be discussed and are brought to the readers' attention in this text. Without overarching candor, discussion, and explanation, my narrative would risk being classified as simply more consultant-conceived generalities. I did not want this to be the case.

The segment on root cause failure analysis makes a point made often before. It alerts readers to the fact that thousands of process machines are plagued with repeat failures. I follow up by again bringing a bit of simple commonsense logic to their attention; there are only two possible explanations why a facility experiences repeat failures: reason (1) might be that the troubleshooters or failure analysts involved have not found the real cause of the failure and reason (2), the root cause of failure is in fact known, but nobody cares to address it. Long-held biases and unspecified agendas get in the way. That said, I would like readers to consider a sobering definition: a mistake repeated more than once is called a decision. There are only two types of decisions: right ones and wrong ones.

In essence, this book tries to teach how mistakes can be remedied or avoided. As you grasp the underlying principles and come to appreciate the near-impossibility of fully separating the multitude of interwoven reliability topics, you will understand how and why so very many topics cross over or overlap into different subject categories. That understanding will get you well on your way to making good decisions, the right decisions. Summing up: throughout the time it now took to collect, condense, and update what I had published in the years since my first article appeared in a Processing Journal in 1974, I tried to synthesize facts and the principles that I observed. I related and commented on what I learned from, and during, my long and rewarding employment by a multinational petrochemical corporation. In this book and whenever possible, I expressed what I had absorbed while teaching over 500 workshops on six continents. And rather late in the game, I thought of a very famous wise man's proverb: "the king finds pleasure in a servant who acts with insight." Well, this book should assist the reader in doing just that—acting with insight.

Subject Category 1

Agitators

Chapter 1.1

Your Agitator Just Failed: What Now?

INTRODUCTION

As the reader will see later, this text repeatedly uses the terms "best of class" and "best of competition" (BoC) in our discussions of procedures, approaches, management routines, and training issues. Neither the term "best of class" nor "best of competition" is to imply that only one company deserves the attribute. We may consider one particular plant as the leader in vibration and condition monitoring of the major machinery systems; another plant might tower above the rest in terms of mechanical seal selection and so forth. Here, we share how a process plant in the Upper New York State excelled in systematically and successfully tackling their equipment distress events.

CONSISTENCY IS ONE OF THE KEYS TO PERFORMANCE

This 1600-employee chemical manufacturing site was blessed with a mid-fortyish reliability engineering manager who knew his craft like few others. The key to his own and also to his employees' strong performance was simple: together, they subscribed to the belief that tangible and lasting reliability improvements are achievable only if management and wage earners view every maintenance or downtime occurrence as an opportunity to find the true root cause. In essence, each event triggers efforts to upgrade the equipment, taking time for thoughtful examination (Fig. 1.1.1), rather than just fixing it and risking that the same failure repeats itself. If upgrading was feasible, could it be cost-justified? If it was cost-justified, who was in charge of getting the upgrade done? Are we laying the groundwork to avoid future repeat failure issues? Are we taking the right measurements (Fig. 1.1.2) and are we filing them in the right retrieval system?

Petrochemical Machinery Insights. http://dx.doi.org/10.1016/B978-0-12-809272-9.00001-3

FIG. 1.1.1 Whenever equipment fails, our reliability focus leads to taking steps toward future failure avoidance. Enlisting competent outside firms goes a long way toward being prepared. *(Courtesy Hydro Inc., Chicago, IL, www.hydroinc.com.)*

FIG. 1.1.2 Measuring a preexisting spare part and asking if it could be upgraded. Recording the dimensions in at least two different computerized data storage systems makes sense. *(Courtesy Hydro Inc., Chicago, IL, www.hydroinc.com.)*

Whenever equipment distress was reported at this Upstate New York facility, the reliability manager or designated staff convened a meeting of key individuals: the clout, the chairperson, the data source, the client, the recorder, and the knowledgeable worker are either present or, at least, represented. These functional responsibilities were delineated and defined whenever an occasion arose.

Their meeting room was equipped with a flip chart and a recorder stepped up to this chart to write down the task headline, in this case *create standard checklist for agitator system major component failure—job plan.* Every meeting participant contributed. Up went the listing—essentially a mere memory jogger for the purpose of keeping everyone focused on the task:

1. Create failure assessment—what gave way and what seems to be wrong.
2. Prepare failure analysis—failure mode observation, how, why, sequence, and root cause (only seven possibilities exist: design, assembly or installation, fabrication or manufacturing, material defect, maintenance deficiency, off-design or unintended service conditions, and improper operation).
3. Determine production history—run time and product properties (viscosity, etc.).
4. Review equipment history (computerized event log).
5. Review/copy all applicable system and component prints.
6. Obtain all current manufacturers' operating and maintenance manuals.
7. Identify and contact inplant/outside experts, if necessary (manufacturer/contractor/other specialists (Figs. 1.1.1 and 1.1.2)).
8. Develop job time line—identify production window.
9. Check parts availability—swap options and discrete components.
10. Establish and maintain production communication interface.
11. Consult with safety personnel—arrange for the presence of safety coach or have them address all concerns (lockout, tagging, rigging, etc.).
12. Comprehensively check other related components and/or systems to avoid compound/recurring/supplemental problems (lube system, filter, relief valves, coolers, controls, hydraulic systems, pumps, filters, and motors). Create fluid analysis (water, dirt intrusion passageways, and sources).
13. Thoroughly inspect all critical components including gearboxes, seals, bearings, wiring insulation, relief valve settings, and critical instruments.
14. Identify manpower availability/requirements.
15. Assign tasks and accountabilities.
16. Check all interlocks and production control sequences.
17. Review all applicable design specifications.
18. Create start-up checklist, including acceptance criteria.
19. Key up necessary preventive maintenance routines not yet in system.
20. Identify future similar job assignment responsibilities.
21. Conduct post-job debriefing.
22. Create formal procedure and checklist for similar jobs *now*.

And keep the momentum going; capitalize *now* on lessons learned.

The entire approach used by this company was and still is worthy of note. Failure assessment and analysis replaced the traditional reaction of scheduling a work crew with instructions to go for the customary quick fix. A deliberate thinking process was being pursued by this company. Production and equipment histories were scrutinized for clues as to why a repeat failure might have occurred. The equipment manufacturer's manuals were not disregarded. Before reinventing or guessing, they determined if someone else had applicable expertise. Spare parts and safety considerations were being aired. Potential contributing influences were listed. Inspection assignments were made and chain of approval and accountability clarified.

In any well-managed facility, data collection is part of the job. Data retrieval is simple, because pertinent data are stored and analyzed with the help of a dedicated moderate-cost software package that makes equipment surveillance and analysis orders of magnitude simpler and faster than would be achieved by work order-driven maintenance management computer systems. A good equipment surveillance and analysis program are essentially a cost and product control stewardship system designed for operating machinery populations. Various forms of such a system are in use worldwide and have been responsible for dramatic savings in equipment maintenance costs. The computerized segment of this system is a tool designed to facilitate the collection and interpretation of data (Fig. 1.1.2). It identifies the burden of some repeat failures, monitors progress, encourages proactive team participation and worklist compilation, tracks costs and savings along the way, simplifies the reporting function, and provides a compact, centralized file system.

Benchmarking starts with knowing where you are and how your failures stack up against those experienced by the competition. This implies that you are serious, accurate, and consistent with your failure analysis, data logging, and data retrieval efforts. There has to be a structured, repeatable approach to all this; half-hearted or occasional efforts will not allow a company to be among the pacesetters or best-of-class performers. Remember this approach when you receive a call from your manager.

Subject Category 2

Alignment

Chapter 2.1

Alignment Choices Have Consequences

INTRODUCTION

Modern laser alignment methods take far less time and are more precise than the old-style methods they replace. Common sense should point toward exclusively using precise, modern methods. Inadequate alignment still causes major calamities (Fig. 2.1.1), whereas the results of sound alignment approaches typically show up as improved mean time between repairs (MTBRs) and a reduction in maintenance outlay.

FIG. 2.1.1 Fluid machine failure that started with misalignment, high vibration, and bearing distress. *(Courtesy Murray & Garig Tool Works, Baytown, TX—in remembrance of my colleague Malcolm Murray.)*

Petrochemical Machinery Insights. http://dx.doi.org/10.1016/B978-0-12-809272-9.00002-5

ALIGNMENT RULE OF THUMB EXPLAINED

A likely consensus among reasonable people holds that precision alignment typically lowers vibration to one-half of the value of "conventional" alignment. Fig. 2.1.2 represents an estimate of bearing operating life extension due to reduced vibration severity for typical process pumps.

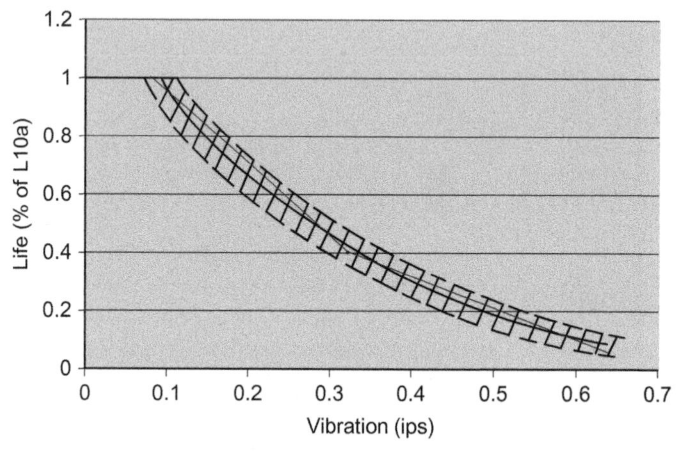

FIG. 2.1.2 Bearing housing vibration velocity vs. bearing life for process pumps [1].

FIG. 2.1.3 Major machinery shaft alignment in progress. *(Courtesy Ludeca, Inc., Miami, FL; www.ludeca.com.)*

FIG. 2.1.4 A medium-sized machine being aligned. *(Courtesy Ludeca, Inc., Miami, FL; www.ludeca.com.)*

Of course, coupling type and shaft speed affect these ratings. Reliability-focused users have at their disposal alignment tools that keep shaft misalignment at tangents of 0.001 and lower. Miami-based Ludeca Inc. has marketed suitable instruments since the 1970s, and highly accurate laser optic alignment tools used in Figs. 2.1.3 and 2.1.4 are now widespread among competent users in every conceivable industry.

Industry feedback indicates that precision alignment *alone* would result in a pump MTBR multiplier of somewhere between 1.4 and 1.7. The general observation is that best-of-class performers inevitably implement additional upgrades and will seldom confine their reliability focus on better alignment alone. That is why Bloch and Budris [1] puts the pump MTBR of a very marginal performer at 1.6 years, while best performers often get 9 or more years between process pump failures.

Finally, Fig. 2.1.5 gives an indication of how a major bearing manufacturer rates the effects of misaligned bearings. Most styles of rolling element bearings fit somewhere between the two rule-of-thumb-derived (empirical) curve boundaries. At tangents below 0.001, bearing life is thought to exceed a relative rating of 1 and can reach as high as 1.5 [2]. However, severe misalignment tangents can rapidly lower the relative rating to a mere fraction of 1.0.

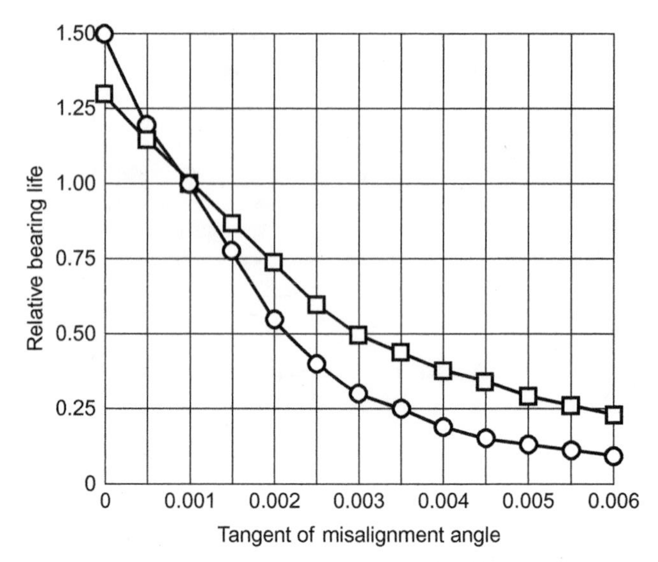

FIG. 2.1.5 How tangent of misalignment angle affects bearing life.

REFERENCES

[1] H.P. Bloch, A. Budris, Pump User's Handbook: Life Extension, fourth ed., Fairmont Press, Lilburn, GA, 2013 ISBN: 0-88173-720-9.

[2] R.L. Leibensperger, Look beyond catalog ratings, Mach. Des. April 3 (1975).

Chapter 2.2

Pump Alignment Reduces Power Demand

INTRODUCTION

Having an awareness of energy efficiency is one of the minimum job qualifications for reliability engineers. In the summer of 1994, Jack Lambley, then an intern at Imperial Chemical Industries (ICI) Rocksavage site in the United Kingdom, was assigned the task of quantifying the effect on power consumption of misaligned process pumps. A surplus pump was overhauled and new bearings were fitted; the pump was then reinstalled and water was used in a suitably instrumented closed-loop arrangement. A groundbreaking German manufacturer loaned Jack Lambley a modern laser optic alignment instrument, and Lambley went to work.

LAMBLEY'S WORK SUMMARIZED

As an undergraduate student, Jack Lambley had learned how misalignment affected bearing load and how bearing load increases caused exponential decreases in bearing life. His supervisor, Steve Moore, had asked Lambley to read the engineering sections of SKF's general catalog, which stated that a 25% increase in bearing load caused rated bearing life to be cut in half.

Jack Lambley investigated alignment accuracy and methods in use at that time. He found that straightedge methods were inappropriate for refinery pumps. Rim and face alignment methods were judged difficult and generally unreliable. Properly executed, reverse dial indicator methods required consideration of bracket sag and would take more time than modern laser techniques. Still, from the data available at Rocksavage, he calculated that typical misalignment consisted of 0.02 in./0.5 mm vertical and horizontal offset and 0.002 in./in. vertical and horizontal angularity. In 1994, lasers were already known to be inherently more accurate (Lambley believed 10 times more accurate) than the best competing techniques.

Lambley constructed several graphs and tabulations; they are reproduced here, duly acknowledging ICI. The resulting recommendations were to align machinery to within 0.005 in./0.12 mm shaft offsets and to limit deviations in hub gap to 0.0005 in./in. of hub diameter. Lambley further documented that adhering to these recommendations would reduce ICI's power consumption by about 1%. He confirmed that laser alignment was fast and superbly accurate. He determined that laser alignment technology was bottom-line cost-effective; he deserves credit for establishing facts instead of repeating the opinions of others. He summarized his findings in Figs. 2.2.1 and 2.2.2, giving the effects of parallel offset (Fig. 2.2.1) and angular misalignment (Fig. 2.2.2) on a pin coupling. The power consumption of pin and toroidal couplings was at 3000 rpm. In Figs. 2.2.3 and 2.2.4, we see the effects of parallel (Fig. 2.2.3) and angular (Fig. 2.2.4) offset on a toroidal or tire-type coupling, again operating at 3000 rpm.

As we then use data from a midsize refinery, power savings alone should make us take note of the following:

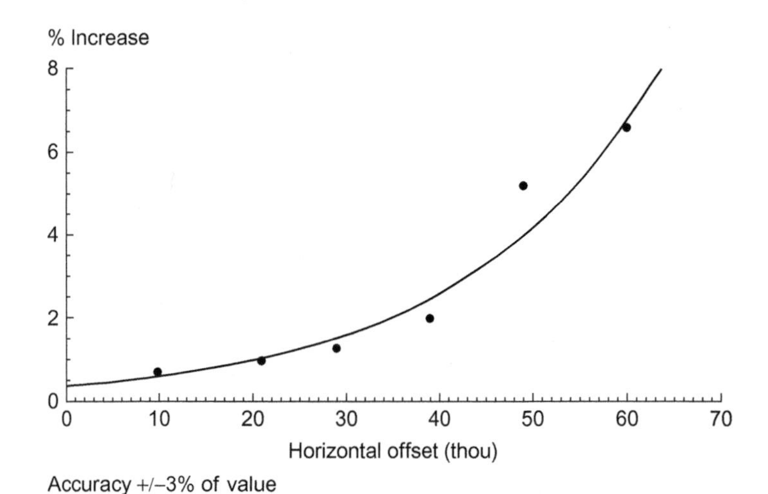

FIG. 2.2.1 Graph 1: effect of parallel offset on power consumption of a pin coupling at 3000 rpm.

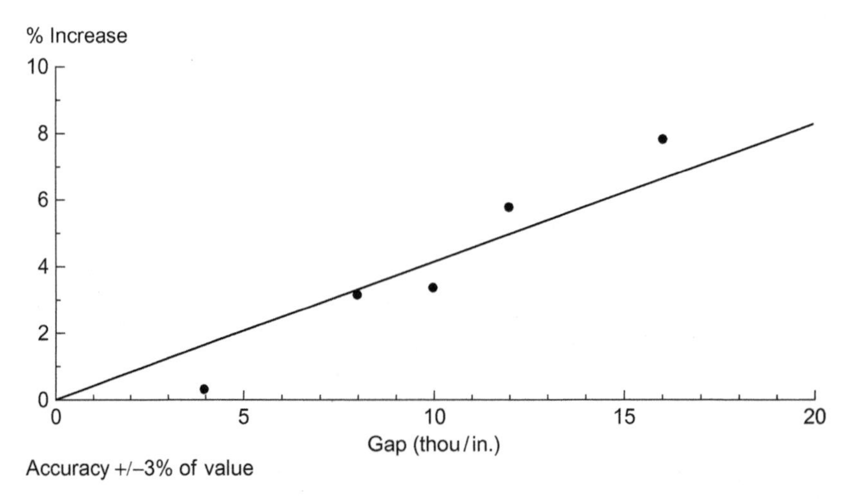

FIG. 2.2.2 Graph 2: effect of angular misalignment on power consumption of a pin coupling at 3000 rpm.

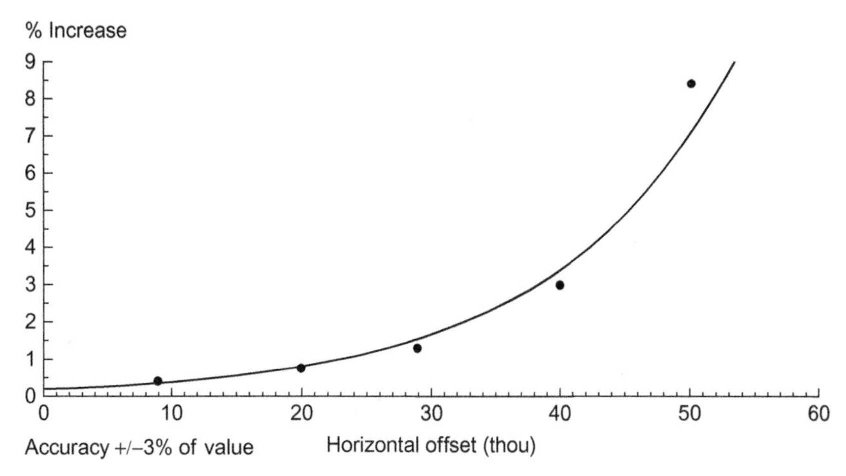

FIG. 2.2.3 Graph 3: effect of parallel offset on power consumption of a toroidal (tire-type) coupling at 3000 rpm.

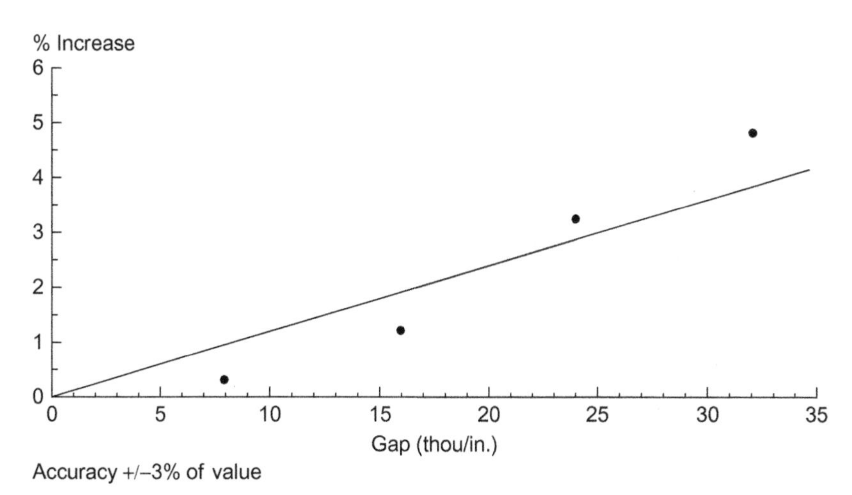

FIG. 2.2.4 Graph 4: effect of angular misalignment on power consumption of a toroidal (tire-type) coupling at 3000 rpm.

Average demand: 27 kW/pump × 8760 h/year × $0.1/kWh × 1000 pumps × 0.01 = $236,520/year. And, with 1000 pumps operating at any given time, this location would save close to $.25 M/year in avoided power consumption.

We leave it to the reader to obtain current cost of laser alignment instruments and to add the cost of training. Take into account seven man-hours of time-saving credit per alignment job. For even more data, read about thermography and infrared monitoring techniques. These have been used to quantify the significant temperature increase in a coupling located between misaligned pump and driver shafts. You might then compare the energy wasted by increasing the temperature of a coupling with the energy loss described by Jack Lambley. You will find that, regardless of calculation method, laser alignment is for the reliability-focused users and will result in surprisingly rapid payback. Last but not least, in all your reliability improvement endeavors, never let somebody's opinion get in the way of sound science and facts. Developing numerical answers is usually not difficult.

Notice how "alignment" spills over into our next subject category, "Asset Management." And who would have thought there is a connection....

Subject Category 3

Asset Management

Chapter 3.1

Alignment Specifics Contribute to Effective Asset Management

INTRODUCTION

Over the past 20 years, a large number of surveys dealt with rankings of maintenance practices and the effectiveness of different reliability organizations [1]. Industry specialists and other observers and practitioners also attempted to rank technology trends, but their projections and predictions were greatly disrupted by economic trends in the 2008–2012 timeframe. Few people anticipated these adjusted trends. The rankings of control-related trends had included industry moves toward wireless technology, enterprise asset management (EAM), enhanced predictive maintenance, stronger networking, smarter alarm management, and more effective security.

In machinery reliability improvement, a number of rankings prior to 2008 anticipated accelerated conversions to oil mist lubrication, synthetic lubricants, better energy utilization (power efficiency), and, finally, smarter programs to more accurately identify spare parts needs. Experience showed that an upward trend slope can unexpectedly shift to become a downward trend slope. Sustained upward trends require sponsorship by motivated individuals who can ascertain and also obtain whatever financial backing is needed.

As seasoned reliability professionals, we should welcome trends that support increased asset reliability. Early on, we should convey to higher management our conclusion that making good use of these developments in the future calls for lots of preparation and action today. Specific training, gaining knowledge, and being focused and diligent are needed *now* (in 2016) to translate any of these existing or future trends into safe and profitable uptime at industrial facilities.

We are fully aware that the preceding three paragraphs differ little from what we have so often criticized as "consultant-conceived generalities." Such

Petrochemical Machinery Insights. http://dx.doi.org/10.1016/B978-0-12-809272-9.00003-7

generalities add no value unless one knows and then acts on relevant specifics. Technology implementation requires knowledge-based action on seemingly minor items. Getting the "big picture" and embracing the concept of asset management (AM) are commendable, but much more is needed for sustainable success. So, please, read on.

AN EQUIPMENT ALIGNMENT PROBLEM POINTS TO THE NEED TO UPGRADE

In 2014, a machinery engineer on temporary assignment overseas struggled with equipment alignment issues. While shaft alignment seemed like a very "low-technology" issue at best, he realized that it can have demonstrable impact on availability and reliability. With that in mind, the engineer alluded to Ref. [2] and wrote

Suppose you have very precisely aligned the shafts of pump and driver; nevertheless, shims placed under the equipment feet in order to achieve this precise alignment caused the shaft system to slant 0.005″ or 0.010″ per foot of shaft length. As a consequence, the brass or bronze oil ring (slinger ring) will now exhibit a strong tendency to run "downhill." While bumping into other pump components thousands of times per day, the oil ring gradually degrades and sheds numerous tiny specks of the alloy material. These specks of metal cause progressive oil deterioration and, ultimately, bearing distress.

He then relayed more information and asked a few questions:

We are currently installing a couple of hundred motor-driven pumps on steel modules. The modules are being built in a Pacific Rim country and will be shipped to another continent when completed. I have a concern now, after reading your book. The pumps are all installed on skids from different manufacturers. After setting the flatness of the skids' machined surfaces to the requirements spelled out in an applicable standard (API-686, 0.25 mm/m), we come back the next day or week and find the surfaces out-of-tolerance due to the sun's orientation and changes in ambient temperature. The equipment stays in a common plane, but not within the guidelines of API-686.

Do you have any recommendations? Can you shed more light on the expected equipment or component life reduction if the pump orientation is out-of-flatness?

Are pumps used on ships different from API-compliant pumps?

We replied that we might not have all the answers, but shaft misalignment reduces the expected trouble-free operating time (Fig. 3.1.1). Pumps on shipboard are often grease-lubricated and the regreasing frequencies on well-managed ships tend to be better than those practiced on land. Regardless of where installed, bearing life—in oil-lubricated pumps equipped with loose oil rings dipping in the oil sump—will be influenced by oil replacement (preventive maintenance) frequency. Oil cleanliness, degree of immersion in the oil, variation of oil viscosity

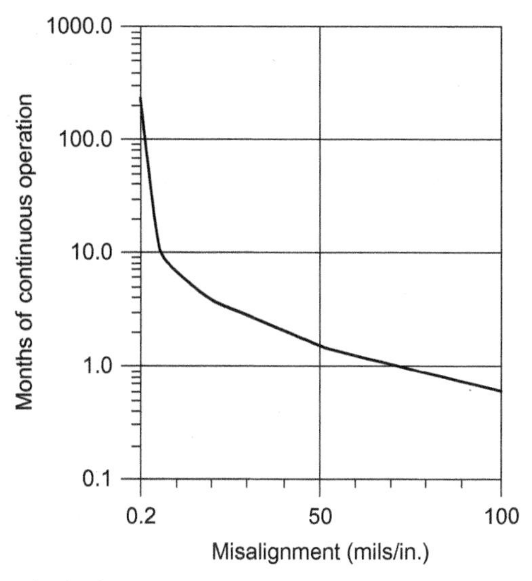

FIG. 3.1.1 High shaft misalignment versus expected trouble-free operating time [3].

from an ideal as-designed value, bore roughness (RMS) of the oil rings, and the degree of horizontality of the shaft system are factors that influence bearing life.

The out-of-roundness of loose oil rings is very important, and some researchers have asked for concentricity within 0.002 in./0.05 mm [4]. In 2009, at a facility in South Texas, we measured malfunctioning oil rings that were more than 0.06 in./1.5 mm out-of-round. For loose oil rings to remain within the asked-for concentricity is difficult. Therefore, flinger disks clamped to the shaft (see index) are much preferred over loose oil rings. But if, for some unexplained reason, the reader must use loose oil rings, these should be manufactured with stress relief annealing, which is just one of the several required manufacturing steps. Cheap oil rings very often omit stress relief from their various manufacturing steps. They are disallowed by reliability-focused purchasers. Working with a reputable pump rebuilder (their competence usually shows up in impeller design and fabrication know-how; Fig. 3.1.2) will go a long way toward reducing the risks associated with "lowest bidder."

For companies and individuals with reliability focus, working with competent OEMs and non-OEMs usually involves much more than simple preventive or restorative maintenance. In many cases, there are opportunities to remove or upgrade the weak links in a component chain. Reverse engineering an existing part and pre-engineering a superior or more energy-efficient part are often possible. We can do this by being proactive and by making competent vendors as our technology resource. Our Figs. 3.1.2 and 3.1.3 make the case and Subject Category 22 (later) gives additional input.

FIG. 3.1.2 Select a competent pump rebuilder to design and manufacture upgraded impellers for fluid machines. *(Courtesy Hydro Inc., Chicago, IL, www.hydroinc.com.)*

FIG. 3.1.3 Large API-compliant skid-mounted pump set. *(Courtesy Emile Egger, Cressier/NE, Switzerland.)*

Coming back to our point, reliability professionals must advise their owners, employers, project managers, and bosses on these seemingly small matters. In essence, cheap equipment will require more maintenance, and reliability professionals must bring these facts to light. If it is too late for our overseas reader to insist on flinger disks, he/she could now lay much groundwork for upgrading via suitable retrofits in the future. Upgrading is part of the dictionary definition of AM. So, whenever the first one of the reader's many skid-mounted pumps fails and is taken to the shop, the needed flinger disk adaptations (see Ref. [2]) would be retrofitted. A designated responsible implementer would be involved

FIG. 3.1.4 Multistage horizontal pump and driver "train" on the manufacturer's test stand. *(Courtesy Dengyosha Machine Works, Tokyo, Japan.)*

in this follow-up. Carrying out such upgrades is rarely considered optional by reliability-focused user plants. For them, upgrading is mandatory because only the reliability-focused users will run plants safely and profitably. Using oil rings and expecting highest possible equipment reliability are contradictions.

Attempts to live with contradictions (such as demanding lowest price and expecting highest reliability) will ultimately cost more than implementing best-available technology at the inception of a project. Best-technology API-style process pumps (Figs. 3.1.3 and 3.1.4) are typically mounted on a base common to pump and driver. Once provided with suitable instrumentation (if sold as a standard "package"), the pump set is shipped as a skid, ready to be aligned at the installation site and secured to the foundation. Responsibility for the entire skid should be assigned to the pump manufacturer.

The message: Effective AM deals with specifics. Not knowing these specifics and thus not implementing suitable upgrades will be very painful. The need for having a substantial amount of expertise will be ever more evident and is the common thread of this book and its various subject categories. The need for training is glaringly evident throughout.

Of course, experience demonstrates that "average" facilities will also run but will get locked in a never-ending cycle of repeat failures or random repairs. From the start, these facilities will be repair focused and notably less profitable than their reliability-focused competitors. Ideally, an owner-operator should engage only design contractors who know, specify, and insist on obtaining lowest failure risk components—right down to parts such as flinger disks in locations where "average" users will accept loose oil rings. If the design contractor does not have these insights, the timely and consistent application of machinery quality assessment ("MQA"; see index) becomes more important than ever.

REFERENCES

[1] H.P. Bloch, F.K. Geitner, Maximizing Machinery Uptime, Elsevier Publishing, Oxford, UK and Waltham, MA, 2006. ISBN 13: 978-0-7506-7725-7.

[2] H.P. Bloch, Pump Wisdom, in: H.P. Bloch (Ed.), Pump Wisdom: Problem Solving for Operators and Specialists, John Wiley & Sons, Hoboken, NJ, 2011.

[3] J. Piotrowski, Shaft Alignment Handbook, third ed., Marcel Dekker, New York, NY, 2006.

[4] D.F. Wilcock, E.R. Booser, Bearing Design and Application, McGraw-Hill Publishing Company, New York, NY, 1957.

Chapter 3.2

Dealing With Asset Management and Life Extension

INTRODUCTION

The business manager for the Asset Management Solutions ("AMS") firm in the Middle East faced a big task. He had been asked to implement a major project for an oil and gas producer and had questions on remaining rotating equipment life at existing client sites. The manager's charge was to analyze large oil and gas plants that had been operating for far more than 25 years, although they were originally designed for 20–25 years only.

The aim of his AMS firm was to secure the future of the client's assets for another 20 years. The AMS manager now had to determine what the client needed to do to stay in business for another 20 years without undue risk of production loss and without jeopardizing the high level of safety they had achieved for their human and physical assets.

FIG. 3.2.1 NDT and measuring being carried out as part of combining maintenance with upgrading. *(Courtesy Hydro Inc., Chicago, IL., www.hydroinc.com.)*

CHARTER AND PLANT DATA EXPLAINED

Before meeting with the manager and his staff, we had to set the stage for a productive week of meetings. Once we agreed that his firm's charter was to quantify the remaining life of the client's turbines, compressors, pumps, and other equipment, the deliverables for a reliability consultant had to be delineated. The consultant defined his work effort to defining key parameters and spelling out what the formula or approach would be for calculating the remaining life of each rotating machine. We believe the key ingredient of any useful endeavor to determine remaining life of machinery is hidden in the client plant's own past failure history. Where such history exists and where the root causes of failures have been analyzed, authoritative answers on remaining life are possible. Conversely, where these data are lacking, applicable data from others would have to be substituted.

On stationary equipment and piping, corrosion data should be available from coupons or from nondestructive testing (NDT) readings; Fig. 3.2.1 shows NDT and measuring being used by a highly qualified pump rebuilder (competent pump rebuild shop, or CPRS). If no such test data are available from a particular facility, the AMS firm was advised to examine third party surveys. The firm was encouraged to look at corrosion rates experienced at comparable industries and under comparable or scalable conditions. This would be an effort that takes time and money; there was no way we could pull these numbers out of a hat.

Because our specialty is rotating machinery, we wanted to look first at process pumps. In the hydrocarbon processing industry (HPI), these simple machines (a simple pump is shown later; see Chapter 4.4, Fig. 4.4.1) suffer many thousands of unexplained repeat failures every year. We outlined to the AMS how, on pumps, the audited plant's own failure history, and past repair data must be reviewed first. To the maximum extent possible, plant data and pump configurations must be compared to upgrade measures taken by successful "best-of-class" organizations. Advanced lube application strategies are used by best-of-class plants. Lubrication is one of the strategies that must enter into the comparison, as will the extension of oil replacement intervals made possible by better lubricants and superior bearing housing protection measures (see index words "bearing protector seals"). Mechanical seal life must be assessed and compared to best available sealing technology. This requires liaison with the most competent mechanical seal suppliers. It requires the seal supplier's active cooperation and divulging of what some claim (albeit without real good justification) to represent proprietary information. For instance, the extent to which superior dual-sealing technology is of value will have to be determined on a service-by-service or even pump-by-pump basis.

In like fashion, the extent to which superior bearings (e.g., ceramic hybrids available from the most experienced bearing manufacturers) would lengthen pump life or avoid bearing failures will have to be determined on a pump-by-pump basis. Lubricant application and standby bearing preservation (see Subject Category 17) are especially important in humid coastal and tropical environments, as well as in desert climates. Oil mist lubrication (Subject Category 23) extends the life of general-purpose machinery, and AMS should consider it.

PIPING AND FOUNDATIONS AFFECT REMAINING LIFE

Then there is the issue of piping for all types of machinery. Just as residential sidewalks and the walls of houses move and settle, pipe supports and equipment foundations will settle. The effect of such settling on pipe connections and equipment nozzles can be visualized and must certainly be taken into account. Examining the grout support under base plates will be quite revealing.

On steam turbines, blade stresses and water quality at the client site must be compared with those in successful long-running installations elsewhere [1]. This is a time-consuming endeavor that requires an investigator's time; mere guessing will not suffice. Of course, if comparable experience exists elsewhere, the investigative effort may take less time.

On geared units, remaining gear life must be examined by calculating tooth loading (stresses on tooth face) and from temperature measurements. In all instances, synthetic lubes from experienced oil formulators will greatly extend gear life. The right oil additives are needed for life extensions. They drive maintenance cost and affect gear life; oil cleanliness ranks next on the investigator's priority list.

Certain warehouse spares (gears, electric motors, etc.) should be upgraded if important. Upgrading spares is likely to speed up equipment recommissioning after an unanticipated future shutdown.

COMPRESSORS: ALL OF THE ABOVE ARE IMPORTANT

For compressors, one looks at all of the above. Valve technology and piston velocity are important comparison-worthy parameters on reciprocating compressors; onstream performance tracking and observation of prior sealing experience are important for centrifugal and axial compressors. This performance tracking and a review of the client's present sealing technology determine seal system upgrade potential. Even the compressor internal seal materials must be examined in detail (Fig. 3.2.2) and judgments made as to their failure potential. Couplings and the work procedures associated with attaching couplings to shafts should not be overlooked and neither should shaft alignment quality and philosophy. They all tell a lot about remaining equipment life and failure risk.

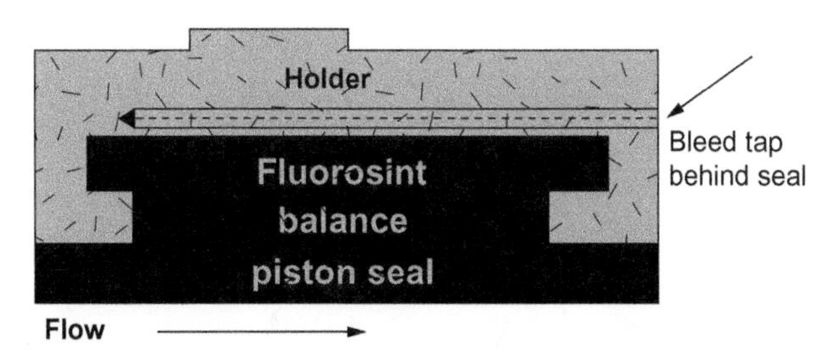

FIG. 3.2.2 Cross section of an internal compressor sealing component [2].

Whether a facility ultimately receives guidance from an established expert or whether an AMS puts its trust in someone else with similar experience is of no consequence, so long as the expert working for AMS

(a) authoritatively spells out recommended measures,
(b) thoroughly explains the recommended upgrade steps,
(c) identifies recommended vendors that should do the upgrading, and
(d) defines the deliverables that should be contractually agreed on between the upgrade provider and client.

REFERENCES

[1] H.P. Bloch, M.P. Singh, Steam Turbines: Design, Applications and Re-Rating, second ed., McGraw-Hill, New York, NY, ISBN: 978-0-07-150821-6, 2009.

[2] S. Quance, Using plastic seals to improve compressor performance. Turbomach. Int. (1997) January/February.

Subject Category 4

Bearings

Chapter 4.1

Picking the Right Oil Viscosity for Your Machines

INTRODUCTION

From our early experience with automobiles, most of us recall that we selected thicker oils (such as SAE 30) in the warm summer months and switched to thinner oils, perhaps SAE 10, in preparation for winter driving. Table 4.1.1 illustrates where these motor oils fit in as we compare them with the industrial oil designations in use today.

Thick oils are more viscous and may not readily flow into the bearings. We can heat up the oil or avoid oil rings and other risk-inducing lube application methods by using smarter and more reliable means of assured lubrication. We could use a jet of oil (oil spray) or could convey the oil, mixed with compressed air, in the form of an oil fog—also called oil mist. Whatever we do, we must guard against using the thinnest oil found on the market because we are concerned about potentially inadequate oil film strength and oil film thickness. But we will also not benefit from excessively thick oils, as we will see.

The MRC (Marlin Rockwell Corporation) "Engineer's Handbook" [1] gives general guidance as it states: "In general, the oil viscosity should be about 100 SUS [100 Saybolt Universal Seconds equates to approximately ISO viscosity grade (ISO VG) 32] at the operating temperature." If for some reason we had a bearing operating at 210°F, Table 4.1.1 would apparently call for a lubricant with an ISO VG somewhere between 220 and 320, but that would be unrealistically thick for most process pump bearings. Oil rings designed to function in ISO VG 32 would definitely slow down and malfunction in such excessively viscous oils. Oil overheating may be an additional concern.

With that in mind, we might wish to consult literature issued by bearing manufacturers with years of process pump experience. First and foremost,

Petrochemical Machinery Insights. http://dx.doi.org/10.1016/B978-0-12-809272-9.00004-9

TABLE 4.1.1 Comparing Typical Oil Viscosity Grades and Popular Designations [2]

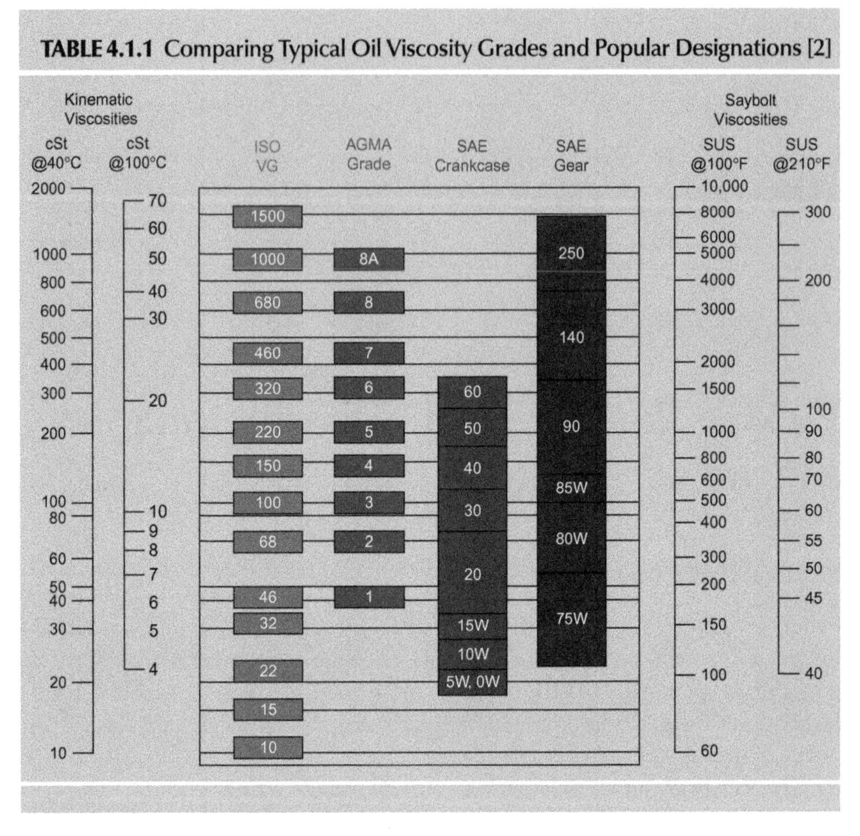

among these is SKF; their graph (Fig. 4.1.1) is simple, time-tested, and widely applicable [2,3]. Fig. 4.1.1 depicts the required minimum (rated) viscosity "ν_1" as a function of bearing dimension and shaft speed [2]. A bearing with a mean diameter of 390 mm at a shaft speed of 500 rpm would require $\nu_1 = 13.2$ cSt.

Here's one more example: if we had a bearing mounted on a 70 mm shaft rotating at 3600 rpm, we might start by assuming that the bearing's outside diameter is twice its inside diameter (ID) or 140 mm. The bearing's mean diameter would then be 105 mm. To simplify, we might call it 100 mm and (in Fig. 4.1.1) travel up from 100 to a location midway between the 3000 and 5000 rpm lines. In this instance, one could operate with a lubricant that, at the bearing operating temperature, is somewhere between 8 and 9 cSt.

Note, however, that we would have to know the operating temperature of a bearing to determine what ISO VG we needed to use here. The operating temperature derives its combined thermal input from bearing load and from lube oil frictional drag. Unnecessarily viscous oils will become hot. Fig. 4.1.2 is

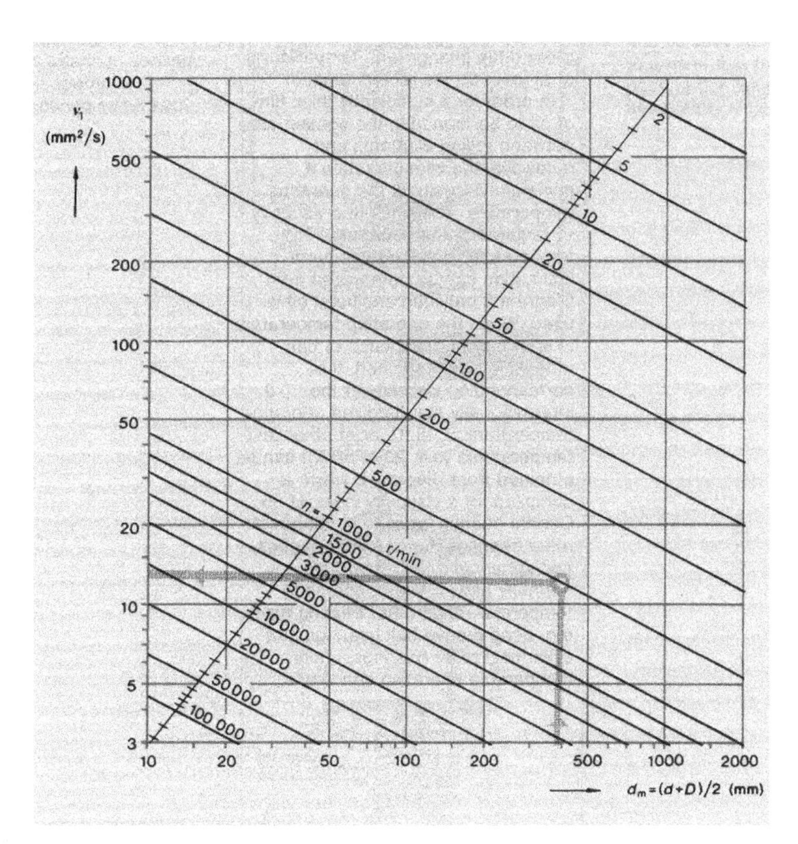

FIG. 4.1.1 Required minimum (rated) viscosity "ν_1" as a function of bearing dimension and shaft speed [2]. A bearing with a mean diameter of 390 mm at a shaft speed of 500 rpm will require $\nu_1 = 13.2$ cSt.

helpful in this regard, but note that Figs. 4.1.1 and 4.1.2 were drawn years ago and apply to mineral oils. If we today use premium-grade synthetic oils, we will enjoy a sizable safety factor in our lube applications. Again, working with the application groups of a reputable bearing manufacturer may simplify our life and greatly improve equipment reliability. The payback we get from making a solid bearing manufacturer our technology resource can be remarkable and is often expressed in weeks.

We would enter the vertical scale in Fig. 4.1.2 near 9 cSt and move toward the right, where we now intersect with oils ranging from very low to very high viscosity. Suppose we chose ISO VG 32; we might start the pump and verify that its oil temperature had leveled off at no >75°C. Alternatively, we might choose ISO VG 68 and verify that its operating temperature does not exceed 100°C (212°F).

Using lubricants with viscosities in excess of those actually needed may generate excess heat and actually work against us. However, thicker oils

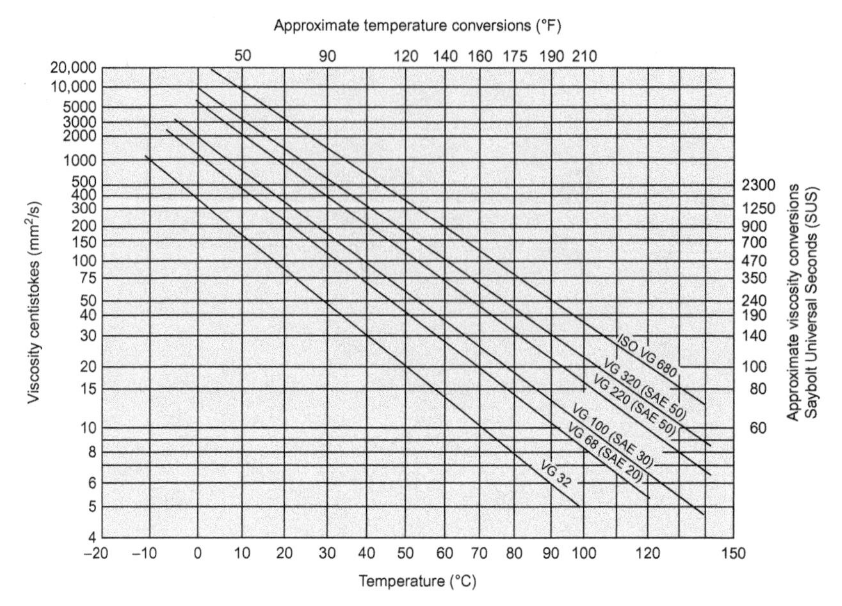

FIG. 4.1.2 For a required viscosity (*vertical scale*), the permissible bearing operating temperatures (*horizontal scale*) increase as thicker oils are chosen (*diagonal lines*). *(Courtesy SKF Corporation.)*

quite obviously have their place, and we must remember that MRC (now part of SKF's global brand) with their 100 SUS rule of thumb had to cover all the bases. That said, a large bearing (200 dm) in a slow speed gearbox (200 rpm) requires an operating viscosity of 40 cSt. From Fig. 4.1.2, we could easily see that ISO VG 100 (or higher) oils would be needed here. However, it's important to remember that highly specific oil application methods may be required for thick oils. Oil rings are very rarely (if ever) designed for thicker than ISO VG 32 lubricants.

A more recent publication by SKF (Fig. 4.1.3) allows estimation of actual oil viscosity (ν) at a range of operating temperatures with a typical viscosity index (VI) of 95 and a mineral oil. The required viscosity is estimated from Fig. 4.1.4; actual (ν) and required viscosities (ν_1) must be reconciled.

But there are many considerations when thick oils are applied. Again and to reemphasize, whenever oil rings are involved, verify the design intent. Oil ring weight and dimensions often assume operation in considerably lighter oils, say, ISO VG 32. As we then attempt to use such rings in thick oils, they often tend to stick, slip, and undergo a pendulum-like motion. Because oil ring-related topics cross several subject categories, additional entries will be found in the index.

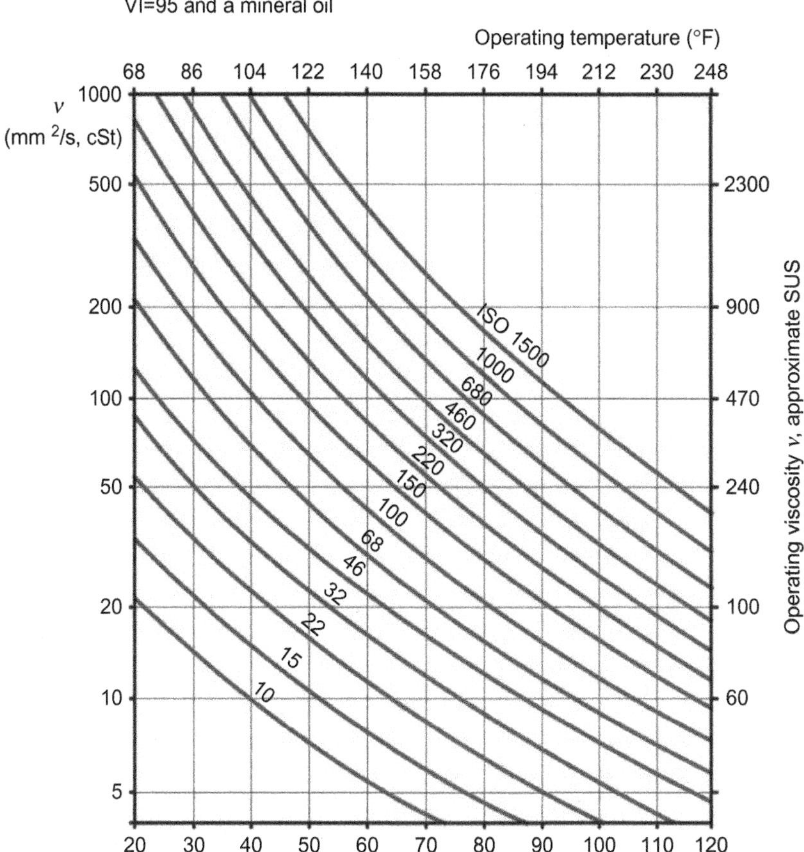

FIG. 4.1.3 Estimation of oil viscosity (ν) at a range of operating temperatures with a typical VI of 95 and a mineral oil.

LEARNING FROM A RECENT CASE HISTORY

In a 2015 case history, ISO VG 100 was applied to a large pump where ISO VG 68 mineral oil or its equivalent (synthetic) ISO VG 32 would have sufficed. (An ISO VG 32 synthetic is the "bearing life equivalent" of an ISO VG 68 mineral oil. The synthetic ISO VG 32 makes bearings run considerably cooler than the mineral oil equivalent.)

With ISO VG 100 mineral oil, the oil-misted radial bearing ran a few degrees lower in temperature than it had with conventional sump and oil ring lube. The user was pleased but expressed disappointment at a triple-row thrust bearing running as hot as before, 190°F. It ran hot because the oil was too thick or not enough oil was applied to carry away the heat.

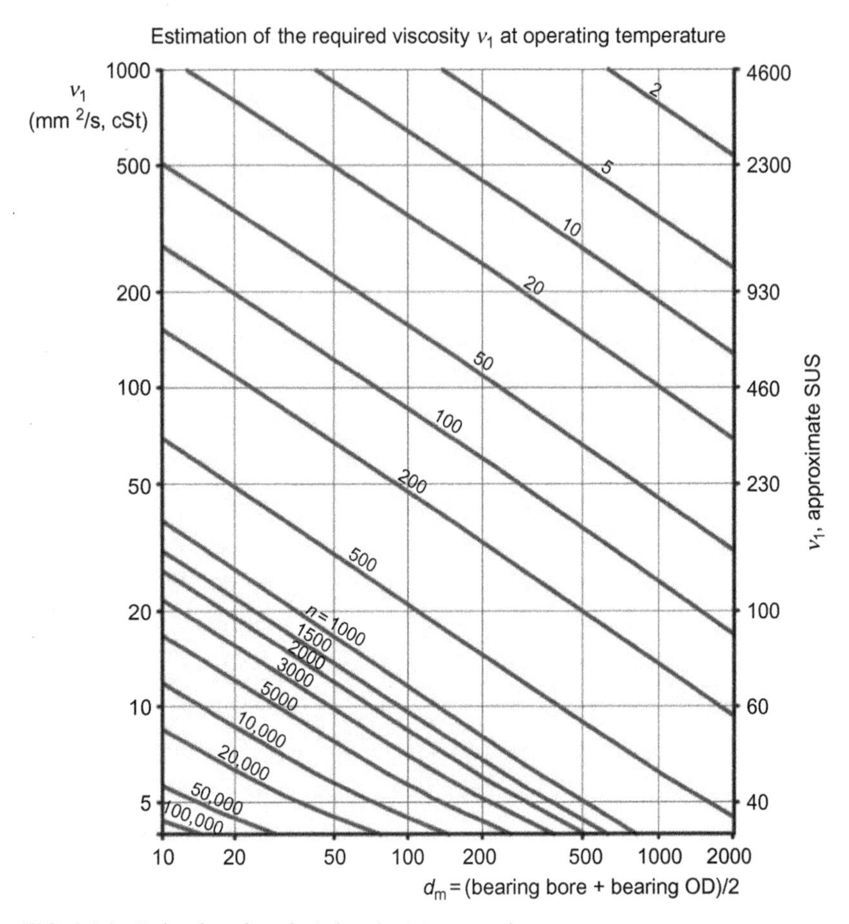

FIG. 4.1.4 Estimation of required viscosity (ν_1) at operating temperature.

To make the long story short, a premium formulation synthetic ISO VG 32 would have been sufficient and would have given the user everything a solid reliability professional could have asked for. Reliability pros would like to see pump bearing housings with no oil rings, no need for constant level lubricators, and few if any repeat failures. They start with the right lubricant and realize that neither too low nor excessively high oil viscosities are conducive to long bearing life (Fig. 4.1.5).

Why, with all that, is the radial bearing nice and cool? After all, it too is presently surrounded by the thick ISO VG 100? It's cool because it's got no load on it [3,4]. The load is in the triple-row thrust bearing, and that creates temperature, in addition to the frictional temperature we mentioned earlier in this chapter. Next, we should perhaps examine what temperatures are reasonable,

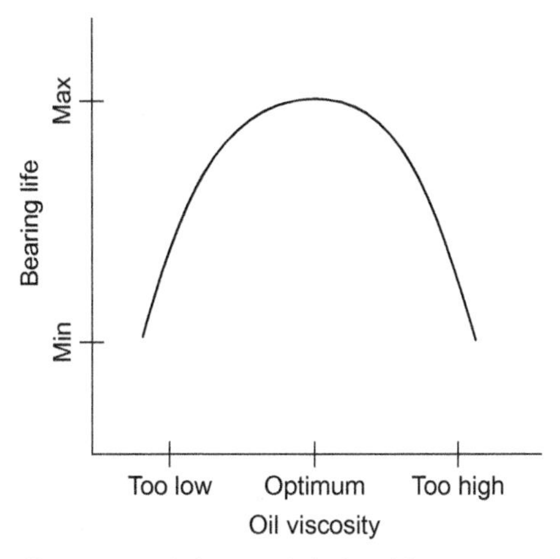

FIG. 4.1.5 Bearing Life Trends and influence of oil viscosities (per author's empirical data).

are high, and are out of allowable range for rolling element bearing housings and pump bearing housings. The findings may surprise us.

REFERENCES

[1] MRC, Engineer's Handbook, General catalog 60, TRW, Jamestown, NY, 1982. Copyright TRW.

[2] SKF America, General Catalog, SKF America, Kulpsville, PA, 2000.

[3] H.P. Bloch, Pump Wisdom: Problem Solving for Operators and Specialists, John Wiley and Sons, Hoboken, NJ, 2011.

[4] H.P. Bloch, Improving Machinery Reliability, third ed., Gulf Publishing, Houston, TX, 1998, ISBN 0-88415-661-3.

Chapter 4.2

Bearing Styles and Configurations

INTRODUCTION

Questions are often raised about rolling element bearings, and the reader is directed to the various references in articles grouped under this subject category. Recall, however, that many hundreds of books and thousands of articles and conference papers deal with this subject. When exploring bearing issues, try to link up with a bearing manufacturer that has an application group and gives you access to these experienced individuals. Very few bearing manufacturers meet these two criteria. This text makes a number of important points and relates experiences by highlighting pump and electric motor bearings. Recall, however, that the lessons and observations very often apply to other machines as well. When it comes to bearings, work with one of the leaders in application engineering. Cultivate access to this leader; shun those who are only selling bearings but cannot provide application know-how. Learn from their know-how; you will almost certainly have greater equipment availability and far fewer unscheduled downtime events.

BEARING TYPES

There exists a profusion of different bearing types, and the left-to-right images in Fig. 4.2.1 show fewer than one percent of available types and configurations at best.

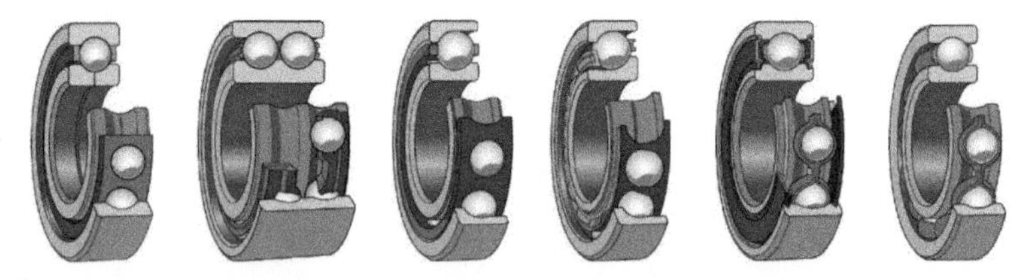

FIG. 4.2.1 Bearing styles, from left to right: split inner ring bearing, double-row bearing, light-duty unidirectional angular contact thrust bearing, heavy-duty unidirectional angular contact thrust bearing, double-shielded radial bearing, standard radial bearing.

Process pump bearings are of considerable interest to the petrochemical industry. These pumps generally incorporate a radial bearing and a thrust bearing. A radial bearing is usually located near the mechanical seal, whereas the thrust bearing is fitted close to the drive end. Fluid pressures acting on impellers cause a net axial force on the shaft system; we call it axial thrust. In operation, thrust acts in one direction that would make the two "tandem-mounted" bearings in Fig. 4.2.2A suitable for thrust acting from right to left. If they are

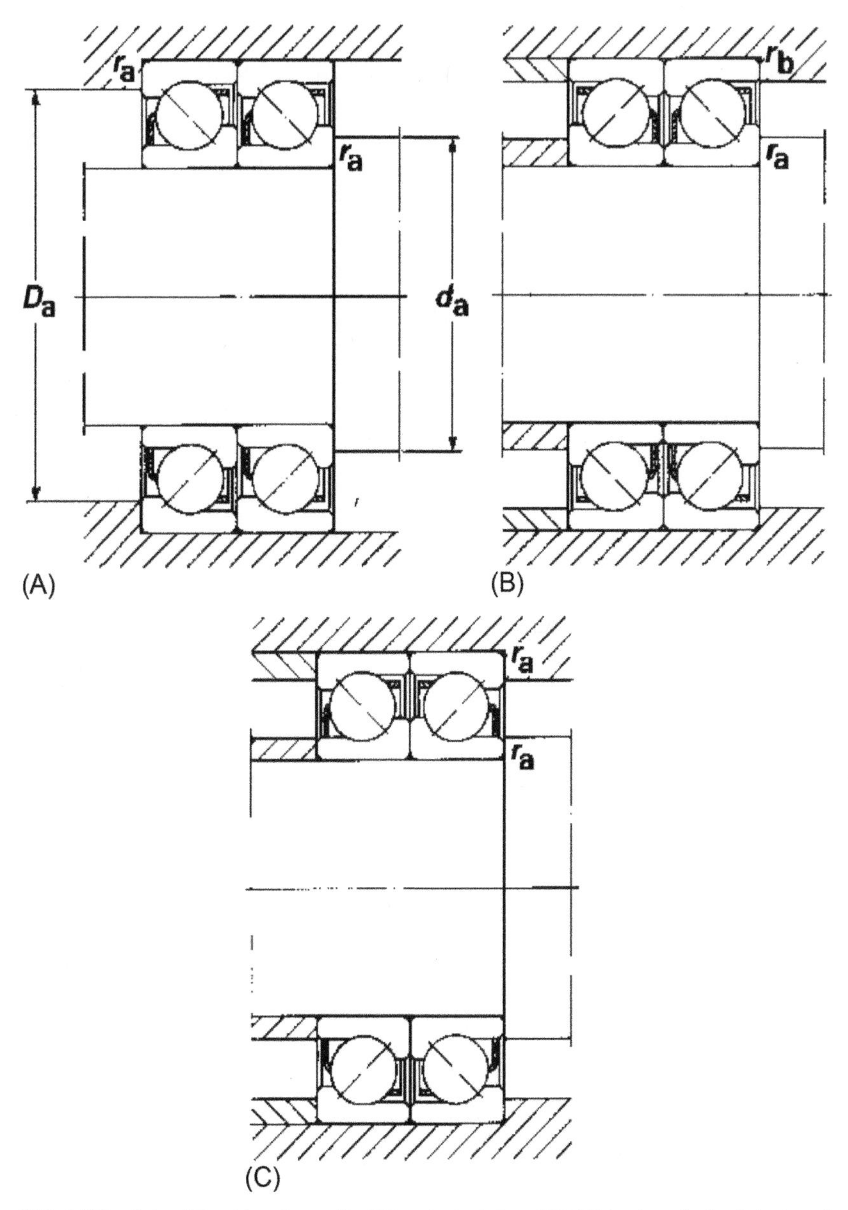

FIG. 4.2.2 Sets of thrust bearings with different orientations [1]: (A) tandem, for load sharing of a pump shaft thrusting from right to left; (B) back-to-back, the customary API 610 recommended orientation with shafts possibly exerting axial load in each direction; (C) face-to-face, rarely desirable in centrifugal process pumps.

precision-ground, the bearings will share the load and each will carry 50% of the total. If they're not precision-ground, there will be issues.

Expect the thrust action to briefly shift from the normally active to the opposite, or inactive, direction when process pumps are started and accelerating to operating speed. This concern favors using two angular contact thrust bearings mounted in back-to-back fashion, as depicted in the "back" of an angular contact bearing that is the wider outer ring land and the narrower outer ring land that is the "face." For the sake of completeness, Fig. 4.2.2C is a "face-to-face" mounted set. If, in Fig. 4.2.2C, the inner ring became hotter than the outer ring, the resulting thermal growth would load up the bearing. This is but one of a number of reasons why face-to-face bearings are rarely used in process pumps.

At all times, care must be taken not to allow interference at the bearing and shoulder radii r_a and r_b. Also, precision-ground bearings are, understandably, more expensive than bearings with more liberal manufacturing tolerances. However, as we think through how the thrust bearing set in Fig. 4.2.2B works, we realize that with one bearing heavily loaded, the other one might be unloaded. When that happens, the unloaded bearing may skid and wipe the lubricant off the raceways. Metal-to-metal contact on the skidding bearing creates much heat and accelerated bearing failure.

API 610 asks for the contact angles in each bearing making up a set to be equal, which is why we showed them as equal in Fig. 4.2.2B. The API recommendation is influenced by the desire for standardization and by the economics of initial cost and the desire to simplify training of technicians. Economics aside, two angular contact thrust bearings with *equal* load-carrying capacities are not necessarily best for pumps that briefly experience thrust reversal at start-up. An unloaded bearing may skid (unintended) while the rolling elements in the loaded bearing are rolling, just as intended. Skidding bearing elements tend to wipe off the oil film, in which case there will be metal-to-metal contact. This contact can initiate failures and often manifests itself as high bearing temperature.

FIG. 4.2.3 Triple-row thrust bearing set.

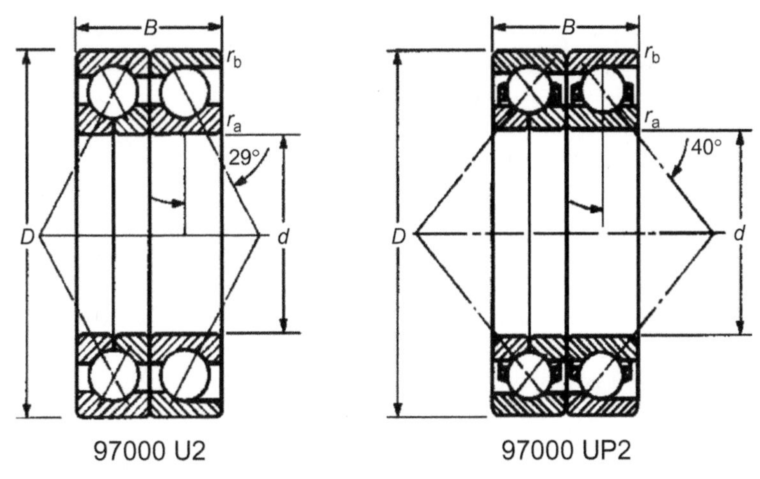

FIG. 4.2.4 Thrust bearing sets with dual inner rings [3].

Special sets of back-to-back mounted thrust bearings may be a better choice in certain applications. These sets would use dissimilar angles; as an example, instead of the customary 40/40 degrees, a 40/15- or 29/15-degree design might be used to avoid skidding [2]. Even an occasional set of triple-row bearings (Fig. 4.2.3) can be found in multistage pumps. Like Fig. 4.2.2B and related sets, they must be carefully matched and precision-ground at all contacting faces and surfaces.

Finally, there are thrust bearing sets with matched/mated split inner ring (separable ring) and angular contact bearing geometry (Fig. 4.2.4). Because they accommodate the maximum number of rolling elements, split inner ring bearings can impart favorable load capabilities but are more expensive to produce and acquire.

BEARING CAGE MATERIAL

The cage (or ball separator) material showing the most promise in angular contact ball bearings is polyetheretherketone, commonly called PEEK. Bearing manufacturer SKF has been using PEEK cages in angular contact bearings for several years in both super precision bearings for machine tool applications and industrial angular contacts for higher speed centrifugal compressor applications. PEEK cages have also been used in angular contact bearings in compressors exposed to sour gas. In all cases, the PEEK material has proved to be an excellent cage material. SKF has used both cages machined from PEEK tube and cages molded from PEEK. Both have worked, but in higher volume, the molded cage is much more economical.

The advantage of PEEK is that it is very strong, inert to most chemicals, very temperature-resistant, and lightweight. Especially when molded, the cage pockets conform well to the balls spreading the already minimal weight of the cage over a large area. The cage material is easily strong enough to support the forces necessary to pilot the cage and separate the rolling elements. Design-wise, the

PEEK cages are nearly identical to the molded cages made of polyamide. The polyamide cages typically perform very well, but the limitations of polyamide limit the temperature capabilities of the bearing and the material ages over time when exposed to moderate temperature. PEEK does not limit the temperature capabilities of the bearing nor does it age over time. Therefore, it is an excellent cage material allowing the bearing to operate over the full range of capability of the metallic components.

Bearings with PEEK cages have not been used extensively for pump applications. This is not because such cages are incapable, but rather because the market's preference is machined brass. While a machined brass cage is an excellent cage, it does have some drawbacks. The cage is heavy, and for ball-guided cages, that weight is supported by the balls. Especially in situations with marginal lubrication, that force can be enough to cause pocket wear. Misalignment or pinching can also create conditions where cage pocket wear happens. The wear debris is deposited in the oil speeding the creation of black oil, plus the particles are run over by the balls creating debris dents throughout the bearing eventually leading to wear or spalling. The PEEK cages are more flexible and

FIG. 4.2.5 A typical sleeve bearing on a large process pump. *(Courtesy Hydro Inc., Chicago, Illinois, www.hydroinc.com.)*

lighter so they will not exert as much force on the balls. The flexibility of the cage also helps prevent any cage-generated damage in misaligned or pinched situations.

Overall, PEEK represents the state of the art for cages in angular contact ball bearings. Many of the applications where it has been used are very demanding and the service records of those cages have been excellent. For typical centrifugal pumps, there is no reason to believe PEEK cages wouldn't perform superbly there as well.

As an aside, sliding bearings (sleeve bearings) are used on larger equipment, generally machines at over 500 kW (Fig. 4.2.5). With very few exceptions, sleeve bearings require relatively thin oils, such as ISO VG 32. The strengths and limitations of sleeve bearings are well understood. A maximum load of 250 psi is considered a safe upper limit. However, unusually light loading (<20 psi) must be avoided as well.

REFERENCES

[1] Pump Bearings, SKF America Publication, Kulpsville, PA, 2000.

[2] P. Eschmann, L. Hasbargen, K. Weigand, Ball and Roller Bearings, John Wiley & Sons, Hoboken, NJ, 1985.

[3] H.P. Bloch, Lubricant delivery advances for pumps and motor drivers, in: Proceedings of 31st International Pump Users Symposium, Houston, September, 2015.

Chapter 4.3

Optimized Lubricant Application for Rolling Element Bearings

INTRODUCTION

Lube application methods include grease, oil flooding, and oil rings or slinger disks used to bring lubricant from an oil sump to the bearing elements. As of 2015, there existed about 3000 plant-wide oil mist systems, and hundreds of these have been in highly successful use since the early 1960s. As of 2015, approximately 150,000 process pumps and 27,000 electric motors are lubricated by oil mist. Some oil mist providers consider this a low estimate. Either way, the message is clear: check into oil mist lubrication (see index for additional coverage).

RANKINGS EXAMINED

Only jet oil application (Fig. 4.3.1) is ranked higher than oil mist. Jet oil lubrication applies the filtered lubricant precisely where needed and dispenses with constant-level lubricators. The oil jet is directed at the space between the outside diameter of the bearing inner ring and the bore of the cage. Metallic cages are

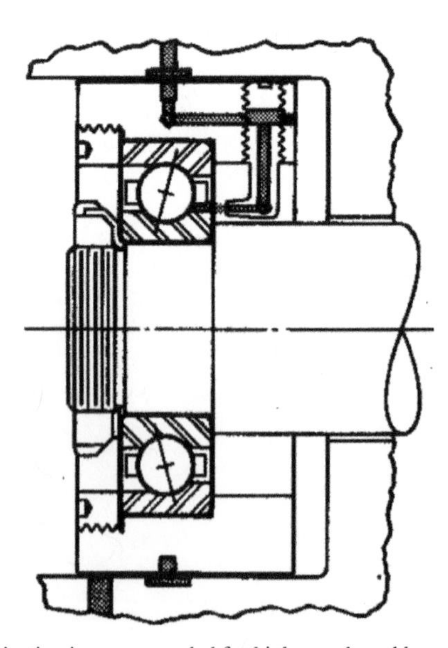

FIG. 4.3.1 Jet oil lubrication is recommended for high speeds and heavy loads [1].

used because they have the highest allowable temperature capability. Means for scavenging the oil should be provided on both sides of the bearing, allowing oil to collect, and overheat in those spaces must be avoided. Note how the oil system may be used to allow free axial floating of the bearing cartridge in the housing on a thin pressurized oil film [1]. A clearance of 0.0005–0015 in. is typically recommended between the cartridge and the housing.

Reliability-focused users are generally disinclined to view process pumps with old-style, failure-prone oil rings (see index) as reliable as they would view pumps with circulating lube systems. Over 100 technical publications agree that oil rings are sensitive to shaft horizontality, depth of ring immersion in the oil, material selection, and machining and roundness parameters.

Envision, therefore, how the lube application of Fig. 4.3.2 will open a window of opportunity for reliability-focused users and/or innovative pump manufacturers. Think of a small oil pump, either internal to the process pump bearing housing or incorporated in a small assembly screwed into the bottom drain of your present process pump. This upgrade can provide filtration, metered flow, and proper pressure downstream of the oil sump and upstream of the spray nozzles [3].

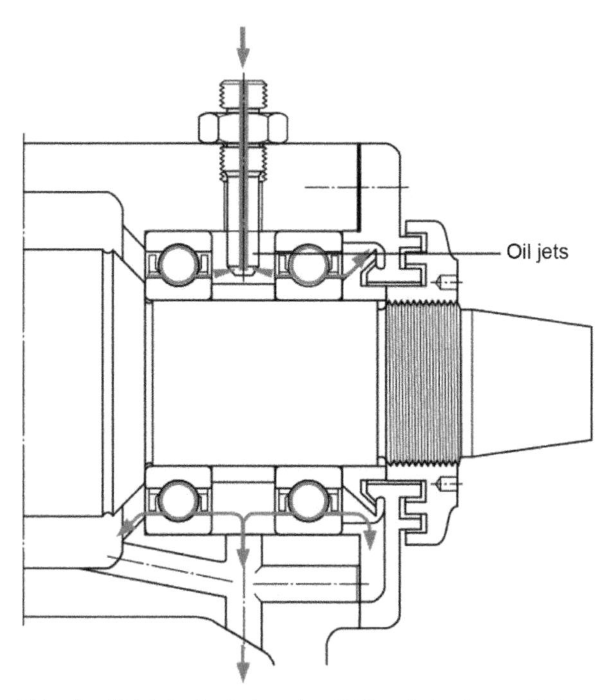

FIG. 4.3.2 Oil jet (jet oil) lubrication is the only reliable solution for very high speeds and heavy loads [2].

Motivated users have written this preferred lube application approach into their pump specifications, and as of 2016, many are actively pursuing this pump reliability enhancement. It will be needed for high-reliability installations where oil rings are either deemed too maintenance-intensive (and preventive maintenance is definitely required to lower the failure risk) or too failure-prone (in installations that refuse to spend time and money for preventive maintenance).

REFERENCES

[1] MRC, Engineer's Handbook, General catalog 60, TRW, Jamestown, NY, 1982. Copyright TRW.
[2] Advertising literature issued by SKF America, Kulpsville, PA, 2014.
[3] H.P. Bloch, Lubricant delivery advances for pumps and motor drivers, in: Proceedings of 31st International Pump Users Symposium, Houston, September, 2015.

Chapter 4.4

Pump Bearings and Allowable Temperatures

INTRODUCTION

Lubrication and bearings are of greatest importance to machinery. Lubricants provide a separation film between stationary and moving bearing surfaces; they also remove heat. Viscosity is the most important property or "indication," as some lube providers call it. Once a lubricant is selected—usually by the machinery manufacturer—the equipment owner/user usually monitors performance on the basis of operating temperature. However, users/owners are rarely correct in their understanding of allowable lubricant temperature in general purpose machines, such as process pumps. Fortunately, best-of-class (BoC) companies are more detail-oriented, and the next few paragraphs summarize important highlights traceable to BoC experience.

ROLLING ELEMENT VERSUS PLAIN BEARINGS

As we of course know, hundreds of millions of horizontally oriented process pumps are moving fluids in modern industry. The larger of these, typically pumps in ranges above 500 kW, incorporate sliding bearings. Except for a relatively small number of pumps with product lubrication or with exotic and appropriately expensive bearings, horizontal process pump bearings (Fig. 4.4.1) need oil lubrication. In the majority of cases, these pumps have a set of ball bearings in the thrust position; a symmetrical ball groove or cylindrical roller bearing supports the pump rotor in the radial direction.

Among the many references linking bearing health and life expectancy, both in the plant and in the field, Fig. 4.4.2 stands out because of its general applicability. The illustration pertains to the thermal range of a typical rolling element bearing. Note that bearing metal temperature is often higher (10–25°C) than the oil temperature in the bearing within an oil circulation system. In noncirculating systems, bearing metal temperature is more likely 5–10°C higher than the oil temperature.

The central zone in Fig. 4.4.2 represents the optimum zone for bearing and lubrication temperature; operating in the adjacent zones will reduce lubricant and bearing life. If bearing temperatures are found in the extreme end zones, expect both the bearing and a mineral oil lubricant to be severely compromised. It is of particular interest to note that pump bearing housing temperatures around 160°F are quite normal. This temperature is to the left of 100°C, comfortably

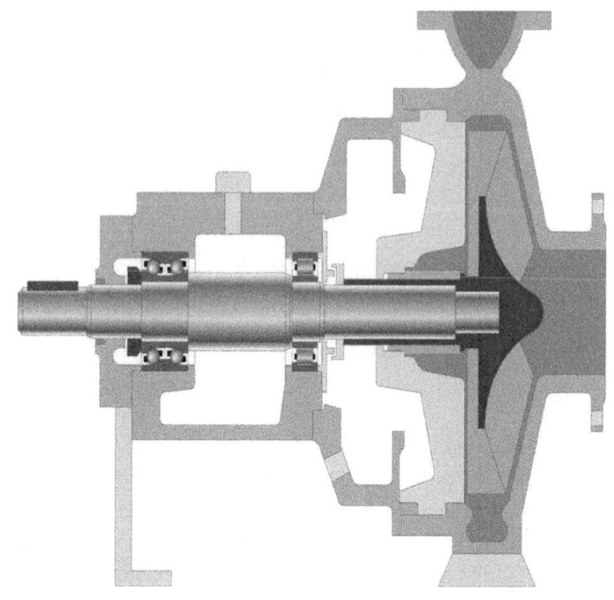

FIG. 4.4.1 Generic representation of a back pull-out process pump shown here with back-to-back thrust bearing (left) and cylindrical roller radial bearing (right). *(Courtesy NSK Bearing Corporation.)*

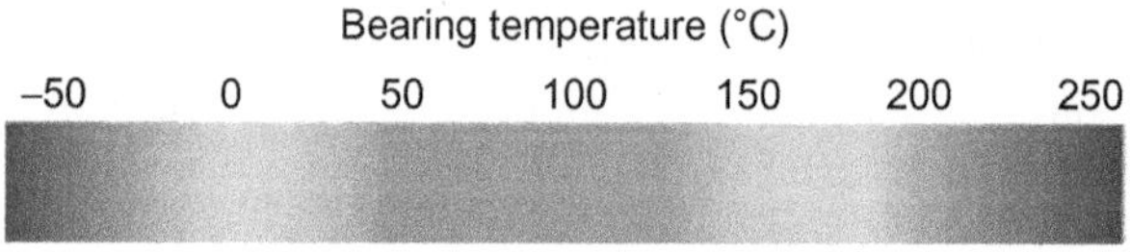

Bearing temperature (°C)

−50 0 50 100 150 200 250

FIG. 4.4.2 Heat ranges of bearings. The lighter shades of gray represent favorable temperatures where bearing and lubricant life are most favorable [1].

near the center of the lightly shaded zone. Yet, bearing housing temperatures of 190°F are allowable and will likely occur in process pumps if an unnecessarily high oil viscosity is selected. Bearing manufacturer MRC states that its brand of rolling element bearings will run in continuous duty at a temperature of 121°C [2]. Therefore, operating at 190°F, while perhaps wasteful if the oil is unnecessarily viscous, is permissible. Be sure to add a personnel protection guard, though.

Of course, operating with a lower viscosity (properly selected) synthetic oil would often be a suitable alternative. In fact, operation with temperatures in the elevated risk zone of Fig. 4.4.2 can usually be avoided by switching to a lighter oil, preferably a premium-grade synthetic lubricant. It should be noted that premium-grade synthetics tend to remove more heat than equivalent grades of mineral oil. Synthetics run cooler, but they, too, will become hot if unnecessarily high viscosities are chosen or if the rate of oil flow is insufficient for effective heat removal.

Rolling element bearings will start degrading once a heat treat-annealing temperature, probably slightly over 130°C (266°F), is exceeded. However, no process pump bearing housing designed or built by reputable manufacturers has ever even come close to this temperature while properly lubricated with the as-purposed quantity and quality of lubricating oil. In a recent case, copious amounts of an excessively viscous lubricant applied as an oil mist leveled off and stabilized at 190°F.

An interesting lesson is again centered on the misunderstanding that the higher the lubricant viscosity, the better the bearing protection. That is far from correct, because rolling elements plowing through thick oils at high speeds will generate considerable frictional heat. Using oil in the correct overall viscosity range is always important; the recommendations of major pump manufacturers and/or manufacturers of rolling element bearings are backed by decades of experience, and we should not disregard them. Moreover, bearing geometry and oil residence time can influence heat generation. Visualize a few "old" drops of thick oil stagnating in the same spot. Now visualize the same spot but a few drops of "new" oil displaced the old oil every second of the day. The respective temperatures will differ.

Table 4.4.1 illustrates the extended temperature ranges available from modern synthetic lubricants. Properly formulated synthetic lubricants allow

TABLE 4.4.1 Temperature Ranges of Modern Synthetics

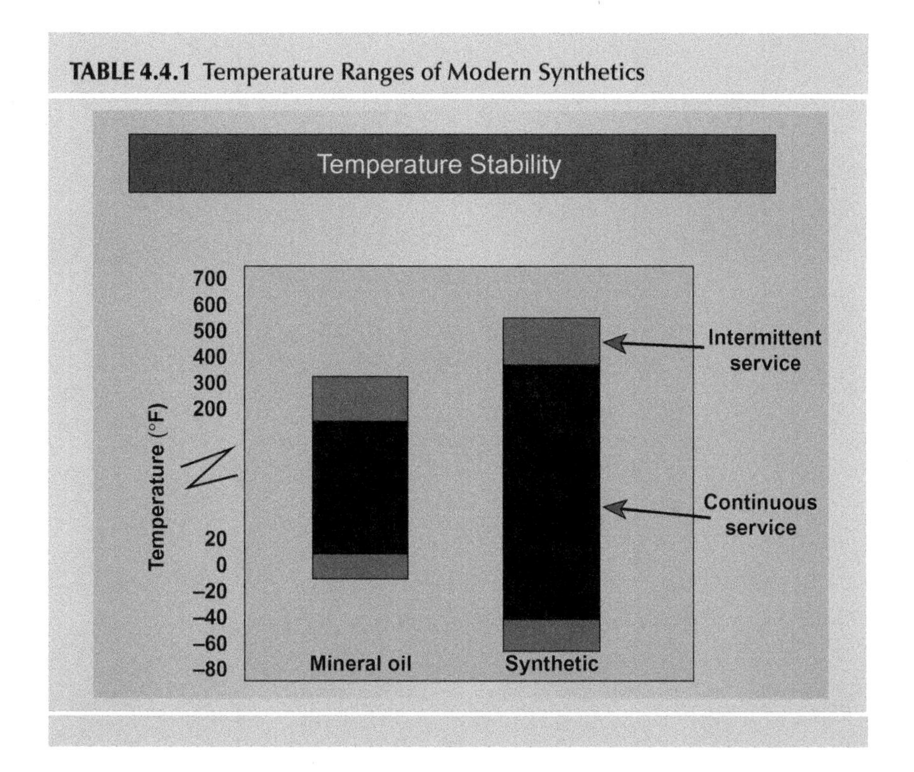

continuous operation at 400°F (~205°C). Although no rolling element bearings in process pumps are ever exposed to this temperature, we would be correct in pointing to the information conveyed by Table 4.4.1. It supports the findings that premium synthetic formulations can be used with every conceivable bearing style or configuration found in process pumps.

REFERENCES

[1] D.F. Wilcock, E.R. Booser, Bearing Design and Application, McGraw-Hill, New York, NY, 1957.
[2] P. Eschmann, L. Hasbargen, K. Weigand, Ball and Roller Bearings, John Wiley & Sons, Hoboken, NJ, 1985.

Chapter 4.5

Ranking Bearings and Lube Application Methods for Process Pumps and Drivers

INTRODUCTION

An authoritative text ("Ball and Roller Bearings," [1]) recommended the relative rankings per Table 4.5.1 for general guidance. But Table 4.5.1 addressed bearings and lubrication applied in a very wide variety of machines, whereas this chapter of our text is making observation relating exclusively to process pumps and their electric motor drivers. When published in 1985, Ref. [1] represented a general consensus. Its relative accuracy has been reaffirmed and largely corroborated by an additional 30 years of experience. We find it particularly reassuring that Ref. [1] ranked circulating filtered oil and/or oil spray into the cages of rolling element bearings at the very top of the list.

As of 2016, all world-scale manufacturers of rolling element bearings continue to support oil spray as the best lubricating choice. These manufacturers noted that, after first filtering, oil is sprayed into the bearing so as to remove heat and to provide an oil film of proper thickness. With the proper oil film, no undue amounts of frictional energy are created. Therefore, Ref. [1] gave oil spray 10 out of 10 possible points. The information and guidance in this chapter of our texts will revert back to these recommendations; it will harmonize with the advice of bearing manufacturers. In other words, it will answer the question "so what?" Sound answers are of importance and interest for many conscientious reliability professionals.

MAKING CHOICES

This particular narrative is based on a tutorial presentation made by the author at the Texas A&M University International Pump Users Symposium in Sep. 2015. It does not infringe on, or limit, the choices made by pump manufacturers and/or owner-purchasers. It is well known that many kinds or styles of lubricant application are presently available. However, in developing the original tutorial, the author set out to address the needs of individuals and groups searching for better solutions. Pump users often suffer unexplained repeat failures and have not found adequate redress for some of these. Therefore, this chapter of our text will be of interest to many of the affected pump users.

TABLE 4.5.1 Relative Ranking of Lube Methods Dating Back to the Early 1980s[a]

Decreasing service life[b] →	Oil	Grease
	Rolling bearing alone	Rolling bearing alone
	Circulation with filter, automatic oiler Oil-air Oil-mist	Automatic feed
	Circulation without filter[c]	
	Sump, regular renewal	Regular regreasing of cleaned bearing
		Regular grease replenishment
	Sump, occasional renewal	
		Occasional renewal Occasional replenishment
		Lubrication-for-life

[a]*Note that oil-air and oil mist are ranked near the top of the list. The list pertains to a large number of typical machines and includes automotive and appliance equipment.*
[b]*Condition: lubricant service life $<$ fatigue life.*
[c]*By feed cones, bevel wheels, asymmetric rolling bearings.*
(Courtesy P. Eschmann, L. Hasbargen, K. Weigand, Ball and Roller Bearings, John Wiley & Sons, Hoboken, NJ, 1985).

As a reference point, we start with lube application rankings as published in the early 1980s (Table 4.5.1). Later, in Subject Category 17, we will explore and explain the proposed new ranking.

OIL VERSUS GREASE

As this text emphasizes in many places, the primary purpose of lubrication is to separate stationary from rotating parts by placing lubricant molecules between the components. A flow of lubricant also serves to carry away heat. Oil has advantages over grease because it removes more heat. Grease has an advantage because it is more easily confined. Both oil and grease can be applied in many different ways.

Grease is normally used in electric motor drivers ranging from fractional HP (horsepower) to approximately 500 kW; at over 500 kW, oil lubrication often represents an overall cost advantage. This is because grease can be readily introduced in the small-to-medium electric motor sizes where motor end caps readily accommodate grease.

With occasional exceptions, experienced petrochemical plants often prefer oil—properly applied—over grease, once 500kW is exceeded. However, there are many different oil application details; likewise, there are many different grease application details. Each merits further elaboration and will be discussed later. Depending on equipment speed and power delivered (or absorbed), well-designed oil-lubricated sleeve bearings may be favored in the 3600rpm/1000kW and higher categories. However, rolling element bearings will give generally superior overall service in the lower and medium-power applications. As an example, an electric motor at 3600rpm/300kW will rarely benefit from sleeve bearings. Many factors enter into the decision-making process. Limited end float couplings will be needed for sleeve bearing applications. Motor magnetic centers will have to be observed by the maintenance craftsperson. Competent maintenance execution becomes important. Such maintenance effort costs money.

OIL LUBRICATION EXPERIENCE

Oil applied as a static sump is often called an "oil bath." Static sumps—oil baths—are acceptable for relatively low bearing velocities. With a static sump, the oil level would be at or near the center of whichever rolling element passes through the 6 o'clock (bottom) position. Oil bath lube is feasible for low-to-moderate shaft velocities. Once bearing elements plow through an oil bath at "high" velocities, heat generation will be of concern. Elevated bearing temperatures can degrade lubricant oxidation stability.

A widely used approximation suggests a "DN value" (inches of shaft diameter multiplied by revolution per minute) of 6000 as the threshold where bearing elements should no longer move through the oil bath and where, instead, lube oil is introduced into the bearings by other means. Traditionally, these other means have included oil rings (Fig. 4.5.1), flinger disks (Fig. 4.5.2), "jet oil spray" (shown a few pages earlier in Chapter 4.3, dealing with application methods), and oil mist (Figs. 4.5.3 and 4.5.4).

Flinger disks must be carefully engineered for the intended hydraulic performance; they also must be securely fastened to shafts. Experienced European manufacturers often offer them as standard components. The disks allow moderate deviation from precise horizontality of shaft systems; they make contact with the oil level or are partially immersed in the bearing housing oil sump. Insertion into the bearing housing is usually made possible by removing the thrust bearing assembly shown in Fig. 4.5.2.

As of 2016, oil mist was being used on an estimated 140,000–170,000 process pumps and close to 27,000 electric motors worldwide. API 610 gives application details very similar to Fig. 4.5.3, as do Refs. [2,3]. The key point is that oil mist is introduced between a long-life bearing housing protector seal (see Figs. 4.5.2 and 4.5.3) and a vent location downstream

FIG. 4.5.1 An unrestrained oil ring can touch portions of the inside of the bearing housings and suffer abrasive damage.

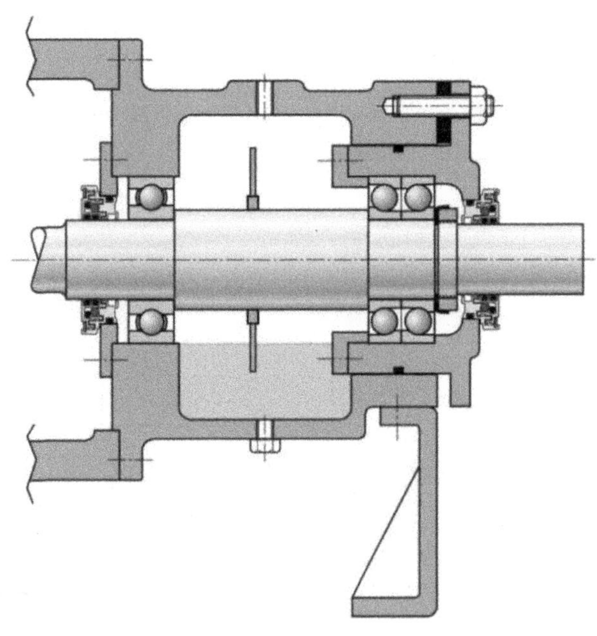

FIG. 4.5.2 Flinger disks that dip in the oil avoid issues with oil rings, but these disks must be carefully engineered. Note that dimensional access (assembly) will require that thrust bearing sets be placed in a cartridge.

of the bearing. As the mist flows through the bearing and while shaft rotation creates turbulence, atomized oil globules combine and form larger oil droplets. The coalesced oil then coats and cools the bearing. Because the bearing housing is at slightly higher than atmospheric pressure, inward migration of atmospheric contaminants is avoided.

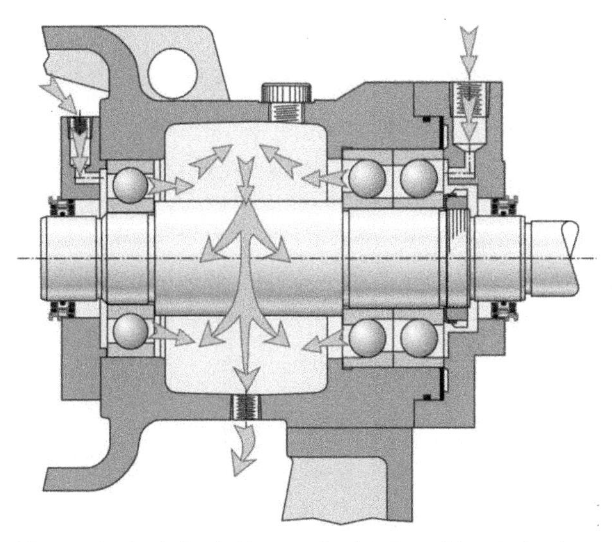

FIG. 4.5.3 Oil spray (or, in similar fashion, an oil mist) directed into the bearing cage through a nozzle or metering orifice. To overcome the "fan effect" (or windage) created by angular contact bearings at high speeds, the nozzle exit may have to be located close to the bearing cage (ball separator).

FIG. 4.5.4 Pure oil mist on a process pump. After passing through the bearings as a mist, coalesced oil is collected in a small sight glass-equipped container (lower left).

CONSTANT-LEVEL LUBRICATORS

Traditional lowest first-cost application of oil usually involves using one of many available constant-level lubricators (CLLs). A widely used version is shown in Fig. 4.5.5; in fact, one is shown on each of the two bearing housings. But note that side-mounted CLLs or "oilers" are unidirectional. They should be mounted on the up-arrow side of the bearing housing. The oiler on the left side of Fig. 4.5.6 shows what happens if the oiler is mounted on the down-arrow side of a bearing housing: It will allow air to be pulled in and the viewer may not see the true oil level. Therefore, we will have to observe the directional sense of the pump shaft in Fig. 4.5.5 and relocate one of the two CLLs. They cannot both be correct.

FIG. 4.5.5 Traditional liquid oil application (oil bath). Because CLLs are unidirectional, one of the two must be repositioned; both lubricators must be on the same side.

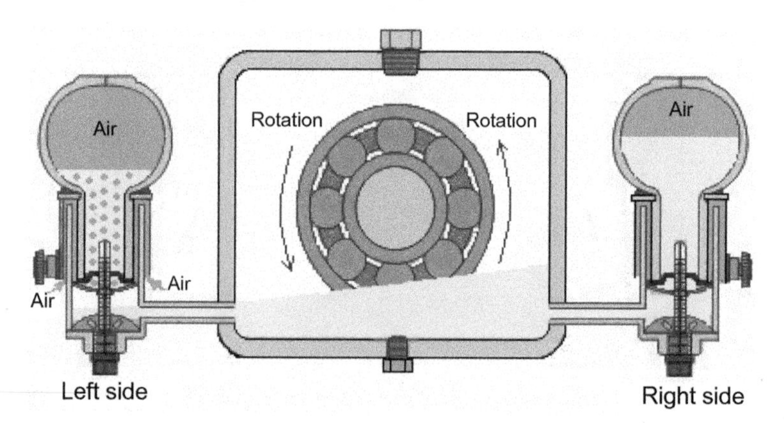

FIG. 4.5.6 Traditional liquid oil application with static sump (oil bath). The lubricator on the left side should be removed.

Visualize how in some instances a small oil level decrease may deprive a bearing of lubrication. Similarly, if the pressure in a closed bearing housing is elevated because of a slight temperature increase, the oil level in the closed bearing housing will go down. If the oil level was initially close to the bearing outer race rim, oil may suddenly no longer flow into the bearing. Black oil will form and the bearing will start to fail. Pressure-balanced lubricators (Fig. 4.5.7) are preferred over unbalanced types shown in Figs. 4.5.5 and 4.5.6.

As a matter of general concern, note that a puttylike caulk is often used on CLLs where the glass bottle meets the cast metal support. Over time, this caulking will craze, meaning it will develop fissures through which rainwater enters the oil via capillary action. Therefore, CLLs often have a finite life—perhaps 5 years—depending on temperature cycles and extremes. They must be replaced as part of periodic preventive maintenance.

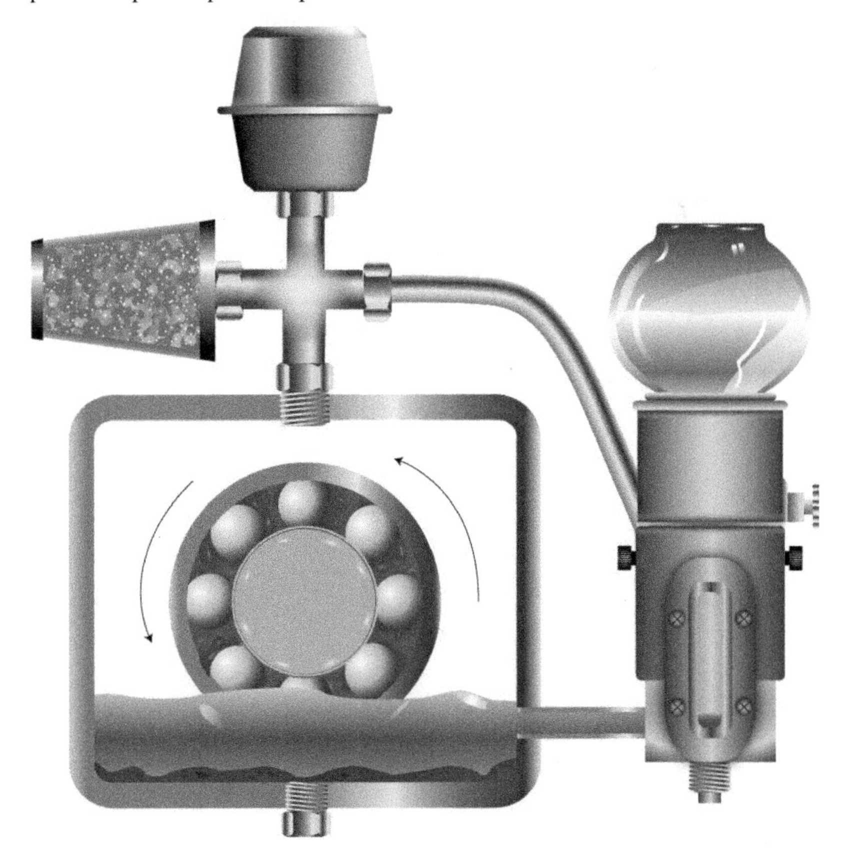

FIG. 4.5.7 CLL with design features that allow installation of a pressure-equalizing line (balance line) between the CLL and the particular bearing housing it serves. *(Courtesy TRICO Mfg. Co, Pewaukee, Wisconsin.)*

VULNERABILITY OF OIL RINGS

Oil rings will work and are found in many machines [4]. It should be noted that oil rings have to be installed on a truly horizontal shaft system and are not allowed to make contact with housing internal surfaces. Their vulnerability is further described a little later in this book (see Subject Category 24). Be certain they cannot get wedged under the long "oil ring travel limiter" screw shown to the left of the thrust bearing in Fig. 4.5.8. Maintain depth of immersion and lube oil viscosity within acceptable ranges. Also, ascertain that bore eccentricity stays within 0.002 or 0.003 in. recommended in Ref. [4].

Oil rings or no oil rings: problems result if oil can get trapped and overheat because no oil return slot was provided. In Fig. 4.5.8, only one oil return slot is shown (note slot below the radial bearing). For some reason, no such slot was provided below the thrust bearing in this illustration. Therefore, the oil getting trapped between the thrust bearing and the housing end cover may overheat and be converted to carbon powder. No bearing housing protector seals are shown in Fig. 4.5.8, but beware of adding just any such bearing protector seal. Doing so may result in somewhat higher pressure to the right of the thrust bearing compared with the pressure in the large, often well-vented, space near the center of the bearing housing. A low-risk bearing housing design will incorporate interior balance holes or other features that ensure equal pressures exist in the center of the bearing housing and also on either side of the bearings.

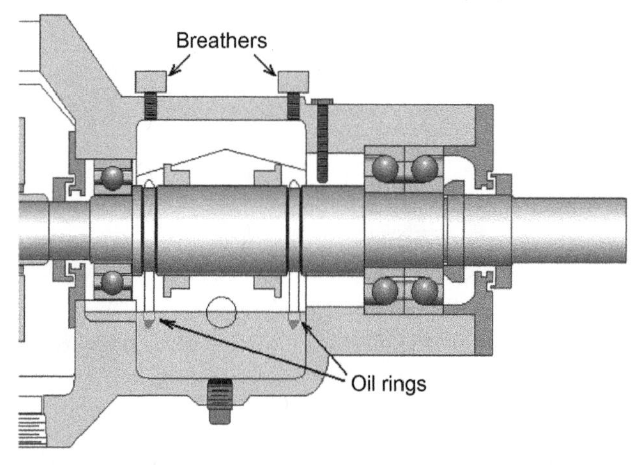

FIG. 4.5.8 An appropriately sized oil return slot (or channel) is shown below the radial bearing. Because the designer overlooked the need for an oil slot below the thrust bearing, oil could get trapped to the right of this bearing. Trapped oil tends to overheat and oxidize behind the thrust bearing unless an oil return slot is provided.

Technical personnel in various job functions with reliability impact have sometimes asked for more detail on oil-air, also called "jet oil lubrication," briefly mentioned in Chapter 4.3 (Fig. 4.3.1) [5]. Jet oil lube represents the highest-rated application method. Envision, therefore, how jet oil—widely used in aerospace applications since the mid-1940s—will open a window of opportunity for reliability-focused users and/or innovative pump manufacturers. Think of a small oil pump, either internal to the process pump bearing housing or incorporated in a small assembly screwed into the bottom drain of your present process pump. This upgrade can provide filtration, metered flow, and proper pressure downstream of the oil sump and upstream of the spray nozzles. Motivated users have written this preferred lube application approach into their pump specifications in the 2010 time frame and have since then actively pursued this pump reliability enhancement. We were told, in 2015, that unless the market demands it, pump manufacturers will not be offering this upgrade.

BRIEF OVERVIEW OF GREASE LUBRICATION

There are isolated instances when bearings should be fully packed with grease. A boat trailer is such an isolated instance. As they back their trailer onto the boat ramp and launch a small boat, owners wish to keep water from entering the trailer's wheel bearings. While towing the boat trailer on a highway, the bearings usually rotate at no more than 900 rpm. We will assume that, on average, the owner tows the trailer 200 hours/year.

Compare this with the average electric motor bearing. Its shaft diameter is twice that of the boat trailer axle. The electric motor bearing rotates at twice or even four times the speed, and we expect the bearing to last 24,000 hours—three or more years. Packing the motor bearing full of grease would create excess heat and reduce bearing life. Therefore, the grease in electric motor bearings should take up only 30–40% of the space between bearing rolling elements. We must be careful when making comparisons. As the old saying goes, do not compare apples and oranges.

Lubrication Charts

Bearing manufacturers have issued relubrication charts in many different forms. The one shown in Fig. 4.5.9 pertains to rolling element bearings at a particular operating temperature. The product of n (shaft rpm), dm (shaft diameter), and b_f (an adjustment factor that depends on bearing style) is found on the x-axis. The user will have to find the appropriate b_f factor from vendor literature or application engineering groups. Depending on shaft size, speed, and load, the recommended intervals can be read off on the vertical axis as hours between relubrication. While these intervals were conservative and pertained to standard greases, Fig. 4.5.9 was often used to envision where the use of lifetime lubrication (meaning fully sealed, non-regreasable bearings) should be discouraged. If a bearing cannot be regreased (and fully sealed bearings cannot be regreased), the indicated interval will at least provide a general guide on expected bearing life. Keep in mind that grease path is of major importance in greased bearings and that overgreasing will cause temperature rise and accelerated bearing failure.

SHIELDS VERSUS NO SHIELDS IN ELECTRIC MOTOR BEARINGS

The advent of entirely different nonhydrocarbon greases means that Fig. 4.5.9 will have to be modified for PFPE-PTFE greases. More information on these

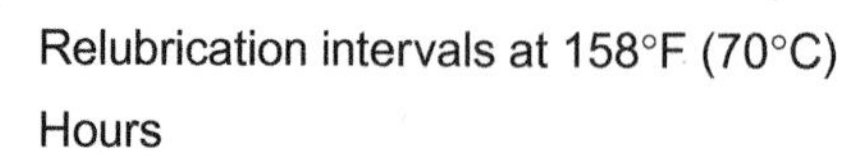

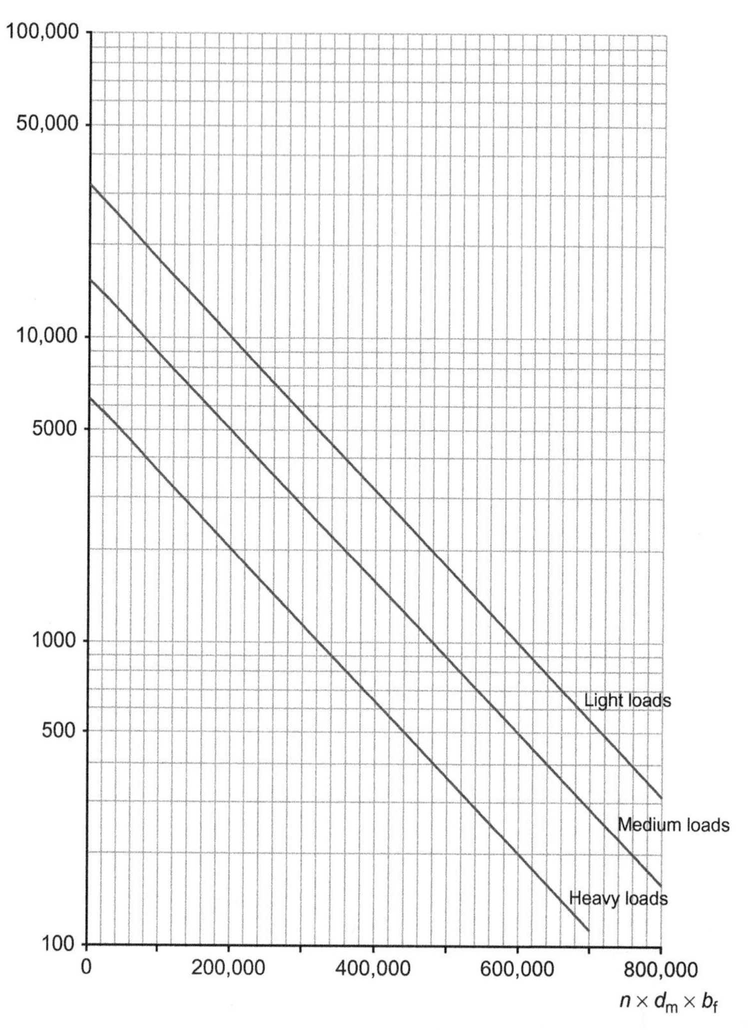

FIG. 4.5.9 Bearing grease relubrication chart intended for a particular bearing style (e.g., tapered roller bearing) as a function of shaft speed and shaft size. *(Courtesy SKF America.)*

greases is discussed later in Subject Category 17. There is also Subject Category 23 with considerable detail on oil mist lubrication of electric motor bearings. Oil mist on electric motors is totally proved technology; it has been widely used by best-of-class petrochemical companies in the United States since 1975. The fact that it seems to have been overlooked is difficult to explain.

These are among the questions to which competent reliability engineers seek answers. If no answers are found, the facility may find that random repeat failures occur from then on. These and other experience factors interact; they admittedly shape and skew rankings in the eyes of field-experienced individuals. Their backgrounds differ, their perceptions differ, and we must leave it to our readers and experienced professionals to judge where to place their trust. We have observed that many process pumps in industry are experiencing repeat failures even as you read this narrative.

Finally, and by way of simple reaffirmation and reassurance, what you presently have in your pumps and electric motors will work most of the time. Still, what you *could* have in your pumps would work better and more reliably. We just thought we might bring it to the attention of those wishing to add value to their enterprise; look for more information in the many subject categories and chapters that are found in this text.

REFERENCES

[1] P. Eschmann, L. Hasbargen, K. Weigand, Ball and Roller Bearings, John Wiley & Sons, Hoboken, NJ, 1985.

[2] H.P. Bloch, A.R. Budris, Pump User's Handbook: Life Extension, fourth ed., Fairmont Press, Lilburn, GA, ISBN: 0-88173-720-8, 2014.

[3] H.P. Bloch, Pump Wisdom: Problem Solving for Operators and Specialists, Wiley & Sons, Hoboken, NJ, ISBN: 978-1-118-04123-9, 2011.

[4] D.F. Wilcock, E.R. Booser, Bearing Design and Application, McGraw-Hill, New York, NY, 1957.

[5] H.P. Bloch, Lubricant delivery advances for pumps and motor drivers, in: Proceedings of 31st International Pump Users Symposium, Houston, September, 2015.

Chapter 4.6

Bearing Housing Protector Seals and the Value of Reducing Lube Oil Contamination Risk

INTRODUCTION

Excluding moisture and solid contaminants from bearing housings and other machine shaft penetrations can have significant economic benefits. Bearing housing seals are contaminant exclusion components that appeal to reliability-focused users. They include inexpensive lip seals that are, however, quite prone to wear. The choice is frequent lip seal replacement or else loss of sealing capability.

The shortcomings of lip seals led to the development of rotating labyrinth seals, generally called "bearing isolators" [1]. These, too, can include makes and models that deserve closer scrutiny. Upon closer examination, one finds low-cost models that, on occasion, incorporate not only an air gap but also a single O-ring that can easily make contact with the sharp edges of an O-ring groove. As the O-ring degrades, such protector seals may still allow airborne contaminants to reach both lube oil and bearings. (Note that Subject Category 14 expands on the topic "Bearing Protector Seals.")

ROTATING LABYRINTH SEALS

The ultimate design intent of rotating labyrinth seals is to gain extended operating life for a machine equipped with bearings. Because bearings depend on lube oil, long operation requires a clean operating environment. Translation: keep dirt away from the lubricant.

Three generic categories of rotating bearing housing protector seals are depicted in Fig. 4.6.1. In the upper left is a dynamic, radially moving O-ring. Unfortunately, such a component runs in close proximity to the sharp edges in one of the two opposing O-ring grooves. A drag-inducing elastomer is used in the type illustrated in the upper right. The bottom image shows a rotating labyrinth assembly with an axially outward-moving, optimized, design.

It should also be noted that axial action permits O-ring contact with a considerably wider area (Fig. 4.6.2) than lip seals. Wide area contact greatly decreases both contact pressure and the resulting wear risk. There are also the well-known spring-loaded mechanical seals, but these either are too expensive or take up too much space. Certain applications favor the use of magnetically closed dual-face bearing housing seals. All are discussed later (see Subject Category 14).

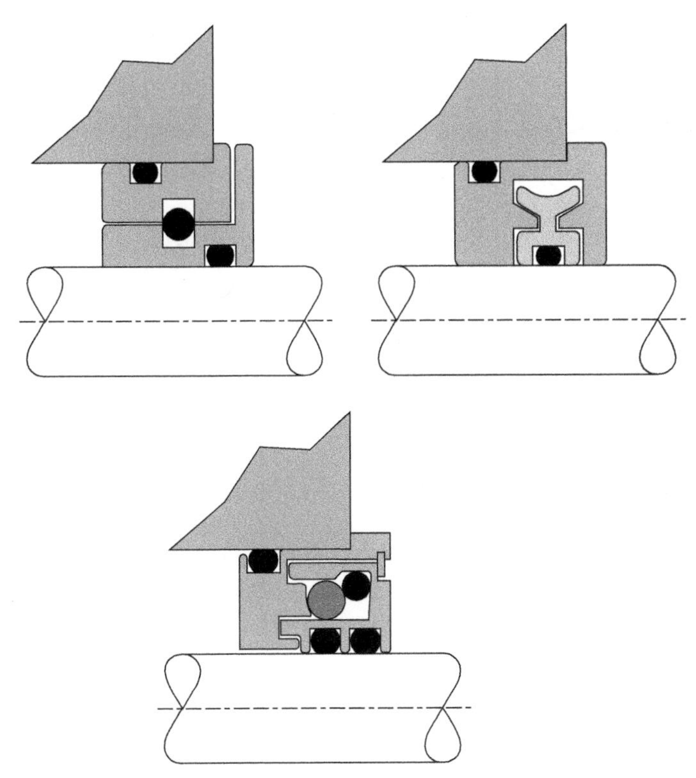

FIG. 4.6.1 Three generic categories of bearing housing protector seals. Upper left—dynamic, radially moving O-ring in proximity of sharp edges. Upper right—stylized image of a high frictional drag-inducing contoured elastomer. Bottom—axially outward-moving, optimized, design. *(Courtesy AESSEAL Inc., Rotherham, the United Kingdom, and Rockford, Tennessee.)*

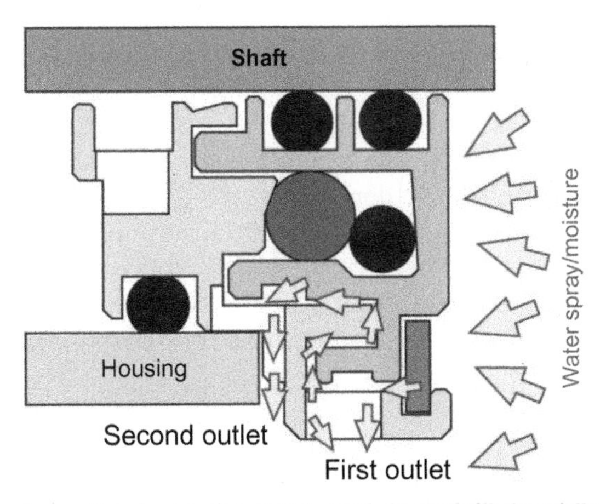

FIG. 4.6.2 An advanced bearing housing protector seal. At standstill, the axially outward-moving O-ring is contacted by a somewhat smaller radial O-ring. The combined O-ring action ensures that water spray cannot reach the bearing housing interior. *(Courtesy AESSEAL Inc., Rotherham, the United Kingdom, and Rockford, Tennessee.)*

JUSTIFYING THE COST OF MODERN BEARING HOUSING PROTECTOR SEALS

Assessing the true value and cost-justifying modern bearing housing protector seals is not difficult. Diligent research has made it possible to quantify the negative effect on bearing life of both moisture and solid contamination of lube oil. Researchers have published life extension factors (LEFs) that accompany changing from moist oils to drier lubricants. Others have estimated the effects of solid contaminants on bearing life and published reasonable estimates as LEFs. Using these tables, cost-justifying modern bearing housing seals are greatly facilitated [2].

LEFs make it easy to calculate with reasonable accuracy and desirable conservatism the intrinsic value of modern bearing housing protector seals. In one example, actual plant failure statistics indicated installed and presumably "contaminated" bearings to last 1.9 years, although data given by the bearing manufacturer showed a load- and speed-dependent L-10 life of 84,000 hours (or 9.6 years) with very clean oil. This user decided to reasonably expect that with well-sealed bearing housings, that is, bearing housings with plugged vents and oil cleanliness assured by retrofitting either modern rotating labyrinth or dual-face magnetic seals (see Subject Category 14), bearing lives of 9.6 years would be achieved. Bearing life would thus be extended fivefold and several repairs would be avoided. The user researched suitable advanced bearing protector seals and magnetic seals. Because even the most advanced bearing protector seals cost considerably less than one single pump repair, such upgrades have always proved cost-justified.

Just as water in bearing lube oil is detrimental to long-term, reliable equipment operation, introducing solid contaminants (particulates) will degrade lube effectiveness and bearing life. Experience shows that most oils are simply not clean. They may not look dirty, but they are full of small particles in the 2–30 μm range that the eye cannot see. These particles consist mainly of fibers, silica (dirt), and metals. In the presence of water, they form sludge. It is easy to cost-justify superior means of keeping out contamination and reducing bearing distress. All that remains to be done is comparing the cost of a set of superior bearing housing protector seals with the value of just one avoided equipment outage or repair event. Singling out the best manufacturers of clean lubricants and equipping fluid machinery with the best available bearing protector seals make economic sense, and many subject categories in this book reach this conclusion as well.

REFERENCES

[1] H.P. Bloch, Pump Wisdom: Problem Solving for Operators and Specialists, John Wiley & Sons, Hoboken, NJ, 2011.

[2] Noria Corporation, Tulsa, OK. www.noria.com, ; also H.P. Bloch, A.R. Budris, Pump User's Handbook: Life Extension; fourth ed., Fairmont Publishing Company, Lilburn, GA, 2014.

Subject Category 5

Blowers

Chapter 5.1

Modern Blowers Can Be Reliable

INTRODUCTION

More than least-cost equipment is usually needed and desired to become a best-of-class user. But when very high reliability equipment is also available at least cost, both owner-purchaser and equipment manufacturer will benefit. Reasonably priced, reliable, modern blowers are available to informed owner-purchasers. Here, we relate an example where all parties found themselves on the winning side.

HOW BEST-OF-CLASS PERFORMERS GROOM PROJECT EXPERTISE

As a general rule, responsible engineers at best-of-class companies see to it that cost estimating manuals reflect the outlay for uptime-optimized equipment. They verify their findings and disseminate all relevant information. Doing so will at times require standing up to the misguided folks who confuse the highly desirable sharing of knowledge with an illegal practice called restraint of trade.

As reliability professionals, we could support the better vendors and manufacturers by publishing our successes. We could do it by teaching others what we have learned from the success of others, and we should be emulating their best practices. Perhaps those who disagree with these sentiments might contemplate how they would fare if medical doctors refused to share with others the lessons they learned in practicing their skills!

These thought processes are at work when real good companies assign the specifying engineering team (project engineers) to operations duties right after completing the project's start-up phases. Carrying over project work into operational assignments is a powerful incentive for project engineers to learn more

Petrochemical Machinery Insights. http://dx.doi.org/10.1016/B978-0-12-809272-9.00005-0

about an asset than mere cost and schedule. Assigning the specifying engineering team to operations duties during and after commissioning allows companies to groom good engineers; it encourages them to investigate thoroughly and choose wisely. In short, their future career in project engineering and project management is linked to how well their first project performs in the two or three years after initial start-up.

PRACTICING RELIABILITY THINKING AND CHECKING THINGS OUT

Important detailed feedback was obtained from a senior corporate reliability expert in the Middle East who made it his professional obligation to thoroughly acquaint himself and others about available upgrade options and alternatives. In an equipment experience survey and during subsequent field visits dating back to 2001, he had been searching for a high-quality multistage centrifugal blower manufacturer capable of replacing the lower quality, lesser reliability multicasing, tie-rod-type blowers he had seen at some facilities. The best mean time between failures (MTBFs) for the old tie-rod type blowers averaged two years, with a mean time between overhauls (MTBO) of three years. During one of his field visits, he noticed three blowers, Figs. 5.1.1–5.1.3, at one of the corporation's rather remote locations. These were air blowers rated at 600 HP (450 kW) each; note that they did not incorporate the usual tie-rod construction features that often "sandwich" together the different blower stages.

The corporate reliability expert teamed up with a maintenance foreman assigned to the facility's sulfur unit. Together, the men ascertained and verified that the reliability of these blowers was well recognized at other projects. Looking at all of the corporation's sites, the blowers ranged in size from 350 HP (260 kW) to 1200 HP (900 kW). At one remote location, the blowers had operated for 20 years without failures. In other words, the blowers achieved a 20 year MTBF.

It was established that the vendor who contributed Figs. 5.1.1–5.1.3 also produced a variety of highly reliable horizontal and vertical centrifugal pumps, as we will see later in this text. We can safely conclude that reliable equipment is not only still being built but also can be affordable as well. Highly reliable manufacturers can compete because they keep scrap rates low. They keep their costs down by having infrequent warranty claims against them.

We also know of Swiss manufacturers that have very few warranty issues to contend with. They too are often price competitive in world markets. They are certainly not among companies that pay substandard wages to their craftsmen. Their success is attributable to consistently high product quality and consistently low scrap rates (Fig. 5.1.4). The equipment so manufactured generally has "eye appeal"—it looks good. This promotes the desire to keep the operating environment clean.

FIG. 5.1.1 Process blower package leaving factory. *(Courtesy Dengyosha Machine Works Corporation [DMW], Tokyo, Japan.)*

FIG. 5.1.2 Modern process blower. *(Courtesy Dengyosha Machine Works Corporation [DMW], Tokyo, Japan.)*

FIG. 5.1.3 Modern process blower. *(Courtesy Dengyosha Machine Works Corporation [DMW], Tokyo, Japan.)*

FIG. 5.1.4 Process blower ready for installation. *(Courtesy Dengyosha Machine Works [DMW], Tokyo, Japan.)*

This is not to infer that others with similarly high reliability do not exist. It is just that we cannot identify them if they are held back by legal concerns or for unspecified reasons are unwilling to share information on reliability performance. Reliable components (Fig. 5.1.5) can be found through the machinery quality assessment (MQA) process explained in several locations throughout this text (see index word "MQA").

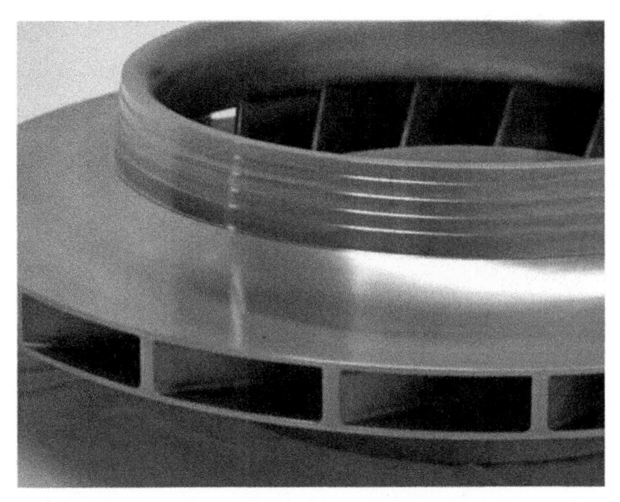

FIG. 5.1.5 Modern blowers often incorporate well-designed, highly efficient closed impellers. *(Courtesy TURBOCAM Corporation, Barrington, NH.)*

Meanwhile, we are indebted to Dengyosha ("DMW," the contributing blower manufacturer) and also the corporate reliability professional in the Middle East who collected and reported these details for everyone's benefit.

Chapter 5.2

Positive Displacement Blowers With Bearings Using Both Pure and Purge Oil Mist Lubrication

INTRODUCTION

Positive displacement blowers (Fig. 5.2.1) are often used to move air ranging in pressure from slightly negative to about 15 psig (~1 bar) positive. These rotary lobe machines are typically used for polymer powder or pellet transfer and come in different sizes. A number of models with 10, 12, and 14 in. shaft center to shaft centers are among the most widely used versions. Drives for these blowers include direct-connected motors and a number of gearbox and belt arrangements. Many positive displacement blowers are oil splash lubricated, while some of the larger ones are forced-feed lubricated. Questions might arise when selecting lubricant application and lubricants. Vendors react to market forces, and traditional designs often win unless customers demand innovation.

FIG. 5.2.1 Positive displacement blower with oil mist lubrication arrangement.

COOPERATION IS ESSENTIAL TO SUCCESS

When several positive displacement blowers were installed at one US facility, there was, at the time of purchase, an understanding between the manufacturer and purchaser. The two parties agreed to consult with an oil mist provider [1] to ascertain and determine in what manner oil mist could be used for the blower bearings and the timing gear area at a later date. Consensus was reached on a wet sump (purge mist) arrangement to be the initial, as-provided lube method for the timing gear/oil sump side of these splash-lubricated units. (The two primary oil mist methods are dry sump and wet sump [2]. For more detail, view Subject Category "oil mist lubrication.")

Years later, the user-owner wanted to implement an upgrade project and researched if it was acceptable to eliminate forced-feed lubrication from the larger blower models in his plant. Although not disallowing it, the original equipment manufacturer (OEM) had indicated having insufficient experience converting forced-feed units to best-practice oil mist lubrication.

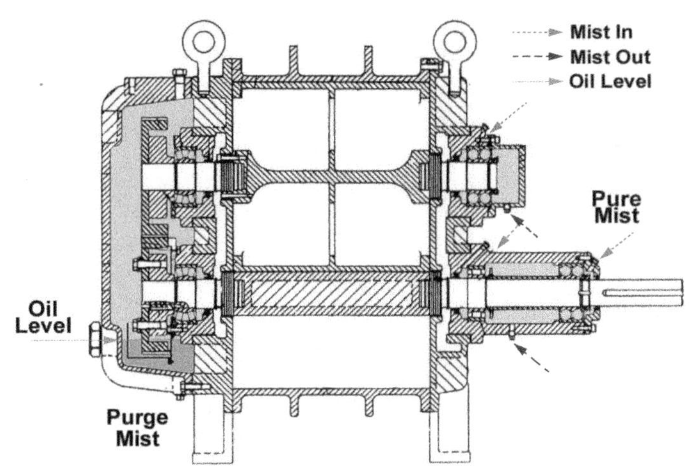

FIG. 5.2.2 Roots-type blower cross section with oil mist purge (*left*) and pure oil mist (*right*). *(Courtesy Colfax Industries/Lubrication Systems Company, Houston, Texas.)*

THE QUEST FOR CASE HISTORIES

Regardless of the OEM's position, we know that oil mist lube on Roots-type blowers exists and has been quite successful on many sizes. Competent oil mist providers can point to case histories and would be pleased to share these with any prospective client. More typically, many Roots-type blowers use a combination of dry sump and wet sump, as shown in the cross-sectional view of Fig. 5.2.2, and conversions from forced feed are generally feasible. However, first and foremost, the dry sump oil mist is intended for antifriction bearings, as shown on the right side of the illustration. Wet sump is shown on the left side, where the gears pick up oil and fling the lubricant into the two spherical roller bearings. The exact point where the mist is applied is of critical importance. If routed per API 610 (late editions), pure oil mist will protect bearings better than anything short of an oil jet impinging on the rolling elements.

Oil mist can work even if not applied per API 610. Yet, a facility doing it the way it used to be done in the 1960s cannot, 50 years later, claim to be reliability-focused. For best effectiveness, dry sump oil mist connections must be arranged for mist flow *through* the bearings. The wet sump (purge mist) side requires no such porting, since purge mist must merely represent a region of elevated pressure and serves no lubrication-related purpose.

As spelled out in a separate Subject Category in this text, the only reason for purge mist is to keep out atmospheric contaminants. Although oil jet lubrication would be the one superior form of lubricant application to bearings, jet lube would be difficult to cost justify. Dry sump oil mist—properly applied—is an attractive lubrication method for rolling element bearings in virtually all industries and for rolling element bearings in all types of rotating machinery. It also serves as a protective environment for stored or nonrunning standby machines.

When utilizing a suitable synthetic lubricant either as a liquid or an oil mist, the maximum allowable temperature is usually set by bearing internal clearance and bearing metal considerations (generally limited to 230°F maximum). But clearances are rarely an issue as long as bearing inner and outer rings are within 60 or 70°F of each other. As long as premium grade synthetic lubricants are used, lubrication matters do not enter the picture. The maximum permissible operating temperatures of synthetic lubricants certainly exceed 300°F, and many suitable lubricants are available.

Some potential user of oil mist wish to engage in experience surveys, and surveys are a commendable effort. Nevertheless, try not to put blind faith in what someone said, regardless of whether they claim good or bad or "just so-so" experiences. Sometimes, a nonexpert's word-of-mouth experience is not relevant at all. Since the nonexpert rarely does root cause failure analysis (RCFA), any feedback or opinion must be linked to several variables and would mandate that we knew these variables. In any event, there is never a substitute for understanding how parts work and how they fail.

So, a simple review of the cross-sectional configuration of a given positive displacement blower would be helpful. Anything short of such a review is just guesswork; it rarely adds value but always adds some risk. As long as this diligent review includes all components and fully explains the machine's inner workings, an owner-operator will prosper. Such a review is as important for oil mist on blowers as it is for the bearing housing protector seals on these machines. Examine these seals and be weary if half the O-ring is contacted by a groove in the stationary part and the other half is contacted by a groove in the rotating part. Did you ever wonder what happens to the O-ring at slow roll or when there is axial movement of the two parts relative to each other? As you think about these matters, you may perhaps realize that you did not purchase the best available bearing protector seal [3,4] or the most cost-effective bearing or the most suitable lubricant. Understanding how components work and then making smart buying decisions will surely contribute to downtime avoidance.

REFERENCES

[1] H.P. Bloch, A. Shamim, Oil-Mist Lubrication Handbook—Practical Applications, Fairmont Press, Lilburn, GA, ISBN: 0-88173-256-7, 1998.
[2] H.P. Bloch, Improving Machinery Reliability, third ed., Gulf Publishing Company, Houston, TX, ISBN: 0-88415-661-3, 1998.
[3] H.P. Bloch, Consider Dual Magnetic Hermetic Sealing Devices for Equipment in Modern Refineries, Pumps & Systems, 2004. September.
[4] H.P. Bloch, Counting Interventions Instead of MTBF, Hydrocarbon Processing, 2007. October.

Subject Category 6

Compressors, General

Chapter 6.1

Upgrade Your Compressor Antisurge Control Valves

Compressor "surge" is a low-flow phenomenon known to exist in "dynamic," that is, centrifugal and axial compressors found in petrochemical plants, liquefied natural gas (LNG) facilities, and pipeline compressor stations. A simplified definition would describe surge as a series of rapidly occurring reversals of the pressurized gas flow.

It is well known that surging can cause serious damage to compressor internals; surge events must be prevented by applying a suitable control strategy. These control strategies include sophisticated hardware and software. Both, or either, can be incorporated in process control computers. In typical antisurge systems, a quick-acting valve recycles a portion of the compressor discharge flow back to the compressor suction. The compressor itself is thus always exposed to a gas volume sufficient to ensure forward flow of the gas. However, while simple in terms of operation, this recycle application requires antisurge control valves to satisfy a multitude of factors. These include providing up to 40 decibels of noise abatement, high flow capacity, and the ability to fully open the valve in less than 2 s. But, energy conservation also mandates operating large compressors close to their respective surge lines, and this places strenuous control demands on valves with 24 in. (600 mm) strokes.

Needless to say, the importance of these valves is well understood by reliability-focused professionals. They not only pursue and implement the most up-to-date solution to protect the compression equipment but also strive to maximize efficiency by safely operating compressors as close as possible to their surge limits.

Petrochemical Machinery Insights. http://dx.doi.org/10.1016/B978-0-12-809272-9.00006-2

IMPROVEMENTS ARE POSSIBLE

Much advancement has been made in noise-attenuating technologies in recent years. These technologies address the potential for damaging noise and vibration; also, some of the most dramatic improvements can be traced to improved valve actuation and positioning technologies.

In many existing antisurge valves, old-style accessory configurations are used. Meeting the emergency stroking time requirements, old-style accessory designs often ignore how well the unit should—or could—be controlled when not in emergency conditions. Until recently, poor control strategies required dynamic process gas compressors to operate a safe and generous distance away from the surge line. In those installations, efficiency was not optimized, and issues arose with valve stability during small step changes.

The picture brightens for facilities capitalizing on advancements in digital valve positioners and extensive dynamic testing by modern provider companies. New and existing antisurge valves can now be tuned to provide accurate response for fast, yet as needed, small or large step changes, without impeding valve stroking speed.

Other benefits include ease of tuning during start-up and commissioning. Older technology required a multitude of accessories to provide the required stroking speed. High stroking speed often meant repeated tuning of bypasses and needle valves that occasionally took as much as 3 days per valve.

With the latest implementation strategies, suitable adjustment of the pneumatics is all that is required during initial installation. This solution also eliminates the drawn-out process of tuning the accessories at least once a year as the equipment begins to degrade and drift.

ON-LINE MONITORING AND DIAGNOSTICS

Using the latest technology, the performance of recycle valves can also be monitored online, in real time. This allows the user to identify any potential issues without requiring a shutdown. With now 8- and 9-year turnaround intervals in modern ethylene plants, it often proved difficult to isolate potential valve performance issues. However, the latest software developments allow users to realize the expanded scope of predictive diagnostic software.

Valve-related software has the ability to detect and identify more than 200 fault conditions that can occur in control valves. Superior software can also spell out and recommend corrective action. This enables reliability professionals to actively preplan any necessary or available improvements or equipment upgrades well in advance of a shutdown rather than reviewing each and every installation during a shutdown.

IMPROVED VALVE PACKING SYSTEMS

Avoiding valve trouble requires packing systems that manage stress level, ensure proper stem and shaft alignment, contain the correct amount of packing material, and offer packing containment. As you link up with a competent surge control provider, be sure to consider the benefits of modern technology. Become familiar with the merits of Belleville washers to provide constant load over the life of the packing material, lined packing followers, an optimized amount of packing, and antiextrusion rings that allow only a predetermined amount of packing deformation.

Finally, compare potentially significant performance differences between modern, advanced graphite compositions and such traditional materials as expandable graphite, wedge-shaped graphite, and flexible graphite over a carbon core. Get the right stuff when you upgrade. Work with experienced companies and identify them through your informal networking contacts. Spend time with potential suppliers and scrutinize their reference lists. As a responsible reliability professional, do what corporations engage in before aligning themselves with a takeover target or a potential corporate partner. It is called due diligence.

Chapter 6.2

Consider Single-Point Responsibility for Compressors

INTRODUCTION

Equipment purchasers are always interested in meeting the cost and schedule targets for projects. There could, however, be conflicting goals when purchasing compressor drivers and/or auxiliaries from vendors other than the compressor manufacturer. While cost targets might initially be met, missed schedules and future warranty disputes present risks that must be considered. It is worth noting what experienced owner-purchasers do to avoid these conflicts. And rest assured that smart owner-purchasers are not limiting their strategies to process gas compressors only.

AGREEMENT NEEDED ON "TRAIN RESPONSIBILITY"

Whenever best-of-class companies separate the procurement of a major fluid machine from the procurement of its driver, one of the two suppliers is given train responsibility. The owner-purchaser then pays the responsible vendor—usually the compressor manufacturer—a modest but equitable sum for accepting single-point or equipment train responsibility.

When this rule was disregarded by an overzealous project person at a company (usually) priding itself as a sophisticated best-of-class performer, a lot of unproductive finger pointing resulted. A little later, the issue received renewed attention during a performance appraisal. The project person's employer-boss considered it a career-limiting oversight to not assign single-point responsibility for compressor trains.

EXCEPTIONS TO THE RULE

Of course, giving a vendor single-point responsibility does not mean that all risk has been removed. As a case in point, reciprocating compressors had been purchased from a manufacturer who had just acquired another company's product line. Unfortunately, the owner-purchasers quickly encountered glitches in spare parts identification and timely availability of spares. In a second instance, a manufacturer of reciprocating compressor frames and main running gear tried to adapt another manufacturer's pistons and cylinders. Part of the assembled machine thus originated as subassemblies for high-speed compressors; the rest of the machine consisted of API 618-compliant slow-speed subassemblies. The marriage had a real rough time early on. Initially, there was much consternation as to who was at fault when the compressors suffered from insufficient capacity, and different components started to break.

Soon after these hybrid machines were installed on a floating production/storage/offloading unit (commonly called FPSO), things really went wrong. The owner-purchaser initiated litigation against as many parties as possible, and years were wasted in tedious and contentious arguments. In fairness, the problems were ultimately solved, so keep these machines on your approved bid invitation list for future consideration.

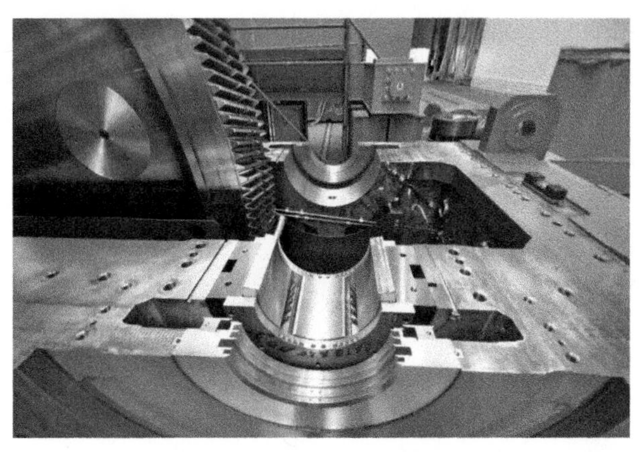

FIG. 6.2.1 One of several pinion bearing locations in an integrally geared multistage process gas compressor. *(Courtesy MAN Turbo, Oberhausen, Germany.)*

Regrettably, both of these experiences and many other case histories are often hidden from our view. The parties sign protective agreements because neither side wants their reputation blemished. All involved are reluctant to publish detailed articles or papers about these events, and lessons learned turn into lessons forgotten.

FIG. 6.2.2 Main bull gear in an integrally geared multistage process gas compressor with single-vendor responsibility. *(Courtesy MAN Turbo, Oberhausen, Germany.)*

FOCUSING ON THE BRIGHT SIDE

In contrast, much less risk exists when a purchaser opts for entire compressor systems from manufacturers with a proved record of long-term quality and as-promised performance. A case in point are modern process compressor packages (Figs. 6.2.1–6.2.3). The lower half of such a compressor is shown in Fig. 6.2.2 where the drive gear (bull gear) is clearly the largest major component. A fully assembled machine is shown in Fig. 6.2.3.

Quality machines are built in accordance with applicable portions of the different American Petroleum Institute (API) standards covering integrally geared centrifugal compressors. The design intent is to fully and seamlessly meet the demands of the petroleum, chemical, power, and gas industries.

FIG. 6.2.3 Assembled view of an integrally geared multistage process gas compressor. The vendor accepts full responsibility for providing a turnkey compressor-driver package. Note that part of the support base serves as the lube oil reservoir. *(Courtesy MAN Turbo, Oberhausen, Germany.)*

To date, integrally geared compressors have been built for services including gas turbine fuel gas (methane), hydrocarbon refrigeration gases, ammonia, carbon monoxide, carbon dioxide, syngas, gas field gathering, and many other services. Conventional wisdom used to suggest or assume that integrally geared compressors are somewhat smallish machines. Today, however, we see flowrates ranging to 400,000 m^3/h, molecular weights from 6.5 to 58 MW, stage pressure ratios as high as 2.6:1, inlet pressures from ambient to 45 bar, and discharge pressures to 220 bar. Integrally geared compressors exemplify the best in single-point responsibility for major machinery assets.

One of the most experienced manufacturers of integrally geared compressors has designed these for drivers up to 60,000 kW. Another manufacturer specialized in considerably smaller machines has, over a number of decades, produced 12,000 integrally geared machines. As of 2016, up to eight (8) high-speed stages are available per gearbox, and even more stages are being considered for future development. Today's efficiencies can approach 90%. Seal designs have kept pace with modern dry gas sealing technology, as have their well-proved intercoolers.

In any event, there are many opportunities to use multistage integrally geared machines where mere tradition may have steered us in the direction of positive displacement machines. The design of these integrally geared late-generation dynamic compressors overwhelmingly favors single-point responsibility. As an additional feature, modern integrally driven compressors often represent the best possible blend of standardization and customized design. These machines certainly merit consideration and invite comparisons with every other type of compression machinery. Upon closer examination, we may have to adjust our thinking about packaged fluid machines for process plants.

Chapter 6.3

Toroidal Chamber Rotary Compressors

INTRODUCTION

Although toroidal chamber compressors represent proved technology, they are not widely used in process plants. Then again, we want to briefly describe them for the simple reason that machinery engineers may wish to know about the general parameters of this machine type. In essence, these rotary compressors offer the same performance as traditional low-flow compressors but do so at considerably lower speeds: 500–6000 rpm. The flow is continuous and surge-free.

OPERATING PRINCIPLE

The compression chamber is a toroidal channel in which an impeller rotates. Patent application illustrations available on the Internet depict construction details for these compressors. As a rule, the gas trapped between the vanes is centrifugally forced to the periphery of the chamber and swirls around the core before it is caught by the next vane on the wheel, repeating the process all around the channel and transforming the kinetic energy into pressure. A stripper separates the inlet from the outlet ports and helps in guiding the gas flow from suction to discharge. The clearance between the impeller and stripper is very tight to limit the gas slippage.

The location, orientation, area of inlet/outlet ports, the geometry of the vanes (bending radius, height, penetration, and angle), the shape and area of the chamber, and the shape of the channel impart different compressor characteristics (flow rate, pressure, and efficiency). Accordingly, they offer multiple solutions for varied process conditions.

APPLICATION SUMMARY

In 1970, the first of five toroidal chamber rotary compressors entered service in nuclear test loops. Each of these machines was used to circulate 2240 scfm (standard cubic feet per minute) of CO_2 at a temperature of 660°F; the suction and discharge conditions were 870 and 942 psia, respectively. Since then, many of these single or multistage rotaries have seen service on gases ranging in molecular weight from 2 to 44, flows from an extremely low 15 scfm to as high as 39,000 scfm, and power inputs from 15 to over 600 hp.

Several manufacturers offer toroidal chamber rotary compressors for boosting and recycling of hydrocarbon gases and hydrogen-rich mixtures. These machines serve in gas-phase reactor circuits, regeneration and molecular sieve applications, fuel gas boosting duties, and vent/purge gas recovery. Efficiencies can range as high as 60%, and pressure rise up to 15 bar/220 psi is not unusual on three-stage machines. Both dry and wet sealing arrangements are in service; high reliability and attractively low maintenance cost are claimed. Within their particular operating ranges, toroidal chamber compressors may merit attention in the future.

Chapter 6.4

Vent Gas Compressor Selection

INTRODUCTION

Vent gas compressors fit in a category of positive displacement machines often used in low-cost petrochemical plants. As a general rule and compared with other available options, sliding vane compressors are selected for vent gas duty by user-purchasers whose emphasis is on low initial cost. If emphasis is placed on maintenance cost savings and long uninterrupted run length, screw compressors or even reciprocating machines are often preferred. However, this chapter of our text highlights certain principles relating to experience checks and follow-up that carry over into asset selection in general. Keep in mind that reliability principles are universally applicable.

EMPHASIS OFTEN DIFFERS

Energy conservation is of ever-increasing importance to process plants from the smallest to the largest. Gas streams that were flared a decade ago are now often recompressed and recycled back to the suction vessel or sent to furnace heaters and other energy consumers. Depending on flow and compression ratio, a wide range of compressor configurations, both positive displacement and dynamic, could be considered. As a side comment, twin-screw "wet" rotary positive displacement machines are in contention for many of these duties.

It should be noted, however, that the economic justification and return on investment calculations for such projects are often based on cost estimating manuals that list the least expensive compression machinery. We have always expressed the view that only a thorough life-cycle cost analysis can determine which equipment makes most sense. Yet, even the reliability-focused will admit that economics will occasionally favor certain less expensive machines. Therefore, they are in contention or "in the running" if future maintenance is not a factor in the decision-making process.

In essence, sliding vane compressors are sometimes offered for vapor recompression services. They are rotary-element gas compressors in which spring-loaded sliding vanes are evenly spaced around a cylinder. As this cylinder revolves off-center in a surrounding chamber, the vanes pick up, compress, and discharge the gas being processed. In fact, these compressors had somehow been recommended to an overseas reprocessor of used lube oil where the available vapors contained naphtha, gas oil, and a host of other gases.

EXPERIENCE CHECKS ARE IMPORTANT

A thorough experience check and open communications with competent vendors will assist in leading the purchaser in the right direction. Sliding vane compressors from "A" might be better than sliding vane compressors from "B." Service competency could favor one over the other, as might maintainability, accessibility, space occupied ("footprint"), operating efficiency, downturn (reduced capacity), and overload considerations. Prevailing monetary exchange rates might make certain US-built machines attractive for overseas purchasers. However, the sliding vanes of these compressors (Fig. 6.4.1), by definition, make rubbing contact and will be experiencing more wear than certain compressors with lower—or zero—sliding contact. Gas composition and cleanliness also make a difference in the selection process, and the sliding vane machine should be assumed to experience lower availability and higher maintenance cost. This must be weighed in making the buying decision.

FIG. 6.4.1 Cutaway view of sliding vane compressor often used in vent gas service. *(Courtesy GE Oil & Gas, Houston, Texas, USA.)*

FOLLOW-UP IS NEEDED

In any follow-up communication with the vendor, the potential buyer should ask for references and should expend the necessary effort to check these out via phone calls or e-mail or even by making a visit to the present user's plant site. The question of vane tip velocity, heat generation and dissipation, suitable oil type, oil-to-process compatibility, and even oil consumption and lubricant cost must be included in the assessment. For higher overall reliability and availability, one might look at twin-screw rotary compressors. Dry screw and wet screw machines are used in these services, and only a thorough life-cycle cost analysis will tell which one might be appropriate.

The same analysis is needed if certain (typically higher) compression ratios point to reciprocating machines. There is much favorable experience with labyrinth piston reciprocating compressors. These too are available in many sizes and will certainly require less maintenance than seemingly equivalent reciprocating compressors with piston rings and rider bands.

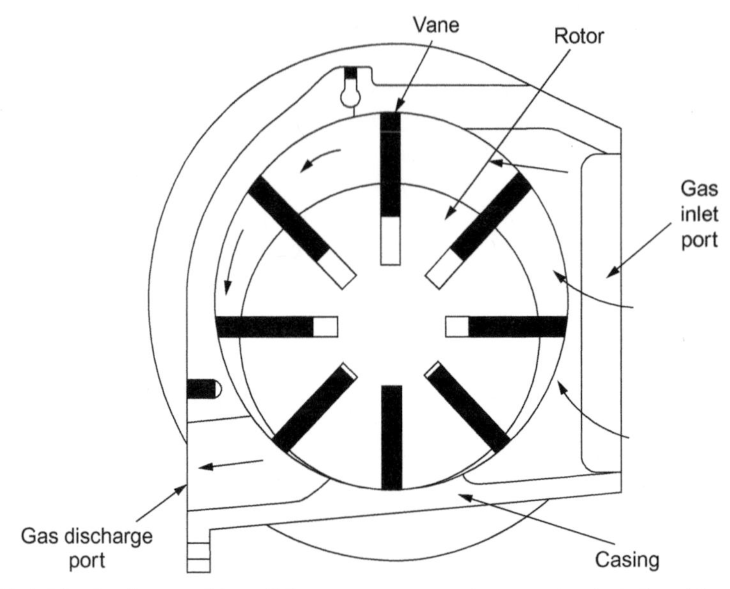

FIG. 6.4.2 Small motor-driven sliding vane compressor in a petrochemical plant. *(Courtesy GE Oil & Gas, Houston, Texas, USA.)*

Sliding vane compressor interiors are relatively simple; major parts are shown in Fig. 6.4.2. These machines are often advertised as a package. "Package" implies that the compressor manufacturer or another entity has mounted the compressor and its support system on a skid. These skids facilitate installation, and the purchaser benefits from installation cost advantages.

But whenever you buy a "package" and regardless of its physical or power-indicated size, be sure that the overall turnkey responsibility is clearly defined in the contract. Beware of situations where the party supplying the compressor refuses to warrantee performance of the lube supply or lube injection modules or where supervisory instrumentation becomes an undefined "gray" zone. Frankly, the downstream concerns are overwhelming in those instances. Wrestling with these concerns can be costly and time consuming. They often end in litigation.

Subject Category 7

Cooling Tower Fans and Drives

Chapter 7.1

Consider Innovative Cooling Tower Fan Motor and Drive Technology

INTRODUCTION

Since 1986, a cooling tower engineering firm ("CTE," for short) has been designing and installing high-quality cooling towers for the large institutional market, including hospitals, universities, and airports (Fig. 7.1.1). The systems use an air-conditioning approach that requires a cooling tower to exchange heat and return cooled water back to the chiller. However, for two or more decades, CTE had been searching for a better method of driving fans in cooling towers. In about 2008, they found what they had been looking for in Baldor's RPM AC Direct Drive Cooling Tower Motors (Fig. 7.1.2).

THE SEARCH FOR A BETTER SOLUTION

In the arrangement of Fig. 7.1.2, the same very compact electric motor and control technology that is used to power today's most sophisticated hybrid automobiles have been adapted for cooling tower drive applications. Advances in motor power density using laminated steel frame construction are combined with high-flux strength neodymium iron boron (NdFeB) permanent magnet salient pole rotor technology. This resulted in the full commercialization of a high-torque, slow-speed, low-profile motor that is mounted directly to the fan and operates at variable speed to maximize system efficiency. The permanent magnet rotor design allows these fan drive motors to excel in many ways. The motors produce high torque at superior efficiency, low weight, low noise, and very high reliability. Power is supplied via a cooling tower drive module made by the same company.

Petrochemical Machinery Insights. http://dx.doi.org/10.1016/B978-0-12-809272-9.00007-4

FIG. 7.1.1 Traditional cooling tower configuration with right-angle gear and fan.

FIG. 7.1.2 Baldor variable speed direct drive cooling tower fan arrangement.

A MATURE PRODUCT

It took little time for these products to mature, and many are now installed in petrochemical plants and oil refineries. Sizes in the 150kW range are not uncommon, and Saudi Aramco oil company is installing low-speed direct drive motor fan units on over 75 cooling tower drives and heat exchanger fin fans on a grassroots gas plant project scheduled for completion in 2017. In Baldor's RPM AC Direct Drive Cooling Tower Motors, this major user found a product which neatly sidesteps all of the issues of a traditional system. "If you don't mind the phrase, I think it's a simple and elegant solution," said a seasoned cooling tower professional with CTE. From his vantage point, it is elegant in the sense that a user will have traded all of the old-style components for one moving part.

Finally, many retrofits have been implemented because retrofits are neither difficult nor time consuming. The Baldor engineers and designers succeeded in creating a low-profile motor design that fits in the same space and a mounting footprint identical to the traditional gearbox. Indeed, it is almost just a drop-in replacement. In all, we believe that permanent magnet-based variable speed low-profile salient pole motors deserve our consideration in original projects and retrofit applications. The reasons can be summarized in just four important findings:

(1) Elimination of drive shafts, couplings, and gearboxes in cooling towers.
(2) Elimination of gearbox lubrication and maintenance.
(3) Elimination of belts, pulleys, drive shafts, bearings, and their maintenance in forced draft process heat exchangers.
(4) Variable speed drive motor is standard drive, can pay back costs in 1–2 years' operation by energy savings alone. And if there is any question on the reliability of permanent magnet salient pole motors, the manufacturer can be required to furnish a 2- or 3-year motor warranty.

It will be interesting to see the cost breakdown for the various fan drives in contention for a particular application. Be on the lookout for developments as time goes on.

Subject Category 8

Couplings

Chapter 8.1

Coupling Types and Styles

INTRODUCTION

Couplings serve to connect the rotating shaft of a driver to the shaft of the driven machine. Although the shaft centerlines are usually well aligned (see earlier text chapter on alignment), maintaining perfect alignment is difficult or near-impossible. A flexible component is thus needed between the coupling hub fitted to the driving shaft and the coupling hub fitted to the driven machine's shaft. Flexible couplings can accommodate a small amount of parallel and angular misalignment; a flexing element reduces both vibration amplitude and bearing loads.

FLEXING ELEMENTS MADE FROM ELASTOMER MATERIALS

The basic components of elastomeric couplings with a deformation-tolerant element (usually a block or star shape) in *compressive load* come in dozens of different styles. Their use in special purpose (meaning nonspared, extended run length) petrochemical machinery is usually confined to large slow-speed machines. Here, the primary purpose of an elastomer block or contoured insert is to avoid torsional resonance conditions.

Resonant frequencies are rarely an issue in small pumps. However, couplings incorporating flexible elastomer elements in compression are frequently applied in somewhat smaller close-coupled process pumps. In these, yearly maintenance and parts replacement are the norm.

In *elastomeric couplings* with tire-like flexing elements, the "tire" undergoes twisting; *tensile loading* prevails. There is similarity between this flexing element and certain vehicular tires. Rim geometries are suitable for securing the sides of the "tire" to an adjacent coupling hub. Other brands of elastomeric flexing elements are designed with a cross section resembling "omega," the last letter in the Greek alphabet. Regardless of cross section, some models or styles

Petrochemical Machinery Insights. http://dx.doi.org/10.1016/B978-0-12-809272-9.00008-6

of flexing elements have each rim fused into a metal ring. In turn, the metal ring has to be fastened to the adjacent coupling hub and the fasteners must be selected with great care. Torque wrenches must be used to securely and properly tighten the various bolts.

Because elastomeric couplings with tire-like flexing elements are designed with the flexible element twisting or pulling, they are rarely used in large fluid machines. They are, however, applied in so-called general purpose (as contrasted with "special purpose") small- to medium-sized pumps.

A number of factors come into play whenever elastomeric couplings are chosen:

- Toroidal ("tire-type") flexing elements can exert an axial pulling force on driving and driven bearings. Also, they are difficult or near-impossible to balance.
- Polyurethane flexing elements (or inserted elements in compression) perform poorly in concentrated acid, benzol, toluene, steam, and certain other environments.
- Polyisoprene flexing elements (or elements in compression) do poorly in gasoline, hydraulic fluids, sunlight (due to aging), silicate, and certain other environments.
- Components made of natural rubber may be susceptible to ozone attack.

OBTAIN USER FEEDBACK

User experience feedback is of great importance in selecting toroidal couplings. Well-trained reliability professionals always establish that a particular offer meets their company's long run length or low failure risk expectations. These professionals compare warranty duration and coverage, specific features, and power capabilities, including installation details, tooling, and allowable combined parallel and angular misalignment.

Occasionally, pump and coupling users have encountered sales or marketing personnel who will deflect user questions in the hope of making a sale. The primary focus of such marketing strategies may be low initial cost and attractive wear part replacement cost. A certain size elastomeric coupling may indeed cost "X" dollars less than a competing longer-lasting coupling. But what if the initially less expensive coupling has to have its flexible elements replaced far more frequently? What if running a coupling to failure can cause parts to become airborne? What if a precautionary shutdown prevents failure, but costs a (hypothetical) "7X" dollars? Be sure you know as many answers as possible before buying couplings. You may even look at past litigation involving potential vendors and verify that a particular causal event has since been addressed in suitable redesigns. Alternatively, changes in maintenance or related work procedures should have been implemented.

Gear couplings require grease replenishment and are certainly more maintenance-intensive than nonlubricated disk pack couplings. If gear couplings are selected, they should be on the plant's preventive maintenance schedule. Only approved coupling greases should be used for periodic replenishment. Realize that standard and multipurpose greases are unsuitable because their oil and soap constituents are getting "centrifuged apart" at typical pump coupling peripheral speeds.

That usually leaves a reliability-focused user with *nonlubricated disk pack* (Fig. 8.1.1) or other variants of alloy steel membrane couplings (e.g., diaphragm couplings) as the preferred choice for lowest life cycle cost, or LCC. The LCC design must be such that the spacer piece or center member is "captured," meaning retained between opposing coupling hubs. In the unlikely event of a disk pack failure, a "captured center member" will not leave the space between the two coupling hubs.

API-610-compliant disk couplings are designed to be fatigue-resistant for a theoretically infinite service life, as long as the bending stresses applied to the coupling disk packs do not exceed their rated limits. The disk packs are mounted in a cartridge that is removable without any need to disturb the driving and driven hubs, allowing for easy access to pump seals. The various coupling elements have overlapping features ("captured center members") to mitigate the risk of the spacer tube being thrown from the machine in case of coupling failure due to overload. This particular coupling design transmits rotary power by pure face friction, eliminating shear loading and stress concentration in the assembly bolts, making the system more reliable.

It should be noted that many brands of inexpensive couplings do not have captured center members. Avoid using them.

FIG. 8.1.1 A widely used version of a reliable high-quality disk pack coupling for fluid machinery, including large process pumps. *(Courtesy R+W Couplings, Klingenberg, Germany.)*

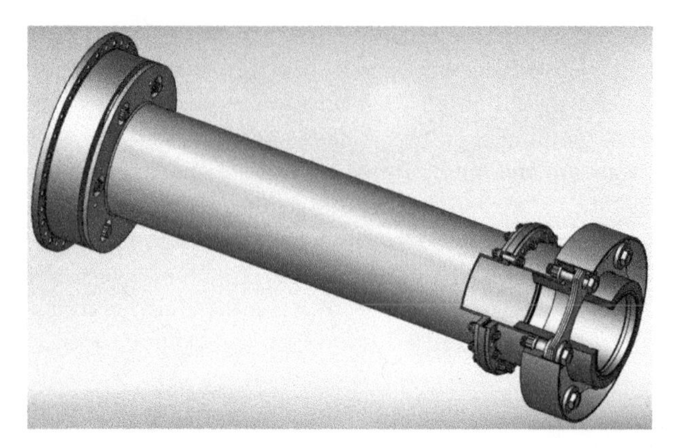

FIG. 8.1.2 A highly reliable special purpose (API-671 compliant) high-quality disk pack coupling for fluid machinery, including large process pumps. Image shows left side configured for gearbox fit-up; the right side is configured to fit the driver's output shaft. *(Courtesy FlexElement Texas, Houston, TX.)*

The thoroughly engineered disk pack coupling of Fig. 8.1.2 is used in numerous upgraded fluid machinery installations. It complies with API-671 and other high-performance standards. Its primary range of application is in 1000–100,000 horsepower turbomachines. Among its important positive attributes are high-quality materials, advantageous machining methods, and optimized disk coatings. Lead times, that is, the time elapse between order placement and delivery, are almost always very attractive.

INSTALLATION AND HUB REMOVAL

Before installing a coupling, examine it for adequacy of puller holes or other means of future hub removal. The coupling in Fig. 8.1.3 was mistreated at disassembly because no thought had been given to future removal.

FIG. 8.1.3 Coupling hub damaged with severe hammer blows at disassembly.

For parallel pump shafts with keyways, use 0.0–0.0005″ (0.0–0.012 mm) total shaft interference. Depending on user experience, several available thermal dilation methods (heat treatment oven, superheated steam, and electric induction heater) are available to mount hubs on shafts.

Disallow loose-fitting keys for coupling hubs because they tend to cause fretting damage at shaft surfaces. During rebuilding or repair, allocate time needed for hand-fitting keys; they should fit snugly in keyway. On all replacement shafts, reliability optimization requires avoidance of sharp corners in keyways. Arrange to have only "radiused" corners (meaning that the corners have fillet radii). Sharp-edged keyways are locations where excessive stress concentrations exist.

Reliability-focused petrochemical plants often modify the key contours to match the radius contour. Others are specifying coupling hub bores with special keyless fit arrangements. Highly advantageous friction-based hub clamping options are available from some of the best coupling manufacturers.

Subject Category 9

Dry Gas Seals

Chapter 9.1

Advances in Dry Gas Seal Technology for Compressors

Centrifugal process gas (and also certain rotary) compressors require sealing elements between the pressurized gas-containing volume and the bearing assemblies that support the compressor rotor. A variety of different seal types and styles have been available for decades; in the majority of these seals, lubricating oil serves as the fluid that separates rotating from stationary sealing elements. The barrier oil for these "wet" compressor seals is typically introduced at pressures approximately 30 psi (2 atm) higher than the opposing compressed process gas. Wet seals require a seal oil supply system that generally includes an oil reservoir, two or more oil pumps, filters, coolers, valves, and control instrumentation.

Already three decades ago, some rather successful dry gas seal (DGS) applications (Fig. 9.1.1) became viable sealing options. Although wet seals will probably remain in contention for a while, a number of considerations have accelerated the development of DGSs. As of this writing, seven years of uninterrupted compressor service are no longer the exception for services with clean gases. This partially explains why, over the past 20 years, DGSs have displaced many of the different precursor seal styles. Another recent development began in about 2006 at a highly innovative UK-based seal manufacturing company [1]. The company has taken on both the design of their own and the successful refurbishment of all kinds of DGS assemblies originally provided by other seal (or compressor) manufacturers. The functional features of DGSs are described in the succeeding text, and this chapter also provides an overview of related developments.

Petrochemical Machinery Insights. http://dx.doi.org/10.1016/B978-0-12-809272-9.00009-8

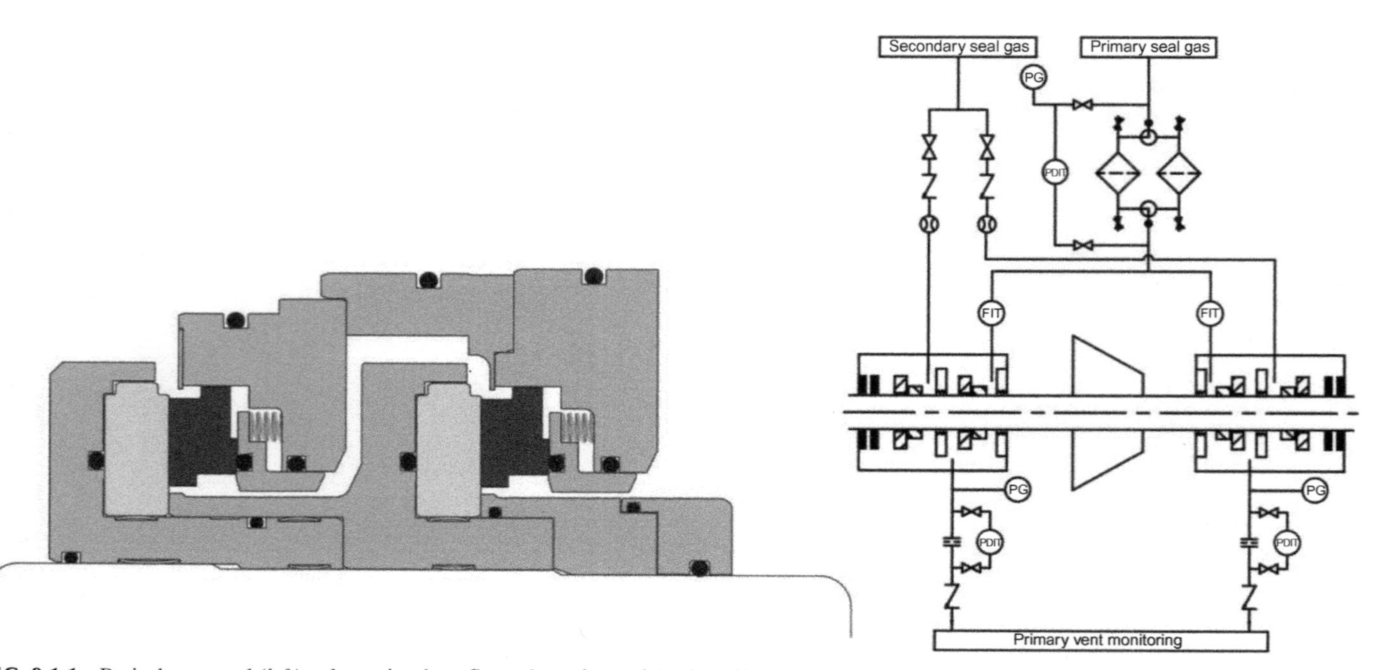

FIG. 9.1.1 Basic dry gas seal *(left)* and associated gas flow schematic, on right side of image. *(Courtesy AESSEAL, plc, Rotherham, the United Kingdom.)*

HOW DRY GAS SEALS FUNCTION

Swwimply put, DGSs operate by creating and maintaining a very thin gas film (<5 μm) between two mating disklike surfaces, one stationary and one rotating [2]. The stationary face is spring-loaded, and maintaining this gas film under all operating conditions is essential to reliable seal operation. The gas film is so thin that the most efficient method of demonstrating its existence is to perform a rigorous running test. For this reason, all DGSs should be dynamically tested at the time of manufacture and also after seal repair (often called seal refurbishment). The basic schematic representation in Fig. 9.1.1 shows a DGS and its associated controls. Depending on the nature of the gas and the criticality of service, the design of gas conditioning and DGS support and monitoring system can be quite sophisticated.

There are many functional similarities between DGSs and their precursor models. The precursor or predecessor oil-lubricated ("wet") seals include many variants of face, bushing, and floating ring seals. Yet, there are component features that differ in wet seals as contrasted with DGSs. For instance, the seal face of the rotating mating ring in a DGS can be divided into a grooved area at the high-pressure side and a dam area at the low-pressure side (Fig. 9.1.2). The shallow grooves are often laser-etched, spark-eroded, or chemically milled.

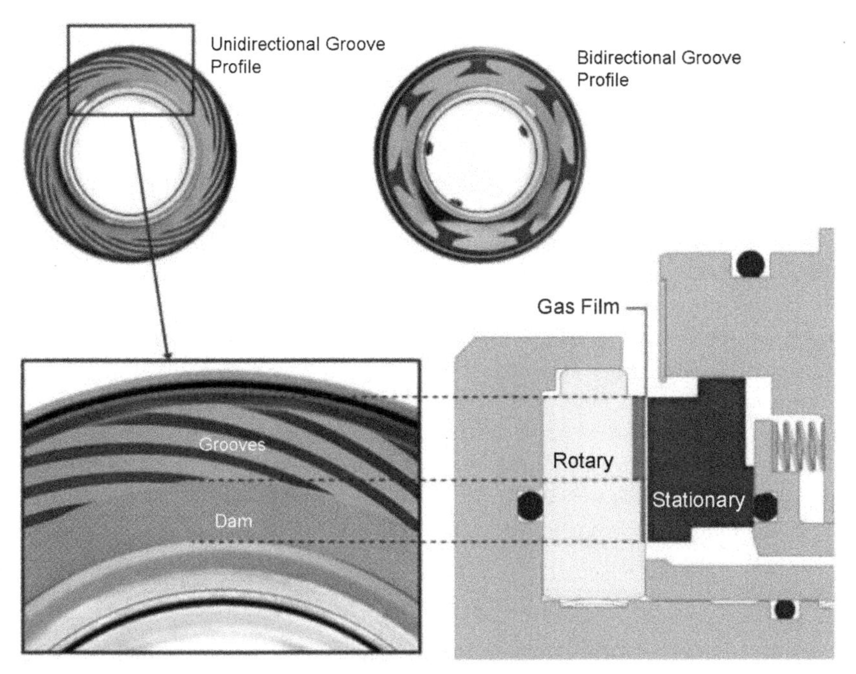

FIG. 9.1.2 Mating rings incorporáte vane-like grooves. Top left depicts a spiral groove; top right is called "swallow tail." *(Courtesy AESSEAL Inc., Rotherham, the United Kingdom; also H.P. Bloch, F.K. Geitner, Compressors: How to Achieve High Reliability and Availability, McGraw-Hill, New York, NY, 2012, pp. 34–40, ISBN: 978-0-07-177287-7.)*

A typical depth is ~0.0003 in. = 8 μ, quite obviously achieved through highly precise machining operations. T-shaped and/or V-shaped grooves are bidirectional. L-shaped (unidirectional) grooves have been produced as well, and each configuration has its advantages and disadvantages. A stationary sliding ring is pressed axially against the mating ring by both spring forces and sealing pressure.

The sealing gap is located between the mating ring and the sliding ring. The needed gas film is achieved by the pumping action of the grooves and the throttling effect of the sealing dam. A suitably designed groove geometry is critical for trouble-free operation of the seal.

Before opting for DGSs in retrofit situations, reliability-focused users ask if an apparently flawed precursor seal really represents the best the vendor was able to offer. Use Table 9.1.1 as your preliminary screening tool [3] and follow up by a more thorough review.

More specifically, the claims of competing vendors and claims made for different styles of seals must be checked against actual experience. In some cases, these checks lead to the purchase of late-generation liquid film ("wet") face seals instead of DGSs. In all instances, the areas of safety and reliability must be given full and special consideration.

MINIMIZING THE RISK OF SEALING PROBLEMS

Upgrading from the traditional compressor seals to advanced DGSs is entirely feasible. There are cases where gas seals are a good economical choice because the user does not have to purchase an elaborate seal oil console. Suppose, however, the compressor train is a few years old and incorporates older, but technically satisfactory, "wet" oil seals. Suppose also that the compressor train's already existing seal oil console has proved reliable. In such instances, it may be difficult to justify the cost of installing modern DGSs.

But experience shows that preexisting "wet" oil seals and their support systems are not always reliable and inexpensive to maintain. These are among the reasons why DGS has come into prominence and why a DGS merits consideration for most process gas compressors [4]. Giving due consideration implies that specification, review, purchasing, and installation of a DGS and its requisite support system cannot be left to chance. An audit of the owner's facility and the particular process unit or gas in which the compressor will be installed is foremost among the steps toward selecting the right seal configuration. These comprehensive and active review steps often lead to the selection of dual seals; dual seals may be needed to maximize DGS life, long-term performance, and equipment and process safety.

Although orders of magnitude less complex and costly than their wet seal counterparts, DGS support systems merit the attention of reliability-focused buyers. In their respective order of importance, the following factors should be considered in examining dry seal support systems for centrifugal compressors:

TABLE 9.1.1 Turbocompressor Seal Selection Guidelines [3]

Experience-Based Recommendations for Compressor Seal Selection			
Application	**Service**	**Inlet Pressure (1) (kPa)**	**Seal Type (psia)**
Air compressor	Atmospheric air	Any	Labyrinth
Gas compressor	Noncorrosive	Any	Labyrinth
	Nonhazardous		
	Nonfouling		
	Low value		
Gas compressor (2)	Noncorrosive or corrosive	69–172 (3)	Labyrinth with injection and/or ejection using gas being compressed as motive gas
		(10–20)	
	Nonhazardous or hazardous		
	Nonfouling or fouling		
Gas compressor	Noncorrosive	≥25,000	Gas seal (4)
	Nonhazardous or hazardous	≥(3600)	Tandem preferred
	Nonfouling		
Gas compressor	Corrosive (5)	Any (6)	Oil seal, double gas seal (6)
	Nonhazardous or hazardous		
	Fouling		

Notes:
1. Operating seal pressure range.
2. Where some gas loss or air induction is tolerable.
3. Pressure ranges shown for labyrinth seals are conservative. Manufacturers extend this range upward, resulting in a debit due to power losses.
4. Within state of the art.
5. H_2S is the most common corrosive component in process gas compressors.
6. Dry running gas seals often have pressure limitations below those of oil seals.

Gas composition: Understanding the actual gas composition and true operating condition is essential, yet often overlooked. For example, it is necessary to define when and where phase changes start and that condensed liquids must not be allowed in the sealing gas. A checklist approach is often helpful:

1. Is clean and dry buffer gas available at all anticipated compressor speeds?
2. Is the seal protected from bearing oil?
3. How is the compressor pressurized or depressurized?
4. How is the machine brought up to operating speed and how will the seal react?
5. Are all operating and maintenance personnel fully familiar with the compressor maintenance and operating manual?
6. Is the full control system included and adequately described in these write-ups?
7. Are the key elements of the system design understood and do they include buffer gas conditioning, heating, filtration, regulation (flow versus pressure), and monitoring?

SEAL SAFETY AND RELIABILITY

Positive sealing of the compressor during emergencies must always be assessed. In the event that the "settling-out" gas pressure in the compressor casing exceeds that of the seal gas pressure, advanced face seal assemblies are designed with shutdown pistons. These small pistons are located and dimensioned to exert a force proportional to gas pressure. Their force action keeps the seal faces closed and prevent gas release to the environment.

It is an understatement to note that working with an experienced DGS repair service provider is always helpful. The non-OEM provider's allegiance is primarily with the seal user. Experience-based input may cast light on where DGSs may or may not provide positive gas sealing if the seal faces are damaged or distorted. The backup seal may show reasonable performance under low pressure, but may fail to perform under higher pressure if the primary seal fails. Elastomeric O-rings are used in all DGSs and recall that Fig. 9.1.1 showed no fewer than 14 of these O-rings. They are being replaced and seal faces relapped to perfection during DGS refurbishment.

The alarm and shutdown devices of the seal oil system must be of reasonable range and sensitivity to provide reliable alarm and shutdown characteristics. DGS systems often rely on pressure switches of very low range and high sensitivity. These switches might have a tendency to malfunction or give a false sense of security. Make sure the system will provide a sufficiently high degree of alarm and shutdown performance.

Some seal configurations excel at online monitoring. The favored seal must allow easy onstream verification of sound working condition. Beware of seal systems that are hampered by small-diameter orifices that tend to plug; avoid sensitive pressure switches that often become inoperable. DGS failures are not as easy to detect as wet seal failures.

In all instances, the well-versed user must go through a rigorous cost justification analysis. Fig. 9.1.3 represents an experience-based approximation and may serve as an example.

COMBINING OEM AND REPAIR EXPERTISE

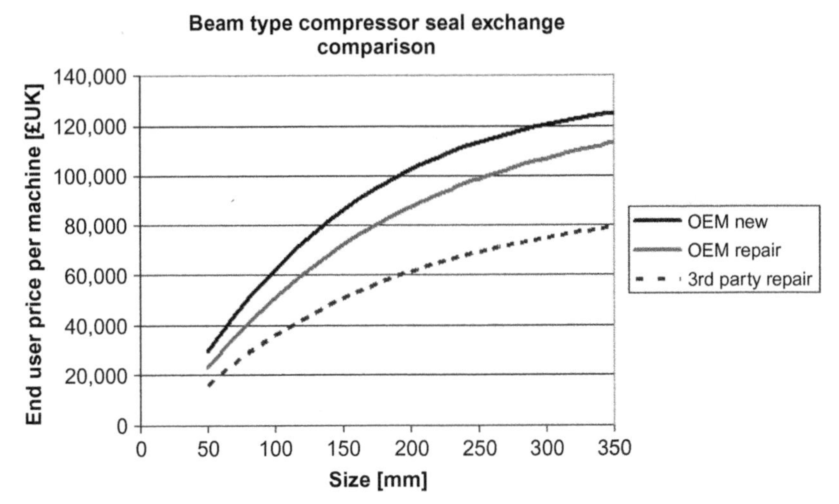

FIG. 9.1.3 An experience-based estimate contrasts likely cost of three different DGS exchange strategies. Similar cost comparison should precede any "buy new" versus "refurbish" decision. *(Courtesy AESSEAL Inc., Rotherham, the United Kingdom.)*

Compressor end users, also known as owner-operators, normally have to face considerable expense when a machine is taken out of service for a routine overhaul. At this point, they may elect to refit brand new seals supplied by the original DGS manufacturer. Given that there is usually no discount, it could make this "go-new choice" an expensive option. Some saving can be attracted by having the seals repaired by the original DGS manufacturer where the associated repair costs are often reaching up to 80% of the price of new seals. In extreme cases when the original DGS manufacturer has full order books and 100% test utilization, they may well elect to price the repair at an even greater cost than a new unit. The original manufacturer may justify this approach so as to improve new seal production efficiency with batch size economics. A more cost-effective option may allow a competent third-party DGS repair specialist to refurbish the seals. Using a typical two-seals-per-machine beam-type compressor as our example, Fig. 9.1.3 illustrates a cost comparison between the three options just outlined.

Regardless of hydrocarbon gas prices, it will be prudent to consider gas leakage rates in order to justify a conversion to gas-lubricated seals. Along these lines, one researcher [4] suggests we study the failure statistics of our oil-type seal and compare these statistics with current estimates. Current failure rate estimates for gas seals are in the vicinity of 0.175 failures per year, meaning that we could expect a problem every six years or so. At least one DGS manufacturer recommends basing maintenance intervals for DGSs on limits imposed by an elastomer's aging process [5]. This manufacturer suggests the following maintenance routine after 60 months of operation:

- Replace all elastomers.
- Replace the springs.
- Replace all seal faces and seats.
- Carry out a static and dynamic test run on a test rig.

MAKING TECHNICALLY SOUND CHOICES

Thus, our final advice to the reader is to make informed choices. Consider DGSs only in conjunction with a clean gas supply. Bottled or industrially available nitrogen is entirely feasible, but could cost considerable money. If your process gas can cause fouling deposits to develop, ask critical questions of anyone offering DGSs for use with that kind of gas. If extensive microfiltration is needed, factor in the cost of maintaining a DGS support system. Look for seals that will survive a reasonable amount of compressor surging. Consider DGSs that incorporate features ensuring start-up and acceleration to full operating speed without allowing the two faces to make contact. If these seals are not available from your supplier, look beyond the usual sources.

Recall again that at least one innovative manufacturer offers modern DGSs for OEM as well as aftermarket applications. Also, this manufacturer has the capability to repair and test DGSs made by others [6]. In the 6 or 7 years leading up to 2015, a considerable number of users and sites have been added to the reference list published in 2009. Today, in 2016, reliability engineers are urged to investigate the extent to which any compressor seal manufacturer can meet the owner-purchaser's reliability requirements. The logical follow-up would be to consider using DGSs from experienced manufacturers.

MAKING ECONOMICALLY SOUND CHOICES

Most compressor OEMs have service and rerate divisions that will upgrade, repair, refurbish, and modify anyone's machines. There are also third-party independent repairers that are often formed, owned, or staffed by former employees of compressor OEMs. That simply means that either the OEM or an independent non-OEM compressor manufacturer will service, overhaul, repair, and even rerate entire compressors regardless of origin.

However, while compressor users tend to accept as standard practice all of these repair options, the same is not true for DGSs. For reasons nobody can explain, DGSs tend to be sent back to seal OEMs. This is clearly a paradox because an entire compressor train is far more complex than a gas seal. Moreover, a highly qualified DGS manufacturer may be eminently more qualified to do DGS refurbishments than (some) compressor manufacturers.

DGS production for new compressors has been ramped up since about 1995; the same is true for the well-planned implementation of dry seal retrofits. As with all industries where demand for new products outstrips the capacity to support existing products, both pricing structure and lead time for component repairs have come under scrutiny. In other words, primary emphasis on the manufacture of *new* DGSs has resulted in service issues related to product engineering and repair support for DGSs that have been in use for several years.

A major UK seal manufacturer has anticipated this issue even as it began to develop. So, in the 2000–04 time frame, this manufacturer started to design and establish facilities that are specifically aimed at the repair and testing of dry gas compressor seals (Fig. 9.1.4). As of late 2015, they have repaired many hundreds of DGS units; they are also engaging in thorough review and updating of seal gas control panels whenever such efforts are called for.

FIG. 9.1.4 One of two dynamic test rigs at a DSG manufacturer-refurbisher's factory location. *(Courtesy AESSEAL Inc., Rotherham, the United Kingdom.)*

TESTING CAPABILITIES ARE IMPORTANT

Unlike some OEM repair facilities that do a shortened test on repaired DGS units, this seal manufacturer-rebuilder fully tests all DGS units, both new and refurbished, as outlined in API 617 (Fig. 9.1.5). Only full testing harmonizes with today's emphasis on asset reliability. It has been said that modern test facilities are the key to competent repairs.

Of course, properly engineered DGSs contribute greatly to overall machine reliability. Advanced seal designs reduce overall maintenance requirements; they also save power and gas consumption and thus contribute to sizable operating cost savings in gas transmission services.

DGS TECHNOLOGY ADVANCES THROUGH GLOBAL REPAIR SERVICES

One of the premier global DGS repair services has upgraded the customary repair and inspection facilities to unparalleled standards of excellence. Since the repair and testing program was aimed at any type, configuration, or specification of DGS, the new facility had to be devised with utmost versatility in mind. It therefore ensured that the repair of large-diameter and high-speed seals is accommodated. In the autumn of 2005, a new electrical substation dedicated to reliably supply power to the company's DGS test facility was commissioned to supply two DGS test rig motors as well as the other services necessary to conduct dynamic tests on these seals. Each of the two test rigs contains a state-of-the-art inverter drive enabling precise speed control of the 106 kW-rated motors up to a maximum speed of 7000 RPM. Each motor provides the input to a planetary gearbox with three output modules that can be interchanged to provide output speeds up to 45,000 RPM. The power and speed capacity of each test cell allow the dynamic testing of large-diameter high-speed seals under full operational conditions.

THE DGS REPAIR PROCESS

The ability to test DGSs to high standards is only one aspect of a full repair service. A leading company has drawn upon many years of experience in the conventional seal industry and transferred the same service-based culture into a first-class DGS repair program. The first element of the repair process is usually an initial price quotation that can also be based upon electronically transferred images or historical repair knowledge of the seal's condition. This allows the service provider to quickly respond to customer demands for reasonable estimates of cost and time to repair; the estimate sometimes leaves the office before a defective seal arrives at the DGS service facility. Once a seal arrives, it is disassembled and mapped for damage assessment. The damage is then documented in a concise examination report that provides the full scope of work needed to restore the seal to original as-designed condition. Firm price and delivery are rapidly offered together with details of the dynamic test protocol, one normally conducted in accordance with API 617.

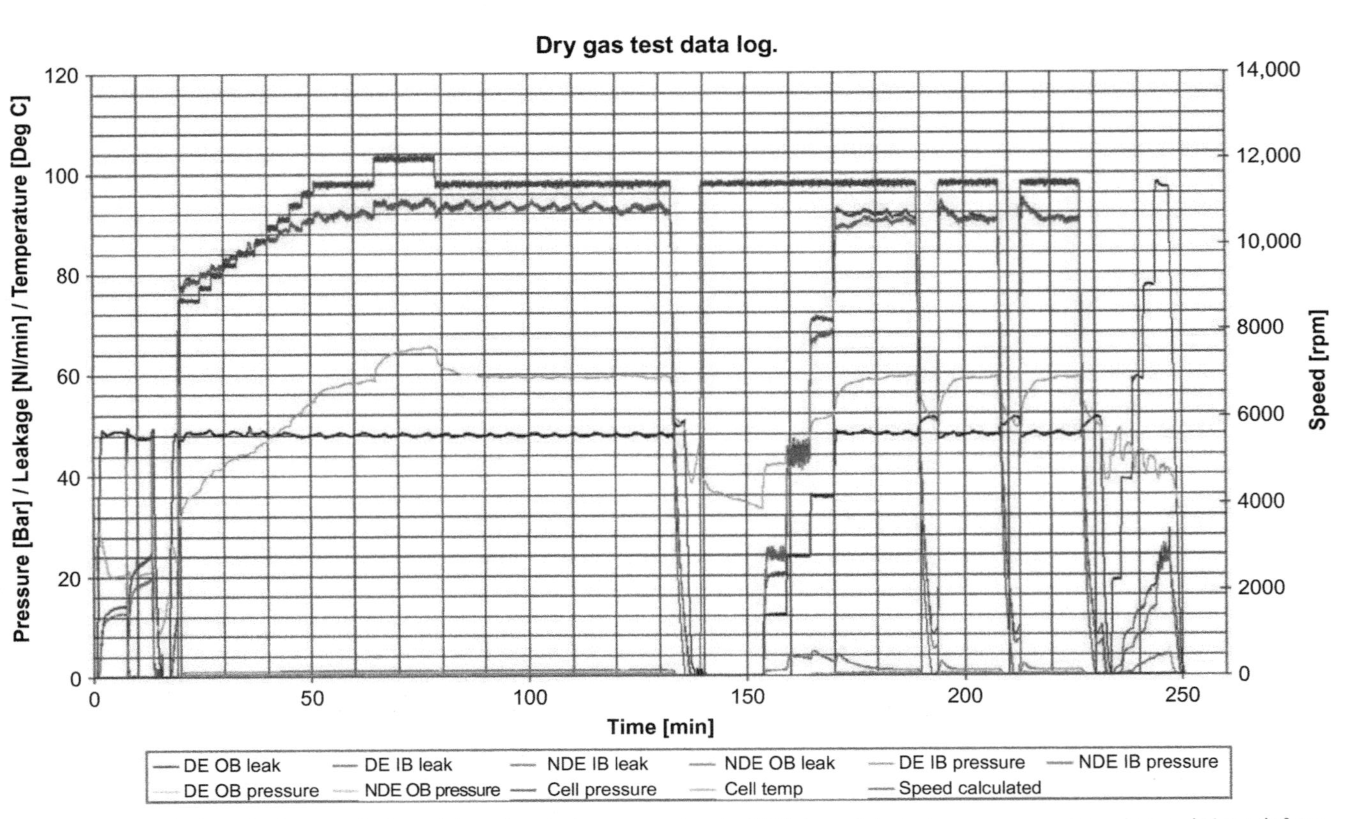

FIG. 9.1.5 A typical graph relating to the logged data taken during a comprehensive DSG dynamic test where pressure and speed are varied to suit future operating conditions. *(Courtesy AESSEAL Inc., Rotherham, the United Kingdom.)*

Already at this stage of the repair, initial design work is conducted. This design work typically covers equipment and fixtures—such as spin test and balance mandrels—needed later in the repair process. Because the facility is geared toward different seal and compressor designs, fixtures must be purposefully designed for a particular repair. They will later be needed to verify the integrity of rotating face materials and, if required, to effect dynamic balancing to ISO Standard 1940. A complete material assessment is done, whereby all materials of construction are fully certified. Providing both rapid repair response and all requisite documentation is facilitated by close links to top suppliers of raw materials. The company has made it a practice to order materials beyond those needed for a particular repair. The resulting buildup of DGS part stock advances the speed of responding to numerous repair requests.

REPLACEMENT AND REBUILDING STEPS

Replacement of damaged faces is an important facet of the repair process. Premium-grade corrosion-resistant tungsten carbide, reaction-bonded and sintered silicon carbide, and silicon nitride seat replacements are offered. Stationary faces can be supplied in blister-resistant carbons and diamond-like carbon (DLC)-coated silicon carbide where additional PTFE (Teflon) coatings are also be applied to seal faces. Steel components can be repaired and recoated as required, and where replacement parts are necessary, full code requirements can be met. In all repairs offered by competent DGS refurbishers, replacing consumable items is standard. Such replacement includes all springs and fasteners; also included are secondary seals, where all O-rings fitted to DGS repairs are explosive decompression-resistant grades. Whenever polymer upgrading is feasible, the original selections are replaced with spring-energized PTFE-equivalent sealing devices.

Since most DGSs are supplied with external barrier devices, these components can also be repaired along with the seals. Labyrinth components and both contacting and noncontacting segmented seals are often supplied as part of the repair program.

Once all of the replacement parts have been produced, reassembly and testing can proceed. A full spin test is conducted at 23% over the maximum speed rating of the seal. The component itself is thus subjected to twice its normal rotational stress. Also, whenever applicable, full dynamic balancing is done on all rotating assemblies. Similarly, all stationary assemblies are subjected to pressure test that exercise the face assembly and serve to identify seal hang-up. Only then are the various assemblies finally fitted into the test equipment shown in Fig. 9.1.4.

Clients can observe DGS testing at the company's main facility. Even if unable to attend, customers can still participate anywhere in the world and view the test progress in real time via a *WebEx* Internet link. A second remote surveillance option involves a 3G mobile phone link, whereby the test is being filmed with a 3G camera. This option is available in the form of live video streaming to any mobile phone with suitable 3G features. Irrespective of whether customers choose to witness the tests, each test is fully recorded and a CD of the full test event is produced for each individual seal. The CD is normally supplied with the information pack that is placed inside the shipping cartons for the repaired seals.

DURATION AND DOCUMENTATION

Although tests are usually conducted to stipulations or compliance in harmony with API 617, at least one experienced GGS refurbisher often accepts and carries out additional customer testing requirements. Dynamically testing each seal typically takes three or four hours, which includes second-by-second data logging.

The company provides an information packet for each repaired seal; the data packet normally includes the prerepair inspection report, a posttest condition inspection report, full details of the dynamic test itself, a full set of tabulated dynamic test results, a graph of the data logging results, and a full set of installation details. It includes a spin test certificate, a balance test certificate, a dynamic test certificate, and full certificates of conformity for any materials used. The repaired seal is shipped in a heavy-duty airfreight case that incorporates carrying handles and a locking mechanism. In addition to the actual seal repairs, an assembly tool replacement service is offered. This is particularly useful when seal assemblies were produced at locations remote from the compressor factory, but the compressors were shipped without the best available DGS assembly tools.

Nevertheless, the success of a seal design and manufacturing company that can provide seals of their own design and point to the successful refurbishment of all kinds of DGS assemblies deserves to be considered here. At least one company's DGS repair service continued to grow in the decade leading up to our 2016 publishing date [1]. As of that time, they had received and completed repairs on hundreds of these seals with a multimillion dollar value covering several OEM product ranges.

In late 2014, one researcher suggested to study the failure statistics of your oil-type seals and to compare them with his then current statistics. At that time, failure rates of gas seals were in the neighborhood of 0.175 failures/year, meaning that users could expect a problem every six years or so [4]. At least one seal manufacturer bases recommended maintenance intervals around gas seals on limits set by the elastomer aging process. He suggests the following maintenance routine after 60 months of operation:

- Replace all elastomers.
- Replace the springs.
- Replace all seal faces and seats.
- Carry out a static and dynamic test run on a test rig.

Clearly, then, compressor owner-operators now have unprecedented and solidly verifiable options [5]. They can decide to continue working with the

OEM or can trust the demonstrated capabilities of a firm whose expertise in seal design is firmly anchored in a totally modern and highly accessible testing facility. Most importantly, the customer is now at liberty to investigate the full extent of this company's commitment to timely repairs and superbly executed testing of any compressor DGS in use today.

REFERENCES

[1] C. Carmody, Dry Gas Seal Repair and Testing, CompressorTech2, 2009. December.

[2] H.P. Bloch, Consider Dry Gas Seals for Centrifugal Compressors, Hydrocarbon Processing, 2005. January.

[3] H.P. Bloch, F.K. Geitner, Compressors: How to Achieve High Reliability and Availability, McGraw-Hill, New York, NY, ISBN: 978-0-07-177287-7, 2012. pp. 34–40.

[4] Mid-East Turbomachinery Consulting Ltd, Dahran, KSA. Personal reviews and professional communications with the co-authors, 2014.

[5] H.P. Bloch, F.K. Geitner, An Introduction to Machinery Reliability Assessment, second ed., Gulf Publishing Company, Houston, TX, ISBN: 0-88415-172-7, 1994. pp. 242–247.

[6] H.P. Bloch, C. Carmody, Advances in Dry Gas Seal Technology for Compressors, Hydrocarbon Processing, 2015. December.

Subject Category 10

Dynamic Compressors

Chapter 10.1

Three- to Eight-Stage Integral Gear-Driven Centrifugal Process Gas Compressor Overview

In the 1970s and 1980s, multistage centrifugal compressors in the 3000 acfm (~5000 m^3/h) and lower flow ranges were largely confined to instrument and plant air duties. Three stages with intercooling were the norm, and few manufacturers concentrated on developing integral gear compressors with four or more stages. In the late 1990s, some manufacturers started to capitalize on decades of know-how with high-speed, low-flow/high head centrifugals. Using API 617 and API 614 as their compliance guidelines, these companies designed, and are now actively producing, multistage compressors (identical or similar to Fig. 10.1.1) for virtually all process gases. The inlet flows and achievable pressure ratios for four stages quite obviously bracket ranges that were previously reserved for positive displacement machines. Eight-stage integral gear compressors reach far into the capacities previously covered by a single rotor between bearing ("beam type") centrifugal compressor. Efficiencies for integral gear-driven multistage process gas compressors have moved into the high 80s, and relatively low maintenance requirements make these integrally geared compressors ever more attractive.

The general layout of these multistage centrifugal compressors is modular (Fig. 10.1.2), and drivers from 40 to over 25,000 kW are available. Impellers are usually three-dimensional and all kinds of machining methods are employed as of 2016.

Technology advances become evident as reliability professionals began to delve into design details such as deflection pad bearings, impeller attachment, and tandem dry gas seals with separation labyrinths. Experienced compressor manufacturers now often delegate the design and manufacture of seal support

Petrochemical Machinery Insights. http://dx.doi.org/10.1016/B978-0-12-809272-9.00010-4

FIG. 10.1.1 An assembly bay for multistage integrally gear-driven process gas compressors. *(Courtesy MAN Turbo, Oberhausen, Germany.)*

systems to the company's own in-house talent. However, the seals are made by competent seal manufacturers (see Subject Category 9, "Dry Gas Seals").

FIG. 10.1.2 Modular construction features allow entire compression stages to be used on different-size compressor frames. The entire "package" shown here will leave the factory fully assembled and ready for rapid installation at the user's site. *(Courtesy MAN Turbo, Oberhausen, Germany.)*

High-quality compressor manufacturers pay considerable attention to the packaging. Look for spotlessly executed welds, intelligently placed overhead air-cooled heat exchangers, piping, and conduit flawlessly configured. A valuable approach to closed-loop testing often differs from the "standard of convenience" frequently found in the compressor manufacturing industry. Rather expensive refrigeration gas mixtures may have to be used in these tests. Using such gases enables a manufacturer to simulate and closely match the purchaser's ultimate process gas operating conditions.

Subject Category 11

Electric Motor Life

Chapter 11.1

Monitoring Electric Motor Vibration and Optimizing Motor Bearing Lubricant Application

INTRODUCTION

Reader feedback is very important to technical columnists. From the many questions that made their way to our desk, we could sense the pulse of industry trends, the level of training, professional competence, and many issues of concern. We usually sent off a reply and later reworked the more interesting correspondence into one of our columns of wider interest. Keep this in mind as you follow our discourse with reliability professionals, feedback obtained at conferences, etc.

COMMENDATIONS FOR A CASE HISTORY ADMIRABLY TABULATED

To begin with, note that this reliability professional (the reader who sent us the question) is employed by an owner-operator with a number of ammonia and urea plants in locations where blinding sand storms often occurred. He deserves credit for making important observations on a double-ended electric motor. A high-pressure carbamate pump and a booster pump were connected to the motor's shaft ends:

- Each motor bearing housing had one vertical (*X*) vibration probe and one axial (*Y*) vibration probe. (There was no horizontal (*Y*) probe.) Both are seismic probes that resolve acceleration into vibration velocity—inches/second or millimeters/second.

Petrochemical Machinery Insights. http://dx.doi.org/10.1016/B978-0-12-809272-9.00011-6

- The motor tripped on high vibration at one of its bearings. Initially, only one vertical (*X*) probe reached the trip value, but the second one did not. After 30 s, both probes reached the trip value of 7.1 mm/s, and the motor was shut down, exactly as intended.
- All bearings were deep-groove style 6317, meaning the bearing bore was 85 mm.
- The failed motor bearings showed bluish discoloration on shafts and bearing inner races, pointing to a lubrication issue. There was no trace of lubricant.
- The original design intent was for these bearings to be lubricated by automatic grease dispensing devices; however, no such automation was in place. The bearings were last (manually) lubricated in Sep. 2014. No regreasing was done until the bearings failed in Jul. or Aug. 2015, after about 10 months of operation.
- After rebuilding the motor, the axial (*Y*) probe was repositioned/relocated to the horizontal (*Y*) location.

Our reader inquired about API 670 fourth edition (2000). This industry standard mentions dual voting logic, which, the reader believes, is adopted by a majority of end users. He noted that the recently released API 670 fifth edition (2014) recommends single voting logic for radial vibrations. Based on their in-house experience, his company favors either monitoring radial vibration excursions without trip logic or, more recently, two-out-of-two voting logic. The reader sought our advice on the best voting logic for radial seismic acceleration/vibration monitoring of motors and asked us to be mindful of ever-present concerns over plant operational availability and machine reliability priorities.

TREAT ROOT CAUSES, NOT SYMPTOMS

Our advice was experience-based. To have this motor "protected" with one transducer per bearing housing is probably cost-justified because the plant already has all the associated electronic modules. However, the facility's objection to using just one transducer was stated in follow-up correspondence; it had to do with concerns about spurious trips shutting down a highly profitable plant. So, researching the probability of spurious trips in modern installations would be appropriate. The reader's memory of failing transducers may have to be updated. Alternatively, if someone in authority demanded two seismic transducers and two-out-of-two voting logic, he might plan to install these in the vertical (*X*) and horizontal (*Y*) directions. He should consider implementing two-out-of-two trip voting logic and install the two probes in readily accessible locations. Both may be at convenient angles or at the 12 and 3 o'clock locations. The first excursion should sound an alarm if one of the two readings exceeds 7 mm/s. Automatic trip activation should be linked to both probes measuring an activity exceeding 7 mm/s. A special caveat was illustrated in the so-called orbital plot generated by the reader's rather sophisticated vibration monitoring system. His plot showed a squashed orbit—call it a rather oval looking noncircle—with the

X-vibration probe showing much less amplitude than the *Y*-probe. Therefore, it is possible that during a vibration excursion, the *Y*-probe could be in a trip state (HiHi) and the *X*-probe showing normal. Over the years, we have found that if one end of a rotor is in distress, the other end should show some change from normal. It might not be in a HiHi alarm state on the nondistressed end, but it should at least be showing a Hi alarm on one of its vibration probes. Of course, more cards have to be installed in the monitoring rack for this kind of exploration.

EVIDENCE OF OUTDATED LUBRICATION TECHNOLOGY

Well, we understand how someone wants to make the case for more vibration monitoring. However, we thought that this plant would do well making reliable bearings and lubrication its priority concerns. Vibration monitoring (hopefully) lets us know when there is a problem; sound best-of-class (BoC) lubricant application strategies prevent problems from happening in the first place. We recommend they adopt only best-available bearing selection and plant-wide automated lubrication strategies. The plant's top technical and mid-level managers should appreciate why dry sump oil mist has been successfully used by BoC companies for the past 40 years.

Unless the bearings are lubricated by oil mist, BoCs disallow rolling element bearings for electric motors above 500 hp; Siemens allows oil mist in motors up to 3000 kW [1]. For those insisting on grease, details on automatically or manually applied grease lubrication are important but will differ with the location and orientation of shields (if any) and drain ports. There is considerable reliability impact depending on the type of grease; moreover, certain grease application methods sometimes result in wrong fill volume, excessive grease pressure (deflecting shields), rust or dust in bearing element paths, and bearing flat spots (in an installed spare pump set) due to shafts not being rotated—to name just a few. Again, proper greasing procedures and lubrication management are far more important than placing/mounting/maintaining more monitors on a rolling element-equipped motor bearing housing.

An electric motor with 85 mm bearings is obviously not a small machine. With grease lubrication, an 85 mm bearing becomes maintenance-intensive by virtue of the fact that it will require grease replenishment at least six and, in some cases, 16 times per year. If rivet heads pop off in a riveted-cage bearing, the motor sometimes grinds to a halt in mere seconds. We referred the reader to an article describing how BoCs are using oil mist on many electric motor bearings (see HP's issue for March 1977—obviously decades ago). As of 2015, an estimated 26,000 electric motors (and 150,000 process pumps) were using dry sump oil-mist lubrication, and some of these have not needed bearing replacements in the time period from 1977 until 2014—at least 37 years. Why the reader's company is not availing itself of oil-mist lubrication is very difficult to comprehend and not even worth speculating about. The one sure thing we know about achieving reliability is that it cannot be obtained with business-as-usual mindsets and lots of manual labor.

Allow us to zero in on the real problems as we saw them. First, the reader is probably only responsible for vibration monitoring and analysis tasks. His assignment may be limited in scope, and he cannot tell higher management that we believe his company is vulnerable in its use of old lubrication technology. Here's how others solved the dilemma: *First*, at least two companies accepted our recommendation to send four or five managers to a three-day off-site technology management update session where subject matter experts (ones without allegiance to either vendors or bosses) candidly briefed them on how companies expressing the *desire* to become BoCs actually *became* BoCs. *Second*, competent reliability professionals must be given more direct access to managers that were willing to listen at their plants. The value of teaching midlevel managers in small groups is far greater than trying to present in-plant seminars to 40 disinterested lower-rung folks. *Third*, reliability professionals must be multifaceted in the sense of wanting to do more than, say, a vibration data collector. They must lead by learning and teaching. They must identify the sources of persistent deviations from long and reliable operational performance of machines. They must learn to not just treat the symptoms of premature equipment distress. Instead, they must make it their goal to firmly establish the root causes of deviations.

In this instance, the root causes of inadequate lubrication had to do with using methods that were discontinued by BoC companies in the early to mid-1960s. Reliability professionals must become change agents, not supplier of temporary solutions, or more and more monitoring. They should eliminate the need for monitoring by implementing whatever it takes to eradicate failures. They must become the providers of lasting solutions. Adding more monitoring instruments is vastly inferior to eliminating the root causes of problems.

REFERENCE

[1] H.P. Bloch, A. Shamim, Oil Mist Lubrication: Practical Applications, Fairmont Publishing, Lilburn, GA, 1998. p. 109.

Chapter 11.2

Electric Motors and Mechanical Efficiency

INTRODUCTION

Reliability-focused subject matter experts know quite a bit about oil mist and many use plant-wide oil-mist systems for their process pumps and electric motor drivers. In fact, some major grassroots olefins plants used oil mist on motors with rolling element bearings as small as 1 hp (0.75 kW) and as large as 1250 hp (925 kW) as early as 1975. What's less known is that oil mist, together with the right viscosity synthetic lubricant, will save considerable energy.

SYNTHETICS FOR MOTORS

An important presentation on the "*Effects of Synthetic Industrial Fluids on Ball Bearing Performance*" [1] described that a readily available synthetic lubricant, having a viscosity of 32 cSt at a temperature of 40C (98 F), offered long-term contact surface protection for process pumps and their electric motor drivers [1]. Although "only" 32 cSt, the protective effect of this *synthetic* lubricant was found to be equivalent to that of a baseline *mineral oil* with the higher viscosity of 68 cSt. The same good wear protection could *not* be achieved with a reduced viscosity *mineral* oil. The tests also showed the lower viscosity *synthetic* lubricant providing energy savings of approximately 4% of the normal power draw of the electric motor.

QUANTIFYING THE ENERGY SAVINGS POTENTIAL

Using both oil mist and synthetic lubes makes economic sense. According to the test described in Ref. [1], the frictional losses in rolling element bearings can be reduced as much as 37%. On the 65 mm bearings typically used in 15 hp (~11.3 kW) process pumps and electric motors, 0.11 kW could be saved (Table 11.3.1, later). While the small absolute value of 0.11 kW per bearing tends to make the savings appear insignificant, petrochemical process pump rotors are typically supported by a double-row radial ball bearing and two angular contact ball thrust bearings. These then represent a total of 4.8 test-equivalent bearings $[(4\times0.7)+(2\times1)=2.8+2=4.8]$. Therefore, the total savings available from an actual motor-driven pump set are 4.8 times the single test bearing energy savings of 0.11 kW; that equals 0.53 kW or 4.7% of 11.3 kW.

Assuming that the average pump operates 90 percent of the time and rounding off the numbers, this difference creates energy savings of 4180 kWh per year. At $0.10 per kWh, yearly savings of $418 should be expected. By using synthetics on conventionally lubricated equipment, oil-replacement schedules are typically extended fourfold; the extended drain intervals more than compensate for the higher cost of synthetic lubricants.

With oil mist, the bearings run cooler and last longer than those typically lubricated by conventional oil sumps. While open oil-mist systems typically consume 12–22 L (3.1–5.7 gallons) per pump set per year, closed oil-mist systems consume no more than 10% of these yearly amounts. Again, at these extremely low makeup or consumption rates and compared with the cost of mineral oils, the incremental cost of synthetic lubricants is relatively insignificant.

Considering annual energy saving per 15 hp pump and driver set to be worth $418, we realize that these savings should be multiplied by the number of pumps actually operating in large refineries—850 to 1200. Again, using $0.10 per kWh, annual savings in the vicinity of $450,000 would not be unusual. A detailed calculation can be found in Ref. [2]; it will prove the point:

- Total pump hp installed at the plant = 15 hp × 1000 = 15,000 hp
- Total pump kW installed at the plant = 15,000 × 0.746 = 11,190 kW
- Total consumption kWh per year, considering 90% of 8760 h/yr
$= 8760\text{h} \times 0.90 = 7884\text{h / yr; then } 7884 \times 11{,}190\text{kW} = 88{,}220{,}000\text{kWh / yr}$
- Total US dollar value of yearly energy consumed, assuming $0.10/kWh:
$= \$8{,}822{,}000$

The total energy savings for 1000 average-sized pump sets would thus equal 0.047 × 8,822,000 = $414,600. That is an amount that should not be overlooked.

REFERENCES

[1] F.R. Morrison, J. Zielinsky, R. James, Effects of Synthetic Fluids on Ball Bearing Performance, ASME Paper 80-Pet-3, New Orleans, LA, February 1980.

[2] H.P. Bloch, Practical Lubrication for Industrial Facilities, second ed., The Fairmont Press, Lilburn, GA, 2009, ISBN: 0-88173-579-5.

Chapter 11.3

Smart Motor Lubrication Saves Energy

INTRODUCTION

Many decades of experience confirm the success of oil mist for rolling element bearings in the operating speed and size ranges found in motors for process pumps. For the past 40 years, empirical data have been employed to screen the applicability of oil mist to pumps and electric motors. The applicability of oil mist is expressed in a rule of thumb that incorporates bearing size, speed, and load. It uses the parameter "DNL" (D = bearing bore, mm; N = inner ring, rpm; and L = load, lbs) and limits oil mist to values below 10^9, or 1,000,000,000. An 80 mm electric motor bearing, operating at 3600 rpm and a load of 600 lbs, would thus have a DNL of 172,000,000—less than 18% of the allowable threshold value.

LONG AND SUCCESSFUL HISTORY

As of 2016, tens of thousands of oil-mist lubricated electric motors continue to operate flawlessly in reliability-focused user plants. Capitalizing on this favorable experience, the procurement specifications for both new projects and replacement motors at many of these plants require oil-mist lubrication in motor sizes 15 hp and larger. The specifying entities generally know that energy savings are possible with smart lubrication strategies, and last month's column gave some highlights. Recall that synthetic lubricants and oil mist are involved across the entire range of electric motors. Although the largest motors suitable for pure oil mist in refineries and petrochemical plants can exceed 2000 hp (~1500 kW) in size, prevailing practice among reliability-focused users is to apply oil mist on horizontal motors, 15 hp and larger, and on vertical motors 3 hp and larger (Fig. 11.3.1). At a savings of approximately 4% (and as much as 5%) of the motor's power draw, these efficiencies are hard to ignore in a cost and reliability-focused plant environment.

While Ref. [1] focused on a diester-based oil and viscosity effects, it also established that equivalency of protection existed for the 32 cSt *synthetic* versus 68 cSt *mineral* oils. Bearing service life matched the theoretically predicted levels with either lubricant, but the lower viscosity together with oil mist saved about 0.53 kW on the average pump set. Since publishing the energy savings potential of oil mist in conjunction with synthetic lubricants in 1980 [1], superior additives technology has also yielded improvements.

FIG. 11.3.1 Single-stage vertical inline pump with pure oil mist lubricating its electric motor since 1976. *(Courtesy Colfax Lubrication Management, Lubrication Systems Division, Houston, TX.)*

Because synthetic fluids are chemically different from mineral oils, one might expect effects that go beyond those attributable to viscosity relationships alone. Indeed, lubricant properties and application methods also affect lubrication effectiveness and the frictional torque to be overcome.

QUANTIFYING THE ENERGY SAVINGS POTENTIAL

The potential cost savings through power loss reduction are quite substantial when we realize that industrial machines consume an estimated 31% percent of the total energy in the United States. Ref. [1] clearly established that power losses in rolling element bearings could be reduced as much as 37% (Table 11.3.1).

The resulting savings with different lubricants and lubrication methods, that is, sump versus pure oil mist, are highlighted in the bar graph of Fig. 11.3.2. Prorating these savings to a refinery with 1000 centrifugal pumps and their respective drivers could save in excess of $400,000 [2].

These are realistic expectations, and the higher cost of synthetic lubricating fluids is compensated by reduced maintenance requirements [3]. While again

TABLE 11.3.1 How changes in lubricant type and application method affect kW power losses [1]

Change	Δ Power loss per bearing	Total reduction
Sump: MIN 68 to SYN 32	0.017	6%
Mist: MIN 68 to SYN 32	0.022	8%
Sump MIN 68 to Mist MIN 68	0.080	29%
Sump SYN 32 to Mist SYN 32	0.085	31%
Sump MIN 68 to Mist SYN 32	**0.11**	**38%**

MIN, mineral; *SYN*, synthetic

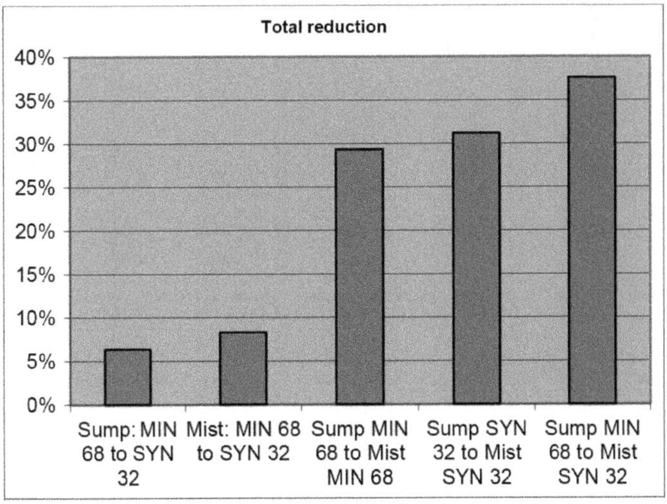

FIG. 11.3.2 Percentage savings in frictional energy when changing lubricant type and method of lubricant application. *MIN*, mineral oil; *SYN*, synthetic oil.

making the case for oil-mist lubrication and adding energy conservation to its other advantages, it should be pointed out that synthetics make longer drainage intervals feasible in conventional sump-lubricated pumps.

In summary, electric motor lubrication by oil mist is highly advantageous. It saves both energy and labor costs. Sealing and mist drainage are well understood and have been thoroughly explained [2]. Although oil mist will neither attack nor degrade the winding insulation on electric motors manufactured since the mid-1960s, mist entry and related sealing issues have been addressed by competent users and suppliers. Still, regardless of motor type, that is, TEFC, X-Proof, or WP II, cable terminations should not be made with conventional electrician's tape. The adhesive in this tape will last but a few days and then become tacky to the point of unraveling. Instead of inferior products, experienced motor manufacturers use a modified silicone system (Radix) that is highly resistant to oil mist. With this rating, modified silicone systems have consistently outperformed the many other "almost equivalent" systems in the terminal boxes.

REFERENCES

[1] F.R. Morrison, J. Zielinsky, R. James, Effects of Synthetic Fluids on Ball Bearing Performance, ASME Paper 80-Pet-3, New Orleans, LA, February 1980.

[2] H.P. Bloch, Practical Lubrication for Industrial Facilities, second ed., The Fairmont Press, Lilburn, GA, 2009, ISBN: 0-88173-579-5.

[3] O. Pinkus, O. Decker, D.F. Wilcock, How to save 5% of our energy, Mech. Eng. 32 (1997) 32–39.

Chapter 11.4

Oil Mist and Electric Motor Windings

INTRODUCTION

Plant-wide oil-mist systems were first commissioned in 1962 or 1963 [1]. At forward-looking user companies, these systems soon included the electric motor drivers. The Reliance Electric Company in cooperation with user plants in Texas and Louisiana had clearly established that oil mist will in no way affect electric motor windings with epoxy insulation. Only the T-lead insulation must be specified as cross-linked polymeric (low swell volume) instead of inferior insulating tape with its high swell rate. Competent motor manufacturers are completely aware of this requirement. Reliability-focused user companies rarely allow uninformed bidders—the ones questioning these facts—to be on their list of acceptable vendors. Cross-linked polymeric tape was introduced in 2003 and has since been an unqualified success.

While oil mist will not harm modern motor, some users desire to minimize oil-mist intrusion into electric motors. Minimum oil-mist intrusion occurs when an advanced bearing protector seal is chosen; you might ask the vendor prove IP66 compliance. IP is the universally recognized Ingress Protection Code; it's discussed elsewhere in this text (see index, also Ref. [2]).

HISTORY

There are other historical facts relating to oil-mist lubricated motors. By the mid-1970s, oil mist had demonstrated its outstanding suitability for lubricating and preserving electric motor bearings [3]. In that decade, petrochemical plants in the US Gulf Coast area, the Caribbean, and South America proceeded to convert in excess of 1000 electric motors to dry-sump oil-mist lubrication. In 1986, there were more than 4000 electric motors on oil-mist lube in the US Gulf Coast area alone. As of this writing (in 2016) and per conservative estimates by total lubrication management (a unit of Colfax Fluid Handling in Houston, TX), about 28,000 electric motors are presently operating on pure oil mist with outstanding success. Many of these electric motors are located in the Middle East.

However, universal acceptance did not come overnight. On the one hand, it seemed logical to extend oil-mist feeder lines from centrifugal pump bearing housings to the adjacent electric motor bearings. On the other hand, concern was voiced that lube oil would enter the motor and cause damage to winding insulation or cause overheating until winding failure occurred. Initial efforts were, therefore, directed toward developing lip seals or other barriers confining oil mist to only the bearing areas. Those efforts date back to about 1975.

FIG. 11.4.1 A worn Vee-ring removed from an electric motor. *(Courtesy AESSEAL Inc., Rotherham, the United Kingdom, and Rockford, Tennessee.)*

When, in the late 1970s, failures of old-style Vee-ring seals (Fig. 11.4.1) were experienced in operating motors, oil mist did in fact enter and coated the windings with coalesced oil. The potential explosion hazard was again investigated on this occasion, and confirmation obtained that the oil/air mixture of a plant-wide oil-mist system remained several orders of magnitude below the sustainable burning point. Experiments had shown the concentration of oil mist in the main supply manifolds ranging from 0.005 to as little as 0.001 of the concentrations generally considered flammable. The fire or explosion hazard of oil-mist lubricated motors is thus no different from that of NEMA-II motors. No signs of overheating were found, and winding resistance readings conformed fully to the initial, as-installed values.

With the introduction of epoxy motor winding materials several decades ago, it was shown that these winding coatings will not deteriorate in an oil-mist atmosphere. This has been conclusively proved in tests by users and motor manufacturers. Among them were Reliance Electric (Cleveland), Continental Electric (Newark), and an oil refinery in the Caribbean where windings coated with epoxy varnish were placed in beakers filled with various types of mineral oils and synthetic lubricants. Next, these windings were oven-aged at 170 °C (338 °F) for several weeks and then cooled and inspected—again with no problems.

Final proof was obtained during inadvertent periods of severe lube-oil intrusion at an oil refinery in Aruba; this facility had been among the first to use oil mist on a wide scale. In one such case, a previously liquid-oil-lubricated, 3000 hp, (−2200 kW), 13.8 kV motor had been converted to pure oil mist (Fig. 11.4.2). The old-style Vee-ring seal was defective, but the motor ran well even after a considerable volume of excess oil was being drained from its interior. The incident caused a small increase in dirt collecting on the windings, but it had not adversely affected winding quality.

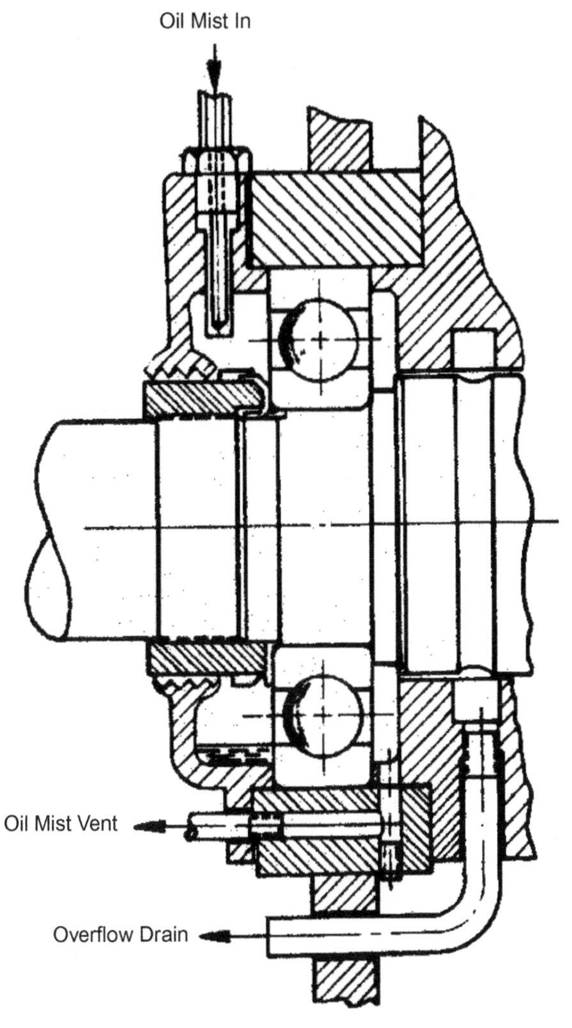

FIG. 11.4.2 A successful oil-mist lubricated motor bearing dating to the mid-1970s.

Decades ago, experimentation with motor winding and cable terminations in conduit boxes showed that a Teflon-based wrap should be used in the conduit box for best results. Other materials, including silicone tape, seemed to exhibit a tendency to swell or become gummy when exposed to oil mist. It was then decided to provide sealant between the motor frame and conduit box to reduce the (in open systems unsightly) mist emissions at the conduit enclosure. Mist supply and condensed-oil-drain ports were made accessible without the need for covers and guards. A simple pipe nipple or similar extension was considered just fine in the 1980s. Today, environmentally friendly "closed circuit" oil-mist systems would be used in industry, but oil mist is still the best way to lubricate.

REFERENCES

[1] H.P. Bloch, A. Shamim, Oil Mist Lubrication: Practical Applications, The Fairmont Press, Lilburn,GA, ISBN: 0-88173-256-7, 1998. p. 103.
[2] H.P. Bloch, A. Budris, Pump User's Handbook: Life Extension, third ed., The Fairmont Press, Lilburn, GA, ISBN: 0-88173-627-9, 2010. pp. 477–478.
[3] H.P. Bloch, Dry sump oil mist lubrication for electric motors, Hydrocarb. Process. 57 (3) (1977) 133–135.

Subject Category 12

Foundation and Grouting Systems

Chapter 12.1

Consider Pregrouted Pump Baseplates and New Grout Systems

INTRODUCTION

Conventional grouting methods for nonfilled pump baseplates are, by their very nature, labor- and time-intensive. Utilizing a pregrouted baseplate with conventional grouting methods helps to minimize some of the cost, but the last pour still requires a full grout crew, skilled carpentry work, and good logistics. To further minimize the costs associated with baseplate installations, a new field grouting method has been developed for pregrouted baseplates. This new method utilizes a low-viscosity, high-strength epoxy grout system that greatly reduces foundation preparation, grout form construction, crew size, and the amount of epoxy grout used for the final pour.

While there may be other low-viscosity, high-strength epoxy grout systems available on the market, the discussion and techniques that follow are based on the flow and pour characteristics of Five Star Fluid Epoxy. This type of low-viscosity grout system can be poured to depths from ½″ to 2″ (13–50 mm), has the viscosity of thin pancake batter, and is packaged and mixed in a liquid container. The material can be mixed and poured with a two-man crew.

CONCRETE FOUNDATION PREPARATION

As shown in Fig. 12.1.1 and irrespective of baseplate style, that is, pregrouted (see also "HP In Reliability," November, 2003) or traditional unfilled, correct preparation of the top of the concrete foundation will have long-term reliability

Petrochemical Machinery Insights. http://dx.doi.org/10.1016/B978-0-12-809272-9.00012-8

implications and is important. The laitance on the surface of the concrete must be removed for proper bonding, regardless of grouting method and material selected. Traditional grouting methods require plenty of room to properly place the grout, and this requires chipping all the way to the shoulder of the foundation (see Chapter 12.3, Fig. 12.3.2). However, utilizing a low-viscosity epoxy grout system will greatly reduce the amount of concrete chipping required to achieve a long-term satisfactory installation.

NEW GROUT FORMING TECHNIQUE

With the smooth concrete shoulder of the foundation still intact, a very simple "2×4" grout form can be used. One side of the simple grout form is waxed, and the entire grout form is sealed and held in place with caulk. While the caulk is setting up, a simple headbox can be constructed out of dux seal. Due to the favorable flow characteristics of the low-viscosity epoxy grout, this headbox does not need to be very large or very tall.

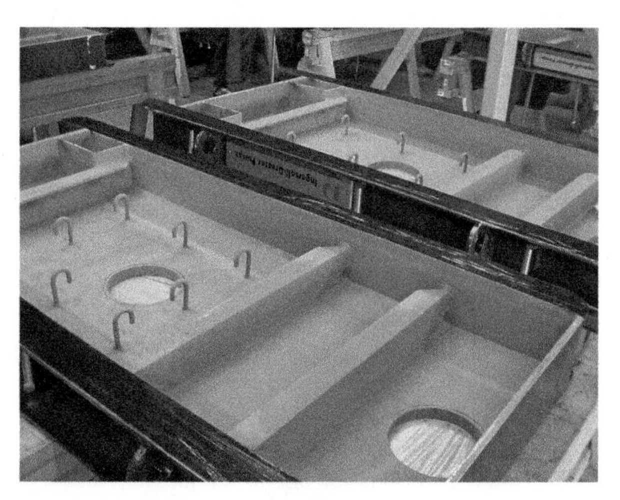

FIG. 12.1.1 Baseplates are inverted, then sand-blasted, and precoated with epoxy primer before being filled with epoxy. *(Courtesy Stay-Tru, Houston, TX.)*

The low-viscosity epoxy grout is mixed with a hand drill, and all the grout is poured through the headbox to prevent trapping air under the baseplate.

This new installation method has been used for both ANSI- and API-style baseplates with excellent results. With this technique, field experience has shown that a pregrouted baseplate can be routinely leveled, formed, and poured with a two-man crew in 3–4h.

FIELD INSTALLATION COST COMPARISON

The benefits of using a pregrouted baseplate with the new installation method can be clearly seen when field installation costs are compared. This comparison applies realistic labor costs; it does not take credit for the elimination of repair costs associated with field installation problems, such as void repair and field machining.

Industry experience shows that eight men are typically involved in the average size conventional grouting job. An actual per man-hour labor cost of perhaps twice the person's take-home wage must generally be used in US installations when employee benefits and overhead charges are included.

A cost comparison can be developed, based on the installation of a typical API baseplate using epoxy grout, for the conventional two-pour procedure and a pregrouted baseplate using the new installation method. The following conditions apply:

Baseplate dimensions, 72″ × 36″ × 6″ (1.8 × 0.9 × 0.15 m)
Foundation dimensions, 76″ × 40″ × 2″ (1.93 × 1.0 × 0.05 m) (grout depth)
Labor cost, $45 h^{-1} (use multiplier relevant to your location)
Epoxy grout cost, $111 ft^{-3} ($3920 m^{-3})

In 2003, a baseplate with the listed dimensions could be pregrouted for $2969, and suitable multipliers should be used for your location and calendar year. This would include surface preparation, epoxy grout, surface grinding, and a guaranteed inspection. The total installed cost for a conventional two-pour installation was $6259, whereas the total installed cost for a pregrouted baseplate, installed with the new installation method, amounted to $4194. Aside from the very obvious cost savings, the reliability impact of this void-free and fully coplanar installation is of great importance to reliability-focused pump users. As you are reading this, an appropriate multiplier may have to be used. However, the long-term experience of what was earlier called a "new" baseplate prefill and grouting system has proved to be as good as had been promised and anticipated many years ago.

Chapter 12.2

Field Erection and Installation Specifications for Special Purpose Machinery: How Detailed Should They Be?

INTRODUCTION

It is not the purpose nor is it within the scope of this text to give detailed field erection and installation specifications for the many rotating machines found in modern process plants. However, field erection and installation specifications are definitely needed by reliability-focused plants. Moreover, they must be reviewed, understood, and approved by a competent machinery engineer who directly represents the owner plant and has a personal stake in its long-term reliable operation.

COMMONALITIES OBSERVED

We found that all of these specifications have a few things in common:

- The scope of a standard is always explained first. For instance, a field erection and installation standard would cover mandatory requirements governing installation and erection for compressors and drivers mounted on baseplates or sole plates.
- Additional information is almost always superimposed on existing industry standards. An asterisk (*) might be used to indicate additional information is required. Here, the contractor may have to specify and the owner's machinery engineer may have to approve information.
- A summary of additional requirements is provided. A separate tabulation of applicable cross-references usually lists documents that have to be used with the particular standard.
- Design requirements are clearly explained. As an example, concrete foundations must be properly sized and proportioned for adequate machinery support and prevailing piping forces. The complete compressor train (compressor, gear, and motor or other drivers) must have a common foundation.
- Foundations must rest on natural rock or entirely on solid earth or good, well-compacted, and stabilized soil. They must be supported on pilings that have a rigid continuous cap or slab cover.
- Foundation must be isolated from all other structures such as walls, other foundations, or operating platforms. They have to be designed to avoid

resonant vibration frequencies at operating speeds, 40–50% of operating speeds, rotor critical speeds, gear meshing frequencies, two times operating speeds, and known, specified background vibration frequencies.

- The temperature surrounding a foundation must be analyzed to verify uniformity so as to prevent any distortion and misalignment. Concrete foundations must also be properly cured (~28 days) before loading.
- Foundation arrangements are described in considerable detail. Anchor bolts must be designed by specialty firms and must be sleeved. In most instances, a civil engineer will provide and/or certify a foundation drawing or separate foundation specification.
- Around the perimeter, a W 8 or larger I-beam must be properly anchored to foundation for supporting small piping, conduit, and instruments. Auxiliary structures, including piping, merit special and separate design. The owner's engineer ascertains that proper attention is given here.
- Typical compressor, gear, and motor foundation arrangements and baseplates must be completely filled with epoxy grout. Sole plates must be completely supported with epoxy grout.
- Reinforcing rods, ties, or any steel members must be a minimum of 2 in. (~50 mm) below concrete surface to permit chipping away 1 in. of concrete without interference.
- A minimum space of 1 in. (~25 mm) must be provided between foundation and chock block for proper grout flow. The maximum distance between foundation and baseplate should not exceed 4 in. (~100 mm). The minimum distance between the top of the foundation and bottom of the baseplate should not be less than 2 in. (~55 mm).
- For epoxy chock applications, the distance between baseplate or sole plate and top of grout should be 1 in. (~25 mm), unless otherwise approved by the owner's machinery engineer.
- Chock block arrangement and installation are described. Chock blocks must be properly sized to distribute anchor bolt and machine loads so as not to exceed 10% of weakest compressive strength material in the foundation structure. (The customary design is 300 psi for concrete.)
- Instructions and appropriate illustrations of field erection and assembly tools must be provided. For instance, a special hydraulic coupling hub to output shaft installation tool will probably be used in virtually any modern plant.

In this day and age, much lip service is paid to reliability concepts and uptime optimization. These two goals often clash with the desire to award plant construction to engineering, procurement, and construction (EPC) contractors who, in turn, are driven by cost and schedule concerns. In those instances, an owner-operator firm would be foolish not to give utmost attention to the existence of specifications that reflect reliability focus and uptime extension goals. Reliability professionals are then assigned to ascertain compliance with these specifications.

Chapter 12.3

How Pump Installation Differs From the Way Pumps Were Shipped to Your Plant

INTRODUCTION

Just because pump sets (pump, motor, and baseplate) are conveniently mounted for shipping, a "package" has often led to the erroneous assumption that the entire as-mounted set can simply be hoisted up and placed on a suitable foundation. However, doing so is not best practice; in fact, how pumps are shipped has relatively little to do with how they should best be installed in the field [1].

MOUNTED FOR SHIPPING

In general, process pump manufacturers are asked to provide pumps as a "set" or assembled package comprising pump, driver, and baseplate (Fig. 12.3.1). After ascertaining correct shaft separation to accommodate the selected coupling and prealigning the two shaft centerlines within perhaps 0.020 in. (0.5 mm), mounting holes are spotted from pump and driver to the baseplate's mounting pads. The pump and baseplate provider then proceeds to thread-tap bolt holes that have a diameter of about 0.060″ (1.5 mm) less than the mounting holes ("through holes") in pump and driver. Mounting bolts are inserted at this stage of the pump assembly, and the complete pump set is now considered ready for shipment as a prealigned "mounted" package to the designated recipient.

FIG. 12.3.1 Small baseplate-mounted pump set installed at a plant in Texas. Note dry-sump oil mist lubrication; coalesced oil is collected in the container mounted at concrete foundation. *(Courtesy Total Lubrication Management, Division of Colfax Industries, Houston, TX.)*

The receiving site sees a conveniently mounted-for-shipping package, which has, over the past few decades, led to the erroneous assumption that the entire package can simply be hoisted up and placed on a suitable foundation. However, doing so is not best practice; in fact, how pumps are shipped has relatively little to do with how they should best be installed in the field [1]. The old OEM field service personnel knew about this issue, but wise old field service folks are no longer employed there. That is why we have to reinvestigate, read, and hope that certain erroneous ways we've become used to are uncovered and then corrected.

AS SHIPPED CONDITION IS NOT YET READY FOR INSTALLATION

Contrary to the understanding of many of today's pump manufacturers' installation or service personnel, best practices companies will not install the equipment as a mounted "set." To ensure level mounting throughout, the baseplate by itself is placed on the foundation into which hold-down bolts or anchor bolts (Fig. 12.3.2) should have been encased when the reenforced concrete foundations were poured [2]. Leveling screws are then used in conjunction with optical laser tools or a machinist's precision level to bring the baseplate mounting pads into flat and parallel condition side to side, end to end, and diagonally, within an accuracy of 0.002 in./ft (~0.15 mm/m) or better. The nuts engaging the anchor bolts are now being secured and the hollow spaces within the baseplate, as well as the space between baseplate and foundation filled with epoxy grout.

Note that the filling of hollow spaces does not apply to baseplates that were prefilled with epoxy, as described earlier in Chapter 12.1. If a baseplate has been prefilled with epoxy, has been allowed to cure, and thereafter had the mounting surfaces for pump and driver machined coplanar and parallel, the whole assembly with pump and driver mounted should be ready for installation. Removal of pump and driver from such a baseplate is optional.

FIG. 12.3.2 The baseplate is being leveled with pump and driver removed. Grout application is the next step. *(Courtesy Perry Monroe, Livingston, TX.)*

After the epoxy grout has cured, pump and driver are aligned to criteria that harmonize with best practices—essentially the workmanship guidelines and reliability-focused practices of modern plants [1,2].

Also, while being aligned, dial indicators monitor soft foot and pipe stress; sensitivity to piping being flanged up is closely monitored. Any dial indicator movement in excess of 0.002 in. will require making corrections to the piping. The pump is not allowed to become a pipe support.

Finally, it should be noted that best-in-class users specify and generally insist on epoxy prefilled steel baseplates. Indeed, consideration should be given to dispense with the labor-intensive conventional grouting procedure described above. You can eliminate much of this by purchasing baseplates prefilled with epoxy. They represent a monolithic block that will never twist and never get out of alignment [1].

REFERENCES

[1] H.P. Bloch, A.R. Budris, Pump User's Handbook: Life Extension, fourth ed., Fairmont Press, Lilburn, GA, ISBN: 0-88173-720-8, 2014.

[2] H.P. Bloch, F.K. Geitner. Major Process Equipment Maintenance and Repair, second ed. Gulf Publishing Company, Houston, TX. ISBN: 0-88415-663-X, 1997.

Subject Category 13

Gasket Designs

Chapter 13.1

Assessing Advances in Gasket Designs

INTRODUCTION

The physical properties and performance of a gasket will vary extensively, depending on the type of gasket selected, the material from which it is manufactured, and the service into which it is installed. In all instances, staying abreast of new developments and working closely with experienced gasket manufacturers will pay rich dividends in failure avoidance and maintenance cost reduction.

Long component service life and maintenance cost avoidance are of great interest to reliability professionals in the hydrocarbon processing industry. The maintenance cost reduction and reliability extension goals are supported by a renowned gasket manufacturer with over 100 years of industrial experience. The manufacturer supplied Fig. 13.1.1 for our use; it illustrates the many available gasket cross sections and recommended flange finishes.

This manufacturer's recent innovation, the "Change gasket"—probably a play on the terms *heat exchanger* and *change of design*—caught our attention in the wider context of advancing technology improvements. The company's metal-wound *Change heat exchanger gasket* uses an unusually thick metal spiral and full-penetration laser welding processes. The product requires no inner and outer ring and is as easy to handle and install as a double-jacketed gasket. Its test performance can be observed from Fig. 13.1.2.

The manufacturer anticipates that the useful life of this gasket makes it suitable for heat exchanger service. But well-made corrugated and Kammprofile heat exchanger gaskets already seal leak-free from turnaround to turnaround. No new gasket can exceed this cycle life, as gaskets are replaced when the exchanger is serviced. So the testing does not "show" that it can last longer than existing designs.

As in virtually all reliability-focused endeavors, we should seriously consider making competent component manufacturers a technology resource [1].

Petrochemical Machinery Insights. http://dx.doi.org/10.1016/B978-0-12-809272-9.00013-X

Surface Finish Requirements

Gasket Description	Gasket Cross Section	Flange Surface Finish Microinch Ra	Flange Surface Finish Micrometer Ra
Spiral wound gaskets		125–250	3.2–6.3
Flexpro gaskets	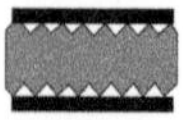	125–250	3.2–6.3
Metallic serrated gaskets		63 MAX	1.6 MAX
MRG		125–250	3.2–6.3
Solid metal gaskets		63 MAX	1.6 MAX
Metal jacketed gaskets	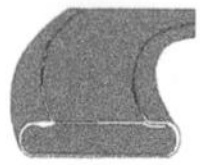	100–125	2.5 MAX
Soft cut sheet gaskets		Mat'l<1.5mm Thick 125–250	Mat'l<1.5mm Thick 3.2–6.3
		Mat'l≥1.5mm Thick 125–500	Mat'l≥1.5mm Thick 3.2–12.5
Change gaskets		125–250	3.2–6.3

FIG. 13.1.1 Gasket cross sections and recommended flange finishes.

Catalog material is of interest and should be requested or downloaded. In many instances, notable manufacturers would be pleased to organize briefing or teaching sessions for interested user-purchasers. However, our obligation to pursue "due diligence" goes a bit further than just taking a brief look. The principles explained elsewhere in this text as MQA ("machinery quality assessment"; see index) apply to critically important components, which certainly include gaskets. In order not to be swayed by overly enthusiastic claims by manufacturers, it is important to carefully examine the information provided. In this instance and at a minimum, the end user should confirm the following:

1. That the supporting data were derived from tests that simulate the actual service for which the gasket is intended.
2. That the claims are not inconsistent with other well-established and well-documented facts.

A careful review of the test data produced in support of the "Change" gasket shows that all the quantified data came from studies of small flange gaskets. This is inevitable, as many of the standard industry performance tests of gaskets are based on 2″ to 6″ raised faced flanges. The end user should carefully consider whether the results obtained by testing on small flanges apply to all heat exchanger services. Expect that in heat exchanger services, the gasket seating stresses can be much higher, the thermal fluctuations are persistent, and differential thermal fluctuations potentially could cause radial shear forces that will damage the joint.

The possibility that test results in Fig. 13.1.2 may *not* directly apply to heat exchangers is suggested by the fact that the test data indicate that double-jacketed (DJ) gaskets perform better than corrugated gaskets (CMGs). While this is undoubtedly true of the test performed by the manufacturer, it directly conflicts with data derived from actual heat exchanger simulations. Radial shear tightness testing is designed and performed by Brown [2]. This researcher demonstrated conclusively that DJ gaskets cannot seal reliably when subjected to shear forces that arise in heat exchanger services. In Ref. [2], we find plots that show pressure increases in the secondary annulus that indicate leakage past the primary gasket. No DJ gaskets passed this test, while CMGs with graphite facings performed better—on the whole—than other gasket styles.

The major point of the testing in Ref. [2] is that it is not possible to make generalizations about the performance of gaskets by gasket type. The author/researcher considered it essential to closely verify the applicability of a gasket's construction and to relate it to full-scale testing or verification of experience prior to final selection. His comprehensive conference paper warns against incorrect assumptions regarding basic gasket performance (such as the level of relaxation and the ability to handle radial shear). He emphasizes that such erroneous assumptions far outweigh other factors (such as room temperature leakage performance) in obtaining leak-free joint operation.

The testing in Ref. [2] prompted its author to strongly advise against the often-practiced "trial-and-error" approach that has been considered the norm in gasket selection. To quote: "It is this trial and error approach that is directly responsible for millions of dollars of lost production and environmental impact due to heat exchanger leakage annually."

The inconsistency between the various data available to us suggests that the underlying tests in Fig. 13.1.2 and in Ref. [2] are not equivalent. We have no reason to doubt that the change gasket performs as reported by the manufacturer in raised face pipe flange gaskets. We may even conclude that, with time, the gasket may prove to be effective in heat exchangers as well. But until such time as the manufacturer provides test data generated from dynamic heat exchanger environments, the end user should carefully determine whether or not he wishes to entrust his heat exchanger reliability to a gasket that has not undergone exchanger-specific testing.

KAMMPROFILE GASKETS

Kammprofile gaskets are among the plotted lines in Fig. 13.1.2. They were developed in Europe and then standardized in the mid-1970s. Their design is relatively simple; a flexible graphite or PTFE (polytetrafluoroethylene (Teflon®) overlay is bonded to a solid metal core with concentric serrations. These gaskets often compete with metal-jacketed and spiral-wound gaskets.

At installation, the soft facing material is forced into the serrated grooves of the metal core. The compressive stress increases the density of the facing material within the grooves. A series of concentric high-pressure sealing regions are thereby created.

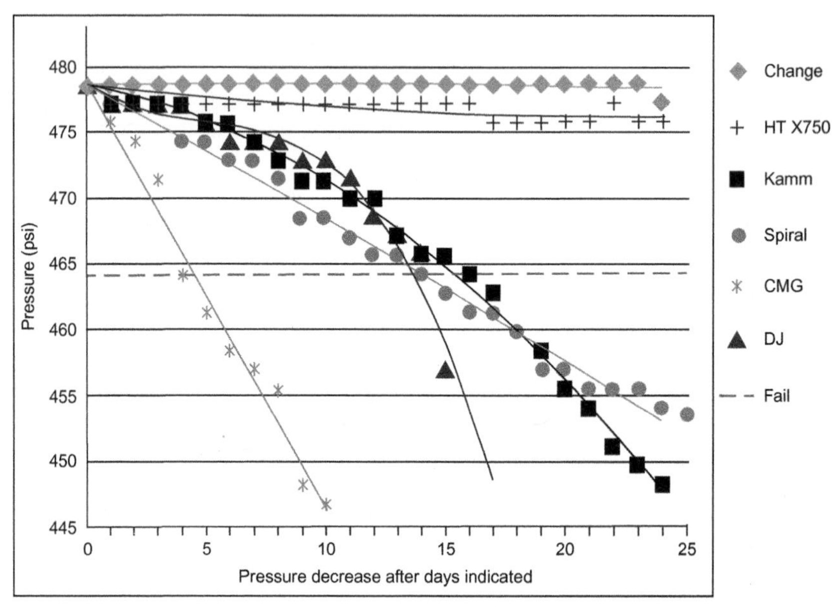

FIG. 13.1.2 Pressure/thermal cycling and how leakage rates differed over 24 days/24 repeats replicating severe industrial conditions. *(Courtesy Flexitallic LP, Deer Park, TX.)*

REFERENCES

[1] H.P. Bloch, Pump Wisdom: Problem Solving for Operators and Specialists, John Wiley & Sons, Hoboken, NJ, 2011.

[2] W. Brown, The suitability of various types of heat exchanger gaskets, in: Proceedings of ASME Pressure Vessel and Piping Conference, Vancouver/BC, Canada, 2002, pp. 45–52. Paper Number PVP 2002-1081.

Subject Category 14

Gear and General Bearing Housing Protection

Chapter 14.1

Sealing Gearboxes Against Water Intrusion

INTRODUCTION

Gear speed reducers and increasers, collectively labeled "gearboxes," have probably been around for more centuries than any other machine. Gears perform duties in every industry today. They are very widely used in municipal waterworks, locks, and flood control systems. Of course, gears require lubrication and lubricants serve us best if they remain free of water and other contaminants. Provision must therefore be made to prevent both lubricant leakage (egress) *from* the gearbox and contaminant ingress *into* the gearbox. Gears and bearings will suffer if sealing is inadequate.

FIG. 14.1.1 Shaft damage started in the lip seal region of this gear shaft and allowed water to enter. *(Courtesy AESSEAL Inc., Rotherham, the United Kingdom, and Rockford, TN.)*

Petrochemical Machinery Insights. http://dx.doi.org/10.1016/B978-0-12-809272-9.00014-1

LIP SEALS AND OTHER SOLUTIONS

For over a century, lip seals—similar to the defective one shown in Fig. 14.1.1—have been used to keep the lube oil from escaping and contaminants from entering into machines such as gearboxes. Lip seals have the advantage of being both relatively inexpensive and easy to install. However, a tight-sealing lip seal may cause shaft damage (Fig. 14.1.1), whereas a worn or loose-fitting lip seal is likely to allow outward leakage of lubricant and will permit external contaminants to find their way into the gearbox or machine.

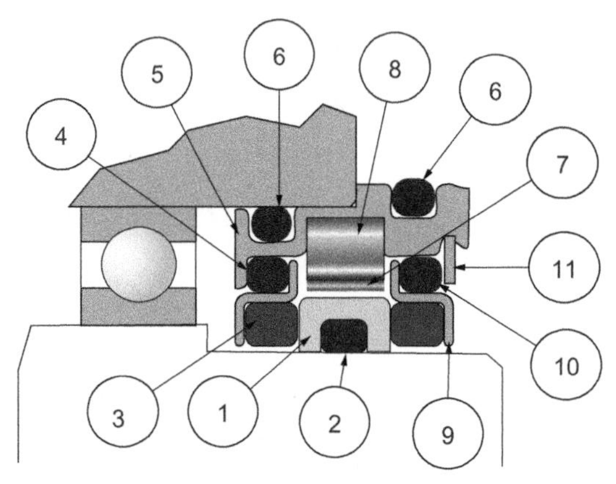

FIG. 14.1.2 Dual-face magnetic bearing protector seal and O-rings (2, 4, 6, and 10), snap ring (11), rotating face (1), stationary faces (3), stationary face holder (9), stationary magnets (8), magnet carrier (7), and outer body (5). Parts 1 and 2 are the only parts that rotate. *(Courtesy AESSEAL, Inc., Rotherham, the United Kingdom, and Rockford, TN.)*

A generic representation of bearing protector seals ("bearing isolators") was depicted earlier in Chapter 4, Fig. 4.6.1. Whenever this important reliability-improving component is specified or purchased, pay attention to its O-ring component. In some models, a dynamic, radially outward-moving O-ring can be cut by sharp corners. In advanced designs from innovative manufacturers, a large O-ring moves axially and two smaller O-rings clamp the rotor to the shaft. These features are highlighted in Fig. 14.1.3; its axial O-ring movement is considered an advantage. Experienced reliability professionals often disallow designs where radially outward-moving O-rings compromise long-term performance and component life.

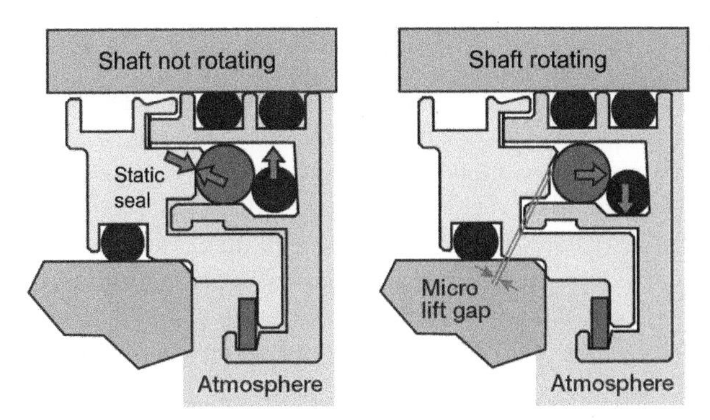

FIG. 14.1.3 Modern protector seal primarily used in premium process pump bearing housings. Note advantageous axially moving O-ring. *(Courtesy AESSEAL, Inc., Rotherham, the United Kingdom, and Rockford, TN.)*

DUAL-FACE MAGNETIC BEARING PROTECTOR SEALS STILL IMPORTANT

Introduced in about 2008, the axial O-ring-style bearing protector seals of Fig. 14.1.3 have accrued an excellent performance record. Moreover, they are less expensive than the dual-face magnetic bearing protector seals ("DFMBPS"—Fig. 14.1.2) that were introduced to the marketplace as early as 2003.

But the more expensive DFMBPS still has its place. It, too, has an outstanding performance record, and hundreds of reliability-focused users (including many gear users in municipal waterworks) line up in an impressive reference list for this long-life sealing solution. The leading manufacturers of DFMBPS will readily share their installation lists and other references with interested gear users and gear manufacturers. It can be reasoned that more and more industries demand extended run length and component life for gear units in 2016 and beyond.

EARLY HISTORY OF MAGNETIC BEARING HOUSING PROTECTOR SEALS

Prior to 2003, only single-face magnetic bearing housing protector seals were marketed. For these precursors of DFMBPS products, they were used primarily in aerospace applications. The application record shows that they were conservatively designed for face speeds approaching 20 m/s, a speed that is well below the operating speed of the great majority of gearboxes in service today. Regarding general aerospace experience and dating back several decades, many commercial and military aircraft used magnetic seals, albeit the single-face version. One precursor-style brand is applied in thousands of planes, including Boeing 707 aircraft introduced in the late 1950s, 747 models since the 1960s, and the more recent Boeing 777 commercial airliners [1].

OPERATING AND APPLICATION LIMITS EXPLAINED

When a hydrodynamic oil film is maintained between the DFMBPS faces, the coefficient of friction can be less than 0.05. (This coefficient is normally higher during early periods of operation when the film and interface are being developed. Once developed, the film helps control wear, reduces frictional heat, and stops bounce or chatter.) In any event, the commercial literature of a long-established US manufacturer of single-face magnetic seals explains applications and case histories at 300°F/5 psi/21,000 rpm and 392°F/14.5 psia/14,407 rpm at temperatures ranging from 65°F to +350°F/20 psi/28,018 rpm and many more [1].

For maximum life, especially at high speeds, face loads are usually kept low; about 15 psi is typical. Proper cooling and/or lubrication must be provided to remove frictional heat from a DFMBPS. Lubrication, of course, is present in the overwhelming majority of oil-lubricated bearings in centrifugal pumps, gear speed reducers, and other machines operating in process plants. In the very few instances where bearing lubrication would not reach the adjacent DFMBPS, it might be wise to question the adequacy of bearing lubrication as well.

TECHNICAL REPORTS ON MAGNETIC SEAL EXPERIENCE

A search of technical reports and other literature relating to magnetic seals shows that serious designers and manufacturers have, since 1971, made successful use of mechanical seal face technology in nonaviation industries. Their successful designs have performed well over the past decade because they make use of superior materials and ascertain seal face flatness.

As an example, in the design of Fig. 14.1.2, only parts 1 and 2 are clamped to the shaft; all other parts are either stationary or free-floating. In the highly unlikely event of the nonrotating stationary carbon "wearing off," the Z-shaped magnetic ring surrounding the carbon (item 3) would contact the rod magnets (item 8). This self-limiting contact would occur before the ring (item 1) could begin to scrape away at the Z-shaped ferrous material. If total wear would ever occur, the seal would simply start to leak.

A business-as-usual approach is at odds with today's reliability performance and best-practices goals. With so many DFMBPS in use today, the facts are easy to obtain. One major gear manufacturer has demonstrably extended oil changes fivefold; upgrading from the lip seals in Fig. 14.1.1 to dual-face magnetic bearing protector seals (Fig. 14.1.2) has extended seal life at least fourfold.

The DFMBPS in Fig. 14.1.2 complies with applicable American Petroleum Institute (API) standards. The European *Health and Safety Executive* also tested and certified the sparking and hazard-related performance of these dual-face bearing protector seals [2]. Face-type magnetic seals can be cost-justified for use in large electric motors. Like any other part ever designed by man, these products must be applied within their respective design envelopes. Peripheral speed and splashed oil lubrication requirements apply.

In the case of grease-lubricated gear and/or electric motor bearings, cost considerations favor a noncontacting bearing protector (see Chapter 4.1; Fig. 4.1.4). The model shown requires no lubrication because the large O-ring can move axially away from its seat well before the equipment shaft approaches operating speed. It is certified by ATEX, the European testing authority for components operating in explosive atmospheres, and is rated IP66 on the ingress protection ("IP") code.

Upgrading to magnetic face seals or well-designed rotating labyrinth protector seals (Fig. 4.1.3) makes economic sense if gearbox reliability is of prime importance. The return on incremental investment is usually measured in months, not years.

TESTING PROVES COMPLIANCE

At the time of initial development of the advanced bearing protector seals of Fig. 14.1.3 around the year 2006, one of the company's many tests was for 1005 h duration, with automated stops and starts. All parts were measured before and after the test, because there will be some contact, on start-up and shutdown between what the manufacturer calls the Arknian shutoff valve and the Arknian elastomer. (To draw a distinction, the O-ring that in the standstill condition contacts the stationary ring contour is called Arknian shutoff valve; the adjacent O-ring has different properties and is called Arknian elastomer.)

The test setup comprised an oil seal (lip seal) at one end of a double-ended motor and the advanced rotating bearing protector seal of Fig. 14.1.3 at the other. Initially, both devices passed the IP55 water ingress test. However, within 5 h, the lip seal had started to leak on a 1.750″ shaft, running at 1800 rpm. After 500 h and once again at 1000 h, the lip seal failed the IP55 water ingress test, whereas the rotating labyrinth seal passed with no leakage. More importantly, at the end of the test, the Arknian elastomer wear was immeasurable, although a slightly shiny surface was visible. There were no measurable wear and only a polishing effect on the Arknian shutoff valve.

LIFT AND WEAR OF ADVANCED BEARING PROTECTORS

Lifting off and the microgap in the typical O-ring-equipped bearing protector occurs at or around 345 rpm, but what happens if there is full contact? To answer that question, the manufacturer provided test data for 1600 h, with a 100 mm (4.000″) device rotating at just 60 rpm. The temperature reached only 7.2°C (13°F) above ambient, which indicates there was almost no friction, even though there had to be full contact at starts and stops. There was also no oil egress.

After 1600 h, there was *no wear* of either the metal components, the Arknian elastomer, or any other part. This is a tribute to the Arknian shutoff design and the Arknian elastomer material selection. The device easily met the IP66 and

5-year duration requirements of IEEE 841-2001. Many thousands were sold in the 9 years since the component first entered into the marketplace; a large number of them had been in continuous service well beyond 8 years when this manuscript went to press in 2016. Although the manufacturer sells an O-ring replacement kit, few (if any) replacement kits have ever been sold.

REFERENCES

[1] Magnetic Seal Corporation, Warren, Rhode Island, commercial literature, c.1971.
[2] European Health and Safety Directive, 89/391/EEC.

Chapter 14.2

Why Bearing Housing Protector Seals are Needed

INTRODUCTION

At the risk of stating the obvious, let's be sure the lube in a pump's bearing housing is kept clean. Even the most outstanding lubricant cannot save a bearing unless the oil is kept clean. This is where bearing housing protector seals are of value [1,2].

Lubricant contamination originates from a number of possible sources and can also be a factor in "unexplained" repeat failures. Unless process pumps are provided with suitable bearing housing seals, an interchange of internal and external air (called "breathing") takes place during alternating periods of operation and shutdown. Bearing housings "breathe" in the sense that rising temperatures during operation cause air volume expansion, and decreasing temperatures at night or after shutdown cause air volume contraction. Open or inadequately sealed bearing housings promote this back-and-forth movement of moisture-laden and dust-containing ambient air. But, simply adding bearing protector seals could change windage or housing-internal pressure patterns in unforeseen ways. This, too, we must recognize as a potential source of "unexplained" failures in housings without internal balance holes, explained later in Chapter 30.5 (see Fig. 30.5.5).

BEARING HOUSING ISSUES

Ideally, housings should not invite breathing and the resulting contamination. There should be little or no interchange between the housing interior air and the surrounding ambient air. The breather vents typically found on the top of pump bearing housings can often be removed and plugged. Don't be shocked by that statement. Many hundreds of millions of refrigerators and automotive air conditioning systems operate with neither vents nor breathers. Conceivably, old-style bearing housing seals allow an O-ring to contact an O-ring groove, as depicted on the left image in Fig. 14.2.1. Although the sketches do not replicate actual product, they assist in visualizing component damage risk.

The illustration on the left highlights what can happen with bearing protector seal designs that incorporate sharp-edged grooves. Contact with sharp-edged grooves invites dynamic O-rings to scrape. That's another disclosure that should not shock us; after all, none of us would think that sliding our fingers over the edge of a knife is without risk. Abraded elastomer shavings can contaminate the lubricant and cause oil to change color. Also, using only a single O-ring for clamping the rotor to the shaft makes the rotor less stable than if two rings are used for clamping duty.

The illustration on the right reminds us of products with parts that make excessive frictional contact. Both invite us to become familiar with how parts work and how they might fail. It can be reasoned that the product on the left of Fig. 14.2.1 is risky in terms of O-ring life reduction and potential rotor wobbling. It competes by being less expensive than the superior product in Fig. 14.1.3 described earlier.

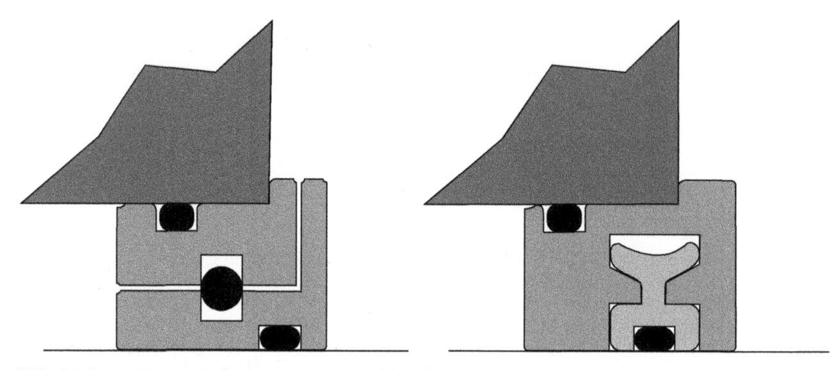

FIG. 14.2.1 Dynamic O-ring in contact with a sharp groove *(left)*; wedge-like elastomer contours undergo lots of friction *(right)*.

Visualize this rotor instability by mentally removing the stationary component in the left illustration of Fig. 14.2.1. The rotor pivots around the clamping O-ring and destructive vibration would occur at high speeds. We could study the rotor dynamics of such a situation and spend a nice sum on doing research. Or we might reach the same commonsense conclusion by giving it some thought. Two clamping O-rings will provide more stability than one single clamping O-ring. And the VEE-ring on the right illustration will cause trouble if this configuration is chosen.

In essence, bearing housing protector seals can greatly improve both life and reliability of rotating equipment by safeguarding the cleanliness of the lubricating oil. However, these protector seals add little value if oil contamination originates with oil ring wear, if pressure-unbalanced constant-level lubricators are used that allow air and moisture to intrude, if the oil is not kept at the proper oil level, if the bearing housing design disregards windage concerns, or if water enters into the oil.

Checking for O-ring degradation by adding see-through containers to the bottom of a pump bearing housing is sometimes being suggested. However, by the time slivers of O-ring material or water becomes visible in such a "sludge cup container," the saturation limits of oil in water will have been exceeded, or an oil ring is shredded to pieces and much damage could have been done to the bearings. We can deduce that free water in the oil is a symptom of not having the right bearing protection. Our reliability focus should be on treating the root cause, not the symptom. We should prevent water from reaching the bearings in the first place. These simple proactive and precautionary thought processes are at the core of pump failure prevention.

REFERENCES

[1] H.P. Bloch, Pump Wisdom: Problem Solving for Operators and Specialists, John Wiley & Sons, Hoboken, NJ, 2011.

[2] H.P. Bloch, A.R. Budris, Pump User's Handbook: Life Extension, fourth ed., Fairmont Publishing Company, Lilburn, GA, 2014.

Subject Category 15

Grounding Technology

Chapter 15.1

Grounding Ring Technology for Variable Frequency Drives (VFDs)

INTRODUCTION

While there are compelling reasons to specify insulated (actually, aluminum oxide-coated) or ceramic ("hybrid") rolling element bearings for variable frequency drives (VFDs), there may be instances where bearing protector rings are well justified and might further reduce the risk of shaft current-induced bearing distress. When specifying such shaft grounding rings (SGRs), steer clear of knockoff products that use carbon fibers and mounting methods in a manner that compromises long-term reliable service.

FUNDAMENTALS OF SHAFT GROUNDING RINGS (SGRS)

At the simplest level, an AEGIS SGR provides the "path of least resistance to ground" for VFD-induced shaft voltages. If these voltages are not diverted away from the bearings to ground, they may discharge through the bearings and cause damage known as electrical discharge machining (EDM), pitting, and fluting failure in bearings. These SGRs can be adapted as an integral part of the motor design. Good products meet both spirit and intent of the NEMA MG1 Part 31 specification, aimed at preventing bearing fluting failure in electric motors as well as their attached equipment. NEMA MG1 identifies induced shaft voltage in VFDs as a potential cause of motor failure and recommends shaft grounding as a solution to protect both motor bearings and attached equipment. Fig. 15.1.1 shows a late version of a well-designed grounding device.

Properly designed SGRs must provide a large number of small-diameter fibers to induce ionization; they must discharge voltages away from motor bearings and to ground. Selecting carbon fibers of specific mechanical strength

Petrochemical Machinery Insights. http://dx.doi.org/10.1016/B978-0-12-809272-9.00015-3

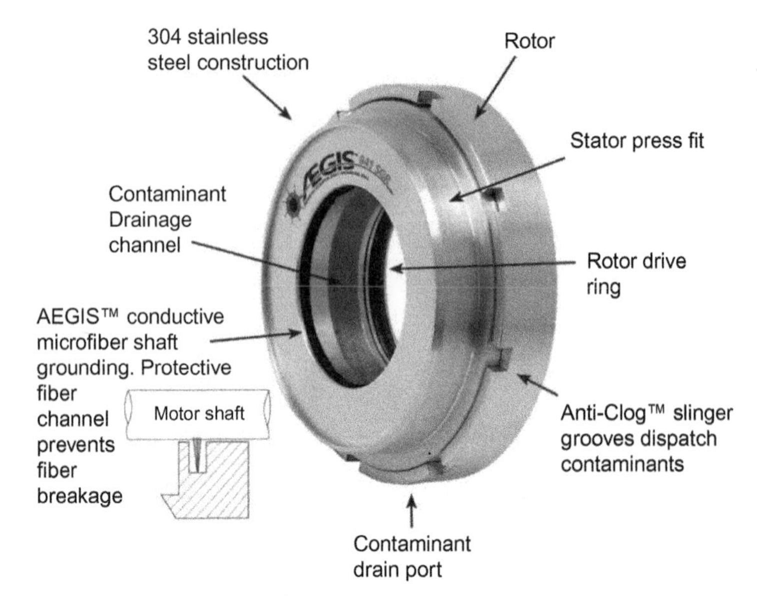

FIG. 15.1.1 SGR grounding device. *(Courtesy Electro Static Technology (EST), Mechanic Falls, ME.)*

and electrical characteristics is critically important to providing break-free and nonwearing service. The fibers must be allowed to flex within their elastic limit and while contacting the shaft with the proper overlap. For long-term reliability, they must be placed in an engineered holder that protects against breaking and mechanical stress. Chances are that, without placement in protective channels, the reliability of SGRs is severely compromised.

CIRCUMFERENTIAL ROWS OF FIBERS

The best available designs will optimize fiber density so as to maintain the required fiber flexibility. If too many fibers are bundled together (as may be the case in less-than-optimal designs), the fibers will break. A soundly engineered SGR has two full rows of fibers. The continuous circumferential "ring" design and fiber flexibility allow them to direst small amounts of oil film, grease, and dust particles away from the shaft surface. It was noted that EST's patents prevent others from copying technology that arranges one or more rows of fibers in a continuous fiber ring inside a protective channel completely surrounding the motor shaft. This design ensures that there are literally hundreds of thousands of fibers available to handle discharge currents from VFD-induced voltages at the various prevailing high frequencies. The fibers can then flex inside the channel while maintaining optimal contact with the motor shaft.

SELECTION STRATEGY

Reliability-focused VFD users would involve both VFD manufacturer and bearing suppliers in issues dealing with bearing failure avoidance strategies. Also, reliability-focused users would endeavor to become familiar with sound grounding ring technology.

A company that initially focused on mitigating static charges in the printing and imaging markets, Electro Static Technology (EST), has—since about 2005—been producing conductive microfiber grounding rings for rotating equipment. EST's Bearing Protection Ring products are marketed as AEGIS products, representing proprietary technology said to provide a reliable and essentially maintenance-free SGR. Such rings may be needed to mitigate the issues of electrical erosion in motor bearings when electric motors are controlled by pulse width-modulated (PWM) VFDs.

Subject Category 16

Innovation Versus Unnecessary Developments

Chapter 16.1

Innovations that are Duplications of Effort

INTRODUCTION

An old saying reminds us that we should not reinvent the wheel. We intuitively understand that this admonition covers anything from machines to components and work processes to procedures. Know what already exists and expend effort on improvement, if possible. Three case histories explain the issue.

Our *case 1 experience* dates back to about 2003, when a major mechanical seal manufacturer started to put effort into a mechanical seal assembly to replace segmented carbon rings in the gland areas of small steam turbines. Over 20 years earlier, this kind of steam turbine seal gland upgrade cartridge (STSGUC) had been pursued and fully implemented by a user company in Texas [1]. Several manufacturers of small- and medium-size steam turbines were contacted by the Texas-based STSGUC users, but the manufacturers showed no interest in commercially developing cost-justified STSGUC retrofits.

Then, in 2004, a major seal manufacturer began marketing STSGUC-like products. As things turned out, the seal manufacturer never acknowledged anyone else's prior contributions. We might, therefore, assume its development teams never read what had been reported on and successfully installed decades earlier. Could the manufacturer have saved money by reading about the earlier developments? Could they have saved time and effort by researching the prior releases that dated back to 1985? Suppose the seal manufacturer's design team had, in fact, researched the matter. In that case, what made them decide not to acknowledge the preexisting product or conference paper and article that had pointed the way?

Petrochemical Machinery Insights. http://dx.doi.org/10.1016/B978-0-12-809272-9.00016-5

Our *case 2 experience* is also of interest. In this instance, a company spent considerable research effort developing means of retaining rolling element bearings in a mounting arrangement accommodating vertical shafts. Ostensibly, the effort was in response to bearing failures at user companies with air-cooled heat exchangers and associated vertically oriented cooling fan shafts. Here, too, the developers seemed oblivious to the fact that well-established major bearing manufacturers had, for a long time, been providing spherical roller bearings with thrust load and angular misalignment capabilities. Knowledge of that fact might have saved money for developments that largely duplicated prior art.

Finally, there is the noteworthy *case 3 experience* involving incipient failure detection (IFD) systems. Fig. 16.1.1 shows IFD transducers—piezoelectric sensors—mounted on an electric motor driver and a pump bearing housing. In the early 1980s, several publications had highlighted the IFD technology and explained why a major multinational petrochemical company in the United States had decided to discontinue its plant-wide IFD program [2]. In 2009, a start-up company began commercializing technology aimed at capturing bearing degradation before it had progressed to the point of failure. It's a predictive maintenance (PdM) approach that might best be described as bearing housing metal stress propagation sensing. When we asked for more data, it became evident the company staffers had not read preexisting conference proceedings, articles, or chapters in books dealing with a pump PdM method known as IFD. The short explanation: each baseline electronic signature displayed on a computer monitor in the control room was different from the next. In the mid-1970s, there seemed to be no cost-effective way to accurately predetermine an intervention threshold. Until such thresholds are defined and incorporated in cost-effective wireless signal transmission technology, the developments will meet with low user interest at best. Then again, good books may shed much light on similar issues where a small amount of time devoted to reading may illuminate little known solutions to problems plaguing pump users. For the time being, IFD is not one of the solutions [3].

FIG. 16.1.1 Piezoelectric transducers mounted on pump and motor bearings in the mid-1970s.

GOOD INNOVATIONS

So as not to leave the impression that innovation is of no value, we want to point to just a few of the many hundreds of examples where innovation creates advantages for manufacturers and users. We point to dual mechanical seals with bidirectional tapered pumping rings and bearing housing protector seals with axially moving O-ring seals (Subject Category 14 and index words at the end of this text).

We point to exemplary non-OEM rebuild shops that routinely use computerized scanning (Figs. 16.1.2 and 16.1.3; see also Subject Category 22),

FIG. 16.1.2 Computerized scanning in progress. *(Courtesy Hydro Inc., Chicago, IL, www.hydroinc.com.)*

FIG. 16.1.3 A scanning arm and instantaneous display. *(Courtesy Hydro Inc., Chicago, IL, www.hydroinc.com.)*

upgrade design, patternmaking, and simulation capabilities that were unheard of a few decades ago. They are anything but duplication of innovation; instead, exemplary non-OEMs qualify as companies that bundle seemingly isolated stand-alone innovation into value-adding equipment rebuilding and upgrading strategies.

REFERENCES

[1] H.P. Bloch, H. Elliott, Mechanical seals in medium-pressure steam turbines, in: Presented at the ASLE 40th Annual Meeting in Las Vegas, Nevada, May 1985, 1985. Also Reprinted in Lubrication Engineering, November 1985.

[2] H.P. Bloch, R. Finley, A decade of experience with plant-wide acoustic IFD systems, in: Presented at Vibration Institute Conference, Houston, Texas, April 19–21, 1983, 1983.

[3] H.P. Bloch, Pump Wisdom: Problem Solving for Operators and Specialists, John Wiley & Sons, Hoboken, NY, ISBN: 978-1-118-04123-9, 2011.

Chapter 16.2

Process Innovations that Pose Dangers

INTRODUCTION

Because most processes use fluid machinery, process modifications and/or "innovations" can affect equipment reliability, safety, and human lives. An unresolved pump problem may lead process designers to modify a process so as to eliminate unreliable pumps. The effects can be far-reaching and many references explain how seemingly innovative steps can ultimately have the exact opposite result. The reader will find a piping-related incident (see index under "Flixboro"), a highly informative article [1], and a book [2] that should remind equipment specialists of their professional and ethical responsibilities. Theirs is the responsibility to detect, question, and resolve repeat failure events. As they perhaps feel inclined to accept or implement "innovations," they must be ever aware that human lives may be at stake—and thousands of them at that.

Fig. 16.2.1 is a fitting reminder that pumps and questionable sealing were involved in a still incomprehensible tragedy involving serious equipment reliability issues in Bhopal. Erroneous decisions were made in the interest of coping with, and circumventing, a chronic machinery reliability problem. These decisions included devising alternative operating strategies—call them "innovations"—that introduced unknown and potentially unacceptable hazards. Creative alternative operating strategies ("process optimization steps") may seem justified in the interest of confronting legitimate personnel-related safety concerns. However, their impact on overall process safety can be disastrous.

FIG. 16.2.1 The MIC feed pump—a reminder of the most tragic loss of life in the history of process plants. *(Courtesy Bhopal Medical Appeal, www.bhopal.org, used by permission.)*

Early on, reliability engineers must focus on resolving equipment reliability defects. Understanding failure causes and implementing upgrades eliminate failures. Simply replacing parts should not be reliability engineering. A reliability engineer eliminates risk-inducing components or elements. Unswerving focus, early remedial action, and zero tolerance for unexplained repeat failures of rotating machinery are the most valuable safeguards against a potentially devastating sequence of events.

In Ref. [1] ("Unreliable Machinery in the Bhopal Disaster"), two experienced coauthors applied modern investigation techniques, using only information in the public domain. They placed before us the exceedingly well-documented sequence of events that preceded the disaster and/or followed in the wake of several process-related decisions. A comprehensive book [2] goes into much more detail and shows how every "innovation" constituted a deviation and how deviations become the accepted norm.

Within the context and scope of the text you're reading just now, here's the point: engineers should have a questioning attitude and should demand of themselves the same conduct they expect from the medical profession. All must offer authoritative advice, concern for everyone, and complete candor. Meeting this mutual expectation is especially important when there are edicts to purchase from the lowest bidder. Engineers should stand firm in demanding to buy from the *lowest bidder that also fully meets responsibly written specification requirements*. Needless to say, one of the engineer's primary roles must be to write such specifications. That further implies and presupposes that they, the engineers, avail themselves of all reasonable postgraduation training opportunities. It also implies that they should never become a party to withholding information from the ultimate equipment users.

So, we believe that engineering schools and their industry training institutions (or any of their adjuncts that include reliability conferences) should make it their business to teach about the sequence of events that resulted in the worst industrial disaster in human history. Not doing so will deprive new generations of engineers of understanding the importance of good engineering judgment.

English critic, essayist, and reformer John Ruskin (1819–1900) knew what would happen if one blindly purchases from the lowest bidder. He stated basic principles and phrased them in commonsense language long before the business schools of the late 20th and early 21st centuries began dispensing their often misguided and also widely misinterpreted and rarely contested wisdoms. Paraphrasing Ruskin:

> *It is unwise to pay too much, but is worse to pay too little. When you pay too much, you lose a little money—that is all. When you pay too little, you sometimes lose everything, because the thing you bought was incapable of doing the thing it was bought to do.*
>
> *The common law of business balance prohibits paying a little and getting a lot—it cannot be done. If you deal with the lowest bidder, it is well to add something for the risk you run, and if you do that, you will have enough to pay for something better.*

There is no way John Ruskin could foretell or foresee what would happen in Bhopal on 3 Dec. 1984. Bhopal was an unmitigated disaster that involved disregarding warning signs, disabling critical safeguards, and permitting normalization of deviances. The record shows deviations of huge consequences included choices in materials technology and pump component selection. The record shows neither the pump manufacturer nor the seal manufacturer. There is no evidence of their involvement after several seal leakage events should have prompted a search for seal upgrades.

As modern industry benefits from examining the many historical facts relating to process safety failures, be it noted that process safety trumps all manner of process optimization. Both of our references make convincing use of material that is now freely available in the public domain. They use TapRooT as the central core of their root cause failure analysis. The entire Bhopal topic will be fully appreciated by those professionals who, by virtue of seeing it as their responsibility to avoid similar incidents, draw the right conclusions and take advantage of the information presented.

REFERENCES

[1] K.P. Bloch, B. Jung, Unreliable machinery in the Bhopal disaster, Hydrocarbon Processing, June 2012.

[2] K.P. Bloch, Rethinking Bhopal—Investigating and Preventing Disasters, Elsevier Publishing, London, UK and Waltham, MA, ISBN: 978-0-12-803778-2, 2016.

Chapter 16.3

Involve Innovators in Fluid Machinery Upgrading

INTRODUCTION

In best-of-class (BoC) process industries, at least one and often an entire group of professional employees are tasked with plant debottlenecking and equipment life extension. Professionals at BoCs are experienced value-adders; they study process optimization and profitability enhancement. They advise managers on feasibility, cost justification, and implementation strategies. By exploring innovations in seemingly unrelated industries (such as aerospace), they often discover significant opportunities in their own industry segment. Learning from the success of others, informed value-adders—the motivated reliability professionals—steer their companies toward becoming innovators in their own right. As readers of this text, you would be correct in assuming that we speak of new and important opportunities available to petrochemicals and oil-refining companies that themselves want to push technology in the right direction.

An exciting wear materials development. First, we wanted to apprise you of a unique diffusion-conversion-based materials technology that improves both the chemical and physical characteristics of steel. These improved characteristics transform lower cost materials so that they either match or outperform more costly alloys in high-wear applications. Depth and hardness of the diffusion-converted strata are 40–70 µm and hardness values of 67 RC are typical. The converted surfaces exhibit high lubricity without applying oils and antigalling properties without imparting brittleness. By altering the crystalline lattice of steel, diffusion-conversion-based technologies eliminate wear on the nanolevel. The main or bulk mass remains at its original hardness, allowing it to adsorb[1] or dissipate the energy of abrasive particles striking the surface. It is important to note that surfaces of diffusion-converted parts resist spalling.

We find it easy to visualize how use of this innovation will open up previously unavailable upgrade opportunities for the hydrocarbon processing industry. In fact, because of an urgent need to find virtually indestructible bearings for product-flooded positive displacement pumps, our initial focus was on this technology in 2014. This company had commercialized an important diffusion-conversion process a few years earlier. The process imparts unusual hardness and very good lubricity without adding a dimensional buildup. It is being used

---

1. Adsorption is the process in which atoms, ions, or molecules from a substance—it could be gas, liquid, or dissolved solid—adhere to a surface of the adsorbent. Adsorption is a surface-based process where atoms, ions, or molecules "infiltrate" the surface and take their place in the voids between atoms or molecules.

in high-temperature automotive turboexpanders and in metal chip conveyors. It certainly has potential in many process and mining pump applications.

Turbocompressor wheels (impellers). Our interest was heightened when we realized the extent to which the materials technology developers also had manufacturing experience in impellers for turbocompressors and expanders. Virtually any conceivable impeller geometry used in the petrochemical industry has been produced by TURBOCAM, with some impellers suitable for service temperatures in excess of 1000°F (537°C). We saw the names of original equipment manufacturer (OEM) compressor manufacturers and other customers on routing sheets for a number of three-dimensional semiopen impellers (Fig. 16.3.1). Several of these impellers were in the process of being machined on the company's over 150 multiaxis machines. The impellers were either special aluminum alloys or stainless steels; some were made of titanium. We also confirmed several of the company's references to, and capabilities in, producing traditional closed (covered) impellers (Figs. 16.3.2 and 16.3.3).

FIG. 16.3.1 Semiopen impeller being machined. *(Courtesy TURBOCAM, Barrington, New Hampshire.)*

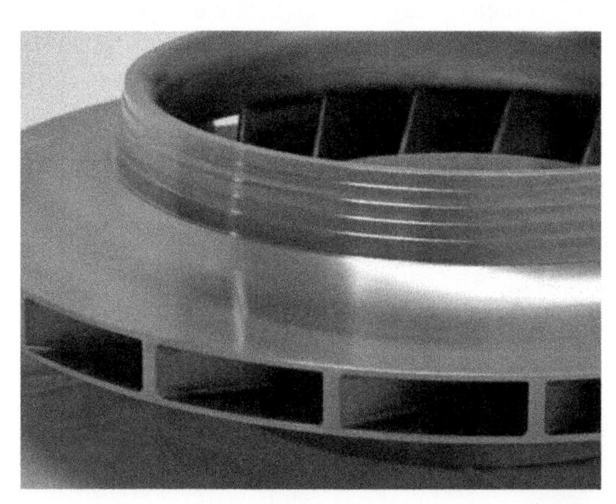

FIG. 16.3.2 An integral cover ("covered") radial turbocompressor impeller produced for both OEM and aftermarket customers. *(Courtesy TURBOCAM, Barrington, New Hampshire.)*

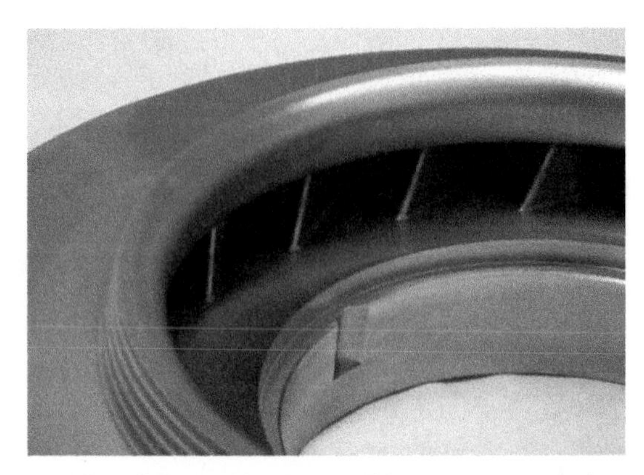

FIG. 16.3.3 Impeller for a fluid machine. Note half-key seat to securely fasten this impeller on an equipment shaft at ultimate assembly. *(Courtesy TURBOCAM, Barrington, New Hampshire.)*

Be prepared to upgrade. From the user's perspective, keep an open mind as to which entities should be entrusted with upgrading your machinery. Make informed choices instead of basing important reliability engineering decisions on mere opinion or past traditions and practices. So, while OEMs are part of the equipment service and component upgrading picture, the very same must be said for competent non-OEMs. We have established that advanced materials technology is rapidly migrating from aerospace to other industries. This migration often bypasses slow-paced OEMs and provides great opportunities for users striving for self-sufficiency and/or ready access to fast-paced OEMs and non-OEMs in particular geographic regions. These users should proceed by ascertaining they have accurate geometric data describing the present internals of their existing fluid machines.

Alternatively, it would be worth it to consider opening "upratable" machines during the next scheduled outage. At that time, a competent service provider will be able to multidimensionally map critical parts of the machine. Investigate if impeller refurbishment would be of benefit and, if the answer is in the affirmative, consider a variety of options. Do some advance planning before a crisis develops and include the true innovators in conscious endeavors aimed at future debottlenecking and reliability improvements. Subject Category 22 sheds additional light on the opportunity.

Subject Category 17

Lubrication

Chapter 17.1

Resolving Grease Lubrication Controversies

INTRODUCTION

In the late 1980s, a major electric power company received a memorandum from one of the premier manufacturers of rolling element bearings stating, in effect, that *neither sealed nor shielded* bearings are regreasable. In a follow-up conversation, one bearing manufacturer's application engineer claimed that double-shielded bearings are no longer considered regreasable because of recently reduced clearances between shield bore and bearing inner ring diameter. Unlike older shielded bearings that had a radial gap of 400 μm (0.016″), the manufacturer's engineer informed that the newer bearings have radial shield gaps of only 250 μm (0.010″).

If the reader finds it difficult to understand why oil would manage to creep across a 0.016″ gap, but not a gap of 0.010″, he's not alone. Most competent reliability professionals, mechanics, and maintenance technicians share his skepticism.

CAPILLARY ACTION EXPLAINED

Rest assured: The claim that shielded bearings cannot be regreased is incorrect. It is generally known that grease consists of approximately 85% oil, the rest being soap (the "matrix") and other additives. There is a body of literature explaining that the oil portion of the grease "bleeds" or "seeps" through the gap between the shields and the bearing inner ring. In other words, capillary action causes the oil to leave the soap and wet out, which is to say, "coating," the rolling elements.

Petrochemical Machinery Insights. http://dx.doi.org/10.1016/B978-0-12-809272-9.00017-7

To say there's confusion is an understatement. In mid-2001, a major bearing manufacturer's marketing communication mentioned that "test data support the finding that little or no grease can be *forced* through the shield-to-bearing-inner-ring gap." To a reliability-focused reader, that should not come as a surprise. The original design intent for shielded bearings was to keep the user from *overgreasing* the bearing and to depend on capillary action to primarily bleed *oil,* not soap, through this gap. It was also the original design intent to use shielded bearings (Figs. 17.1.1–17.1.3) for light to medium loads and moderate speeds and to periodically regrease shielded bearings in these applications.

RELUBRICATION IS INTENDED FOR SHIELDED BEARINGS

One well-known bearing manufacturer infers relubrication of shielded bearings by stating, in Ref. [1]

> *Relubricatable, double sealed (probably meaning double shielded—Author's comment) bearings should be sparingly relubricated at recommended intervals with the identical or, at least, a compatible grease approved by the bearing manufacturer.*

The majority of electric motor bearings fit the light-load criterion. Using another bearing manufacturer's interactive engineering catalog [2], we found the projected bearing lives for many electric motors to be somewhere between 500 and 600 years! These, then, are quite obviously light loads. Lightly or moderately loaded bearings of either the double- or single-shielded variety are widely used in industry. In Ref. [3], the same competent manufacturer shows the grease chamber of an electric motor equipped with double-shielded bearing. The accompanying narrative states:

Experienced bearing manufacturers agree that single-shielded bearings should be installed with the shield facing the grease cavity [4]; the shield in Fig. 17.1.1 acts as a metering orifice and prevents overgreasing. Double-shielded bearings (Fig. 17.1.2) put an end to the controversy. In cross-flow (open) bearings, Fig. 17.1.3, the drain plug must be removed, while regreasing is in progress.

EXPERIENCE-RELATED DATA OF INTEREST

There is no substitute for experience. Also, it behooves us to ask questions. It is only reasonable to assume that manufacturers have lost experienced personnel or to conclude that the goals and aspirations of bearing providers are not necessarily aligned with those of equipment users and reliability professionals. So, what should be the course of action of this latter group?

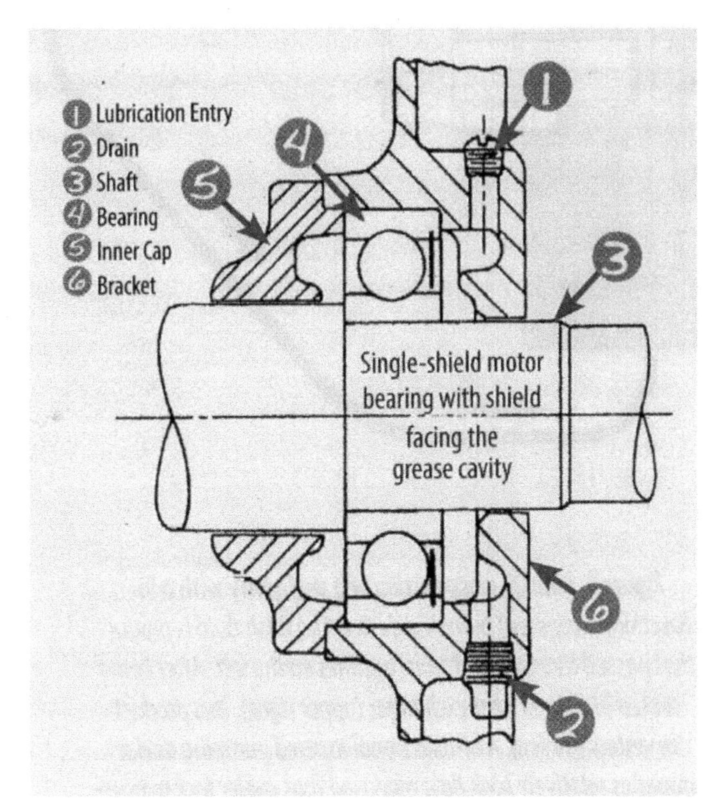

FIG. 17.1.1 A single-shielded bearing; note grease reservoir is next to shield; addition of lubricant to the original supply takes place by seepage of oil from the grease, and a small amount of grease is forced through the clearance in the shield. *(Courtesy J.R. Autenrieth, Motor Lubrication Guidelines, Phillips Petroleum, Sweeney, TX, c.1980.)*

Over the years, guideline values have evolved, and not too long ago, both users and manufacturers shared their experience as follows:

- *Sealed* bearings are preferred in continuous duty applications as long as the product of the bearing bore diameter "*D*," in mm, and speed "*N*," in RPM, does not exceed 80,000. ("Seals" are elastomeric disks fitted to the bore of the bearing outer ring and making rubbing contact with the outer periphery of the bearing inner ring. Sealed bearings are factory prefilled with grease and cannot be regreased.)
- *Regreasable*, shielded bearings are considered the best choice for light to medium loads in the *DN* range from 108,000 to 300,000, and users can opt for either sealed (nonregreasable) or shielded (regreasable) configurations in the "gray" range from 80,000 to 108,000.

- Nonshielded, open, regreasable bearings would be best for high-load applications, regardless of the *DN* value. Consider it a high-load application if the calculated bearing life is below 20 years.
- Finally, once a *DN* value of 300,000 is exceeded, liquid oil lubrication should be much preferred over grease [4,5].

It should be noted that virtually everything you have read and almost every relubrication chart published before 2015 refer to mineral oil-based greases. Mineral oils are hydrocarbons. Check the index in this book for information on PFPE greases that are formulated with substances other than hydrocarbons. Note how PFPEs can extend the life of sealed bearings or allow operating certain fully sealed bearings with larger diameters (or at higher speeds) than what was customary with predecessor grease formulations.

BEARING REPLACEMENT STATISTICS

At one major US petrochemical plant, there were 156 bearing-related repair incidents per 1000 electric motors per year, while an affiliated refinery experienced only 18 incidents on very similar motors lubricated with the exact same grease [4]. Three installations in the Middle East averaged 14 replacement events per 1000 motors per year. Their motor specification insisted on regreasable bearings, and they always removed the drain plug when replenishing grease. Removing the drain plug (item 2 in Figs. 17.1.1–17.1.3) prevents overgreasing and allows spent grease to be expelled. Mixing of incompatible greases can greatly reduce bearing life. Refer to Table 17.1.1 for grease compatibility ratings and verify compatibilities for any grease not presently listed in this traditional table. Conversion to certain high-temperature greases may require removal of even trace quantities of traditional greases from bearings.

At the 156/1000/year location, periodic grease replenishment was done with the drain plug (item 2) left in place. New grease tended to force the spent grease into open or nonshielded bearings; alternatively, new grease under pressure tended to deflect or even deforms the shields of shielded bearings. In contrast, the 18/1000/year location saw to it that drain plugs were removed during regreasing. Using these documented findings and assuming 2000 electric motors at a large refinery or paper mill, proper regreasing could thus avoid 276 bearing replacement incidents. At $2000 per incident, this would save in excess of $500,000 per year.

It is important to note that one US West Coast installation addressed the drain plug removal issue by permanently substituting 6-inch long pipe nipples, followed by pipe elbows and short pipe nipples, Fig. 17.1.4. Newly applied grease forces out the old grease. Spent grease forms the plug, so to speak. This installation and others that use proper selection and grease replenishing techniques can look forward to highly satisfactory bearing lives. Indeed, many electric motor bearings are still in service after 20 years of virtually continuous operation.

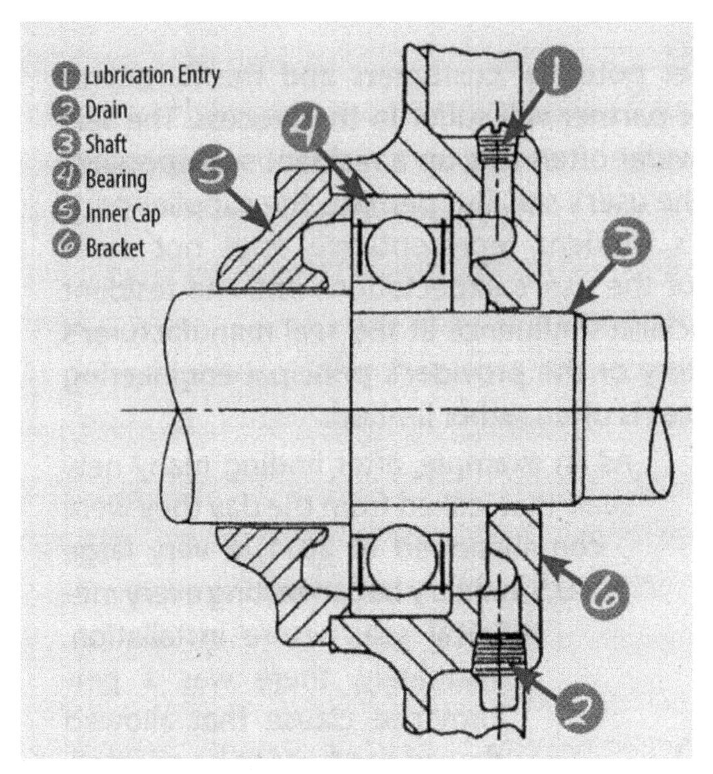

FIG. 17.1.2 A double-shielded electric motor bearing. *(Courtesy J.R. Autenrieth, Motor Lubrication Guidelines, Phillips Petroleum, Sweeney, TX, c.1980.)*

Refer to Subject Categories 11 and 23 for important information on dry sump oil mist that, at best-of-class plants, is extensively used for electric motor bearing lubrication.

For many plants that adhere to grease lubrication, Fig. 17.1.4 shows the best solution from technical acceptability and also from "not wanting to argue with my workers" points of view. Alternatively, we might try accountability and insist on staffers following instructions. We found out how well this admirable approach worked in the United Arab Emirates where a large refinery reported replacing 7 bearings per 1000 electric motors per year. When asked what magic grease formulation they were using, a senior manager explained that his workers simply followed instructions and that grease-related bearing failures are a rarity. In other words, they know what bearings they have, they remove drain plugs, they regrease with the prescribed amount of grease, and they then move on to do the next electric motor. After allowing 2–3 h for grease to settle, a worker returns and reinserts each drain plug. This simply illustrates that there is no substitute for following a proper work execution procedure. Good supervision and managing with integrity prevent failures and generate higher profits.

TABLE 17.1.1 Conventional Grease Compatibility Chart

General Grease Compatibility Chart	Aluminum Complex	Barium	Calcium	Calcium 12-Hydroxy	Calcium Complex	Clay	Lithium	Lithium 12-Hydroxy	Lithium Complex	Polyurea	Silicone
Aluminum complex	X	I	I	C	I	I	I	I	C	I	I
Barium	I	X	I	C	1	I	I	I	I	I	I
Calcium	X	I	X	C	I	I	I	I	C	I	I
Calcium 12-hydroxy	C	C	C	X	B	C	C	C	C	I	I
Calcium complex	I	I	I	B	X	I	I	I	C	C	I
Clay	I	I	C	C	1	X	I	I	I	I	I
Lithium	I	I	C	C	I	I	X	C	C	I	I
Lithium 12-hydroxy	I	I	B	C	I	I	C	X	C	I	I
Lithium complex	C	I	C	C	C	I	C	C	X	I	I
Polyurea	I	I	I	I	C	I	I	I	I	X	I
Silicone	I	I	I	I	I	I	I	I	I	I	X

Royal Purple Ultra-Performance® Grease Compatibility Chart	Aluminum Complex	Barium	Calcium	Calcium 12-Hydroxy	Calcium Complex	Clay	Lithium	Lithium 12-Hydroxy	Lithium Complex	Polyurea	Silicone
Operating temps. <225°F	C	C	C	C	C	I	C	C	C	C	I
Operating temps. 225–350°F	C	B	B	C	B	1	B	B	C	B	I
Operating temps. >350°F	C	I	I	C	I	I	I	I	C	I	I

C=compatible; B=borderline compatible: typically results in a light softening or hardening of the NLGI grade and a lowering of the dropping point of the mixture of grease. I=incompatible: typically results in a softening or hardening of greater than 1 ½ the NLGI grade, a shift in the dropping point, and a possible reaction of additives or base oils.
Note: Ultra-Performance® greases are more stable. This chart is generated from independent lab testing and field experience. Actual compatibility results may vary. It is recommended that bearings be purged of old grease per OEM instructions to ensure proper lubrication and performance.
Source: E.H. Meyers' paper entitled "Incompatibility of Greases" 49th Annual NLGI Meeting.

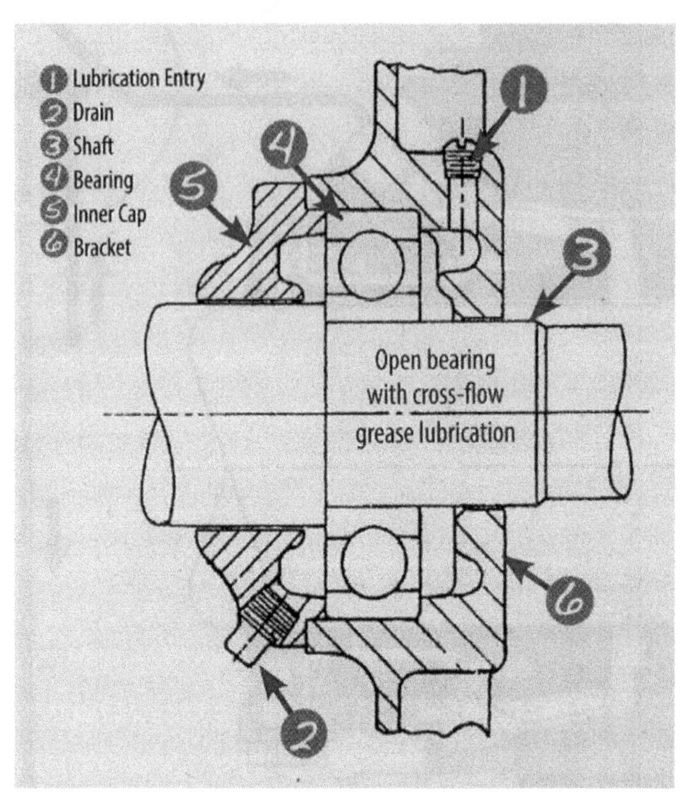

FIG. 17.1.3 An open bearing with cross-flow grease lubrication. The drain plug (item 2) must be removed while regreasing from entry point 1. *(Courtesy J.R. Autenrieth, Motor Lubrication Guidelines, Phillips Petroleum, Sweeney, TX, c.1980.)*

RANKINGS UPDATE

Recall the 1980s' lube application rankings given earlier in Chapter 4.5, Table 4.5.1. In 2014 and based on experience feedback on pumps on motors in US refineries and petrochemical plants, a new ranking was proposed by the author. Except for a rather expensive traditional auxiliary lube pump-around system with reservoir, an integrated lube spray system (Fig. 17.1.5) is the best of all worlds. One would no longer worry about oil rings and their many demonstrated flaws, constant level lubricators, installation accuracy, shaft inclination, and so forth. The same illustration explains optimized through-flow in oil mist lubricated bearing housings. It leans heavily on information presented in Chapter 4.3, Fig. 4.3.1.

Sooner or later, an innovative pump manufacturer will offer a pumping device and attach it to the bottom drain port in Fig. 17.1.5. The oil could get pressurized, filtered, and sprayed into bearing housing end caps.

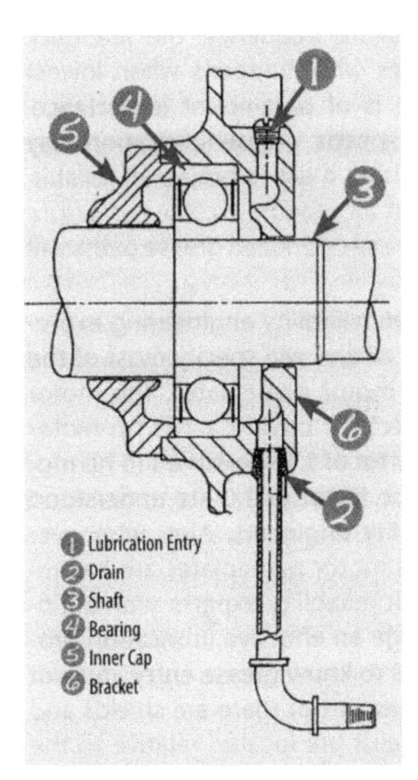

FIG. 17.1.4 A double-shielded bearing; housing plug replaced by hard pipe. When regreasing, new grease displaces spent grease in the pipe, and the permanent drain opening prevents overpressuring. *(Courtesy J.R. Autenrieth, Motor Lubrication Guidelines, Phillips Petroleum, Sweeney, TX, c.1980.)*

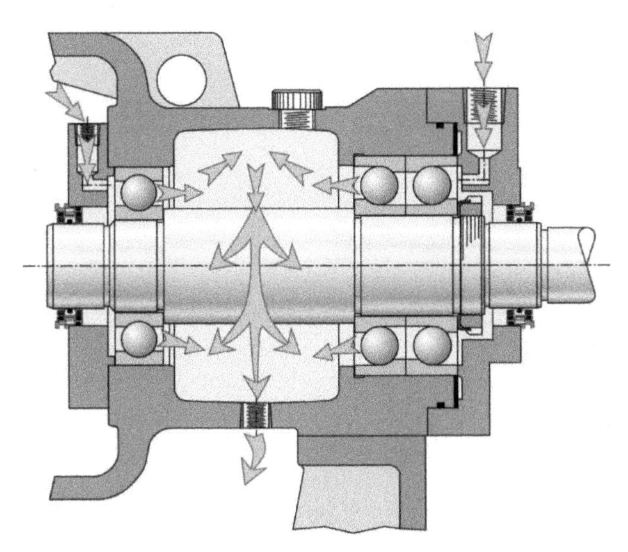

FIG. 17.1.5 An oil spray is given the highest performance ranking, as is oil mist applied from locations between bearing housing protector seal and bearing. *(Courtesy AESSEAL Inc., Rotherham, UK and Rockford, TN.)*

The various other options are given lower rankings. However, the rankings reflect the thinking after obtaining and examining proprietary data from machinery network colleagues (with access to at least 24,000 pump sets) and decades of observing elusive repeat failures (Table 17.1.2).

TABLE 17.1.2 Applicability/Desirability Ranking Sequence for Oil Lubrication and Grease Lubrication

Oil Lubrication	
Oil spray, filtered/pressurized. Also oil mist/coalesced oil coats bearing (highest)	=10
Circulation, filtered/pressurized	=9
Quiescent sump, with debris not stirred up	=6
Sump, with slinger disk (average rank)	=5
Sump and guided oil rings, well within concentricity spec	=4
Sump and nonguided oil rings well within conc. spec	=3
Sump and nonguided oil rings out of concentricity (lowest rank)	=2
Grease Lubrication	
Lifetime PFPE-PTFE (highest rank)	=7
Through-flow, low pressure	=6
Random-flow, low pressure	=4
Lifetime EM polyurea	=3
Random-flow, random press	=2
Random-flow, overpressured (lowest rank)	=1

Oil rings are ranked very low. To function properly, an installation has to be perfectly level, and the oil ring must be stress relief annealed before being machined to exacting tolerances. It must not be touching any stationary part inside the bearing housing but must be immersed into the right lubricant (usually ISO VG 32) and to just the right depth (typically 10mm).

If fluid machine manufacturers disagree, hear them out. A subject matter expert's decades of field observation must be weighed against manufacturers' test cycles of, typically, a few hours. Factory test durations lasting from 2 to 8h were reported by one major pump manufacturer in 2000. It appears they became the basis of advocating plastic oil rings and marginally thicker oils.

However, these changes did not cure the problem of black oil experienced by a disappointed customer, a user company in Canada. Likewise, follow-up talks led nowhere; they left attendees in some of the discussion groups at a major pump conference with considerable frustration. Needless to say, the machines on a pump manufacturer's test stand are properly aligned, and the lubricant is fresh and clean. In contrast, the degree of inaccuracy encountered in many field environments differs greatly from the accuracy found on test stands. Neither the training nor the abilities of crafts and service personnel will always measure up to expectations. In some installations, the piping connected to pumps is pushing and pulling. As a result, bearings are edge-loaded, and the oil film can no longer provide adequate separation of parts [6]. Weigh all factors before making an informed choice.

REFERENCES

[1] Torrington-Fafnir Company, Radial & Angular Ball Bearing Catalog, Form No. 105-10 M-897.
[2] SKF Interactive Engineering Catalogue, CD-ROM, Version 1.2, 1998.
[3] SKF Catalog 140-170, August 1988, Page 40, Figure 17.
[4] H.P. Bloch, Practical Lubrication for Industrial Facilities, second ed., Fairmont Press, Lilburn, GA, ISBN: 0-88173-296-6, 2013.
[5] H.P. Bloch, When to Use Life-Time Lubricated Bearings, Hydrocarbon Processing, 1991. July.
[6] H.P. Bloch, Pump Wisdom, John Wiley & Sons, Hoboken, NJ, ISBN 978-1-118-04123-9, 2011.

Chapter 17.2

Purposeful Grease Specifications Essential to Saving Money

INTRODUCTION

Virtually, any substance known to man is affected by temperature changes. Materials usually expand as they grow hot and contract as they become cold. While a scientist can rightly point to a few exceptions, the rule seems to apply to industrial lubricants. This is where grease specifications take on importance. For a certainty, while a "universal grease" may be suitable for horse-drawn carriages, it will not serve those responsible for keeping machines reliable in an industrial plant.

As if you needed to be reminded, you might say. Nevertheless, industry continues to experience equipment and component failures that could have been avoided by simple root cause analysis and appropriate specification follow-up. A recent motor shaft failure example will illustrate the point.

COLD TEMPERATURE PERFORMANCE OF GREASES

At a power plant in the United States, several electric motors in cold weather experienced severe bearing failures followed, in at least one case, by massive shaft failure. The incidents were attributed to stiff grease not reaching the bearings. In response, a government regulatory agency sent out a message advocating, henceforth, the frequent replacement of electric motor bearings as part of routine preventive maintenance. There did not seem to be any attempts to investigate why thousands of electric motors operate flawlessly in refinery outdoor environments in Canada and other worldwide locations north of the 49th parallel. Nor, so it seems, did anyone bother to ask why many of the world's most profitable paper mills in Finland, Sweden, and Germany use automated periodic regreasing on thousands of pumps and electric motors [1,2]. It would have been easy to find the answer: Numerous successful locations use the right grease and apply it in accordance with well-publicized, good procedures [3,4]. Many good procedures date back a few decades and are readily available to anyone taking the time to obtain these published guidelines.

What, then, was the basis for recent recommendation to treat shielded bearings as nonregreasable, throwaway products? It would only create ill will to speculate and suspect perhaps the desire to sell more bearings or to assume widespread ignorance as to how lubrication really works in greased bearings. The bearing manufacturer may have realized that bearing life is very often cut short by overpressuring because the grease drain plug was left in place. Shop experiments have shown that leaving the plug in place would often yield pressures as high as 15,000 psi (10^5 kPa) in electric motor bearing housings subjected to incorrect regreasing. At these pressures, shields will be pushed into the rolling elements and cause almost instant failure.

Rapid grease deterioration could also be caused by the mixing of incompatible greases [3] or by using contaminated products [4]. All of these fall into the category "incorrect procedures." Replacing bearings would indeed make more sense than allowing *wrong* regreasing practices to persist. On the other hand, by simply implementing correct practices, "best-of-class" plants manage to prevent both downtime and unjustifiable expenditures.

It is thus of immense value to determine if the best bearing option has been selected. Next, one needs to establish if the application meets the load and speed criteria for periodic and correct regreasing and if the correct grease formulation has been chosen. An objective practitioner of lubrication will agree that it should not be difficult to find the right grease formulation and the correct grease replenishing procedure [4]. At all times, diligently searching, finding, and implementing is still the most cost-effective and sensible sequence. Opinions and guesswork have no place here.

SPECIFY WHAT WORKS AND INSIST ON GETTING IT

Let's be honest: We, the users and buyers and consumers of products and services, are part of the problem. We want the least expensive product and still expect it to be of high quality. We expect low telephone rates and yet are surprised if the telephone company's information person doesn't come across as knowledgeable in geography. You might find yourself connected with a party in Berlin, New Hampshire, instead of Berlin, Germany, or Maracaibo instead of Paramaribo.

It's no different with knowledge in industry. Some process plants and utilities and government agencies hire young engineers and leave their further training to video games and what they learn from Bubba Graybeard, the opinionated shop foreman. Buying on faith alone, the purchasing agent tends to forget that vendors and manufacturers have often decided to employ low-cost labor or are perhaps holding on to a loyal but overworked employee. You're in trouble if you depend on someone other than a top quality supplier. The risky supplier expects an employee to do not only his job but also Jack's job. The trouble is that Jack was recently laid off by his employer at age 54, just before he would have been entitled to a small pension.

ENGINEERS AND MANAGERS MUST TAKE AN ACTIVE INTEREST

Good advice is simple and straightforward. *Engineers* must take an active interest in researching facts and uncovering prior knowledge. Establishing an informal network of competent mentors and peers, being resourceful and proactive (i.e., doing lots of reading) will allow them to appreciate what others have learned, uncovered, and specified with the goal of avoiding failures. It would seem that the *proper* specification of grease types and *proper* replenishment methods is the engineer's job. Even if he decides to delegate the job to the maintenance technician, it should still be the concerned professional's job to understand and review the issues involved. *Managers* should see to it that these roles are perfectly understood. And management should also interject themselves into the last point that we will try to make here. It involves a major utility or power plant.

When the power plant did not achieve satisfactory bearing life, it was given the recommendation to modify their procurement specifications for electric motors. A simple add-on paragraph in the specification was to request a cross-sectional view of the motor bearing housing. On it, the motor manufacturer was to show grease path, grease entry fitting, and grease drain provisions. Finally, the manufacturer was to state the grease type originally provided. Since certain greases will enter into a chemical reaction when mixed with other greases, potentially unacceptable performance could thus be avoided.

Unfortunately, the reliability engineer involved took the position that such a modification to the procurement specification was not within the jurisdiction of persons other than the purchasing agent and that the purchasing agent would not agree with the above recommendation. This interchange took place in the twenty-first century, a time when just about everyone claims to be interested in teamwork, high reliability, downtime avoidance, and reduced bottom-line cost. It's also an example of why and where we should make a more serious effort to align our actions with our avowed goals and expectations. As one of the top managers of a well-known industrial electronics company expressed it, "users get what they deserve in the long run, and sometimes in the short run." Let's just ponder over that statement and draw the right conclusions.

If you conclude there is a serious deficiency in the thought processes of managers who allow the clearly unacceptable and wholly deficient interaction between different job functions at this power plant, you are quite correct. Just as safety is everybody's business, so is reliability.

REFERENCES

[1] H.P. Bloch, Automatic lubrication saves money, Chem. Eng. 104 (12) (1997) 125–128.
[2] H.P. Bloch, Best-In-Class Lubrication for Pumps and Drivers, Pumps & Systems, 1997. pp. 36–39, April.
[3] H.P. Bloch, Increasing Pump Reliability, Hydrocarbon Processing, 1992. October.
[4] H.P. Bloch, F.K. Geitner, Major Machinery Repair and Maintenance, second ed., Butterworth-Heinemann, Stoneham, MA, 1994.

Chapter 17.3

Lubrication Misunderstood

INTRODUCTION

The subject of lubrication is rarely, if ever, taught in US universities. That's quite regrettable, because misunderstandings and lube-related errors abound and cost industry millions of dollars each year. A few case histories will make the point rather compellingly. Grooming and nurturing a lubrication specialist is certainly feasible and of value. There are excellent texts that help and even accelerate this grooming; consult the training chapters of this book and become familiar with the various training topics.

SYNTHETIC LUBE SOLVENCY ACTION REMOVES SLUDGE

When a major refinery decided to convert its steam turbine lubricant from mineral oil to synthetic, the refinery procured a premium-grade synthetic lubricant and placed it in service. Soon after, when sludge started to appear in the oil sumps of their small steam turbines, staff members theorized and proclaimed that it had to be the rather regal color in this well-known ISO grade 32 synthetic oil formulation that was responsible for causing sludge. Because neither the maintenance technicians nor the engineers realized that sludge removal is one of the key attributes of many excellent synthetics, the refinery insisted that the oil formulator make future oil shipments without the color.

This lubricant color had never been of concern to hundreds of customers, and thousands of pieces of equipment did not seem to matter. In fact, since the next refill of the now "colorless" lubricant no longer produced sludge, the buyer felt even more strongly that color had been the culprit here. But the refinery was clearly wrong. The original charge of synthetic oil had simply removed all the sludge-like deposits from bearing housings and lube piping. Therefore, color or no color, the second charge no longer liberated any sludge.

As informed engineers well know, some lubricants use color primarily for identifying purposes. So, while color may identify the vendor, different additive formulations often set apart competing vendors and can, indeed, make huge differences. Never mind colors; instead, assist your workforce and administrative staff in understanding that specific additive packages in synthetics are the real key to high equipment reliability and failure avoidance. And good additives cost money—and are well worth paying for!

DON'T SYNTHETICS ATTACK PAINT?

Another facility is being deprived of the benefits of synthetic oil use because an engineer once expressed concern that the (widely used) premium synthetic lubricant might attack the paint inside the plant's gearboxes. However, degrading of paint would be a valid concern only if the gear manufacturer had been foolish enough to supply gearboxes with an internal coating of house paint (house paint in gearboxes would surely set a new precedent). Well, after first ascertaining that the epoxy paint in modern gear units is not subject to attack, one would expect competent engineers to actively consider upgrading to synthetic lubricants. Rigorous life cycle cost calculations taking into account, cooler running, extended oil drain intervals, and reduced frictional power losses will prove that, for self-contained gear units, such upgrading is highly cost justified in the overwhelming majority of cases. So then, to what extent are *you* using synthetic lubricants?

STANDARDIZATION AND LIFETIME LUBRICATION

We've also seen ill-advised (and, in the end, very expensive!) decisions to use a "one-oil-fits-all" approach. Similarly, we know of highly counterproductive moves toward standardizing on the same type of grease for electric motor bearings, gear couplings, and steam turbine valve linkages. Some folks even "standardize" by exclusively using lifetime lubricated, sealed, bearings. Such standardization will indeed work well for vacuum cleaners that are typically in service for about 100 h each year. But using these bearings on your 25 hp and larger electric motors will surely cause excessive maintenance cost outlays for motors that are expected to run without bearing distress for thousands of operating hours.

Another rather sad and costly standardization experience occurred at another refinery. In one process area, which we will call unit "A," the refinery had been using oil mist purge on a large number of sleeve bearing-equipped centrifugal pumps. In another process unit, "B," pure oil mist was being applied on a large number of pumps equipped with rolling element bearings. The pumps in all other process units were conventionally lubricated and had liquid oil sumps; their 32-month MTBF served as the comparison basis.

Properly applied, an oil mist purge will prevent atmospheric moisture and dust particles from entering and contaminating the lube oil. And, if applied per API 610 ninth edition guidelines, pure oil mist will significantly extend rolling element bearing life in pumps. However, when the refinery's unit "A" pump MTBF decreased below that of the conventionally lubricated refinery units and when the MTBF of unit "B" hovered just barely above that of the conventionally lubricated areas of the refinery, doubts were raised as to the viability of oil mist.

The root cause reasons for the surprisingly low pump MTBF at this refinery were found in an unusual step that had been taken by "parties unknown." Concerned that the use of ISO grade 68 mineral oil as a purge in unit "A" would dilute the ISO grade 220 lube oil in a connected gearbox, "parties unknown" had decided to use ISO grade 220 mineral oil in the two oil mist systems. ISO grade 220 oil squarely contradicts the recommendations of knowledgeable bearing and lubricant suppliers. Grades higher than 68 and certainly those in excess of ISO

100 are quite unsuitable for centrifugal pump bearings operating at the *DN* values (D=shaft diameter in inches, times N=rpm) found in modern refinery pumps.

In any event and as customary, the sleeve bearings in the pumps of unit "A" incorporated oil rings and were designed for ISO grade 32 lubricants. Operation on thicker oils caused the oil rings to feed an insufficient amount of lubricant into the sleeve bearings. Likewise, application of the unacceptably high viscosity grade 220 oil to the many rolling element bearings in unit "B" caused the oil to collect in the lowermost spot or "trough" we can envision to exist at the 6-o'clock position of deep groove bearings supporting a horizontal shaft. The rolling elements had to plow through this thick oil. Temperature excursions and skidding of rolling elements resulted, and the bearings failed prematurely. As a consequence, the refinery's pump-related repair and downtime expenses became excessive.

CHECK OUT YOUR PRIORITY CONCERNS

Yes, the list could go on for quite a while. Perhaps, some readers should consider putting a reliability-focused and self-motivated individual in charge of updating and moving the facility's synthetic lube and bearing application knowledge. It certainly is true that there is considerable payback for insisting that reliability professionals implement the many small, yet highly cost-effective maintenance improvement measures. Doing so sets apart the repair-focused old-style practitioners from today's reliability-focused best-of-class performers. If your plant is not as profitable as a competitor's facility, why not ask what it is *the competitor* is doing that *your folks* aren't doing. Update the knowledge base of your personnel by recognizing the likely existence of a costly training deficiency. Perhaps, the prior training of your personnel neglected to emphasize that the consistent application of sound basics must precede looking for high-tech solutions. Without teaching, understanding, and conscientiously implementing the basics, high-tech solutions will be of no value.

So, checking out if a stated concern is—or is not—valid should be a priority task for reliability-focused organizations. Don't allow mere statements of concern to stand in the way of progress. Let your engineers check out and report to you why it is that others thrive on using synthetics for gear lubrication, or why nobody else has apparently ever thought of using ISO grade 220 lubricants in their oil mist systems, or how many years it would take for an extremely small quantity of ISO grade 68 oil mist purge to degrade five gallons of ISO grade 220 lubricant to the minimum acceptable ISO grade 190 level, or why best-of-class performers wisely chose not to install lifetime lubricated, nonregreasable ball bearings in large electric motors.

Trial-and-error solutions are very rarely the right or most cost-effective approach for HPI organizations that employ engineers. Today, sound root cause failure analysis and the implementation of fact-based solutions are needed more than ever before. To paraphrase and apply a "modified" ancient proverb to your machines, "safeguard practical wisdom and thinking ability and they will prove to be life to your process equipment."

Chapter 17.4

How High-Performance Oils and Greases Extend the Application Range for Sealed Bearings

MAIN FOCUS

Being informed on technology advancements is one of the reliability engineer's principal job functions. As regards to grease lubrication for electric motors, sealed bearings (lifetime lubricated bearings) are highly appropriate for the typical appliance motor. Conversely, regreasable bearings were preferred for the majority of larger motors in industrial facilities. In other words, lifetime lubrication was shunned for, say, a 50 kW electric motor driving the average refinery process pump. While these rules of thumb and the regreasing recommendations of major bearing manufacturers may still be in effect overall, it is important to know that certain exceptions are possible because of recent developments. Extending sealed bearing applications by using a very expensive grease may make economic sense.

NEW DEVELOPMENTS

The development of high-performance perfluoropolyether (PFPE) lubricants dates back a few decades. These developments were both necessitated and accelerated by aerospace and aviation markets where lubrication at the extremes of low and high temperature was far more important than it would be in the average industrial environment. Even beyond aviation and aerospace, PFPEs have served admirably whenever higher initial cost was easily overcome by the far more important need to consistently meet and even exceed performance expectations.

To what extent the traditionally lower initial cost of mineral oil-based lubricants has influenced procurement decisions in process industries is of peripheral interest at best. However, solid cost justification for PFPEs has recently become available. Such cost justifications were derived from a large Canadian paper mill [1], which struggled with grease-lubricated electric motor bearings. When the mill opted to dispense with relubrication of electric motor bearings by purchasing and converting to PFPE grease-filled (sealed, lifetime lubricated) bearings, their electric motor bearing life improved drastically.

The purpose of this chapter is to examine PFPE greases and to highlight their applicability in many process lubrication services. The leading provider of these high-performance greases often uses experimental data (Fig. 17.4.1) in its cost justification calculations.

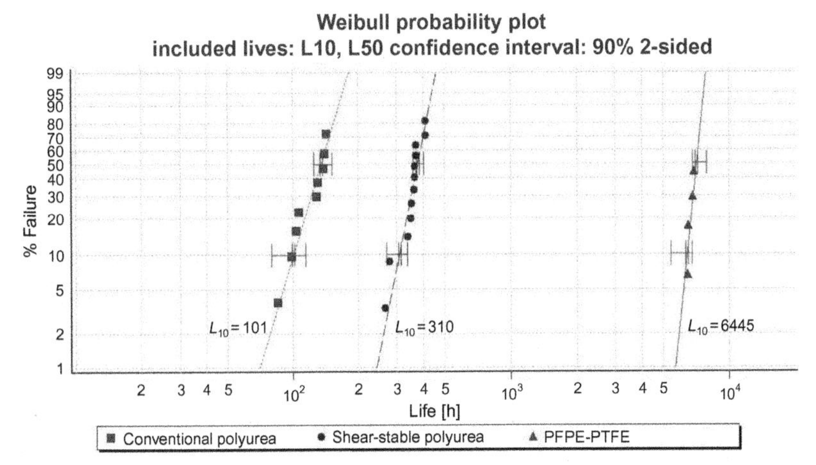

FIG. 17.4.1 Weibull probability plot for modern PFPE-PTFE grease formulations. *(Courtesy Boulden Company, Conshohocken, PA/USA; also Boulden International, Ellange, Luxembourg.)*

PFPE (perfluoropolyether) greases present an interesting lubrication alternative that was studied and fully validated at smaller and/or non-HPI facilities in recent years. It was found that developments in grease technology can greatly extend the application range traditionally associated with "lifetime" lubrication in electric motor bearings.

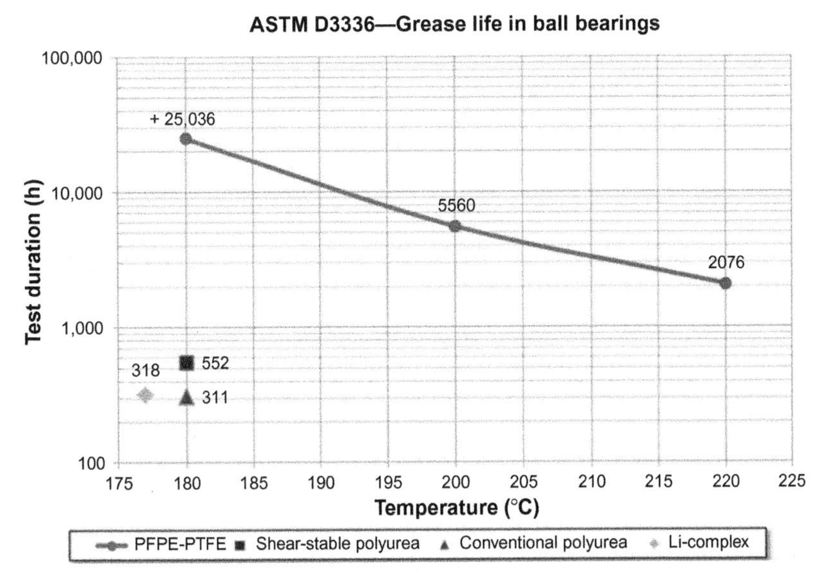

FIG. 17.4.2 Grease life in ball bearings; three widely used grease types are shown in lower left; the plotted line is for PFPE-PTFE. *(Courtesy Boulden Company, Conshohocken, PA/USA; also Boulden International, Ellange, Luxembourg.)*

In one study case, traditional motor bearings were supplied with sealed-in PFPE grease of the proper consistency (its PTFE ingredient is more commonly known as Teflon®) and the resulting life extensions tracked and explained on a comparison plot (Fig. 17.4.2). This detailed cost study (at a Canadian paper mill) showed benefits over periodic regreasing in certain industries and environments. However, the experience with PFPE-PTFE may not apply to every situation, and careful follow-up is always recommended. Also, these greases cannot be mixed with even trace quantities of traditional grease types.

COMPOSITION OF STANDARD PERFLUOROPOLYETHER (PFPE) LUBRICANTS

Standard PFE oils and PTFE (polytetrafluoroethylene, "Teflon®") thickeners contain only three elements: carbon, oxygen, and fluorine. The molecular structure provides thermal and chemical stability to lubricants that are produced in ISO viscosity grades ranging from 2 to 1000. One prominent manufacturer of high-performance chemicals engineered a PFPE molecule with its otherwise degradation-susceptible oxygen atoms fully "encased" by fluorine. The manufacturer's PFPE product bulletins show the degradation temperature or onset of decomposition in air for this grease to be above that of competing products.

A straightforward comparison of PFPE oils to alternatives is available from Ref. [1]; it is reproduced in Table 17.4.1.

From a practical point of view, PFPE lubricants excel and surpass in their capability to form an elastohydrodynamic film—an important oil strength-in-service property that explains effectiveness at all temperatures of interest. The film stays in place under the many operating conditions imposed on—for instance—the rolling element bearings in electric motors. "Staying in place" is a desirable property; it implies both resistance to water washout and the necessity to use special procedures to remove PFPE lubricant from bearings—if that should ever become necessary. Compatibility concerns require that PFPE lubricants be applied to clean bearings only. In this regard, one may take cues from the Canadian paper mill [1] that purchased its electric motor bearings from a competent manufacturer. This manufacturer then prefilled the bearings with the specified grease and applied the bearing seals.

EXAMINING COST VERSUS BENEFIT

Based on experience, polyalphaolefin (PAO) premium grade greases are a baseline competitor of the PFPEs; PAOs are certainly among the leading products presently used in electric motor bearings. The question is: What would be the cost justification for the more expensive PFPEs? If we assume a charge of the PAO grease to cost $1.00 and a certain size bearing sells for $200, the cost of grease equals 0.5% of the total. Based on cost ratio information derived from typical commercial suppliers, the PFPE grease would cost $24 per

TABLE 17.4.1 PFPE Oil Comparison to Alternatives [2]

Property	Mineral	PAO	Diester	Silicones	DuPont™ Krytox®
Thermal stability	Moderate	Moderate	Good	Very good	Excellent
Oxidation stability	Moderate	Very good	Very good	Very good	Excellent
Hydrolytic stability	Excellent	Excellent	Moderate	Good	Excellent
Volatility	Moderate	Very good	Good	Very good	Excellent
Viscosity index (VI)	Moderate	Very good	Good	Excellent	Good to very good
Fire resistance	Poor	Poor	Moderate	Good	Excellent
Seal material compatibility	Good	Very good	Poor	Good	Excellent
Lubricating ability	Good	Good	Good	Poor	Excellent
Toxicity	Good	Excellent	Good	Excellent	Excellent
Cost compared to mineral oil	1	3–5	3–7	30–100	60–120

bearing, although we might assume the bearing manufacturer will charge $250. Purchasing the bearing with PFPE sounds reasonable at this relatively small incremental cost. But we need to make a more detailed comparison. Our projected incremental cost (perhaps $50 per bearing) should convince us to dig a bit further. In a more careful examination, we may wish to know what it really costs to periodically reapply traditional PAO-based greases to electric motor bearings.

The frequency of grease replenishment is determined by the rotational speed, bearing diameter, and the environment in which the bearing operates. We have to look at a number of plausible scenarios and compare these with simply purchasing and installing lifetime, PFPE-prefilled (sealed), motor bearings. Three different scenarios are envisioned below, but others are entirely possible. Our purpose is to show the ease with which such cost justifications can be explored and how the results are easily expressed as payback or benefit-to-cost ratio.

Scenario 1, using bearings with PFPE sealed-in (no regreasing possible). This is the base case scenario. All comparisons will take into account that a set of sealed-in (no regreasing possible) electric motor bearings will cost $100 more than customarily supplied (regreasable) bearings.

Scenario 2, periodic regreasing. A reasonable assumption would assume that the average bearing is being regreased 16 times during its (assumed average) 8-year life. A rather optimistic expectation further assumes that the person doing this type of work will do everything just right. They will ascertain that the grease fitting is clean, will not overgrease, will diligently remove the drain plug while adding grease, and will carefully reinsert the drain plug after greasing is done. That person can do 16 electric motors per day. Counting straight salary, overhead, vacations, training time, administrative costs, etc., of a trained craftsperson costs the employer $800 per day. Therefore, regreasing the bearings routinely found in conventional electric motor will cost its owners $800 over the bearings' 8-year anticipated life. However, the incremental cost of two sealed bearings per motor would be only $100.

Subtracting an incremental $100 from $800 = $700; the motor with sealed bearings leads with a payback of 7:1. That simply means that every year, the owner of the electric motor saves $700/8 or about $87. An installation with 1200 motors would save approximately $100,000 in labor costs per year. Assume further that 10 motors will require bearing replacement each year. Therefore, bearings would be replaced after 8 years of operation, regardless of bearing style (regreasable and being regreased versus lifetime sealed with no need to regrease).

Scenario 3, standard grease with no periodic regreasing. A facility with 1200 electric motors and not doing any regreasing might expect (on average) 200 motors requiring bearing replacement each year. This is to be contrasted against lifetime (PFPE sealed-in) bearings. No labor cost is incurred if standard motor bearings are never getting regreased. However, an incremental number of 190 sets of motor bearings would have to be replaced each year. Replacement bearings and associated labor would cost $2000; 190 × 2000 = 380,000 per year.

It might be prudent to assume there would be a process unit outage event—the cost is anybody's guess. In that case, however, the entire scenario 3 makes even less economic sense than scenario 2.

It simply pays to reconsider "old" regreasing strategies in light of recent experience at a Canadian paper mill. High-performance oils and greases extend the application range for sealed bearings and call for a rethinking of the way things were done before [2].

REFERENCES

[1] R. Aronen, Krytox® Blog, Boulden Company, Coshohocken, PA/Ellange, Luxembourg, 2014.

[2] L.R. Rudnick, Synthetics, Mineral Oils, and Bio-Based Lubricants—Chemistry and Technology, Taylor & Francis, Abington, UK, 2013.

Chapter 17.5

Auditing Your Lubrication Practices

INTRODUCTION

All industrial facilities use machinery, and every one of these machines requires lubrication of one type or another. Modern, profitability-minded plants thoughtfully manage their lubrication practices and reap substantial benefits from the resulting enhanced equipment reliability.

However, not all lubrication practices are cost- and value-optimized. A one-day audit of your lube management practices may uncover near zero-cost improvement opportunities that, if implemented, have paybacks measured in days and may quickly move the plant into the best-of-class grouping.

TYPICAL FINDINGS

During an audit conducted of a world-scale, state-of-art petrochemical plant in the USA, the auditors judged the lubrication program as generally well managed. The plant had selected a competent supplier of both mineral oils and synthesized hydrocarbon ("synthetic") lubricants. Management and the reliability group had engaged an experienced oil analysis laboratory and were certainly aware of the merits of sound lubrication management. Still, they had to be encouraged to make certain changes and improvements. A few illustrations will serve as examples and will convey at least part of the story.

Questionable transfer and storage practices are shown in Figs. 17.5.1–17.5.4. Many lube audits uncover unsatisfactory lubricant dispensing practices, and little oversights can have serious negative consequences. For example, it is important to minimize contamination on lube carts. Galvanized steel dispensing containers are frequently attacked by certain lube oil additives; therefore, good practices mandate the use of plastic dispensing containers. Leaving transfer containers open is simply not acceptable (Fig. 17.5.1).

Storage drums should be located and positioned so that water accumulation (Figs. 17.5.2 and 17.5.3) is ruled out. Changes in ambient temperature cause rainwater on top of the storage drum (Fig. 17.5.4) to be drawn into the drum by capillary action. A drum containing valuable lubricant is thereby rendered unserviceable.

Audit findings and recommendations often involve detection of oil contamination in sump-lubricated equipment and the relative effectiveness of labeling points to be lubricated.

Finally, auditors still find plants that are somewhat arbitrarily "standardizing" on less than optimum grease formulations and are employing incorrect

FIG. 17.5.1 Leaving a transfer container uncovered invites lube contamination and machinery distress.

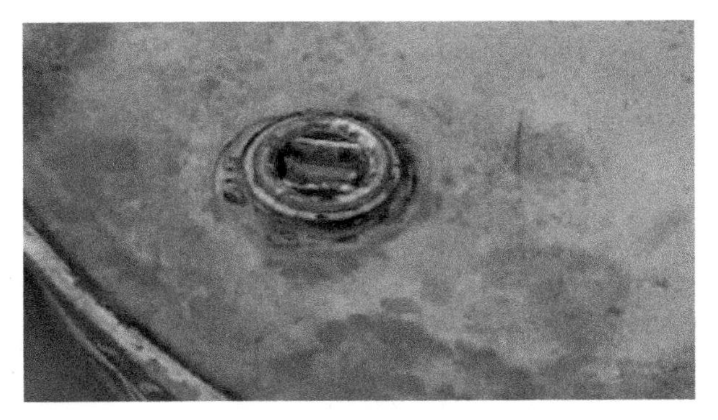

FIG. 17.5.2 Outdoor storage drums are notorious for collecting rain water.

regreasing practices on thousands of electric motors. We have seen superior plants experience as few as 14 motor bearing replacements per 1000 motors per year and average plants with 156 motor bearing replacements per 1000 motors per year. We will spare the reader the statistics of less-than-average plants.

Suffice it to say that being unaware of best available lubrication practices can be expensive. Deviations from best lubrication practices may incur significant, yet readily avoidable maintenance and downtime expenses. Periodic lubrication audits are recommended. Experience shows them to be extremely cost-effective and almost always pointing to areas of improvement. The value of experience-based lubrication audits is intuitively evident.

FIG. 17.5.3 Water accumulation jeopardizes 65 gal of a superior lubricant and risks causing consequential damage to rotating equipment in plants that allow this type of outdoor storage.

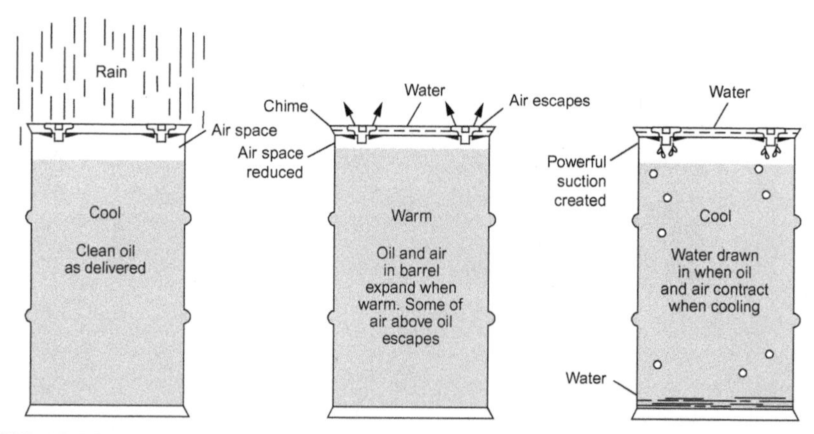

FIG. 17.5.4 Ambient temperature cycling explains the mechanism for water entering into oil drums stored outdoors in the upright position.

DEFICIENT FIELD FOLLOW-UP

Plant management and the facility's reliability group had thoughtfully specified the lubricant to contain a less-than-traditional amount of water. An oil analysis laboratory was on the facility's list of consultants. All were certainly aware of the merits of sound lubrication management.

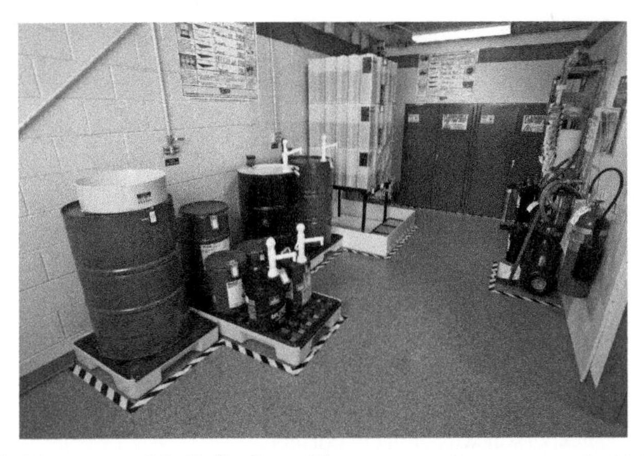

FIG. 17.5.5 The center rack holds 5-gal quantities; a more modern storage method is indicated on the right side of this image. Well laid-out storage rooms can be small, but emphasize accessibility and cleanness.

But things did not trickle down to the field implementation work force members. Indoor storage was sporadic and less than up-to-date. Transfer of oil from the 5-gal buckets in the center of Fig. 17.5.5 to smaller containers was cumbersome and risky. The deficiency was in the process of being upgraded to the storage method shown on the right side of Fig. 17.5.5.

Chapter 17.6

Lubricating Slow-Speed Rolling Element Bearings

INTRODUCTION

A sales engineer visited a customer with different pieces of processing equipment that included an oven. The customer asked a lubrication-related question and received an answer from the sales engineer. Let's see his answer and then decide if the salesperson was entirely correct.

SPEED MATTERS IN BEARINGS

The application is a rolling element pillow block bearing that turns less than 30 rpm, and the temperature radiated from the oven is about 215°F (102°C). The sales engineer recommended an ISO grade 100 or ISO grade 150 (~600 SUS at 100°F) oil for this temperature and made reference to a widely used text. However, he noted that the section in the book dealt with *pumps* that rotated at *3600 rpm* in an elevated temperature environment [1]. So, there was justification for asking if the viscosity recommendation would be different in this slow-rolling pillow block bearing.

Both customer and sales engineer would have found it helpful to communicate with the application engineering department of a world-scale bearing manufacturer. One company's three-dimensional viscosity selection chart, Fig. 17.6.1, indicates the helpfulness in providing a reasonably close answer to the basic viscosity-related question.

To begin with, our illustration [2] conveys the fact that bearing peripheral speed, and not bearing rpm, is important here. In essence, the sales engineer's question cannot be answered without first knowing the bore diameter of the bearing.

For the sake of illustration, suppose this diameter was 300 mm. To use the chart, one would proceed as follows:

1. Determine the bearing *DN* value: multiply the bearing bore or shaft diameter, measured in millimeters, by the speed of the shaft, measured in revolutions per minute (rpm). In this example, $DN = 300 \times 30 = 9000$.
2. Select the proper operating temperature. Although at 30 rpm, temperature increases will probably be very small; with higher speeds, the operating temperature of the bearing may exceed that of the ambient temperature. The temperature scale of Fig. 17.6.1 is meant to reflect the operating temperature of the bearing—215°F is just outside the limits of the chart.

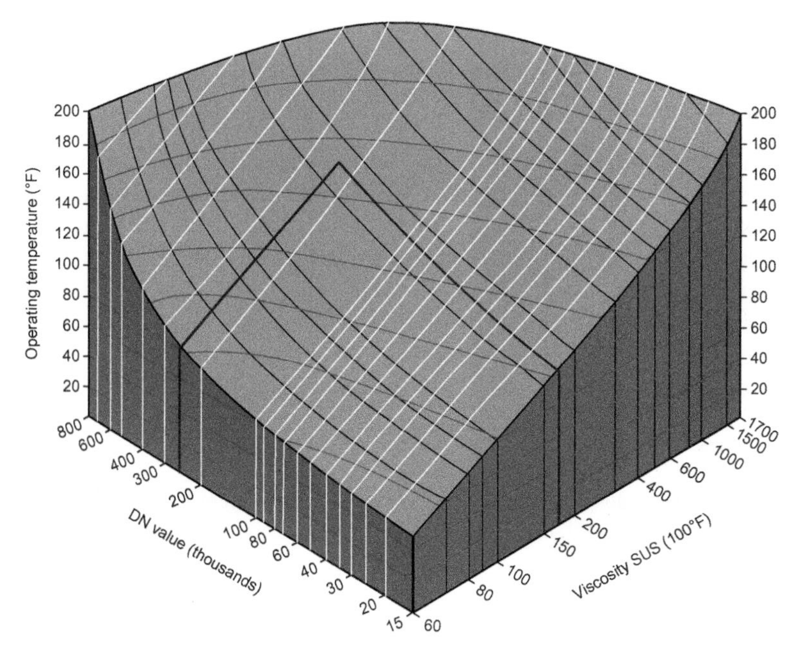

FIG. 17.6.1 Viscosity, speed, and operating temperature relationships. *(Courtesy Torrington.)*

3. Because our *DN* value (9000) is a bit off the chart, enter the three-dimensional graph at the nearest value—*DN* = 15,000, for instance.
4. Follow or parallel the "dotted" line to the point where it intersects the selected "solid" temperature line. The nearest value will be 200°F.
5. At this point, follow or parallel the nearest "dashed" line downward to the right of the viscosity scale.
6. Read off the approximate viscosity value, expressed in Saybolt universal seconds at 100°F. At 200°F, the corresponding value "viscosity SUS at 100°F" would be 1700 or higher. Comparison charts show its international equivalent as ISO viscosity grade 320 [1].

ISO grade 320 would be a rather thick oil, and selecting this not-so-common lubricant might merit additional consideration. Checking with the equipment manufacturer and obtaining written recommendations would be wise, as would be the selection of a properly formulated synthetic lubricant. Because many synthetic formulations have considerably greater temperature stability than mineral oils, their useful life is almost always superior to that of traditional lubricants.

First and foremost, however, is the issue of oil application [2]. Just how is this thick oil applied? A simple oil ring will be unlikely to lift enough of the lubricant into the bearing. Did the manufacturer provide the least-cost application method and leave it to the user to apply the oil in a burdensome and labor-intensive manner? Is this installation a candidate for upgrading? Has the best possible bearing protector seal been installed? Has oil mist, or a directed, recirculating spray of liquid oil, or automatic (timed) grease injection been considered?

All of these are issues that influence lube optimization on low-speed rolling element bearings. The entire scenario attests to uptime improvement opportunities that exist throughout industry. Note that these opportunities would include a fundamental switch to lifetime lubricated bearings prefilled with a PFPE non-hydrocarbon grease. See "Chapter 17.4 and subject index" for more detail on these greases.

REFERENCES

[1] H.P. Bloch, Practical Lubrication for Industrial Facilities, Fairmont Publishing Company, Lilburn, GA, ISBN: 0-88173-296-6, 2000.

[2] H.P. Bloch, A. Budris, Pump User's Handbook: Life Extension, fourth ed, Fairmont Publishing Company, Lilburn, GA, ISBN: 0-88173-720-8, 2013.

Chapter 17.7

How the Best Lubricant Providers Add Value

INTRODUCTION

A very competent lubrication professional shared the many small (but pesky) problems we sometimes encounter in our work experience. And we shared our enthusiasm for those gaining a better understanding of modern industrial lubricants. High-value lubricants are more than ever a fascinating and evolving business. We spoke about this lubricant provider's superior ISO VG 68 “Excel” (2015 vintage) as a worthy substitute for the dibasic ester-based synthetic ISO VG 100 (“DBE 100”) in oil mist systems that both of us had helped commission in the late 1970s and early 1980s. The gist of our conversation was how reliability professionals must recall the rationale we used decades ago to justify and explain lube issues and how, in view of new knowledge and even better formulations, our thinking has to change in order to keep up with progress.

NEW LUBRICANT FORMULATIONS OF INTEREST

A case in point is the presently available “Excel” and how its viscosity versus temperature chart compares to a then (1979) appropriate ISO 100 diester lubricant. Because their respective viscosities are similar at elevated temperature and the Excel's pour point and cloud point are sufficiently low, the Excel 68 will work well in the coldest temperatures that the Texas Gulf Coast can reasonably expect. There's an examination or perusal of the list of properties of each oil, that is, the 1970s dibasic ester (DBE) and the Excel 68. The Excel 68 contains no zinc that, under some conditions, might degrade and form sludge. To be sure, some VG 100 and 150 precursor formulations have served us well in the past; however, given 2015-era knowledge, they represent “overkill” in oil mist systems on the US Gulf Coast. Moreover, they are unsuitable for high-speed equipment with oil rings (slinger rings) [1].

The lubrication engineer also revisited the old argument on running one's “A” process pumps and how frequently to switch over to the “B” set. Anyway, best practice plants on pure oil mist could preserve equipment by manually rotating the “B” shaft one or two turns every month or six weeks. This periodically carried out manual practice will prevent bearing elements from sitting in the same spot. The periodic rotation also decreases the risk of bearing damage due to vibration transmitted from an adjacent operating machine. However, in plants without oil mist and with bearing housings still fitted with open breather vents that allow atmospheric dust and water vapor to wreak havoc on oil cleanliness, it will usually be best to switch and run pumps every 4–6 weeks. Alternate running will drive off some of the moisture, although it will also stir up sludge. Yes—taking into consideration all the facts can really surprise us.

The bottom-line cost and proved efficiency advantages of modern formulations over many traditional synthetics can be readily calculated. Experienced lube application engineers can assist their customers in demonstrating the value of keeping up with changing technology in lubricants. Make these experienced lube application engineers and others your technology resource. Be sure to include bearing, mechanical seal, and gasket engineers in your fluid sealing and bearing protector seal knowledge download efforts. Work closely with their application engineers. Realize that even the best lubricants will suffer if seal components degrade and contaminate the oil. And be prepared to pay for the application engineering know-how that sets apart the lowest as-delivered-per-gallon oil merchants from the expert formulators who, we can be quite certain, will save an industrial user several orders of magnitude in the long run. Shun those who lack application experience or are unwilling to explain their lubrication oil formulations.

A 2015 UPDATE

Informal feedback and networking among specialists are among the most important reliability enhancement actions available to us. As of about 2013, a petrochemical plant in the Texas Gulf Coast region discontinued using constant level lubricators and oil rings on many of their process pumps. The facility discarded their open-to-atmosphere breather caps and plugged the opening. At the housing location from which constant level lubricators were removed, they placed a sight glass. The old-style bearing housing protector seals (ones with a "flying O-ring" making occasional contact with a sharp O-ring groove) were discarded also. In their place, this petrochemical company now uses advanced bearing protector seals that incorporate a large axially moving O-ring. A large multinational lubricant manufacturer provides this particular Texas Gulf Coast user company with high-quality mineral, synthetic, and mixed formulation hydrocarbon oils. Taken together, these reliability improvement actions allowed the user to extend oil change intervals by factors of four or greater compared to previous practice. The labor requirements reported by this user in 2015 have dropped quite significantly, and oil leakage is virtually nonexistent. The takeaway from this experience underscores the value of staying informed and of making a superior vendor/supplier our technology resource.

REFERENCE

[1] H.P. Bloch, Pump Wisdom: Problem Solving for Operators and Specialists, John Wiley & Sons, Hoboken, NJ, ISBN: 9-781118-04123-9, 2011.

Chapter 17.8

Expansion Chamber Sizing

INTRODUCTION

Expansion chambers (Fig. 17.8.1) are designed to restrict pressure increase in closed volumes. From basic physics, we of course know that the pressure in a closed volume increases as the temperature goes up. In case a shaft seal is so tight that it will no longer allow air confined in bearing housings or gearboxes to flow in and out, the trapped air would be pressurized as temperature rises. Creating a larger volume for the trapped air would keep its pressure down.

We find expansion chambers advertised for use on process pumps and gearboxes. They can replace breather vents that are commonly found on pump bearing housings or gear casings and, for the lack of a more precise subject category, often find themselves grouped with matters of lubrication.

FIG. 17.8.1 A thread-on type expansion chamber is sometimes used at the location where the bearing housing vent had been installed originally. The image shows the expansion chamber oriented horizontally; a desiccant breather is at the top vertical location. Note a grease dispenser supplying lubricant to the motor bearing. *(Courtesy TRICO Mfg. Co., Pewaukee, Wisconsin.)*

CONSTRUCTION FEATURES

Most expansion chambers incorporate a rolling diaphragm (usually Viton® and, occasionally, Teflon®). The diaphragm divides the interior volume of the chamber so that ambient air can be aspirated into, or expelled from, the (up-facing) space above the diaphragm. The down-facing surface of the diaphragm is contacted by the air (or air-oil mixture) that exists in, say, a pump bearing housing or gearbox interior.

EXAMINING THE PREMISE IS ALWAYS A GOOD FIRST STEP

However, one might question if an expansion chamber is really needed in some applications or services. An example relates to a shaft seal supplier that started discussions with a wind turbine gearbox manufacturer. The seal supplier wanted advice on calculating the size of expansion chambers for gearboxes with dimensions in the vicinity of 3 m × 2 m × 2 m. A worst-case oil-to-ambient temperature difference of 100°C was anticipated by the two parties.

Well, realistically speaking, temperature differences of 100°C (180°F) are rather unusual in a gearbox. We should always look at the bigger picture and perhaps even challenge the basic premise.

That said, let's be certain to select the right lubricant and to accommodate thermal expansion that, on a 3 m long steel gearbox, might be somewhere around 0.14 in. We obtained this number by multiplying the coefficient of expansion for steel, times the length (inches), times the anticipated temperature change [$DL = (0.0000065)(L)(DT)$].

Anyway, here's the academic exercise. We might elect to work with a base temperature of, say, 100°F, which would be (100 + 460) = 560°R. Continuing in US units, an increase of 100°C (180°F) gets us to (180 + 560) = 740°R.

Charles' law states that the volume of an ideal gas at constant pressure varies directly as the absolute temperature. Thus, $V_2 = V_1(T_2/T_1) = \text{constant}$. For equal pressure, the new volume V_2 would have to be V_1 (740/560) or 1.32 times that of the original air space (or volume) of 1. Hence, 0.32 volume units would have to be added to the original air space or volume unit of 1.

It's also worth noting that the volume of an expansion chamber would have to be larger if we had assumed a lower base temperature, say, 0°F. In that case, the needed volume addition would be based on $[V_1(640/460)] - 1$, and 0.39 volume units would have to be added to an original air space or volume of 1.

Much of the 3 m × 2 m × 2 m overall gearbox volume will be taken up by the gears and the oil. So, if the remaining air volume had been 30% of the gearbox total, that is, (0.3)(12) = 3.6 m^3, one would have to add an expansion chamber with a useable volume of (0.32)(3.6) = 1.15 m^3 in the first instance. In the second instance, the needed addition would be (0.39)(3.6) = 1.4 m^3.

While this might have answered the original question, we might think of all kinds of different scenarios, including seeing a 1.15 or 1.4 m^3 hump on the gearboxes of future wind turbines. Or perhaps we simply will not pursue the expansion chamber idea because someone explains intelligent sealing options to the gearbox manufacturer. An intelligent sealing option would be a balanced seal or a seal that can take the pressure increase that comes with a constant volume.

WHAT IS THE PRESSURE INCREASE WITH A CONSTANT VOLUME?

Suppose we didn't add an expanding volume to the wind turbine gearbox and the temperature rose from 0°F (~−18°C) to 180°F (~+83°C). What would be the pressure increase?

Well, we might just consult the Internet and type in "gas law." We would observe the before-versus-after conditions indicated by the sub-1 and sub-2 characters:

$$\frac{P_1V_1}{n_1T_1} = \frac{P_2V_2}{n_2T_2}$$

This is the same as the expression $P_1V_1 / n_1T_1 = P_2V_2 / n_2T_2$.

The molecular masses n probably will not change, and neither will the volume; they are the same on both sides of the equation. We assume our installation is at sea level; the atmospheric pressure is 14.7 psia.

Therefore, the absolute pressure will change in direct proportion to the absolute temperature, and

$$P_2 = P_1T_2 / T_1 = (14.7\text{psia})(740°\text{R}) / (560°\text{R}) = 19.4\text{psia}$$

So then, the delta-*P* across the seal would be 19.4 − 14.7 = 4.7 psi. Now, we could design a seal for that, even if we needed to divert a slip stream of bearing lubricant to provide a bit of cooling for the seal. And we would perhaps ask ourselves if we needed the expansion chamber in the first place.

Subject Category 18

Maintenance

Chapter 18.1

No Fast Path to Maintenance Brilliance

INTRODUCTION

Many years ago, the vice president of a major steam turbine manufacturing company had to give a legal deposition in Canada. It had been alleged that oversights or engineering malpractice by his firm had caused issues with the company's mechanical drive steam turbines. During the unavoidable back-and-forth questioning, the VP made the statement that the same problems tend to reappear in 30 year cycles. In essence, he believed that corporations have memories that go blank every 30 years.

DESIGNING A MAINTENANCE STRATEGY

A reader request reminded us of the VP's 30-year-cycle comments. The reader said the following:

> *I work for a well-known petroleum company and have a role in a project that focuses on designing our maintenance strategy for 2020 and beyond. It had me wondering whether you could point us in a few right directions. Here are my three questions:*
>
> 1. *Which companies are standing out as the world's leading practitioners of maintenance excellence?*
> 2. *Viewing our present-day involvement and realistic oil and gas industry environment (increasing complexity, safety challenges, catastrophic risks, etc.), are there any specific thought leaders you might point out to me?*
> 3. *Where in US academia is the best research and thinking going on in the maintenance area?*

We knew that the reader's employer—we will call them XYZ—was counted among the very best-run companies roughly 30 years ago. Then, somebody

Petrochemical Machinery Insights. http://dx.doi.org/10.1016/B978-0-12-809272-9.00018-9

discovered the catchy terms "lean and mean" and claimed that reducing the number of subject-matter experts (SMEs) on XYZ's payroll was a smart move. Of course, the folks in charge of XYZ had been misled, to say the very least. But we decided to reply to the reader and also wanted to share our experience-based comments with interested managers and reliability engineers who read this publication:

1. Best and/or leading companies are those that groom their own talent, pay them well, give them a written role statement, and then hold these future SMEs accountable. Accountability means their performance is measured against their stated role. A management sponsor has overall oversight responsibilities, and she or he will insist on compliance throughout.
2. There has never been, nor will there ever be, a good substitute for guided learning. The expense and responsibility for learning must be shared equally between employer and employee.
3. Truly exceptional companies are those realizing that learning begins after graduating from college or university. Therefore, these companies institutionalize a rigorous training program. This training is mapped out in great detail and consists of four phases. Within each phase, there are activities such as (phase I) shared reviewing of trade journals, (phase II) conducting "shirt-sleeve seminars" for your maintenance/technical employees, and participation in structured weekly meetings with operator/maintenance/technical representatives. Training also includes (as phase III) presentations to a plant steering committee, attending local evening events of ASME/STLE/Vibration Institute (or similar) meetings, and involvement in organizing lunch and learn sessions—with the emphasis on learn, not lunch. Most of these training activities cost next to nothing and precede phase IV, attending out-of-state technical conferences. All of these phases are further discussed and described in books, articles, conference proceedings, etc. You might also explore the index words on "training" in this text; training is certainly a recurring theme.
4. There is no "Internet-only" approach, and no "magic bullet" will ever lead to best-of-class (BoC) performance. Only a consistently pursued training approach will yield excellence in the highly interdependent areas of safety, reliability, and maintenance cost-effectiveness.
5. Reliability and maintenance cost-effectiveness always start with a cost estimating manual with prices of *reliable* equipment not just bare bones *cheapest* equipment. Projects use lists of competent vendors and use only sound specifications and a budget, which includes the outlays needed for machinery quality assessment (MQA). At BoCs, SMEs see to it that high reliability and low future maintenance cost are designed into machinery up front. The next best alternative is systematic identification and removal of weak links on existing equipment. Again, only highly motivated SMEs can do that.

6. While there are at least three US universities that offer a maintenance curriculum or degree, there are good reasons why the remaining hundreds of universities are not offering such curricula. When it's all said and done, an employee has to have the desire and motivation to be a top performer. So hire, nurture, groom, reward, and further train those who want to excel. They are the doers, not the talkers.

What ultimately causes a few individuals and a few companies to become leaders in the maintenance and reliability fields is worth passing on. But unless acted upon by responsible, fair-minded managers who will then fight for drastic changes in the prevailing mind-sets, lessons will have to be relearned every 30 years. The VP gave us this feedback after all the attorneys had left the room.

Chapter 18.2

Maintenance Best Practices Involving Pumps

INTRODUCTION

Maintenance best practices (MBPs) are a frequent subject in the literature pertaining to reliability. Maintenance is an ever-moving field of endeavor. Unless maintained, machinery cannot last forever. We use pumps as an example because pumps, next to electric motors, are the most frequently used machine in industrial societies. We ask readers to apply the principles enunciated here to virtually all machinery categories. In particular, we ask you to pay attention to the interweaving of the thinking of industrial engineer-philosopher W. Edwards Deming whose "14 points of total quality management" are found in several of his books. We expanded them here into areas of interest to reliability-focused readers.

MORE ON THE PUMP EXAMPLE

MBPs for pumps in industrial and municipal services deserve more of our attention than we seem willing to routinely give them. Well-explained MBP details for pumps can be found in a wide variety of published literature, including a variety of trade journals, which often take the lead in disseminating useful information on the subject. Generally speaking, relevant details are in the public domain, in plain view of those truly interested in learning the details.

That said, we do our part to encourage reliability professionals and managers to embrace MBPs for pumps. Some reliability professionals and managers need to make changes if the right priority is not presently given to the implementation of true MBPs. Elevating the priority ranking of doing repairs properly often requires a shift in the mind-set of managers and senior professionals. But while managers and maintenance personnel tend to desire quick results and because human nature is what it is, changes in mind-sets are mandatory. Even a subconscious awareness of having to make changes becomes an obstacle to the implementation of best practices. Bear with us and we will explain.

PRECURSOR ACTIVITIES

Best maintenance practices are achieved only after having done due diligence. In other words, precursor activities had to be carried out. By way of analogy, the reliability of a passenger car is dependent upon (a) its design, (b) the quality and diligence of follow-up maintenance, and (c) the competence of the car's driver-operator. It's no different in a pump user's facility: each of these entities, design—maintenance—and operation must do their respective jobs correctly. Unless all three are doing things right, there will be no reliability. While there will be much talk about best practices, little (if anything) is often done to impart lasting value.

PROJECT ENGINEERING MINDFUL OF FUTURE MAINTENANCE

Again reverting to an automobile analogy, planning to market a new automobile model involves up-front design. Design details always create maintenance consequences. It is no different with the assets, installed in a pump user's facility. All the maintenance efforts in the world cannot turn a fundamentally bad or maintenance-intensive design into a technically superior asset. "You cannot buy a sow's ear and make it into a silk purse" was reputed to be one of William Shakespeare's famous utterances. A facility needs a sensible budget, not a bare bones budget. Good projects originate with cost estimating manuals that reflect reliable equipment. A project must use experience-based specifications to describe and properly estimate the cost of reliable equipment. Reliable assets must be purchased and installed. Before the actual commissioning, soundly executed projects move into a training phase. People must learn (i.e., be trained) how to properly operate and maintain this equipment.

RULING OUT THE BLAMING OF OTHERS

A logical realization is to rule out finger-pointing and faultfinding between a maintenance department and an operating department. The smartest organizations are conveying to the maintenance department head that he or she will be asked, at random intervals, to trade places with the operations department head. The operations department head will then become the maintenance department head. That simply means that the two department heads will positively learn to communicate and cooperate; they will adopt a stature whereby they show consideration for each other's issues and concerns on a daily basis. If they do not learn to communicate and cooperate, each will have limited or terminated their own careers.

SEEKING INPUT FROM SMES

Best maintenance practices are put in writing; they require input from subject-matter experts or SMEs. SMEs are not born; they are groomed and nurtured. These experts may report at different levels in an organization but will have to be given access to a sponsor, someone very near the top. SMEs are people with abilities and know-how, which would be perilous for project people to ignore. SMEs are given a role statement, which delineates their role and assigns to them both accountability and empowerment. Their role quite obviously includes project involvement up front, never after decisions are already made by managers whose concerns are inevitably focused on the present and for whom the future is too far out to be of much (if any) concern. Regrettably, in today's environment, reward and progression are often based entirely on cost and schedule instead of long-term lowest cost of ownership.

OWNERS' INVOLVEMENT NEEDED

In our context, "owner" simply means that best practices companies designate an "owner of final decisions." That "owner of decisions" has to live with his rulings, first as the project executive and then as the plant manager. He or she instructs the next layer of managers to use predictive maintenance (PdM) routines, which are tracked so as to identify the optimum maintenance or scheduled shutdown intervals. It's in this particular layer of management where the trading of places between an operating department head and a maintenance department head takes place. In this layer of management, a person sees to it that written instructions are followed, be they maintenance- or operations-related. The managers in this layer are held accountable and, in turn, will hold accountable the supervisory levels reporting to them.

DEMING WAS CORRECT

Following the above road map for several decades will inculcate in an organization what an exceedingly competent observer of the industrial scene, W. Edwards Deming, taught with absolute clarity in the mid-1940s. In his "14 Points of Total Quality Management," Deming described a set of management practices, which bring about quality and productivity. Wildly optimistic maintenance practitioners (and their disciples) sometimes push aside Deming and replace his fundamental principles with a new "flavor of the year." Shortly after World War II, Deming tried to provide highly relevant in-depth consulting services to US automobile makers. When the US auto industry ignored his work, W. Edwards Deming went to Japan and assisted Japan's auto industry to regain and expand both quality and market share. We now know that Deming was right over 75 years ago; we also know that his principles are as true today as ever.

Rediscovering these practices and following Deming's guidance toward MBPs makes much sense. We can certainly apply his thinking to the process pumps we design, manufacture, install, repair, maintain, and operate:

1. *Create constancy of purpose for improvement of product and service.* A reliability engineering group must view every maintenance event on existing equipment as an opportunity to upgrade. The investigation of the feasibility of upgrading is discussed in many of the Subject Categories found in this book. Such investigations are also among the topics relating to MQA or machinery quality assessment. Investigating upgrade feasibility and

performing MQA should have been done before a purchase order is placed; it should be a proactive endeavor.

2. *Adopt a new philosophy that makes mistakes and negativism unacceptable.* Accountability is the key here. If you've taken your car to the dealer on three consecutive Mondays to have a leaking water pump repaired and find the job to be unsatisfactory every time, you might ask some serious questions. Why not ask questions at your plant when a critical process pump repair isn't done right three times in a row? Why the double standard as we shift from our automobile to the plant's process pumps?
3. *Stop being dependent on mass inspection.* While this is not usually a concern for a pump user's facility, invoke the corollary. Ask the responsible worker or maintenance technician to certify that his or her work meets the quality and accuracy requirements stipulated in your facility's work procedures and checklists.
4. *End the practice of awarding business on price alone.* You've probably never bought the least expensive pair of shoes so don't buy the cheapest pump or the cheapest replacement parts for the pump. Understand and redefine the function of your purchasing department. Let them negotiate price and delivery of the mechanical seals and rolling element bearings specified and selected by a competent reliability engineering group, but don't let the seal vendor lobby your buyers. Pay particular attention to the meaning of "specify" and work up detailed specifications for important parts.
5. *Improve constantly and forever the system of production and service.* Understand that in a process plant, improvement will come from the proper daily interaction of operating, mechanical/maintenance, and reliability/technical workforces. Be sure not to let the reliability/technical function become a service organization. They are the support arm that needs to push your company to the top. Don't let them get inbred by depriving them of access to the outside world.
6. *Institute training.* It has been our experience that very few managers know just exactly what type of training is required by their mechanical/maintenance or reliability/technical workforces. Say, for example, the shop is primarily involved in repairing repeat pump failure events. The mechanic or machinist dismantles a pump, finds a defective bearing, and replaces it. For decades, that's been your repair approach, but not so at the best-of-class competition. Rest assured that a best-of-class competitor's approach has been to teach their personnel that bearings fail for a reason, and unless you uncover the reason, you are just setting yourself up for many more repeat failures.
7. *Institute leadership.* What an elusive concept it is! Perhaps we could look at a small slice of it, the one that deals simply with guidance and direction. A leader recognizes that for reliability professionals to be productive, they must be resourceful. The leader must be in a position to outline the approach to be followed by the reliability professionals in, say, achieving extended pump run lengths. The true leader would steer professionals to

publications and mentors, would recommend the development of single-page specifications for acceptable bearing types and configurations and would see to it that inquiries to mechanical seal manufacturers would elicit responses in sufficient detail to acquire the statistical basis for justifying the procurement of better products.

8. *Drive out fear.* In my younger years, we used to have an unwritten contract, which implied that employee loyalty will be rewarded by job security. This contract no longer exists today. But you can initiate guidance and action steps that show personal ethics and evenhandedness, which are valued and respected by your workforce. Their performance will be motivated by your example and by your work habits and character traits. Fear will no longer be a performance motivator.
9. *Break down barriers between staff areas.* You should never tolerate the kind of ill-perceived competition among staff groups that invariably causes them to withhold pertinent information from each other or makes one group shine at the expense of their peers.
10. *Eliminate slogans and exhortations for the workforce.* Just because a catchy safety slogan at the plant entrance serves as a fitting reminder to work safely does not mean that hot-air exhortations will strike a responsive chord. We all subscribe to the belief that actions speak louder than words.
11. *Eliminate numerical quotas.* Think how well your automobile would be repaired if the mechanic had a quota of eight automobile repairs per day. If your reliability professionals work 40h a week and you expect them to solve 20 problems per week, don't be surprised if they do a rather superficial job, at best. If a problem is worth solving, allocate the time needed to do the work. One powerful reason why some process plants are only marginally profitable is because they claim to have neither the time nor the money to do the job right the first time but somehow end up doing it over a second and third time. That's not how the best-of-class competition operates!
12. *Remove barriers to pride of workmanship.* Give your reliability person credit for keeping his or her work environment clean. Don't convey the message that the job must be done so quickly that there are no time to make it look good, no time to accurately monitor bearing temperatures in the induction heater, and no time to calibrate the torque wrench. If reliability could be achieved without pride of workmanship, we would be happy to take delivery of a new car with ill-fitting doors and streaky paint. And the surgical scar on a hospital patient might as well look like the zipper on an old mailbag if nobody cares!
13. *Institute a vigorous program of training and education.* I've asked myself why W. E. Deming seems to repeat himself in listing TRAINING under number six and also number 13. No doubt he wanted to underscore the staggering importance of training to the achievement of consistently high quality, productivity, and profitability. We can be equally certain that Deming understood that engineers leaving colleges and universities

require copious amounts of additional training in order to be productive and proficient contributors. This training may come from such sources as trade journals, vendor seminars, or even the assignment of presentations to management or to certain operations groups. Two of the most costly management misconceptions are that maintenance can always be deferred and that training the workforce is no more important than the (occasionally misguided) training that executives are often receiving from industrial psychologists, wild-water rafters, and golf pros.

14. *Go beyond a plan and take action to achieve this transformation.* Taking action means exercising leadership. Leadership must come from a perceptive, knowledgeable, even-handed individual. Risks must be assessed and roadblocks and impediments must be removed. Leadership is needed because new directives have to be communicated to others. Their cooperation has to be sought and "empowerment" redefined in some instances. Deming used the term "transformation" because he no doubt realized that that's what it may take to escape from the constraints the traditional mind-set is imposing on the pump maintenance worker.

It can be shown that the most productive and highly profitable pump users are ones viewing every maintenance intervention as an opportunity to upgrade. At these best-of-class companies, a designated reliability professional is ready and able to answer two questions without delay and hesitation: (1) is upgrading possible and, (2) if upgrading is possible, will it be cost justified in the particular case at issue. This task is given to professionals who report not only to a maintenance supervisor but also to a management sponsor.

As was mentioned before, the management sponsor makes sure that reliability engineers are shielded from the day-to-day maintenance pressures. In an operating company, reliability engineers are given the task of eliminating repeat failures of equipment. Designating, grooming, and rewarding a reliability improvement expert will be of great value to an enterprise. Proper training and the evenhanded assignment of accountability will promote good reading habits and are sure to drive down all equipment failure frequencies. Proper maintenance management has an influence on the future of a company. We again remind the perceptive reader of the urgency of ruling out finger-pointing and faultfinding between a maintenance department and an operating department.

We want to again remind the reader that, in some of the most profitable and reliability-focused organizations, a high-level executive conveys to the maintenance department head that he or she will be asked, at random intervals, to trade places with the operations department head. The operations department head will then become the maintenance department head. That simply means that the two department heads will practice full communication and cooperation. The two heads will adopt a stature whereby they show consideration for each other's issues and concerns on a daily basis. If they do not learn to communicate and cooperate, their careers will be dead-ended or doomed to fail.

Chapter 18.3

Confused about Reliability-Centered Maintenance?

INTRODUCTION

The process industries are often among the leading practitioners of advanced maintenance techniques. Some of these are adaptations of successful aerospace maintenance practices. While these adaptations clearly have merit in some circumstances, they may not always apply. Here is an example of candid correspondence that makes the point.

RCM QUESTIONS

A trade journalist received a letter from Brazil. "Considering that I am a little bit confused about what you mean by terms such as wear-out, progressive deterioration, wear-out failures, life shortening events, etc.," said its writer, "I would appreciate your thoughts and guidance on these RCM (Reliability-Centered Maintenance)-related matters."

Since the gentleman quoted from some older articles, the journalist's first reaction was to surmise that some things said years ago were perhaps outdated or irrelevant. Our reply reaffirmed that there's no change, although with different job functions and experience levels, clarifications are often needed. Here then are some clarifications and examples, question-and-answer style.

WHAT FAILS?

To begin with, the questioner referred to the statement: "Only those assets which show a clear age-related pattern are subjected to periodic, time-based maintenance (TBM). Equipment exhibiting evidence of random failure, or likely to undergo progressive deterioration is subjected to predictive, i.e. condition-based monitoring (CBM)." He then asked the question:

> **Q**: "What is the difference, if any, between age-related patterns (same as wear-out, as I understand it) and progressive deterioration? Are not all age-related failures a function of progressive deterioration?"
> **A**: Age-related patterns might be progressive hardening (the loss of flexibility) of an O-ring. If the deterioration is progressive, it's obviously possible to plan for a time-based change out. However, wear-out of other components may be related to the number of starts and stops, to name just one. Age alone may thus not be causing progressive deterioration.

Q: You quote: "Only equipment that exhibits wear-out failures will optimally respond to RCM, whereas random failures are best detected by predictive techniques."

A: That's a correct statement. For instance, a random premature failure of the O-ring will occur after the accidental overpressuring of its fluid environment or the inadvertent admission of an incompatible fluid, etc. Knowing that accidental overpressuring has occurred, this out of normal time sequence (hence "random") O-ring failure can now be predicted.

Or, this O-ring typically "wears out" after 100,000 strokes of the piston. Unbeknownst to us, a sliver of metal has contacted it after 20,000 strokes: employing the predictive technique of fluid level monitoring will indicate that leakage flow exists. This knowledge may now allow the vigilant technician or operator to anticipate failure.

Q: You state: "Our assets will virtually always fail in one of the predictable wear-out modes, which is ideal for the application of RCM."

A: In fact, the statement was regrettably out of context. The full paragraph reads: "By all accounts RCM does not respond to random failures. Aerospace [where RCM works] and process industries represent two different worlds. Let us, nevertheless, *suppose* that we, just as Boeing, McDonnell-Douglas and the European Airbus Conglomerate, have installed equipment, subsystems, and components that are life cycle cost optimized. *Assume* further that, just as is the case in the aerospace field, voluminous statistical and experimental data attested to the fact that our assets will virtually always fail in one of the predictable wear-out modes, which is ideal for the application of RCM."

Well, the above clearly stated *assumptions* are of course wrong. In the majority of process plants, we have neither installed life-cycle cost-optimized subsystems and components nor do we have data indicating a preponderance of wear-out failures in our industry. Therefore, investing in procedural RCM *before* implementing widely known cost-effective upgrade measures is both costly and unproductive. To say it differently, RCM is *wasted* if your plant has not yet implemented the component upgrades previously proved effective by best-of-class performers.

Also, a plant that is not implementing the *basics of sound maintenance* will never benefit from RCM.

Q: You have recently written: "Parts and components that are subjected to predictable wear and other life-shortening events exist only in plants that refuse to implement known upgrade measures. Relatively few machinery component wear-out failures exist in *b*est *p*ractices *p*lants." *There* appears to be a contradiction among what is quoted in those paragraphs. Would you please explain this apparent paradox?

A: Hopefully, the preceding Q&A sequence answered the question.

Q: Considering that *relatively few machinery component wear-out failures exist in best practices plants*, there is then no room for the application of RCM in these best-of-class companies. Is that what you mean?

A: First things first. There are far more immediate and revenue-saving opportunities that should be pursued. Many of these opportunities have been the subject of our books and articles since 1974. We believe that these must be understood, explained, and implemented before cost-effectiveness or measurable benefit is derived from RCM.
Q: If TBM should be avoided, why do they, in the field of civil aviation, continue to disassemble the whole aircraft from time to time?
A: I suppose there are many reasons. An aircraft engine alone consists of approximately 8000 parts. Unlike the majority of parts found in rotating machinery in an oil refinery, many of the aircraft components are exposed to thermal fatigue, potential creep rupture, and so forth. Total dismantling overhauls are needed to replace slightly corroded fuselage rivets; also, the entire aircraft is flying and has irreplaceable human cargo on board, etc.

THEORY AND PRACTICE

There is a great lesson in this story. The pursuit of largely theoretical approaches seemed embedded in the engineer's questions. Such pursuits are likely to detract from what we should all emphasize first: "picking the ripe, low-hanging fruit from the trees." And this "picking" is just another way of saying that, first, explore and exhaust the many close to zero cost efforts that are able to *avoid* failures. Until then, defer spending precious resources on predicting *when* parts "wear out."

To use an actual example, decades ago, leading bearing manufacturer SKF distributed a free of charge "bearing selection and life prediction" CD. The CD showed many electric motor bearings to have a theoretical life of *in* excess of 500 years—five centuries. Now, does it make more sense to realistically teach your technicians how to make bearings survive 20 or 30 years or does it make more sense to use statistical and probabilistic methods to predict (by using numerous "assumptions") when these bearings will "wear out"? Efforts put into pointing out oversights or bad work practices that are causing these "wear-outs" will certainly pay more immediate and often long-lasting dividends.

Here's another example. In late 2003, a very prominent RCM expert gave a talk at an International Process Plant Reliability Conference in Houston, Texas. He made the comment that, if a sleeve bearing in a compressor were to fail after a few hours, days, weeks, or whatever, he would call it a "random failure" and simply replace the bearing—no questions asked. A conference delegate decided to respectfully differ with the RCM expert. The bearing failed for a reason, and, unless the reason (note that we could instantly explain twenty different possibilities, symptoms, clues, and remedies) is determined, the user is being set up for a repetition of the failure. The RCM expert didn't volunteer to counter this logic.

Not to belabor the point, aerospace components inevitably operate in a predictable environment. All aviation fuels are identical. Given a certain type of jet engine, all combustion temperatures are the same. All hydraulic fluids in a modern plane are of the same controlled maximum allowable contamination level. All compressors on board a jet aircraft operate on air of a (relatively) narrowly defined temperature. Given the weight of an aircraft, all landings produce a predictable maximum shock load on the landing gear—the list could go on. That's where "wear-outs" are predictable; that's where RCM works.

Contrast this with the profusion of different fuel gas characteristics in the various expansion and combustion engines at the refinery. Understand the different temperatures at which gases and liquids are being processed. See the different maintenance practices and levels of maintenance diligence that influence filter procurement and filter change frequency and filter beta ratio. There are huge differences in the pipe stress and the foundation grout condition and the baseplate stiffness and the shaft alignment accuracy and the choice of couplings and a myriad of other things that influence bearing behavior and bearing life. Suffice it to say, there are precious few parallels in even something as elementary as a fuel pump in an aircraft and a centrifugal pump in a refinery!

Chapter 18.4

Use Selective PM and PdM for Your Compressors

INTRODUCTION

Questions such as whether time-based preventive maintenance (PM) or condition-based predictive maintenance (PdM) should be used for machinery seem to arise periodically. The answers to such questions will differ; a particular situation must be reviewed and understood, and no one answer fits all machines. We perform periodic PM by changing the oil on an automobile but would not do PM on its engine's valves.

LAYING THE GROUNDWORK

Perhaps you too are working for a company that is asking challenging questions or demanding certain implementation strategies for which they have not laid the necessary groundwork. Such seems to be the case at a well-known refinery. One of their staff wrote:

> *Regarding process gas compressors, is it possible to use predictive monitoring and no longer perform preventive maintenance on set intervals? Management has stated that the previous method of performing periodic overhauls during planned turnarounds is not acceptable. We are asked to use state-of-art predictive equipment to determine when a failure will occur and then plan an outage accordingly.*
>
> *Maybe I was mistaken but I thought one did the periodic preventive tasks to ensure that the process would not be affected during planned run times. Available state-of-art predictive routines can still be used to minimize the impact of a premature failure, or to prevent off-design operation such as improper rod loading.*

To provide an answer to the gentleman's questions, we must direct our attention at a number of facts, conventions, and scenarios.

PREVENTIVE (PM) AND PREDICTIVE MAINTENANCE (PdM) EXPLAINED

PM encompasses periodic inspection and the implementation of remedial steps to avoid unanticipated breakdowns, production stoppages, or detrimental machine, component, and control functions. Predictive and, to some extent, also preventive maintenance is the rapid detection and treatment of equipment abnormalities before they cause defects or losses. This is evident from considering lube oil changes. This routine could be labeled preventive if time-based and predictive if done only when testing shows an abnormality in the properties of the lubricant. Without strong emphasis and an implemented PM program, plant effectiveness and reliable operations are greatly diminished.

In many process plants or organizations, the maintenance function does not receive proper attention. Perhaps because it was performed as a mindless routine or has, on occasion, disturbed well-running equipment, the perception is that maintenance does not add value to a product. This may lead management to conclude that the best maintenance is the least-cost maintenance. Armed with this false perception, traditional process and industrial plants have often underemphasized preventive, corrective, or routine maintenance. They have, on many occasions, not properly developed maintenance departments, not properly trained maintenance personnel, and not optimized PdM. Excessive unforeseen equipment failures have been the result.

Correctly executed, maintenance is not an insurance policy or a security blanket. It is a requirement for success. Without effective PM, equipment is certain to fail during operation. However, in today's environment, effective maintenance must be selective. Selective preventive maintenance (selective PM) results in damage avoidance, whereas effective PdM allows existing or developing damage to be detected in time to plan an orderly shutdown.

COMPRESSOR MAINTENANCE IN BEST PRACTICES PLANTS

Four levels of effective compressor maintenance exist. Although there is some overlap, the levels of maintenance are the following:

1. *Reactive or breakdown maintenance.* This type of maintenance includes the repair of equipment after it has failed, in other words, "run-to-failure." Reactive maintenance is unplanned, unsafe, undesirable, expensive, and, if the other types of maintenance are performed, usually avoidable.
2. *Selective PM.* Selective PM includes lubrication and proactive repair. Onstream lubrication of, say, the admission valve control linkage on certain steam turbines should be done on a regular schedule. In this instance, anything else is unacceptably risky and inappropriate.
3. *Corrective maintenance.* This includes adjusting or calibrating of equipment. Corrective maintenance improves either the quality or the performance of the equipment. The need for corrective maintenance results from preventive or predictive maintenance observations.
4. *PdM and proactive repair.* PdM predicts potential problems by sensing operations of equipment. This type of maintenance monitors operations, diagnoses undesirable trends, and pinpoints potential problems. In its simplest form, an operator hearing a change in sound made by the equipment predicts a potential problem. This then leads to either corrective or routine maintenance. Proactive repair is an equipment repair based on a higher level of maintenance. This higher level determines that, if the repair does not take place, a breakdown will occur.

PdM instrumentation is available for both positive displacement and dynamic compressors. It exists in many forms and can be used continuously or intermittently. It is available for every conceivable type of machine and instrumentation schemes ranging from basic, manual, and elementary to totally automatic and extremely sophisticated. Not knowing the size of the questioner's compressors and if the owner employs such sparing philosophies as installing three 50% machines, two 100% machines, or perhaps only one 100% machine, it is not possible to make firm recommendations as the most advantageous level of monitoring instrumentation, shutdown strategies, etc.

However, PdM instruments are available from key vendors in the United States and overseas. An Internet search will uncover competent manufacturers of monitoring equipment; some of these are discussed in *Reciprocating Compressors: Operation and Maintenance* (ISBN 0-88415-525-0).

Certainly, a PdM expert system can monitor machine vibrations. By gathering vibration data and comparing these data with normal operating conditions, an expert system predicts and pinpoints the cause of a potential problem. The trouble is that detecting vibration is quite different from eliminating vibration. An intelligent but highly *selective PM* program may lead to actions that prevent bearing distress and thus prevent vibration from occurring in the first place. Needless to say, a *selective PM* program may well be a more cost-effective program than the program that waits for defects to manifest themselves. This fact establishes that a sweeping management edict disallowing *all manner of PM* does not harmonize with the principles of asset preservation and best practices. Zero maintenance is usually unprofitable and can even lead to disasters.

Traditionally, industry has focused on breakdown maintenance and, unfortunately, many plants still do. However, in order to minimize breakdown, maintenance programs should focus on the appropriate levels of maintenance. These different levels are described in a very wide spectrum of purely maintenance-oriented literature.

EMERGENCY REPAIRS SHOULD BE MINIMIZED

Plant systems must be maintained at their maximum levels of performance. To assist in achieving this goal, maintenance should include regular inspection, cleaning, adjustment, and repair of equipment and systems. Repair events must be viewed as opportunities to upgrade. In other words, the organization *must* know if upgrading of failed components and subsystems is feasible and cost-justified. On the other hand, performing unnecessary maintenance and repair should be avoided. Breakdowns occur because of improper equipment operation or failure to perform basic preventive functions. Overhauling equipment periodically when it is not required is a costly luxury; upgrading where the economics are favorable is absolutely necessary to stay in the forefront of profitability.

Regardless of whether or not PdM routines have determined a deficiency, repairs performed on an emergency basis are generally three times more costly in labor and parts than repairs conducted on a preplanned schedule. More difficult to calculate, but high nevertheless, are costs that include shutting down production or time and labor lost in such an event.

Bad as these consequences of poorly planned maintenance are, much worse is the negative impact from frequent breakdowns on overall performance, including the subtle effect on worker morale, product quality, and unit costs.

EFFECTIVENESS OF SELECTIVE PREVENTIVE MAINTENANCE

Selective PM, when used correctly, has shown to produce considerable maintenance savings. Sweeping broad-brush maintenance, including the routine dismantling and reassembling of compressors and other fluid machines, is wasteful. It has been estimated that one out of every three dollars spent on broad-brush, time-based PM is wasted. A major overhaul facility reported that "60% of the hydraulic pumps sent in for rebuild had nothing wrong with them." This is a prime example of the disadvantage of performing maintenance of industrial machines to a schedule as opposed to the individual machine's condition and needs.

However, when a *selective* PM program is developed and managed correctly, it is the most effective type of maintenance plan available. The proof of success can be monitored and demonstrated in several ways:

- Improved plant availability
- Higher equipment reliability
- Better system performance or reduced operating and maintenance costs
- Improved safety

A plant staff's immediate maintenance concern is to respond to equipment and system functional failures as quickly and safely as possible. Over the longer term, its primary concern should be to systematically plan future maintenance activities in a manner that will demonstrate improvement along the lines indicated. To achieve this economically, corrective maintenance for unplanned failures must be balanced with the planned selective PM program. Every maintenance event must be viewed as an opportunity to upgrade so as to avoid repeat failure.

KNOW YOUR EXISTING PROGRAM

The starting point for a successful long-term selective maintenance program is to obtain feedback regarding effectiveness of the existing maintenance program from personnel directly involved in maintenance-related tasks. Such information can provide answers to several key questions, and the answers will differ from machine to machine and plant to plant. Your in-plant data and existing repair records will provide most of the answers to the seven questions given below. A competent and field-wise consulting engineer will provide the rest:

1. What is effective and what is not?
2. Which time-directed (periodic) tasks and conditional overhauls are conducted too frequently to be economical?
3. Which selective PM tasks are justified?
4. What monitoring and diagnostic (PdM) techniques are successfully used in the plant?
5. What is the root cause of equipment failure?
6. Which equipment can run to failure without significantly affecting plant safety and reliability?
7. Does any component require so much care and attention that it merits modification or redesign to improve its intrinsic reliability?

It is just as important that changes not be considered in areas where existing procedures are working well, unless some compelling new information indicates a need for a change. In other words, it is best to focus on known problem areas.

To assure focus and continuity of information and activities relative to maintenance of plant systems, some facilities assign a knowledgeable staff person responsible for each plant system. All maintenance-related information, including design and operational activities, flows through this system or equipment "expert," who refines the maintenance procedures for those systems under his jurisdiction. He or she reshapes PM into selective maintenance.

MAINTENANCE IMPROVEMENT AND PROGRAM OBJECTIVES

Problems associated with machine uptime and quality output involve many functional areas. Many people, from plant manager to engineers and operators, make decisions and take actions that directly or indirectly affect machine performance. Production, engineering, purchasing, and maintenance personnel as well as outside vendors and stores use their own internal systems, processes, policies, procedures, and practices to manage their sections of the business enterprise. These organizational systems interact with one another, depend on one another, and constrain one another in a variety of ways. These constraints can have disastrous consequences on equipment reliability.

An effective maintenance program should meet the following objectives:

- Unplanned maintenance downtime does not occur.
- Condition of the equipment is always known.
- Where justified, PM is performed regularly and efficiently.

- Selective PM needs are anticipated, delineated, and planned.
- Maintenance department performs specialized maintenance tasks of the highest quality.
- Craftsmen are skilled and participate actively in decision-making process.
- Proper tooling and information are readily available and being used.
- Replacement parts requirements are fully anticipated, and components are in stock.
- Maintenance and production personnel work as partners to maintain equipment.

Following these general guidelines will be sure to give positive results.

Chapter 18.5

Substantive Change Needed

INTRODUCTION

There is a truly simple way to determine if you have an effective reliability improvement program at your plant: monitor the number (percentage) of repeat failures on process pumps.

Repeat failures indicate one of only two possibilities: (a) Your engineers or technicians have not discovered the root cause of a failure, or (b) they know the failure cause and have decided to do nothing about it. Either way, the facility suffers from a management problem. Chances are there is much talk about reliability, but the results speak for themselves. Failures are likely accepted as "normal," and deviations have become the new normal.

SYMBOLIC ACTIVITY

"When reform becomes the status quo" says Frederick M. Hess, executive editor of *Education Next* and resident scholar at the American Enterprise Institute, "then you'll encounter symbolic activity rather than substantive change." Although Hess obviously spoke of the American system of primary and high school education, his statement aptly describes the peculiar situation in which many reliability professionals continue to find themselves. Their plant maintenance and asset preservation functions engage in symbolic activity and reform after reform, but there's rarely any substantive change in mind-set, training, accountability, or even competence of personnel.

The new boss is under pressure to reduce failure frequencies, repair time, maintenance expenditures, or whatever. Managers often hope to achieve these goals by exhortation and slogans or by hiring a retiree from some out of touch, obsolete refinery. The "flavor of the month" is instituted and lasts barely four weeks, only to be replaced by the next reform scheme. This month's magic fix is perhaps a polymer globule-containing oil lube additive that is expected to ward off the ill effects of water in equipment bearing housings. Next month it could be a wireless high-tech data collecting and monitoring gadget or whatever else promises salvation. But there's just no substantive change.

SO, WHERE'S THE PROBLEM?

Chances are the problems at this facility relate not to a scarcity of maintenance philosophies or the lack of high-tech gadgetry. Hence, whatever ails the plant cannot be cured by philosophical changes or by more "benchmarking." All too often, equipment failures and asset unreliability are the result of not implementing and practicing the basics. Also, repeat failures are frequently the result of engineers or maintenance technicians looking for the "quick fix" instead of finding and promoting thoughtfully engineered solutions.

It stands to reason that true profitability goals can only be met by first understanding and then systematically eliminating the many mistakes still made by a plant's operating and reliability staff. Hopefully, your engineers apply basic fundamentals and refuse to endorse gadgetry and implementation strategies that contradict either the laws of physics or plain simple logic or both. A few very brief examples will make the point.

A vendor sells constant level lubricators and shows how these are used on pump bearings. His images and illustrations show ball bearings with rolling elements touching each other; there are no bearing cages or ball separators. Why would you purchase your parts from a vendor who cannot even be trusted to properly depict a ball bearing? Another vendor shows cross-sectional views of bearing housings with no way for lubricants, once they have passed through a bearing, to flow back down into the oil sump. Ask yourself what happens to trapped oil: it will overheat and turn into carbon. How can that be good for the bearing? (Look for illustrations of bearing housings that allow trapping of the oil. You will find at least two such illustrations in this book.)

One enterprising vendor offers a small pressure relief valve for pump bearing housings. Do you really need such a valve? There is usually a near-ambient pressure in bearing housings where the shaft protrudes through simple labyrinth seals or is fitted with a rotating labyrinth, which, by definition, has an air gap. Consequently, a check valve serves no purpose here, and the purchaser has again spent good money for something that neither adds value nor extends equipment life. The message: Ask how parts work and purchase only what comports with physics and scientific principles. Don't get involved in purchasing add-on components devised by clever marketers whose only motive is to sell stuff.

UPDATING MAINTENANCE PERSONNEL'S THINKING

Beware of allowing basic oversights on the crafts and maintenance technician level. A simple example shows the support pads of an equipment baseplate in Subject Category 12, dealing with foundations and grouting systems. Unless the craftsperson either countersinks or counterbores equipment mounting holes in baseplates, the thread tapping operation will create a burr. Precise machine alignment will not be possible under these circumstances.

Coupling hubs installed with reliability-mandated shaft fits require provisions for proper removal at a later date. Depending on coupling size and application, these provisions might range from puller holes to hydraulic fixtures. Also, totally inadequate piping practices still prevail in many plants. They range from allowing excessive pipe strain causing the skewing of bearing outer rings relative to bearing inner rings in centrifugal pumps to not paying attention to mandatory component clearances and optimum interference fits. The details can again be found in books and articles on maintenance technology.

Ask why are your pump shop technicians still heating bearings to 230°F (110°C) if this temperature is 150°F (~83°C) above the ambient temperature in the shop. The maximum allowable bearing-to-shaft interference fit is typically 0.001″ (0.025 mm), in which case a 75 mm bearing bore will be enlarged by almost 0.003″ (0.075 mm) by this temperature differential. A good procedure calls for differential temperatures that will increase the bearing bore by no more than 0.0015″ (~0.04 mm). After thermally growing only 0.0015″, the bearing inner ring serves as a "ring gage," that is, if the heated bearing cannot be pushed on the shaft by hand, the shaft diameter is too large and the ultimate interference will probably be excessive. Excessive interference fits result in greatly reduced bearing lives.

Please understand the point we are attempting to make here: substantive changes are first of all needed in the mind-sets and work processes employed by industrial facilities. Benefits cannot be derived from high-tech approaches alone. If a plant doesn't know the root cause reasons for its repair-focused behavior, it will not reach reliability focus by symbolic activity and reform for reform's sake. We are also saying that, while "benchmarking" may *indicate* a ranking, only the well-defined training and teaching of engineers and craftspeople can *improve* the ranking. There are still many maintenance departments that use nineteenth-century methods and expect best-of-class twenty-first-century reliability results. Also, suppose there is nobody with the necessary background or requisite empowerment to give authoritative guidance. In that case, which person or entity at your plant monitors if best practices are known and are consistently being used?

HOW BEST-OF-CLASS PERFORMERS EXCEL

It is noteworthy that the world's most profitable companies share a number of key beliefs and practices. Among other things, they

- view every maintenance event as an opportunity to upgrade (they will upgrade if the measure is cost-justified),
- practice root cause failure analysis,
- have low tolerance for repeat failures and will not accept mystery explanations,
- always pick the "ripe, low-hanging fruit" first (they make absolutely sure they have the "basic" right and only then will they invest in sophisticated "icing on the cake" or high-tech approaches).

So again, unless the basics are understood and scrupulously observed, a plant will not reach its true potential. To move from the unprofitable repair focus to the vitally important reliability focus, a plant must overcome the profit-limiting effects of certain ingrained, yet unacceptable, practices. Substantive change is needed.

Chapter 18.6

Was That a Failure or "Just a Repair"?

INTRODUCTION

Purists among reliability professionals are sometimes concerned about the accuracy with which one should (or, in fact, does) measure the mean time between failure (MTBF) of an asset. Everyone has their own way of doing it, but monitoring parts replacements on pumps or on electric motors serves a multiplant refining giant quite well. It certainly allows its corporate reliability engineers to make reasonably accurate affiliate-to-affiliate comparisons.

DISTINGUISHING BENCHMARKS

Some petrochemical and oil refining companies make a distinction between repairs and failures. This was alluded to in what one engineer wrote:

I have a philosophical question around the classification of repairs vs. failures when tracking rotating equipment reliability. As I see it, there are basically two structures or philosophies:

- *Asset philosophy*
- *Component philosophy*

In the world of asset philosophy, one views the equipment train as a singularity, and this is how it's being done here, where I work. A motor and pump combination is a single asset. We observe if the asset as a whole continues to perform its intended duty, that is, pumping product. As long as the asset moves fluid, the asset has not failed. In other words, if a particular component fails and needs to be replaced, the asset has not failed. If, as an example, a seal leaks, the action to correct the leaking seal is logged in as a repair. After all, the asset continues to perform its intended function. In this view, all seal leaks are considered repairs. If—and only if—the seal leak causes the asset to shut down (e.g., the seal blows out), this situation is classified as an asset failure.

In the sphere of component philosophy, an asset is seen as a composite unit consisting of multiple and various components (i.e., motor shaft, motor bearings, pump shaft, pump bearings, impeller, wear rings, coupling, throttle bushing, seal, seal pot, etc.). Each of these has its own failure mode. While it is true that a particular component may cease to perform its intended function and not prevent the asset as a whole from functioning (i.e., a small seal leaking does not stop the pump from pumping), the asset must still be taken off-line to repair the defective component. So, in essence, there will be an asset repair due to a component failure.

I believe the component philosophy is the superior form of equipment classification as it pertains to rotating equipment reliability. It gives my plant the ability to classify both asset and components in their respective statistics or catalog. Moreover, it provides the benefit of seeing issues down to the component level. The component strategy is helpful in identifying what particular items are causing all the problems.

What the reader stated may look interesting; however, it also shows how we can play up one side or the other, even with apparently valid statistics. It is not possible to compare statistics based on narrow definitions with statistics based on much broader definitions. Sensible MTBF statistics aim for simplicity. Therefore, some reliability engineers or professionals see merit in making comparisons as long as they do not involve the judgmental ingredient of questioning if and how a particular component defect *could have* caused the asset to shut down.

It must be stated that, for years, many of the best organizations have called it a failure event whenever a component was being replaced. When first collecting relevant statistics many years ago, these companies decided that only two preventive/predictive action steps, data taking and/or performing scheduled oil changes, would escape being called an equipment failure event. The main aim of the reliability professionals in those organizations was to facilitate comparisons based on facts, not on assumptions or projections. These best practices organizations wanted to steer clear of guessing or speculating if leaving a flawed component in place would have, or would not have, led to an asset shutdown.

COUNTING MAINTENANCE INTERVENTIONS

A few years ago, a major oil refinery on the east coast of England emphasized asset outage numbers. Individual process units competed with each other as they attempted to drive down their respective numbers. They did so by performing lots of preventive actions, which primarily included frequent oil changes. As a result, the refinery's overall asset outage frequencies declined, bearing failures declined, and some process units looked real good—but only on paper. The refinery soon realized that maintenance expenditures in the "good" units were higher than in the "bad" units.

From then on, they made it a practice to count "maintenance interventions." An oil change is a maintenance intervention. Counting maintenance interventions shifted the goals from aiming for favorable statistics to optimized operation. This shift in benchmarking made the various process units pay more attention to the facility's key reliability professional. It was this professional's job to look at plant-wide bottom-line performance and long-term profitability. When he again spelled out optimized work processes and procedures to extend intervals between interventions, the rest of the organization finally started to listen.

The author's former employer treated replacing a $3.95 O-ring the same as a $2150 impeller replacement. The various plants or process units might have looked for opportunistic repair dates, but years ago, this company resisted making the "repair versus failure" distinction. There would have been concern that one person's seal blowout was another person's minor leak. So, the quest was to consistently use simple statistics for comparison purposes. As to the original question raised by the reader, what he called "repairs" was included in the expression MTBF.

For some unspecified reasons, certain industry segments keep track of mean time between repairs (MTBR). While this metric could make one's MTBF look good, the organization might then no longer have the ability to compare itself to competitors who lump things together in calculating their MTBFs. Perhaps it doesn't matter much which calculation method we choose as long as we compare apples to apples.

One can play never-ending games with statistics. Statistics often remind this author of an incident that took place perhaps in 1980. At that time, the affiliate of a multinational oil company petitioned its corporate head offices (HQ) that tweaking pump parts was not really a repair. The decision from up high was to henceforth count as repairs a work event requiring the equipment be taken into the shop. "Field work" did not count as an equipment failure. Not long after this decision, work was being performed on a true basket case. The parts were spread out on some kind of pallet right outside the shop. That affiliate's great low failure record (as reported to HQ) remained unaffected; after all, the machine had never entered the shop. It took a while before more realistic heads prevailed, and someone got wiser. From then on, the managers in charge at corporate headquarters reverted to listing as a failure all incidents where components had been replaced.

Chapter 18.7

How Small Deviations Compromise Reliability in Pumps

INTRODUCTION

Parts and components are given nominal dimensions but are allowed to vary within a band of tolerances. Similarly, machine speeds are usually nominal speeds but are frequently allowed to vary slightly within an assigned band of tolerance and so forth. The problem is that deviations can add up. We usually get away with one, two, or three deviations, but too many deviations in the same machine will render it unreliable. Tolerance for more and more deviations led to a major refinery experiencing serious pump distress. Again, a massive thrust bearing failure occurred on one of the refinery's important 3560 rpm process pump.

The pump had been designed and manufactured by a well-known company. Its two oil mist-lubricated 75 mm/40 degree angular contact thrust bearings were arranged back-to-back, as is customary. Also, in compliance with the current API 610 standard, each of these bearings had massive bronze cages. In Subject Category 4, we illustrated a similar bearing and mentioned that some pumps are not reaching lowest possible life-cycle cost with brass or bronze cage bearings. Depending on axial load and bearing speed, carefully selected bearings with nonidentical load angles and bearings with carefully selected high-performance plastic cages may actually be more suitable than what's listed in API 610. The stipulations of API 610 fit the majority of applications yet never claim to represent universal superiority.

DISTRESS SEQUENCE

As to the refinery experiencing pump distress, the call for assistance in understanding and mitigating these failures initially came from a reliability manager. He was an experienced engineer and, coincidentally, had a solid machine shop background. The manager was certain that the bearings were carefully mounted and that all bearing-related dimensions had been verified within their allowable tolerance bands or dimensional limits. We believe he was correct.

Fortunately, the reliability manager passed along some relevant observations. From these, it became evident that the refinery was struggling with the effects of several deviations from best practices in its oil mist-lubricated equipment:

- The process pumps incorporated the typical oil return notch located at the six o'clock position of the bearing housing seat in Fig. 18.7.1. This notch is very important in equipment with conventional lubrication. Without the notch, oil could get trapped behind a bearing and overheat. However, these bearings were lubricated with pure oil mist—an excellent choice whereby the mist is applied between the bearing and the housing endcap in Fig. 18.7.1. With an open oil return notch at the six o'clock position of the housing bore, some oil mist was being bypassed and no longer available for the dual purposes of lubrication and cooling—the first deviation.

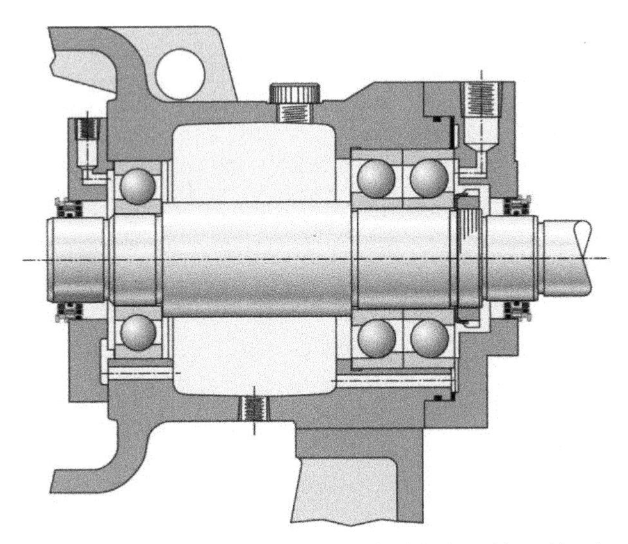

FIG. 18.7.1 Bearing housing with oil return slots at six o'clock position of bearing bore. While quite appropriate on liquid oil-lubricated bearings, plugging these oil return slots represents best practice on oil mist lubrication. Unless plugged, these slots represent "path of least resistance" that would allow oil mist to bypassing the bearings. *(Courtesy AESSEAL Inc., Rotherham, the United Kingdom and Rockford, TN.)*

- An unusually wide bronze cage acted as a restriction orifice for the remaining oil mist. An unusually wide cage became the second deviation from best practices.
- All angular contact bearings create windage. An inclined cage in angular contact bearings acts as a small fan. A fan either *promotes* oil mist flow or *opposes* oil mist flow. With high peripheral cage velocities, this windage must be taken into account. At ~3560 rpm and with a shaft diameter of 75 mm, the peripheral velocity is 2780 fpm. At higher than 2000 fpm, one should use "directed" reclassifiers similar to Fig. 18.7.1, so as to overcome windage. Directed reclassifiers must be mounted with the mist opening no more than 3/8 in (~10 mm) from a bearing cage. (A directed reclassifier was also depicted in one of the author's articles decades ago; see HP March 1977.) Using only a standard, nondirected oil mist reclassifier constituted the third deviation from best practices.

- The standard reclassifiers were mounted a distance of over 24 in away from the bearing housing at the oil mist manifold. Because high windage bearings (shaft surfaces at >2000 fpm) were involved, a large distance from reclassifiers to bearings became the fourth deviation from best practices. And, while placing reclassifiers at the oil mist manifold (24 in from bearings) is normally allowed, doing so will reduce the available factor of safety compared with mounting reclassifiers at the bearing housing. The closer a reclassifier is to the bearing, the easier it will be for the oil mist to overcome the windage, or fan effect, of an inclined bearing cage [1].

Note how, in Fig. 18.7.2, the point of mist exiting from the nozzle/reclassifier fitting is only 3/8 in (10–11 mm) from the bearing cage. Because there are no oil return slots, every tiny droplet of oil mist must travel through bearing, as shown earlier with oil mist routing arrows in Chapter 17.1, Fig. 17.1.5.

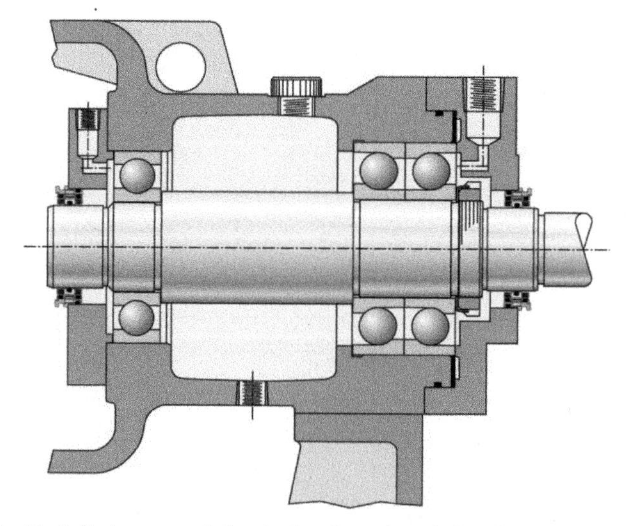

FIG. 18.7.2 Ideal oil mist passage for bearing housings where shaft surface speeds exceed 2000 fpm (610 m/min). *(Courtesy AESSEAL Inc., Rotherham, the United Kingdom and Rockford, TN.)*

Our message is multifaceted, and four points stand out very clearly:

- Small deviations and shortcuts add up, and allowing several risk-inducing practices to exist will cause unacceptably low safety factors.
- Serious reliability-focused users will always use checklists. In particular, oil refineries should make conscious efforts to do things right the first time.
- There is a cost to having uninformed workers in any particular job function at an oil refinery. Staffers may have to overcome a reluctance to read relevant books and articles. Learning never stops, and goals such as becoming above-average performers will elude those for whom it's always "business as usual."

- A series of repeat failures will cost much more than the few hours of reading or listening to a competent consultant who, in this case, could have easily explained how repeat failures could be avoided.

We learned that we will annoy folks by asking them to read and acquire factual knowledge. Expect strong push back when trying to inculcate zero tolerance for deviations from best practice. As you then aim to substitute accountability for today's all-pervasive indifference, you will really ruffle many feathers. Then again, ruffling feathers is a very small price to pay for avoiding a great number of potentially unpleasant outcomes.

REFERENCE

[1] H.P. Bloch, Pump Wisdom: Problem Solving for Operators and Specialists, John Wiley & Sons, Hoboken, NJ, 2011.

Chapter 18.8

Deviations Add Up to Become Exponential Failure Risks in Compressors

INTRODUCTION

If a set of automobile tires is rated for 115 mph and almost new, one can travel clear across the country at 115 mph and have very little probability of a tire failing. The probability increases as the tires experience progressive wear. The failure probability increases again if the tires are underinflated or if we travel at high speed on a road full of potholes. It's no different with process machines. We will usually get away with one, also two, or even three deviations from acceptable dimensions, acceptable speeds, pressures, temperatures, etc. But as more and more deviations are added, the machine will fail.

AN INDUSTRY EXAMPLE

In the early 2000s, an oil refinery experienced a power failure, which caused a large (~2500 kW) three-throw reciprocating compressor (Fig. 18.8.1) to unexpectedly shut down under full load. It was left there, under load, for a few hours. Then, while attempting to put the machine back online, the temperature on the third of its five aluminum main bearings (i.e., one of the crankshaft bearings) ramped up rapidly within seconds, and when the compressor capacity was taken from part load to full throughput, the machine tripped on high frame vibration.

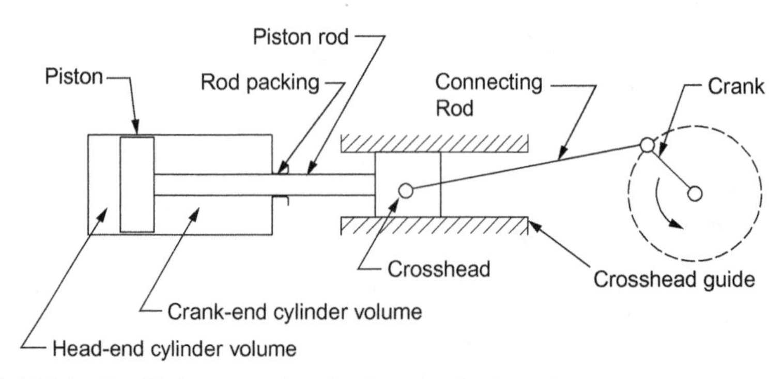

FIG. 18.8.1 Simplified representation of major parts of reciprocating compressors.

The compressor was dismantled and severe damage found on the "hot" main bearing. Substantial extrusion was found on one side; the adjacent connecting rod bearing—also aluminum—had melted. The crankshaft was bent and had to be reconditioned. Since it was also determined that the crankshaft centerline was misaligned relative to the frame, costly regrouting and other repair works had to include line boring of the bearing supports. At first, root cause failure analysis (RCFA) had to be deferred due to other priorities, and "the usual suspects" were hotly pursued. However, although it had initially been theorized that a slug of liquid had entered the compressor, there was neither valve distress nor a bent piston rod. While such collateral damage would be quite typical of liquid slugging, such collateral damage had not occurred here.

Fortunately, some managers later recognized the improbability of their refinery always being the first, or only one, to experience strange events. When the more rigorous RCFA was begun, the effort was hindered because the failed main bearing had been discarded shortly after the incident. In this instance, the far-fetched assumption was that liquid had caused a monstrous bearing failure while leaving all valves, piston rods, and connecting rods in pristine condition.

At this point, a systematic approach was needed to assemble all available data into a logical sequence of events. Reconstructing a logical chain of events must stand the test of plausibility. The most probable sequence is the one supported by all available data, and speculation is inappropriate when it is contradicted by facts or when your company is always the first to experience a calamity.

It was reasoned that the virtually instantaneous stopping of a reciprocating compressor in the loaded condition—while a clear violation of every reputable manufacturer's guidelines—might ordinarily be of little or no consequence to a well-designed and well-aligned machine. However, the machine's alignment was not within specified or acceptable limits. What happens when two violations combine, that is, when internal misalignment and stopping in the loaded condition occur simultaneously?

In this latter case, the force from a loaded cylinder was being transmitted to the main and connecting rod bearing(s) for a much longer period of time. Also, this force was now certain to edge-load the associated bearing(s), whereas in a well-aligned machine, the same force would have been distributed over a much larger projected area of the same bearing(s). With the crankshaft not rotating and the cylinders loaded, the cooling effect of lube oil flow was available only to a small portion of the surface of the crankpin, which added thermal stress and deflection. So it was, with overwhelming probability, the edge-loading that caused metal-to-metal contact and subsequent bearing damage. There is a cross drilled lube oil passage from the crankshaft to the crankpin. As molten metal or debris from a severely damaged main bearing now restricted the oil feed passage to the connecting rod bearing, it too was doomed to fail.

LESSONS LEARNED

The lesson should be that one deviation alone might not be enough to bring on a compressor failure, but when two deviations combine—in this case stopping in the loaded condition and also having a misaligned compressor frame—the failure risk increases exponentially. So, while it might be possible to reduce failure risk by implementing an automatic unloading scheme or adding bells and whistles that annunciate excessive temperatures and rod drops and vibrations and blowby of the gas being compressed, there will never be any substitute for the human brain supplying both logical RCFA and up-front failure prevention processes.

Achieving these processes before calamities and finger-pointing occur requires training, honesty, and accountability. The one accepting a deviation from established practice should somehow be motivated to understand the potential ramifications and should document this understanding in writing. He should sign off after certifying that the deviation is normal and that he can live with the consequences. There can be little doubt that with such documentation requirements being enforced, fewer deviations would be tolerated. And if this documentation requirement was to scrape away part of the veneer of our protective cocoon, so be it. The rewards will be greater than we can imagine.

Subject Category 19

Management and Organizational Leadership

Chapter 19.1

Management's Role in Achieving and Sustaining Reliability Focus

INTRODUCTION

As of 2016 and in US refineries, pump mean times between failure (MTBF) varied from 2.6 years to at least 9 years. In the early 2000s, a West Coast refinery even claimed an 11-year MTBF. Although purists have quibbled over the technically correct definition of MTBF, reasonable people have long agreed that MTBF is the total number of pumps installed, divided by the number of pumps on which work (other than lube replenishment) was performed in a calendar year. With the cost of the average refinery pump repair, inclusive of burden and overhead, approaching (in 2016) $10,000 per event, some repair-focused refineries are spending millions of dollars on pump repairs, whereas their reliability-focused competitors manage to avoid the bulk of these expenditures.

DO YOU KNOW YOUR PUMP MTBF?

The facts mentioned in our introduction should prompt good managers to ask a series of probing questions of their reliability staff. Among these, we would expect to find the following:

1. What is it costing us, per year, to have the (insert your number) pump MTBF that is so obviously inferior to the 8- or 9-year MTBF enjoyed by some of our competitors?
2. What exactly are the procedures and work processes our competitors employ and that we, apparently, are not using?
3. What component upgrades are these competitors pursuing to enjoy these obvious advantages over us?

Petrochemical Machinery Insights. http://dx.doi.org/10.1016/B978-0-12-809272-9.00019-0

4. What deprives my reliability professionals from implementing what the competition has already done years ago?
5. What training or consulting input do you need to give me authoritative answers to my questions?

If you ask these questions, be prepared to receive many answers that, upon closer investigation, make little or no sense. Expect to run into people who will subscribe to purely statistical benchmarking and mathematical reliability prediction methods. Expect to hear that your pumps are old, and theirs are new, or that your labor union is doing this and theirs is doing that. All too often, these attempts at explaining downtime events disregard the human element. In other words, a large percentage of equipment downtime events can be attributed to commissions and omissions and actions and inactions—even plain human error. These include, of course, maintenance neglect and operator oversights—areas where training issues come into play. But some downtime events are indirectly attributable to management styles that are incompatible with the desire to be reliability-focused.

TRAINING AND COMMUNICATIONS

Today, fewer and fewer qualified engineers, technicians, and maintenance workers are practicing the skills that represent the underpinnings of the total productive maintenance (TPM), operator-driven reliability (ODR), and reliability-centered maintenance (RCM) programs and initiatives. We are talking here about such skills as understanding the need to carefully measure the bore diameter of a rolling element bearing and its corresponding shaft mounting diameter. We are referring to the knowledge that component manufacture involves tolerances and that mounting a radial bearing on a pump shaft requires an interference fit that differs from the fit needed for the back-to-back mounted angular contact thrust set. And, of course, we are referring to hundreds of other items that must be carefully observed to achieve longer pump life and hundreds of items dealing with process machinery in general.

Training vs. not training. Even the occasional large multinational petrochemical company has it all wrong when it decides, at the onset of an economic downturn, to suspend training so as to improve the bottom line. Smart leaders make it their business to understand the difference between superficial training and value-added training.

Value-added training is an endeavor conducted by well-informed practicing engineers, not purveyors of philosophical truisms. In other words, good managers steer employees toward value-added training and insist on receiving a briefing on "things learned." Then, they see to it that things learned are translated into action at their plant. The following true example will make this point very clear.

Responding to management's urging to "innovate and do things faster and more efficiently from the way we've done it before," a refinery shift supervisor decided to call for a reduction in steam turbine warm-up time. This caused the rotor to warm up and expand before the surrounding stationary parts had expanded in like manner. The rotating assembly contacted turbine-internal stationary parts and wrecked the machine. Millions of dollars in repair and downtime cost would surely have been avoided had the young engineer consulted his counterparts in the refinery maintenance and reliability groups or if the operators had been encouraged to speak up and express concern.

It's regrettable that incidents of this type continue to occur very frequently and that uncovering and explaining the true root causes of these incidents are frowned upon as finger pointing or as not being a team player. We have even seen such extremes as management rewarding glitzy computer-generated and well-animated presentations. Promotions are often bestowed upon employees who gain high visibility via "computer wizardry," and promotions have been withheld from the diligent readers and "sounders of early warning," since the latter are often perceived as a nuisance.

THOUGHTFUL ACTION IS DESPERATELY NEEDED

Here, then, are action steps that good managers are, or should be, actively pursuing:

- Brutal and often senseless price competition has driven a large number of equipment manufacturers to either lower their product quality, dispense with rigorous quality control, or give shoddy service. Realizing that equipment manufacturers often no longer employ inspectors, the top industrial companies assign the inspection task either to their own or to third-party inspection personnel.
- Good managers realize that buying cheap bearings and other components from importers of surplus and expired shelf-life goods are at cross purposes with their professed reliability focus. They recognize that equipment with reliability impact cannot be purchased on the basis of first cost alone. They, therefore, insist on the development and use of high-grade component specifications.
- Good managers insist that technical employees go beyond guesswork and always substantiate their concerns. In other words, they ask employees to logically explain "concerns" and require follow-up to turn concerns into "nonconcerns." Require staffers to bring you facts, not opinions. Demand solutions, not mere recantations of known problems.
- The best managers abandon the destructive short-range view and take a longer-range approach. They thus make conscious and consistent efforts for knowledge progression and successor planning. They keep their best technical employees from seeking employment elsewhere by implementing a dual ladder of advancement—one administrative and the other technical. They recognize that claims of always being able to hire an outside contractor may be true for your street-paving program, but are fallacious and untrue for many technical issues!

- Top companies stretch the tenure of their in-house experts and future executives. In other words, they keep individuals in a particular position until they are truly proficient. In contrast, cycling a talented individual through fourteen departments in 5 or 6 years is almost certain to produce extremely shallow areas of expertise and superficial thinkers. Good managers groom talent, not arrogant generalists.
- Top managers offset the benefits of past cultures with the need to forge a new culture. They will not allow employees to interact via a flood of e-mails alone. Good managers know that hours spent in posturing and responding to internal e-mail are rarely—if ever—adding value.

Indeed, then, better management is needed at many HP and other industry facilities to achieve long-term equipment reliability and its related plant profitability objectives. ODR, TPM, RCM, and other pursuits will fail where there is no trained workforce. Claims that you can always hire a competent outsider are simply not true. Competent outsiders are a dying breed, and the few that are left will not accept the pay that an unqualified individual will accept. Who, in his right mind, would knowingly select a medical doctor because he knows little and charges less than the competition?

Plants where training is an afterthought, or where training is conducted by individuals who themselves are not up to date, are unable to reach their true reliability and profitability potential. Some of these trainers spend all too much time on discussing maintenance philosophies. What is needed are explanation and implementation of discrete steps to be taken on the component and work procedure level. Regrettably, only the very best HP facilities are implementing the right steps.

Finally, broad communication across functional disciplines is needed. Such seemingly autonomous groups as operations, maintenance, project, purchasing, and reliability/technical can obviously affect equipment reliability, safety, and profitability. Any one of these groups should never be allowed to make far-reaching decisions without input from the related disciplines or functional areas. Holding people accountable is of extreme importance here. Why not start with asking your reliability professionals to explain why the MTBF of pumps at your facility differs so much from that of the competition?

Chapter 19.2

First Priority: Pick the Ripe, Low-Hanging Fruit

INTRODUCTION

Whether it sounds harsh or not, repetitive or not, reliability management is a management function that is truly all-encompassing. William Shakespeare was right when he reputedly wrote that one cannot make a silk purse out of a sow's ear. But whoever talked about the merits of picking the ripe, low-hanging fruit off a tree before doing other things was also correct. Concerning oneself with fruit hidden out of easy reach made sense only after the ripe fruit has been picked. Managers must see to it that improvements that offer themselves in the most obvious fashion should be implemented before tackling the issues that are considerably more difficult or elusive.

THE UNGLAMOROUS BASICS

In the late 1990s, a briefing session was arranged with the reliability team of a well-known petrochemical company—let's call them "WKPC"—to discuss ways to improve the company's reliability performance. In about four hours of discussion, we obtained a reasonably accurate overview of how best-of-class performers achieve their enviable standing and what might be the underlying factors that caused WKPC to fall short of meeting reasonable expectations.

By way of summary, we related that top quartile companies (the upper 25% measured by return-on investment) pay much attention to the often overlooked, generally "unglamorous" basics. We made the point that these high performers emphasize the need to understand when, where, and how appropriate work practices and upgrade measures make economic sense. We, the presenters, attempted to express the belief that WKPC would get more rapid, readily quantified results from the near-term strategy of identifying and picking the "ripe, low-hanging fruit" before embarking on the definition, selection, and implementation of plant-wide TPM (which stands for total productive maintenance) or similar "packaged" programs.

WKPC elected not to go that route. Instead, WKPC engaged the services of a maintenance management firm that did what maintenance management firms do: working on the maintenance process, but not the reliability process. Because we were on WKPC's consultants list, we later received a note from the bankruptcy court advising that the company was reorganizing and was seeking court protection from the demands of its creditors. The moral of the story: don't get hung up in consultant-conceived generalities. Get the basics right; implement them with consistency. Success is in the detail. Groom and nurture people who take pride in doing details right.

CHOOSE THE RIGHT PRIORITIES AND METHODS

It has been our experience that the most successful companies have followed the priority path of identifying the ripe, low-hanging fruit. These successful companies have found this path to accomplish, first, an upgrading of the knowledge base of the maintenance-reliability functions in an industrial facility. Next, this educational uplift will inevitably facilitate the initial acceptance and ultimate success of TPM or well-focused RCM efforts. We have often expressed the fear that many purveyors of "magic bullet" work processes are themselves not sufficiently familiar with what basics we are here referring to nor do they have a grasp for the financial loss that accrues from the lost opportunities.

Knowledgeable professionals have no quarrel with RCM, but are aware that at least 60% of US companies that attempted to adopt it in its generally taught form abandoned it before long. We all know the reasons: extreme cost, manpower being siphoned off from more worthwhile pursuits, years before bottom-line results are beneficially affected, and so forth. A notable exception has been reported by companies that use certain experience-based streamlined RCM (SRCM) processes that wisely single out the approximately 15% of a plant's equipment population for this well-proved and rather effective modified RCM approach.

SRCM methods include visual inspection and qualitative assessments in the field; these are assigned on a daily, weekly, and monthly basis. Relying on the experience of highly experienced engineers or senior technicians, the short frequencies and subjective nature of these tasks make it impractical to schedule, report, and manage the results of these inspections in a conventional computerized maintenance management system (CMMS).

DATA COLLECTOR DEVICES HAVE MERIT

A solution to the problem of *initially* not being able to optimize CMMS involvement is the use of handheld data collectors. Operators can download equipment routines daily; process and visual inspection data are collected in the field and then uploaded—perhaps instantly and without manual intervention—to a host computer. The results of each inspection can be viewed and trended immediately. Linked to the right computer, data collection devices have proved to represent an extremely effective means of successfully managing what amounts to repetitive activities. Through the use of specifically designed software, databases have been populated with hundreds of failure modes, effects, causes, and tasks, all of which permit an efficient analytic process and exceptional level of documentation.

The implementation period of streamlined programs is much shorter and considerably less costly than some competing traditional RCM approaches. The resulting reductions in downtime, lower maintenance cost, and greater availability all contribute to solid return on the investment.

WORKING ON THE RELIABILITY PROCESS

Reviewing a copy of the letter to WKPC, we note that the company was asked to "guard against the pitfalls of conventional wisdom, as was so beautifully explained in a presentation by Fred Logan at a 1996 Process Plant Reliability Conference in Houston." Fred Logan had just retired from a reliability management position with a larger competitor of WKPC and chose as his presentation topic "Abandoning the World-Class Maintenance Approach at a Major Multinational Petrochemical Company."

Fred Logan stated that shutdowns were planned in detail, CCMS was implemented, craftsmen were trained to do it right the first time, the latest P/PM techniques and tools were being used, a test facility was built, and vibration charts were transmitted around the world electronically—everything was in place for the perfect world-class maintenance organization, but the numbers indicated another story. Maintenance costs as a percentage of original plant assets were high—over 50% higher than the competitors' and over twice the 2.5% considered world class.

Matters improved drastically when Logan's employer switched from working on the maintenance process to working on the reliability process. This process is a focus of resources from all areas on improving the reliability of the plant—not just mechanical, but operational reliability of the plant—using data-based decisions incorporating economic evaluations. The areas selected were reliability engineering, root cause analysis, engineering improvements (selective upgrading), and maintenance and operating improvements. The existing maintenance processes were continued even as the need for them diminished. The notion that teaching "precision maintenance" would be the cure-all to deficiencies and false starts was quickly disproved. You cannot buy a cuckoo clock and, using precision maintenance, make it run like a stopwatch. Specifying, manufacturing, and operating highly reliable equipment are the key ingredient to reliable performance. Once that was recognized and accepted as fact, the final outcome was startling. Maintenance costs that had been increasing at 11% annually have been decreasing at 7% per year. What was once considered world-class maintenance productivity—5%—became a reality in 1996, and further improvement materialized later.

Getting back to our original discussion, we had proposed to WKPC an involvement in the reliability engineering, root cause analysis, selective upgrading, and maintenance improvement aspects of their integrated reliability strategy. We had indicated to WKPC that this involvement would start with a 2-day reliability/maintenance effectiveness study.

The 2-day study would have allowed us to see where WKPC stood in or about the year 2000. Study results would be verbally transmitted and would logically lead to presentation of one or more 1-day root cause failure analysis sessions. A few months later, it was thought that a competent specialist might spend half a day presenting a root cause failure analysis refresher session where plant personnel might wish to explore problems that apparently did not respond to the three-pronged approach we have used for the past 20 years.

Subsequent to these endeavors, we thought WKPC might elect to involve a competent specialist in their TPM/RCM contractor selection and reliability engineering/life cycle cost assessment efforts. But none of these recommendations carried. We'll never know if WKPC's profitability would or would not have been substantially better had they accepted our recommendations. We do know, however, that other companies have used the "identify and pick the ripe fruit" approach and they certainly seem to prosper.

Chapter 19.3

Delegating Responsibility Brings Benefits

INTRODUCTION

The most valuable resources at an industrial or process plant are the dedicated employees that are willing and able to respond to balanced and thoughtful direction. Try to make optimum use of this most important asset. Delegating is a sign of maturity and can minimize stress and frustration. Not only will you thus be enabled to concentrate on the most important value-added tasks, but also you will give others the opportunity to gain needed experience. The future of your organization will be given solid underpinnings and all parties come out winners.

HOW A MANAGER VIEWS PARETO'S RULE

After we had written about *Pareto's 80/20 rule* and its applicability to the job function reliability engineering, a top manager contributed his experience-based observations. He explained that he views it as his job to ensure that his company expends the proportional amount of input energy required for the potential output gain. Clearly, if the output gain is very low, he should (a) not be thinking about it or have any involvement in it personally and (b) he should stop any significant efforts of his colleagues if the outcome is likely to have limited value.

He noted that sometimes when you have done a few things and obtained 80% of the gain, it isn't just the effort required to finish some items that needs to be considered. It's also the cost of maintaining them. A perfect example was his company's electronic library. Having spent a few days putting hundreds of documents into it, from which thousands of people have benefited, it became clear that putting additional effort in would be a waste due to the cost of maintenance. In short, it was better for people to ask for specific infrequently used information than it was to try and maintain a huge database.

The thing one needs to know about applying and benefiting from Pareto's rule is that one has to know when to quit as well as when to start. We can liken it all to a big rock of opportunity occasionally becoming but a small pebble of opportunity. One then needs to know when to scale down the resource or move on to a bigger rock of opportunity.

There are exceptions to all numbers or guidelines, one of which is safety. Basically, if we cannot engage safely in an activity and/or if we cannot afford the cost of doing things safely, we must simply stop doing it. Safety is always the most important consideration.

Regarding sound guidance, it appears that top managers share a number of attributes:

- They do not believe that just because a problem is complex, the solution to the problem must also be complex.
- They want people to understand deadlines, but will listen to well-thought-out reasons when and why it might be prudent to extend an occasional deadline.
- They likely favor a competent value-adder to employees who want to shine with brilliance and gifted rhetoric.
- They manage (and reward) people who use analytic skills instead of presenting the top manager with instant compliance; he will change his mind if proved wrong.
- Their inclination is to get people to work smart, not necessarily hard. An employee who sacrifices family to certain achievements at his company will burn out and become unhappy. A top manager will let employees know that he/she is acutely aware of that.
- Instead of demanding or accepting a long list of problems, the top manager looks for a short list of soundly enunciated solutions.
- Top managers discourage employees from sending "cover your back" e-mails to everyone on a long list of people. They request that e-mails be addressed to someone who is given the mandate to act. The sender must make it clear that others be informed in suitable format. Long meetings or big meetings are rarely a suitable format.

DELEGATING

You're probably familiar with texts such as ISBN 0-88415-662-1 ("Machinery Failure Analysis and Troubleshooting") that explore competing approaches to machinery troubleshooting. One of these is the "Mr. Machinery approach," whereby the sole responsibility of a facility's machinery fate rests on a single person, Mr. Machinery. His batting average is usually good. He can make the trouble go away. He is used to calling the shots. He is very valuable to your company.

Unfortunately, there can be a downside to your overdependence on Mr. Machinery. You thought you had placed constraints on his excessive independence by asking him to be part of the reliability team, but Mr. Machinery is known to "go it alone." His presence at the team's meetings still stifles the contributions of others. And you, of course, know that machinery failure analysis and troubleshooting (FA/TS) must be a cooperative effort.

Note that because the three principal job functions represented at your plant (operations, maintenance-mechanical, and reliability-project/technical) are likely to influence equipment reliability in equal measure, all three must share in the plant's FA/TS efforts. Moreover, for plant reliability and profitability to be maximized, continuity of effort must be assured. Others need to be trained and experts should not be burdened with tasks that can be performed by less experienced personnel. Well, perhaps it's time to insist on delegating responsibilities.

DELEGATION IS A DESIRED CUSTOM

The record shows that delegation of authority already existed in the late Bronze Age, some 3500 years ago. When an ancient leader (Moses) took it upon himself to single-handedly judge the people entrusted to him, his father-in-law was concerned and declared: "It is not good the way you are doing. You will surely wear out, both you and this people who are with you, because this business is too big a load for you. You are unable to do it by yourself."

Within downsized and reshuffled modern process plants today, there are many reliability professionals who, like Moses of old, are trying to accomplish more than is reasonable or prudent. It would be in everybody's best interest if they would get help by learning to delegate. Indeed, a person who does not delegate is a poor organizer.

THREE GOOD REASONS FOR TRAINING OTHERS

Delegation of responsibility allows reliability professionals and responsible managers to achieve several important goals. First, when Mr. Machinery or his functional equivalent in other areas of the plant is absent, his delegates will act in his place, and necessary work will not grind to a halt while he is gone. Second, since actions speak louder than words, the "master" could observe the resourcefulness and abilities of the stand-in. Third, the expert gives his delegates an opportunity to gain much-needed experience. An insightful organization or industrial enterprise recognizes the need for continuity of expertise; continuity requires that potential successors be given training.

Engineers may fear that by delegating, they risk losing control. Others may be disinclined to delegate because they feel they can do the job quicker themselves. While this may be true, a thoughtful professional will see much value in training others. Of course, the professional in charge can tactfully make it clear that he plans to exercise an appropriate degree of overview or monitoring of the stand-in's progress. In any event, there is ample evidence that organizations that practice delegation will often prosper and expand at rates that outperform the competition.

Delegating also means getting help with necessary details. Thoughtful delegation allows the trainee to gain experience by being in charge of relevant records such as, say, equipment failure histories. He will thus gain valuable insights into equipment vulnerabilities, maintenance practices, repair costs, and a host of other reliability issues.

THE FOUR-STEP APPROACH TO DELEGATING

The four-step approach to delegating consists of task definition, careful selection of the designee, the provision of adequate resources, and giving support to the decisions of your designee.

Task definition requires making clear what results are expected. Where equipment is involved, the ultimate purpose of your endeavors is undoubtedly asset preservation. More specifically, this entails the safe, reliable, and long-term cost-effective implementation of work processes and procedures. For example, delegating the repair of a centrifugal pump would normally require informing the designee as to whether defective parts are to be replaced in kind or whether component upgrading is desired.

Selection of a designee requires knowledge of the person's competence as well as "people skills." Where a pure "people situation" is involved, an astute boss informs his stand-in that such issues must be dealt with in a manner that doesn't compromise ethics and principles. It is equally important to explain to your designee that he must maintain the dignity and self-esteem of others. To do so, the designee must be equitable, balanced, and considerate.

Whether dealing with the "hard" equipment issues or "soft" people issues, it is important to define not only the scope of the task but also what decisions a person is allowed to make and what matters should be referred to someone else. When you assign responsibilities, try hard to avoid overlap. Whenever more than one person is assigned the same duties, the outcome may be confusion and hurt feelings. Hence, to determine if someone is capable of doing the job at issue, consideration must be given to such factors as personality traits, experience, prior training, and talents.

Assign adequate resources. The one accepting the stand-in assignment will need to have at his disposal certain resources in order to complete an assigned task. Perhaps he needs documentation, access to other contributors, communication equipment, tools, or funds. When assigning responsibilities, inform others that the person is acting in your stead. The authority to act in your place is also a resource. Hence, a thoughtful organization communicates these assignments, where appropriate.

Support the designee's decisions. While the one assigned to act in your place can now get on with the work at hand, it is well to remember that you can be a real source of encouragement to him if you support the good decisions he has made or the good results he has achieved. Delegate the task, not how the task is to be carried out. This point is heeded by best-practices companies favoring role statements where lesser performers sometimes rely on elaborate, and often too detailed, job descriptions. Delegating the task, or outlining the role, has often unleashed refreshing creativity and yielded valuable results.

Furthermore, a stand-in contributor with his feet firmly planted on the ground, so to speak, is often closer to a particular situation and thus better understands the problems associated with it. He will likely respond to problems with solutions that really work. He may also be dealing with factors that are not obvious to onlookers.

Chapter 19.4

Challenges Facing Managers in Industry

INTRODUCTION

When asked about challenges facing the hydrocarbon processing (HP) or, for that matter, any other industry sector, my answer has always been swift and direct: the challenges faced today are bundled up, are comingled with, or can be found in the lessons some of us learned many decades ago. Regrettably, lessons learned and explained decades ago were often disregarded by managers whose focus was short-range. And the focus has not gotten much sharper with time.

THREE GROUPS OF PEOPLE

We can examine some of the reasons for persistent challenges while keeping in mind that all of the people we come into contact with can be divided into three groups:

1. The ones who already know the subject and understand what course of action to take in the best interest of stakeholders.
2. The ones who are not teachable. Trying to communicate with them will be both frustrating and a waste of time.
3. The ones who are presently uninformed and would see merit in you showing them the benefit derived from listening to you and from acting on your advice.

Quite obviously, individuals selling a product may wish to concentrate on group 3. If you work for a company that sells products, consider yourself a stakeholder because you really have a stake in your company being prosperous. But many of us consider themselves user-purchasers.

As we think of the user-purchasers, they, too, are stakeholders. Their stake is in seeing their companies prosper. Needless to say, if both sides prosper, you will have a win-win situation. You will be guaranteed a win-win situation if you consistently practice the "three Cs":

CCC = communication, cooperation, and consideration. Obeying the CCC rule is good for business. It's good for developing products. It's good for developing people. It even keeps marriage together and makes marriage mates happy.

Regardless of one's job function, employer or employee, manager or non-manager, persons fixing their eyes on today or on the future, I urge you to give attention to the rest of the story.

THE REST OF THE STORY

- Today, we deal with a largely uninformed workforce. This adjective is not to be misunderstood; even a genius can be uninformed.
- Some countries that ranked high in technology, mathematics, and science skills decades ago have since slipped in these rankings. This should be of interest to today's managers because the folks who are now in school or have just started work are our present or future technicians and operators. They need to be approached with much forethought.
- *Root causes*. The root causes of an uninformed workforce go back very far and can be traced to generally shallow leadership. Shallow leaders cannot (or will not) give guidance, or mentoring, or provide a nurturing environment. In-depth guidance, appropriate mentorship, and nurturing take time, effort, and dedication. As the decades go on, all of these attributes shrink into a "supply shortage."
- In some locations and environments, few workers are motivated to learn. The fault is absolutely not on one side only; both sides are responsible. Learning is not always rewarded. Industry leaders are not rewarding the one who brings them the facts; all too often, not enough time is allocated to capture and convey facts. Worse yet, facts and opinions are comingled. Strong opinions reap rewards, even promotions. Majority opinions may vastly outnumber factual findings.

There are numerous examples to which we can relate and where facts and opinions and misinformation are often comingled. In one country, a major hurricane was blamed for devastating a large city. But it was the levies that kept back a large lake that broke and the Federal Government and the original construction decisions of entities rarely named that were in fact responsible for the widespread flooding.

APPROACHES TO PROBLEM SOLVING

There are many texts that teach and explain problem solving. They have in common that one must first

(1) *identify* a problem,
(2) *outline the options*,
(3) *recommend a solid and well-researched solution.*

Experience-based remedies exist and should be considered. They represent a sound prescription. As we follow that prescription and examine industry trends, we often see an unhealthy risk-and-reward system. Accountability (or the lack of it) becomes a huge challenge. Intellectual dishonesty exists on a widening scale.

It is not unusual today to see engineers trying to imitate the conduct and behavior of lawyers. The (understandable) aim and job of lawyers is to explain a client's limited responsibility or to make a compelling argument in favor of nonculpability of their clients. In sharp contrast, it should be a reliability engineer's aim and job to clearly define and outline safe and sustainable asset management. Substantive asset management is a detail task that, regrettably, either is *un*appreciated or remains *un*rewarded. It should be no surprise that such detail tasks are, therefore, widely shunned. In *some* companies, an incompetent manager is far better paid than a highly principled and well-rounded engineer.

PROBLEM AND SOLUTIONS

Budgets should not be defined by the lowest possible monetary outlay. The cost estimating manuals at EPC (engineering/procurement/construction) firms often only show least-cost equipment. If these manuals showed operating/maintenance/catastrophic failure-optimized equipment, the budget would need a multiplier of, say, 1.17. Offering to build plants at 1.17 times someone else's offer, the EPC would lose out to the competition. Why? Because EPCs are selected on the basis of bid price or some other yardstick that has little to do with how reliable your plant will be 5 years after it starts producing.

Problems and solutions are still our subtopic. Industry-wide, much lip service is paid to asset reliability. However, asset reliability and lowest initial cost of assets are almost always opposites. They can only be reconciled/optimized by experienced and well-informed professionals. Assuming these experts are still around, they must be given early access to management. Few, if any, are granted that access. Access is one of the solutions.

Because true experts are no longer groomed and nurtured, or because they are brought in far too late in a project definition and execution sequence, we are now stuck in an endless cycle of reinventing "new" initiatives. Grooming and nurturing is one of the solutions.

So, we've made some observations—observations that can be broken down into hundreds of pages of detail. Indeed, an entire branch of publications, trade journals, and conferences has sprung up to engage in debates on the matter reliable solutions. The debates rarely lead anywhere because, as we just heard, actionable implementation strategies require lots and lots of detail. Details will need to be learned, conveyed, and supervised. You don't get what you expect; you get what you inspect.

AN INADEQUATE REWARD SYSTEM

Let's talk about the reward systems we see in place. A supervisor with precise detailed experience will be dissuaded in his pursuits by an inappropriate and often *unjust* risk-and-reward system. Someone will send him the signal: "We don't need you. We don't have enough failures here to justify paying you more than we pay so-and-so." This supervisor quickly learns that only the quick fixers are rewarded. Those who *prevent things* from breaking are often viewed as laggards or sluggish performers.

These are all harsh and unpopular judgments. They cause great annoyance because they allude to the need to drastically change course. I've made you uncomfortable. Verbalizing an unpopular judgment is like telling a mother her baby is ugly—usually not a well-received message, regardless of "facts in evidence." But you will gain nothing by telling a mother your opinion. In contrast, a plant asset can be valued on the basis of facts, carefully leaving off unsupported opinions. You, your employer, your clients, and society as a whole may gain immensely if they're consistently told the truth about your company. Oh yes, changes would be required—and making changes would have to start with conceding the futility of continuing on a course that years ago was already recognized (by unbiased observers) as leading to complications down the road.

Nevertheless, as imperfect humans, we are biased in many ways. Forty years ago, I was biased in favor of keeping my good job. Today, I'm biased in favor of quickly coming to the point because there's so little time left and I don't want to waste anyone's time. I've also come to realize that one can only teach those who want to be taught. They would have to be good listeners, and good listeners are now clearly in the minority.

So, we need remedies for a serious problem: No sweeping initiatives are needed, but the prevailing mind-sets must change. Accountabilities must be defined and adhered to and many of the present reward systems must change.

SOLUTIONS

But as we wind up this chapter of our text, the reader should not be left with nonactionable generalities. We need substantive, experience-based recommendations; we need solutions. These include six steps that have been successfully implemented by leading companies and that you can copy or duplicate and implement without hesitation:

1. Shared educational responsibilities; grooming of successors. Pick people with potential and treat them well. They will no longer be uninformed.
2. Leading by example: be intellectually curious and be very resourceful. The day still only has 24 hours—delegate! Again, the employees will become informed!
3. The end of reliability and maintenance being subservient to operations (we strongly advocate "switching hats"). Note how both sides will be fully informed if a plant manager makes it known that at an unspecified date in the near future, the head of the operations department will become the head of the maintenance department and vice versa! Both sides will start to cooperate with each other. There will be communication, cooperation, and consideration—the three Cs. There will be no more finger pointing and blaming the other side.
4. Specifications developed for reliability, for life cycle cost. Entire budgets must be governed by reliability thinking. It follows that one must…

5. Disallow operation in the safety margin of machines. Finally, one can achieve something very valuable. One can …
6. Nurture absolute accountability by assigning project managers to live with their decisions. Assign a project manager to be involved in the plant start-up. Then make him the plant manager and let him learn from his mistakes before allowing him to be involved in another project. The results will be truly astonishing!

Please recall that these *challenges facing managers* have been taken into consideration and have been overcome by best-of-class companies. Many such companies have, decades ago, implemented what you have just read. We know best-of-class companies that are structured exactly along these lines. They consider these the key ingredients of their lasting success.

Chapter 19.5

Organizing for Continuous Improvement

INTRODUCTION

A colleague of ours and coauthor of a series of books on machinery reliability improvement took a look at North American refineries. He wanted to see how they are typically organized to achieve high reliability and found that most have structured their mechanical technical support (MTS) groups or departments along discipline lines; that is, they have designated machinery, static equipment, I&E (instrument and electrical), and other functional reliability specialists for operational and plant support. The term "support" is very important here, as will be seen.

TECHNICAL *SERVICE* VS. TECHNICAL *SUPPORT*

The majority of North American refineries have abandoned the old technical *service* concept. While technical *service* implies *reacting* to a call for assistance, technical *support* enhances operations when economic and risk reduction considerations favor such *proactive* support. Thus, the service concept with its reactive maintenance connotations and broad (but shallow-depth) expertise and responsibilities is no longer advocated. Only when workload, cost, and reliability levels demonstrated by key performance indicators (KPIs) are sufficiently matured as compared with those of best-of-class (BoC) companies may a blending of discipline activities occur. To put it in different words, there is compelling evidence that many refineries and process plants benefit from the specialization of their engineers. Likewise, many others could yet benefit from organizing for continuous improvement by giving weight to *proactive* expert input.

Common insight has been that in less mature plant organizations, there will be a "firefighting" function or a frontline involvement of mechanical specialists. However, to achieve real gains and continuous improvement, there has to be a segregation of this firefighting function from another important activity. This "other activity" is a reliability engineer's role; it is defined by its objective of addressing and eliminating chronic, repetitive, or "bad actor" problems. It is essentially a separation of day-to-day activities from long-term improvement and upgrading endeavors.

Accordingly, the best refineries in North America have now typically at least one machinery ("rotating equipment") reliability specialist (engineer or technologist) in maintenance backed up by either a senior machinery reliability specialist in an on-site (e.g., new projects) section or a group of specialists in their corporate headquarters (HQ) organization.

NEW PROJECT COVERAGE EXPLAINED

It is vitally important that new projects of any size are adequately supported and accompanied by knowledgeable and experienced specialist engineers to assure reliability problems are not being built into the new plant as the project completes its various cycles. The total life cycle cost (total LCC; top curve in Fig. 19.5.1) is the sum of capital/acquisition costs and maintenance/operation costs. Preplanned and preallocated capital outlays should include the cost of machinery quality assessment (MQA)—typically 5% of the cost of machinery for an oil refinery or petrochemical plant. Experience shows that with intelligently executed MQA, the future maintenance and operating costs are lower than the projected expenditures without MQA. The resulting total LCC will be lowest with diligently executed MQA.

The well-trained and highly experienced reliability professional must have input here. Needless to say, this professional must be trained, groomed, and nurtured to become an effective contributor for the ensuing specification, evaluation, and cost justification tasks. All of these tasks are part of MQA.

We have always taken the position that organizational alignment is unimportant as long as there are two key ingredients that are indispensable for ultimate success:

1. The reliability person must be a voracious reader, resourceful, and "a networker." He/she will have to collect books, proceedings, papers, articles, etc. He/she has to believe in (and single-mindedly practice) root cause failure analysis (RCFA). He/she must loathe repeat failures with a real passion (because such incidents are manifestations of his/her failure to uncover the root cause).

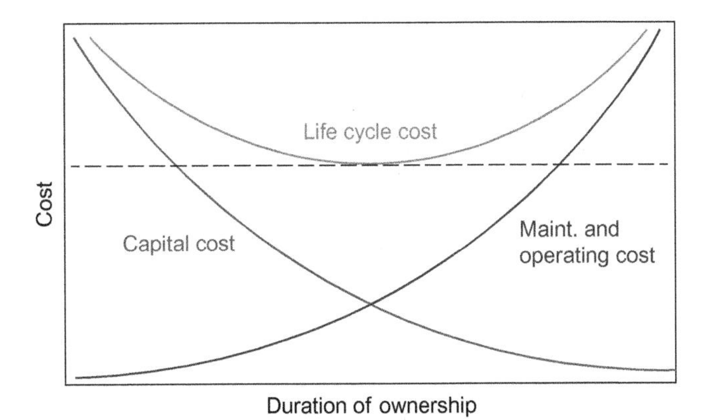

FIG. 19.5.1 Simplified life cycle cost. The cost of MQA is included in capital cost.

2. He/she cannot be told by a maintenance department that all that's important is "firefighting."

It is up to corporate and/or plant management to figure out what it takes organizationally, or motivationally, or financially, or from the point of view of decade-long training and nurturing (career plan!) to comply with these two fundamental and nonnegotiable requirements.

In essence, then, the reliability engineers' role statement should include the following twelve items or duties [1]. There could be more; there could be fewer. The main point is that these are deliverables, expectations, or assigned roles—they must be spelled out clearly.

1. Develop and coordinate reliability maintenance programs. Evaluate reliability maintenance (RM) effectiveness.
2. Develop a system of economic evaluation of the RM effort by publishing and stewarding to KPIs.
3. Review new equipment specifications and drawings with project engineers for reliability and maintainability.
4. Make recommendations for improvements before equipment is ordered.
5. Perform or guide equipment condition and performance monitoring.
6. Perform or guide on-stream analysis of major equipment problems. Solve these problems in a timely manner by whatever means required.
7. Prepare inspection, maintenance, overhaul, and repair (IMO&R) checklists for major machinery trains.
8. Develop scheduled IMO&R (inspection/maintenance/operation and repair) requirements for major machinery trains and accompany/monitor them.
9. Develop (scientific) recommendations for stocking spare parts.
10. Develop "bad actor management" policies.
11. Provide engineered designs for making repairs, replacements, and design changes and upgrading on maintenance projects.
12. Consult with other departments, other refineries, professional associations, and corporate HQ engineering (if available) concerning equipment problems.

REFERENCE

[1] H.P. Bloch, F.K. Geitner, Maximizing Machinery Uptime, Gulf Professional Publishing, Houston, TX, 2006.

Chapter 19.6

Managing Our Time

INTRODUCTION

We may think that we are efficient reliability implementers, but most of us could save time, accomplish more in the long run, and reduce stress if we would have skills in managing time. Effort is required to be productive and an extraordinary measure of resourcefulness is needed. We must balance personal lives, offsetting work with a measure of rest and relaxation. In all of our working endeavors, we should strive to become contributors instead of mere consumers.

GOAL SETTING STARTS EARLY IN THE DAY

On their way to work, serious contributors often set goals for the day ahead. Goals are important; our sense of accomplishment will reward us if we meet daily goals. On our way home from work, we might ask ourselves if we have met the day's goal and if we have contributed by adding value.

Value-adders make, set and readjust their priorities every day. Many work on tasks that require concentration while they're alert, perhaps by midmorning. They organize their tools to be close at hand. A hammer is a tool, but so is a micrometer, a checklist, or a good technical text.

PLANNING ONE'S DAY

If phone calls are necessary, it's perhaps best to make these when we are most likely to reach the person on the other end of the connection. Better yet, we may opt to communicate by e-mail and keep our message as brief as possible. We should be sure to give e-mails and written summaries the distribution they deserve. It's sound advice to do no more and to do no less. Anything beyond "just right" contributes to clutter, confusion, or both.

If we are in charge of meetings, we should strive to keep them brief and stay on schedule. We desire to start on time and finish on time. Sticking to an agenda is important. If we're attending a meeting, let's never miss out on the opportunity to say nothing. Saying nothing may actually add value by not repeating what has already been said or what can be said in a written summary without taking time away from others.

We might also consider delegating some of our work if we're in a supervisory or managing position or if training others is important. Delegating makes it possible for us to accomplish more; it conveys to others that they are being valued. Ask for "buy-in" from the ones to whom you wish to delegate the work and firmly commit to a mutually acceptable delivery schedule.

BEWARE OF PAPER SHUFFLING

Productive professionals try to handle each piece of paper only once. They resist the temptation to move it from one temporary pile to another. Don't be like the freelance writer who recently spent 8 weeks on massaging a four-page article. The freelancer's staccato pace of work adversely affected the efficiency of many others.

So, assume that today is just one more day of seeing before us a seemingly endless number of tasks. The thought of all these duties may cause us to dread the day. Where should we begin? Years ago, many of us learned to keep a written list of things to be done. As new tasks materialized or were assigned, we added them to the list. It always felt good to cross off completed items or tasks. Likewise today, many start by placing a "to do list" on their computer screen; it can certainly make it easier to stay on schedule.

SET REALISTIC PRIORITIES

Priorities are often an issue. Let the boss assist you in setting and resetting priorities. Break down large tasks and list them as smaller ones that don't intimidate or overwhelm you. Be realistic in timing task duration. Do not accept every social invitation coming your way and resist being distracted by the co-worker who always corners you at the water fountain with lengthy recollections of—whatever...

Of course, you can set priorities by numbering each item on your list according to importance. But it's difficult to clearly distinguish between "urgent" and "important." There may even be some slack time—or so you think. What some consider slack time at work, true reliability professionals consider precious. They fill this time (actually, it's the employer's time) with broadening their knowledge base. They work on tool making, or repair tasks, or specification updates. Perhaps they develop a draft version of a technical article that will add value to others or scan some technical articles into their computers.

Generally, and to the extent possible, we handle each activity in its priority order. We set deadlines and don't procrastinate. We stay in control and keep our boss informed if schedule changes cannot be avoided. Naturally, there will be times when we may choose to make an exception and not handle a matter in strict priority order, according to given circumstances and preferences. We want to stay flexible in this regard. Our objective is to stay in control so that what we actually accomplish each day is by choice, rather than by chance.

Seasoned pros do not rush from job to job or worry about doing everything that they have listed. Time management consultant Alan Lakein stressed that one rarely reaches the bottom of a to do list. Alan Lakein very famously said "Time=Life; therefore, waste your time and waste your life, or master your time and master your life." He reminded his students that it's not completing the list that counts, but making the best use of one's time. Along that line of reasoning, we will have accomplished this if the bulk of our time was directed toward what is truly important. As for the unfinished items, see if they can be delegated to others or transferred to tomorrow's list. A hard look at lower-priority items sometimes reveals that they do not need to be done at all. On the other hand, an item at the bottom of today's list may have a higher priority tomorrow.

But how do we go about determining which activities on our list are of high priority? After all, when looking at a long list of duties, many things may appear to be equally important. We must distinguish between "urgent" and "important." Ask if finishing the job produces significant benefits. If not, it may not be a high-priority task.

DISTINGUISH BETWEEN "URGENT" AND "IMPORTANT"

Some tasks yield better results than others, and so, when looking over a list of duties, consider the *results* each one will bring.

True, at first glance, everything on the list may seem urgent. Still, we should ask if urgent matters are always important, deserving a major time investment. Michael LeBoeuf, a professor of time management at the University of New Orleans, makes this observation: "Important things are seldom urgent and urgent things are seldom important. The urgency of fixing a flat tire when you are late for an appointment is much greater than remembering to pay your auto insurance premium, but its importance [the tire's] is, in most cases, relatively small." Then he laments: "Unfortunately, many of us spend our lives fighting fires under the tyranny of the urgent. The result is that we ignore the less urgent but more important things in life. It's a great effectiveness killer."

No doubt you will agree that it is more rewarding to work at something that yields important results than it is simply to be busy at whatever activity happens to be at hand. Try to direct as many of your efforts as possible toward activities that result in true accomplishment.

THE 80/20 RULE

A number of time management experts believe that, in many cases, we can narrow the top-priority items down to about 20% (see index item "Pareto's rule"). These experts cite, as a guide, the 80/20 rule. This principle was formulated by the 19th-century Italian economist Vilfredo Pareto; it states that only about 20% of the causes produce about 80% of the results. If we are alert, we may discover that there are a number of situations in everyday life where Pareto's principle applies. But how can the 80/20 rule be applied to your use of time? Analyze the items on your to do list. Perhaps you can be 80% effective by accomplishing two out of ten items listed. If so, those are the two most important items on your list. Also, analyze a project before diving in. How much of it is truly important to your objective? What part of the job will produce the most significant results? This portion of the task is priority.

Time management consultant Dru Scott, after discussing Pareto's principle, explains how to make it work for you. She says: "Identify the vital ingredients necessary to achieve your objective. Do these things first. You will get the most results in the least amount of time."

One refinery's statistics dating back to the 1970s established that 7% of its 3200 process pumps experienced a disproportionate share of pump outage events. Slightly over 60% of the money spent on pump maintenance went into these 7%. They quite obviously were problem machines that deserved much more attention than average equipment.

Applying the principles discussed thus far, what percentage of your day's activities would you expect to categorize as top priority? Of course, that will depend upon your specific responsibilities.

ENJOY THE BENEFITS

Perhaps at this point we can better appreciate that being the master of your time is not a matter of being preoccupied with never wasting a minute or rushing from crisis to crisis. Rather, effective time management means selecting the appropriate task for right now. It means discerning what activities yield the best results and spending our time on these whenever possible.

There are no fixed rules for personal organization of our time. To benefit from the suggestions in this chapter, be flexible. Experiment, but do so judiciously and with the buy-in of others in your organization. Adapt and if necessary discover what works best for you. By getting better control of your time, think what a sense of accomplishment we will have at the end of each day! Though more duties likely remain for tomorrow, we have the satisfaction of knowing that we directed our efforts toward the most important things. We may even get to realize there's finally enough time for the things that really matter. Then, we will not be the victims of hectic circumstances, but will be the master of our time.

Chapter 19.7

How Flawed Reward Systems Cost You Dearly

INTRODUCTION

Although mentioned elsewhere, deficient reward systems are a major flaw worth exploring in greater detail. In the almost three decades since 1990, many trade journals have observed that successfully implementing a "lean strategy" requires reliable equipment. Reliable equipment is specified and justified by SMEs—subject matter experts. Smart implementation strategies together with excellent equipment increase production capacity, decrease downtime/maintenance spending, and maximize impact and results. That sounds quite logical. Translation: "We want to maximize equipment uptime, have no environmental incidents and make lots of money while keeping everybody happy." At issue is always how to get there. The present reward system is deeply flawed and change is needed.

A NOTICEABLE GAP

There is a disturbing gap between what is being said and what is being done in the fields of equipment maintenance and reliability, and industry does not seem to be closing this gap. An honest appraisal of the facts has long ago convinced us that too many omissions and commissions and a lack of accountability on many levels have prevented managers from achieving tangible equipment reliability. Of course, not all managers are oblivious to the situation and many expend honest efforts to support sensible remedies. We, for our part, try to gauge the intensity and relevance of issues (and answers) from an exposure to all kinds of media, also technical conferences, reader feedback, course presentations, and plant visits. What we discern and feel obligated to report should be of interest. It affects the employer and employee alike. Reporting on why certain professed goals are difficult to reach is quite enlightening.

WHAT MAKES YOU A "RELIABILITY PROFESSIONAL"?

A reader with assignments in exploration and production (E&P) asked that question by noting that his management believes reliability is a function of any branch of engineering and that it is not a "stand-alone" or recognizable discipline. He went on to say that few universities offer "reliability" even as a graduate elective and that, at his corporation, the discipline engineers are generally focused on projects. Even production or operations engineers do a large amount of small project work at his facilities. He commented that these project-focused engineers usually have neither the time nor the experience to assist with failure analysis and troubleshooting. In other words, except where reliability engineer positions exist, reliability seems to be an unfocused topic at best (except tracking production losses, as a lagging indicator).

He wanted us to advise if full-fledged reliability engineering positions existed on the E&P side of major petrochemical and refining companies. He believed that his managers considered reliability engineering an adjunct function to be handled by discipline engineers strictly on an occasional basis within their specific areas of expertise. Also, he was interested in knowing if many of the "majors" recommended or required certification as a reliability professional.

START BY DISTINGUISHING BETWEEN MAINTENANCE AND RELIABILITY POSITIONS

Since maintenance spending was mentioned, let's start with the *definition of maintenance*. We believe that its role is simply to maintain equipment in the as-designed and as-built state. It stands to reason that a certain skill set is required to be a competent maintenance person; indeed, skills and motivation are always needed to add value to any enterprise.

In our view, the *definition of reliability engineering* is considerably more specific. The role of reliability engineering is to

(a) specify so as to design-out maintenance;
(b) define if upgrading is feasible at the bid or proposal stages and also on equipment already operating at the owner's facility;
(c) if "yes" on (b), define if this upgrading is cost-justified;
(d) if "yes" on (c), take full charge of upgrade implementation planning;
(e) help inculcate a mind-set that will not tolerate repeat failures, because repeat failures are indicative of not having found the root causes of a problem or, even worse, knowing the root cause and deciding to not do anything about it.

While it would be desirable for the reliability professional to be fully versed in the maintenance skill set, to be proficient in reliability engineering requires supplementary skills and abilities.

SUBJECT MATTER EXPERTS ARE PERFORMING A RELIABILITY FUNCTION

Within major "best-of-class" petrochemical companies, one will find subject matter experts in piping and vessels, machinery, instrumentation and controls, welding, metallurgy, etc. Sometimes, these subject matter experts reside at a central engineering or headquarters organization. In other instances, the machinery specialist resides at and gets paid by affiliate "A," but spends 40–50% of his time on issues that affect all affiliates. The piping and vessel specialist resides at and gets paid by affiliate "B," but spends time on and communicates issues that affect all affiliates.

At best-of-class companies, subject matter experts spend a large percentage of their time on failure analysis and prevention. From our role definition, failure prevention implies intelligent specification, accurate determination of weak links, and immersion in systematic upgrading. That said, subject matter experts are so valuable that they cannot be burdened with routine maintenance involvement. On the other hand, they may serve as occasional advisors on resolving special maintenance issues.

As long as one accepts the aforementioned and implements job functions and assignments along those lines, there is no need to use the formal title "reliability engineer." However, it would be implied that every one of these subject matter experts is performing exactly as we would expect from a reliability engineer. Recall our definition—one must accept it without equivocation.

We believe that our term subject matter experts and the reader's term discipline engineers describe individuals with identical educational backgrounds. There is, however, a huge difference; it exists because the reader's employee-associates devote most of their time to project work. It is our position that, to be worthy of ultimately contributing to a project, a subject matter expert must have considerable prior plant-related knowledge. Therefore, the importance of making this expert a resident at facility "A," "B," etc. cannot be overemphasized. A close substitute for continued in-plant involvement would be intense prior exposure to multiple plant start-up assignments.

CERTIFICATION AND MEMBERSHIP IN PROFESSIONAL ASSOCIATIONS

A few comments apply to how we might view the pursuit of certification as reliability professionals or simply membership in a professional society such as ASME (American Society of Mechanical Engineers). We see these and other endeavors as evidence that an individual is reaching out and is investing time in gaining a measure of professional recognition. While this "reaching out" would be quite commendable, it is rarely worthy of reward in and of itself. While they might *provide tests* aimed at determining if in-depth knowledge exists, we do not believe that most of these professional associations or entities are able to *impart* the in-depth knowledge that is sorely needed. Case in point, the home facilities of most members of certain professional organizations have repeat equipment failures and are making few inroads toward eradicating these repeat failures. Others have for decades been immersed in the computerized collection of elementary statistical data ("bearing replaced," "bearing failed," "bearing replaced," etc.) without ever establishing the root cause of a particular failure event or replacement need. What does that tell us?

In any case, the reliability job function can be covered by discipline engineers if management understands the issue and is willing to let certain experts share their expertise. We certainly believe that an individual should be rewarded on the basis of performance and contribution. An association member with certification but no motivation is worth much less than a self-starter who reads and implements and applies, although not being a member of a professional association. It's pointless to argue otherwise.

GROOMING AND MENTORING

We certainly are familiar with the grooming, mentoring, and professional development practices of a leading multinational petrochemical and refining corporation. Each year, this corporation reimbursed its employees for two professional membership association fees. The expectation was that this acted as a catalyst for professional growth, but such growth didn't always occur. It was also hoped that the professional would share his or her knowledge by communicating and networking with other potential beneficiaries. While it was not mandatory to be a member of anything, a noncommunicator had trouble rising to the top of his performance group.

When an engineer with a bachelor's, master's, or PhD degree was hired, his or her initial advancement was based on the salary curves for a given performance grouping plotted for people with those degrees. After 7 years of employment, all salary curves merged and salary treatment was based on performance rankings alone. We believe that's what should be done throughout industry.

A perceptive observer is aware of the pitfalls of benchmarking against work order backlog, work orders closed, and so forth. It rewards managers for accepting as "job done" certain maintenance endeavors that, sadly, set the organization up for repeat failures. Rewarding these managers on the basis of failure avoidance would force them to understand that multimillion dollar failures cannot be avoided by a cursory look at failed parts. Time and expertise are needed.

Smart organizations don't just maintain; they upgrade and avoid repetition of errors. Even more important, they groom and mentor professionals. Finally, these organizations reward managers on the basis of leadership in eradicating repeat failures instead of pushing to meet work order backlog benchmarks and other numbers that, with all due respect, are of very little value.

Subject Category 20

Materials Technology

Chapter 20.1

High-Performance Polymeric Wear Materials

INTRODUCTION

Fluid processing industries have embraced the use of current-generation composite materials in centrifugal pumps to increase efficiency, improve mean time between repair (MTBR), and reduce repair costs. One such material that has been used successfully by major refineries is Vespel CR-6100, which is a perfluoroalkoxy (PFA) carbon fiber composite. CR-6100 has replaced traditional metal and previous-generation composite materials in pump wear rings, throat bushings, line shaft bearings, interstage bushings, and pressure-reducing bushings. The properties of CR-6100 reduce the risk of pump seizures and allow internal rotating-to-stationary part clearances to be reduced by 50% or more. Composite wear materials are included in the latest editions of API 610, the widely applied centrifugal pump standard from the American Petroleum Institute.

AVOIDING PUMP SEIZURES

CR-6100 has proven to greatly reduce pump seizures, provide significant dry running capability, and mitigate damage from wear ring contact. Users do not experience pump seizures during temporary periods of suction loss, off-design operation, slow rolling, or start-up conditions. When the upset condition has been corrected, the pump continues operation with no damage or loss of performance. Conversely, when metal wear components contact during operation, they generate heat, the materials gall (friction weld), and the pump seizes. This creates high-energy dangerous failure modes, which can result in extensive equipment damage and potential release of process fluid to the atmosphere. Under most conditions encountered in process pumps, CR-6100

Petrochemical Machinery Insights. http://dx.doi.org/10.1016/B978-0-12-809272-9.00020-7

wear components do not gall or seize. Generally speaking, this eliminates damage to expensive parts, reducing repair costs and mitigating safety and environmental incidents.

Reducing wear ring clearance by 50% increases pump performance and reliability through increased efficiency, reduced vibration, and reduced NPSHR. The efficiency gain for a typical process pump is 4–5% when clearance is reduced by 50% [1]. Minimized wear ring clearance also increases the hydraulic damping of the rotor, reducing vibration and shaft deflection during off-design operation. It can be reasoned that lower vibration and reduced shaft deflection increase seal and bearing life and help users achieve reliable emissions compliance. This reduction in clearance also reduces the NPSHR on the order of 2–3 ft, which can eliminate cavitation in marginal installations [2].

Users have had excellent success installing CR-6100 to achieve all of the earlier benefits. One refinery installed CR-6100 wear rings and line shaft bearings to eliminate frequent seizures in 180°F condensate return service. The condensate return pumps have subsequently been in service for 6 years without failure. Another user improved the efficiency and reliability of two gasoline shipping pumps by installing CR-6100 wear rings, interstage bushings, and throat bushings. In one installation, the gasoline shipping pumps have been in service for years without failure or loss of performance. Hundreds of other applications have benefited from composite wear components. These include light hydrocarbons, boiler feedwater, ammonia, sour water, and sulfuric acid.

As with many other components, proper installation techniques will have to be used. The press fit for PFAs in services with high-pressure differentials across the component has to be just right. Excessive press fits may cause clearances between bushing bore and rotating part to become too small, in which case frictional heat can elevate to the point of destruction.

Close cooperation with knowledgeable vendors will benefit all parties. While it is quite reasonable to anticipate competitive products showing up in the marketplace as time progresses, factual investigations are needed. Do not act on opinions; act on facts. There will always be *someone* who can make things a bit cheaper and sell things for less. Once the pump parts become airborne, that "someone" is nowhere to be found.

REFERENCES

[1] H.P. Bloch, F.K. Geitner, Major Process Equipment Maintenance and Repair, Gulf Publishing, Houston, TX, 1985. p. 32.

[2] V.S. Lobanoff, R.R. Ross, Centrifugal Pumps, Design and Application, second ed., Gulf Publishing, Houston, TX, 1992. p. 96.

Chapter 20.2

FRMP and High-Pressure Fluoropolymer Throttle Bushings

INTRODUCTION

Since the late 1990s, multistage process pumps have often been upgraded with a high-performance plastic produced by the DuPont spin-off company Chemours™. The material is now widely known as Vespel CR-6100. This fiber-reinforced fluoropolymer material (FRFM) provides excellent chemical resistance in refineries or chemical processing. It offers very unique and desirable high-creep resistance for seals, enhanced run dry capability for wear rings for pumps, and easy machinability for tight-tolerance parts like ball valve seats. Additional FRFM components for fluid machines have been designed and manufactured over the years. In general, the user industry can benefit from adopting these components early in the specification and design cycle for modern fluid machines.

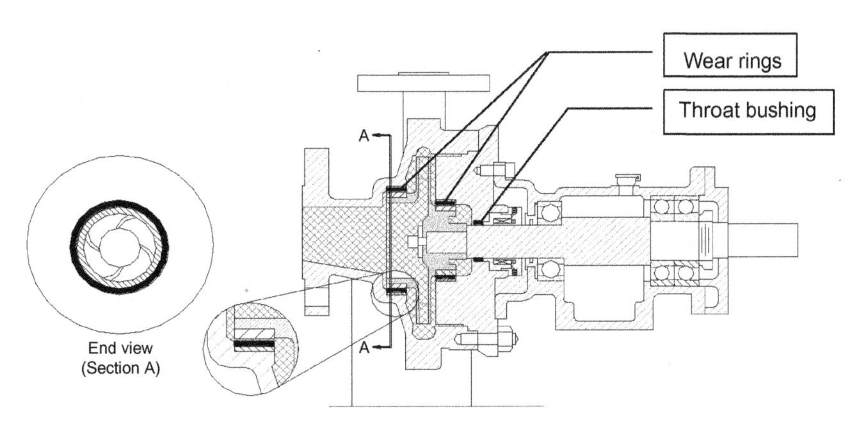

FIG. 20.2.1 Application points (locations) for FRFM parts in process pumps.

CENTER STAGE AND THROTTLE BUSHINGS

Pump users have taken note and virtually every reliability-focused user is now well acquainted with fiber-reinforced fluoropolymers. Typical application points in process pumps are shown in Fig. 20.2.1. More recently, however, the leading provider of this unusually wear-resistant and dimensionally stable material has been investigating the differential pressure limitation of Vespel CR-6100 in center-stage and throttle bushing applications due to questions from users and pump OEMs.

In response to these questions, Pennsylvania-based Boulden Company developed the PERF-Seal design (see Figs. 20.2.2 and 20.2.3) for high differential pressure components like center bushings and throttle bushings. This seal increases the differential pressure limit of Vespel CR-6100 bushings and improves bushing performance. With a solid bushing, high differential pressure can potentially enter the press fit interface, leading to radial distortion of the bushing. Boulden's design utilizes an engineered pattern of radial holes to equalize the pressure between the OD (outside diameter) and ID (inside diameter) of the bushing. Finite element analysis shows that this design nearly eliminates radial distortion of the bushing and dramatically increases the component's differential pressure capability.

In fact, there can be additional benefits of improved throttle bushings: the hole pattern in an engineered PERF-Seal promotes hydraulic effects that are beneficial for centrifugal pumps. The hole pattern discourages tangential whirl, which increases the ratio of beneficial damping to destabilizing cross-coupling forces. This can reduce any tendency toward damaging rotor instability as the pump running clearances wear. (Maintaining as-designed running clearances tends to reduce rotor vibration when operating near critical speeds.)

FIG. 20.2.2 PERF-Seal for use as the balance bushing of a BB5 multistage centrifugal pump. Full view is shown here. *(Courtesy Boulden Company, Conshohocken, PA.)*

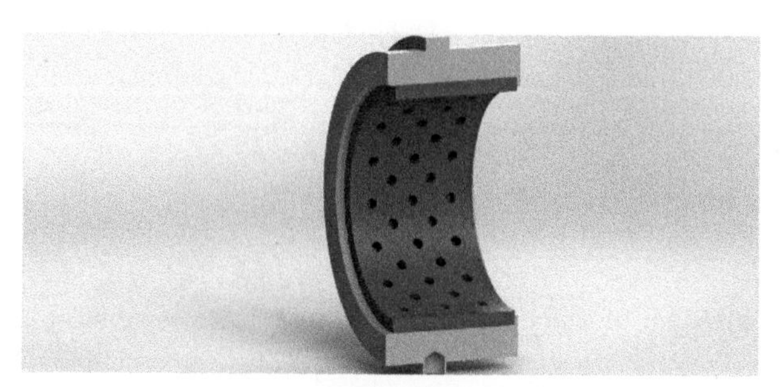

FIG. 20.2.3 PERF-Seal for use as the balance bushing of a BB5 multistage centrifugal pump. A cutaway view is shown here. *(Courtesy Boulden Company, Conshohocken, PA.)*

The holes in the ID surface are also likely to restrict axial flow across the bushing, contributing an incremental efficiency gain beyond the increase in efficiency already experienced from using Vespel CR-6100 wear rings with reduced clearance. Lomakin effect shaft support stiffness is retained, which helps to keep critical speeds out of the pump running speed range. And the design can be used on long or short bushings, throttle bushings, center bushings, or wear rings.

If you are planning to overhaul a multistage pump in the near future, you would do well to work with companies that systematically upgrade the reliability of your pumps. Among your upgrade opportunities, consider utilizing center bushings with an engineered pattern of radial holes to equalize the pressure between the OD and ID of the bushing. In some applications, the upgrade will pay for itself in less than 4 months.

Chapter 20.3

Consider Casting Salvaging Methods

INTRODUCTION

Repair welding of cast components of process machinery is frequently not possible. Fortunately, however, there are salvaging methods that do not involve welding:

1. controlled-atmosphere furnace brazing,
2. application of molecular metals, and
3. metal stitching of large castings.

Braze repair of cavitation-damaged pump impellers is an adaptation of a braze repair method originally developed for jet engine components.

The first step requires rebuilding the eroded areas of the impeller blades with an iron base alloy powder. The powder is mixed with an air-hardening plastic binder and used to fill the damaged areas. Through-holes are backed up with a temporary support and packed full of the powder/binder mixture. After hardening, the repaired areas are smoothed with a file to restore the original blade contour.

A nickel base brazing filler metal in paste form is then applied to the surface of the repaired areas, and the impeller is heated in a controlled-atmosphere furnace. In the furnace, the plastic binder vaporizes and the brazing filler metal melts, infiltrating the alloy powder. This bonds the powder particles to each other and to the cast iron of the blade, forming a strong, permanent repair.

After the initial heating, the impeller is removed from the furnace and cooled. All nonmachined surfaces are then spray coated with a cavitation-resistant nickel base alloy, and the impeller is returned to the furnace for another fusion cycle. After the treatment, the impeller will last up to twice as long as bare cast iron when subjected to cavitation.

Because the heating is done in a controlled-atmosphere furnace, there is no localized heat buildup to cause distortion and no oxidation of exposed surfaces. Unless the machined surfaces are scored or otherwise damaged, repaired impellers can be returned to service without further processing. An average impeller can be repaired for less than a third of the normal replacement cost.

Molecular metals have been applied successfully to the rebuilding and resurfacing of a variety of process machinery components. Molecular metals consist of a two-compound fluidized metal system that after mixing and application assumes the hardness of the workpiece. The two compounds are a metal base and a solidifier. After a prescribed cure time, the material can be machined, immersed in chemicals, and mechanically or thermally loaded.

Molecular metals have been used to repair pump impellers by recontouring or replacing lost material. These repair methods have also been used for centrifugal compressor diaphragms and engine and reciprocating compressor water jackets damaged by freeze-up.

Metal stitching is the appropriate method to repair cracks in castings. One reputable repair shop describes the technique:

1. The area or areas of a casting suspected of being cracked are cleaned with a commercial solvent. Crack severity is then determined by dye penetrant inspection. Frequently, persons unfamiliar with this procedure will fail to clearly delineate the complete crack system. Further, due to the heterogeneous microstructure of most castings, it is quite difficult to determine the paths the cracks have taken. This means that the tips of the cracks—where stress concentration is the highest—may often remain undiscovered. This also means that cracks stay undiscovered until the casting is returned to service, resulting in a potential catastrophe. It takes an experienced eye to make sure that the location of the tips is identified.
2. To complete the evaluation of the crack system, notice is taken of the variations in section thickness through which the crack or cracks have propagated. This step is critical because size, number, and strength of the locks and lacings are primarily determined by section thickness. Where curvatures and angularity exist, the critical importance of this step is further increased.
3. Metallurgical samples are taken to determine the chemical composition, physical properties, and actual grade of casting. This enables the repair shop to select the proper repair material. And this, along with the cross-sectional area of the failure, determines how much strength has actually been lost in the casting.
4. After these decisions have been taken, the actual repair work is started:
   a. Repair material is selected. This material will be compatible with the parent material, but greater in strength.
   b. The patterns for the locks are designed onto the casting surface.
   c. These patterns are then "honeycombed" using an air chisel. This provides a cavity in the parent metal that will accept the locks. Improper use of these tools produces a cavity that is not properly filled by the lock. The result is a joint that lacks strength and from which new cracks may emanate.
   d. Assuming the lock is properly fitted, a pinning procedure is now undertaken. This consists of mating the lock to the parent metal by drilling holes so that one half of the hole circles are in the parent metal and the remaining half are in the locks. High alloy, high-strength, slightly oversized mating pins are driven into these holes with an air gun. This produces a favorable residual stress pattern: in the immediate area of the lock, tensile stresses exist, which change to desired compressive stresses

as one moves out into the parent metal. This is to prevent future crack propagation. Additionally, these pins prevent relative movement between the locks and the parent metal.

e. The final repair step aside from dress-up is the insertion of high-strength metallic screws into previously drilled and tapped holes along the crack paths in between the locks. To clarify, it should be noted that the orientation of the locks is such that the longitudinal axis of the locks is perpendicular to the path of the crack. Thus, between locks, the lacing screws are used to "zipper up" the crack. Care must be exercised to make sure each lock is properly oriented.

Care must also be exercised so that, when the lacing screws are driven to their final positions, a harmonious blending with the parent metal is achieved. An amazing variety of machines have been successfully repaired using metal stitching techniques. Metal-stitched sections are often stronger than the original component or machine. An Internet search will be revealing.

Chapter 20.4

Straightening Carbon Steel Shafts

INTRODUCTION

The feasibility of straightening slightly bent shafts depends largely on the material from which a shaft is made and on the heat treatment it has received as part of the manufacturing process. As in the case of casting repairs, having an "overview knowledge" of available techniques is one of the many roles assigned to reliability professionals.

DO THIS FIRST

Before attempting to straighten a shaft, try to determine how the bend was produced. If the bend was produced by an inherent stress, relieved during the machining operation, during heat proofing, on the first application of heat during the initial start-up, or by vibration during shipment, then straightening should only be attempted as an emergency measure, with the chances of success doubtful.

The first thing to do, therefore, is to carefully indicate the shaft and "map" the bend or bends to determine exactly where they occur and their magnitude. In transmitting this information, care should be taken to identify the readings as "actual" or "indicator" values. With this information, plus a knowledge of the shaft material available, the method for straightening can be selected.

STRAIGHTENING CARBON STEEL SHAFTS

For medium carbon steel shafts (carbon 0.30–0.50%), three general methods of straightening the shaft are available. Shafts made of high alloy or stainless steel should not be straightened except on special instructions that can only be given for individual cases.

The peening method. This consists of peening the concave side of the bend at the bend. This method is generally most satisfactory where shafts of small diameters are concerned—say, shaft diameters of 4 in. or less. It is also the preferred, and in many cases the only, method of straightening shafts that are bent at the point where the shaft section is abruptly changed at fillets, ends of keyways, etc. By using a round end tool ground to about the same radius as the fillet and a 2½ lb machinist's hammer, shafts that are bent in fillets can be straightened with hardly any marking on the shaft. Peening results in cold working of the metal, elongating the fibers surrounding the spot peened, and setting up compression stresses that balance stresses in the opposite side of the shaft, thereby straightening the shaft. The peening method is the preferred method of straightening shafts bent by heavy shrink stresses that sometimes occur when shrinking turbine wheels on the shaft. Peening the shaft with a light (½lb) peening hammer near the wheel will often stress-relieve the shrink stresses causing the bend without setting up balance stresses.

The heating method. This consists of applying heat to the convex side of the bend. Applying heat is generally the most satisfactory strategy with large-diameter shafts, say, 4.5in. (~110mm) or more. It is also the preferred method of straightening shafts where the bend occurs in a constant-diameter portion of the shaft, say, between wheels. This is generally not applicable for shafts of small diameter or if the bend occurs at a region of rapidly changing shaft section. Because this method partially utilizes the compressive stresses set up by the weight of the rotor, its application is limited and care must be taken to properly support the shaft.

The shaft bend should be mapped and the shaft placed horizontally with the convex side of the bend placed on top. The shaft should be supported so that the convex side of the bend will have the maximum possible compression stress available from the weight of the rotor. For this reason, shafts having bends beyond the journals should be supported in lathe centers. Shafts with bends between the journals can usually be supported in the journals; however, if the end is close to the journal, it is preferable to support the shaft in centers so as to get the maximum possible compression stress at the convex side of the bend. In no event should the shaft be supported horizontally with the high spot on top and the support directly under the bend, since this will put tension stresses at the point to be heated, and heating will generally permanently increase the bend. Shafts can be straightened by not utilizing the compressive stress due to the weight of the rotor, but this method will be described later.

To straighten carbon steel shafts using the heating method, the shaft should be placed as just outlined and indicators placed on each side of the point to be heated. Heat should be quickly applied to a spot about 2–3in. (~50–75mm) in diameter using a welding tip of an oxyacetylene torch. Heat should be applied evenly and steadily. The indicators should be carefully watched until the bend in the shaft has about tripled its previous value. This may only require perhaps 3–30s, so it really is very important to observe the indicators. The shaft should then be evenly cooled and indicated. If the bend has been reduced, repeat the procedure until the shaft has been straightened. If, however, no progress has been made, increase the heat bend as determined by the indicators in steps of about 0.010–0.020in. (0.25–0.5mm) or until the heated spot approaches a cherry red. If, using heat, results are not obtained on the third or fourth try, a different method must be tried.

Heat applied to straighten shafts causes the fibers surrounding the heated spot to be placed in compression by the weight of the rotor. This compression is enhanced by expansion on the diagonally opposite side and the resistance of the other fibers in the shaft. As the metal is heated, its compressive strength decreases so that ultimately the metal in the heated spot is given a permanent compression set. This makes the fibers on this side shorter, and by tension, they counterbalance tension stresses on the opposite side of the shaft, thereby straightening it.

The heating and cooling method. This is especially applicable to large shafts, which cannot be supported so as to get appreciable compressive stresses at the point of the bend. This method consists of applying extreme cold—using dry ice—on the convex side of the bend and then quickly heating the concave side of the bend. This method is best used for straightening shaft ends beyond the journals or for large vertical shafts that are bent anywhere.

The action is that the shaft side having the long fibers is artificially contracted by the application of cold. Then, this sets up a tensile stress in the fibers on the opposite side, which, when heated, lose their strength and are elongated at the point heated. This now sets up compressive stresses in the concave side that balance the compressive stresses in the opposite side. Indicators should also be used for this method of shaft straightening—first bending the shaft in the opposite direction from the initial bend, about twice the amount of the initial bend—by using dry ice on the convex side—and then quickly applying heat with an oxyacetylene torch to a small spot on the concave side.

Shafts of turbines and turbine generator units have been successfully straightened by various methods. These include several 5000 kW turbine generator units, one 6000 kW unit, and many smaller units. Manufacturers of turbines and other equipment have long used these straightening procedures, which have also been used by the US Navy and others. Exercising sufficient care, a shaft may be straightened to a centerline deviation of 0.0005 in. or less. This would show up as a total indicator reading (TIR) of 0.001 in. (slightly over 0.02 mm) or less. With very few exceptions, this amount of runout is generally satisfactory.

Subject Category 21

Mechanical Seals

Chapter 21.1

Reconciling Requirements in API-682 Dual-Seal Design Configurations*

Mechanical seal designers face challenges when developing dual seals, and users must weigh available options with input from competent manufacturers. Although originally aimed at the hydrocarbon processing industry, the American Petroleum Institute's seal standard (API-682/ISO 21049) is now very widely used because it lists many universally applicable requirements. Among them is the intent to optimize seal face cooling, which must also address the need to provide a seal that tolerates pressure reversal and abrasive or congealing fluids. Poor seal face cooling can reduce equipment reliability. The seal standard, therefore, offers users a choice of three different configurations, and as can be expected, each version has advantages and disadvantages.

Although the stipulated cooling and other requirements can appear to be at odds with each other, innovative, yet well-proved, designs will satisfy the diverse needs. Instead of trial and error, computational flow dynamics (CFD) and extensive testing have led to uncompromising best available solutions. Field studies have fully validated the underlying design concepts.

DUAL-SEAL ARRANGEMENTS

Dual seals are increasingly important in the hydrocarbon processing and other industries. Plant hazard safety requirements, reductions in allowable fugitive emissions, and the quest for increased equipment uptime are the main drivers. API-682/third edition [1] collectively describes pressurized seal geometries as "Arrangement 3." This arrangement comprises two seals per cartridge assembly,

*Images in this chapter were courtesy of AESSEAL Inc., Rotherham, UK, and Rockford, TN.

Petrochemical Machinery Insights. http://dx.doi.org/10.1016/B978-0-12-809272-9.00021-9

and an externally supplied pressurized barrier fluid is used to provide a beneficial seal environment. The different configurations are described by API-682 as:

- Face-to-back dual seal in which one mating ring is mounted between the two flexible elements and one flexible element is mounted between the two mating rings or seats (Fig. 21.1.1A)
- Back-to-back dual seal in which both of the flexible elements are mounted between the mating rings (Fig. 21.1.1B)
- Face-to-face dual seal in which both of the mating seal rings are mounted between the flexible elements (Fig. 21.1.1C)

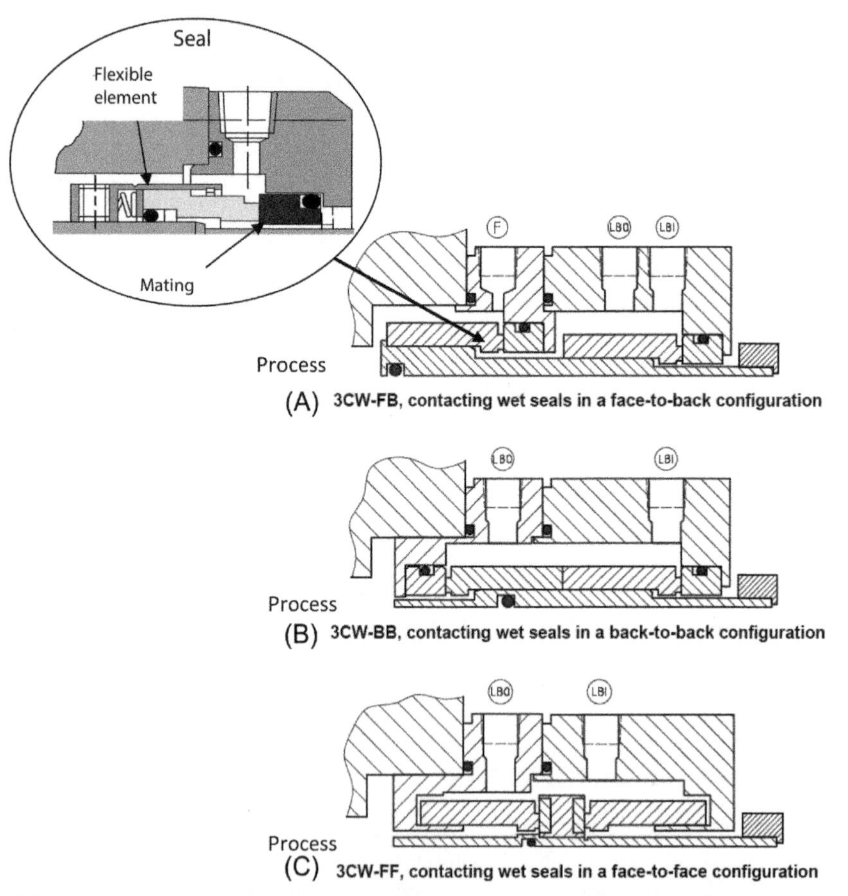

FIG. 21.1.1 Dual seals come in three basic orientations FB (A), BB (B), and FF (C).

The description is, in the opinion of the coauthors of this chapter, incomplete. The principal attribute of the face-to-back configuration (Fig. 21.1.1A) is that the process fluid (pumpage) is on the outer diameter of the seal face. In the other two configurations, the process liquid is on the inside diameter.

Face-to-back arrangements are preferred [2], although back-to-back and face-to-face orientations are offered as purchaser's options. In API-682, non-pressurized dual seals are called "Arrangement 2," and face-to-back is the only arrangement option available in the standard [3]. However, a more thorough comparison of different arrangements is of interest here.

BACK-TO-BACK VERSUS FACE-TO-FACE CONFIGURATIONS

In the hydrocarbon process industries, back-to-back and face-to-face configurations are widely represented [4]. They can potentially offer higher levels of performance—in large measure attributable to the cooling effect of barrier fluid flowing over both inner and outer seals. But there also are disadvantages.

The main shortcoming of back-to-back and face-to-face configurations is that the process fluid is on the inside diameter of the seal faces. Centrifugal force action tends to throw any entrained abrasive solids toward the seal faces, which increases the potential for damage. The "dead zone" formed by the small volume of process fluid underneath the inner seal creates susceptibility to trapped fluids congealing or solids accumulation. The secondary O-ring will tend to then move over this deposit-affected region of the sleeve (Fig. 21.1.2), and "hang-up" is likely to be encountered.

The application of reverse balance can be more difficult with back-to-back designs. Upset conditions such as loss of barrier fluid pressure or increases in process pressure can adversely affect such configurations. Positive retention of the inner seal mating ring can be difficult to accomplish since there will always be certain unavoidable dimensional constraints of associated hardware and seal chambers. Accordingly and unless properly retained, thrust forces during pressure reversal may cause a ring to become dislodged. Also, with some designs, reverse pressure loading will apply a hydraulic force to the inner seal spring plate that then tends to open the seal faces. This has prompted the chemical process industry, since about 1995, to move to face-to-back designs; overall seal reliability has improved as a result.

ADVANTAGES OF FACE-TO-BACK CONFIGURATIONS

Face-to-back configurations overcome virtually all the weaknesses of other designs with pumpage on the outside diameter of the seal faces. The preference for this configuration is noted by API-682 [5] and is summarized below:

> *'The advantages of the series configuration are that abrasive contamination is centrifuged and has less effect on the inner seal.' A further note supports the case [6]. 'Liquid barrier seal designs arranged such that the process fluid is on the OD of the seal faces will help to minimize solids accumulation on the faces and minimize hang-up.'*

Dual balance is easier to incorporate in this configuration, and the seal O-ring can be located so as to permit it to move to either side of the groove [7]. This then supports a closing force regardless of the direction of pressure. Mating rings can be simply retained either positively or, with more modern designs, hydraulically (Fig. 21.1.3). Pressure reversal capability provides for greater safety; it increases degree of tolerance for many process upset or loss of barrier fluid conditions. Again, API-682 reinforces this point in a note [5]:

> *In the event of a loss of barrier fluid pressure, the seal will behave like an Arrangement 2.*

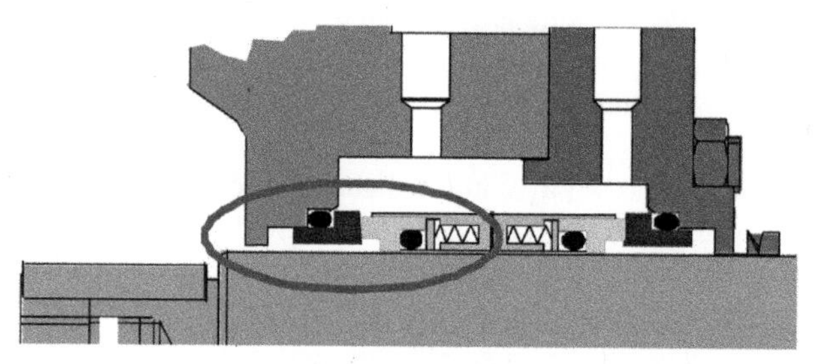

FIG. 21.1.2 Back-to-back process fluid on ID of face with potential for hang-up inner mating ring unretained.

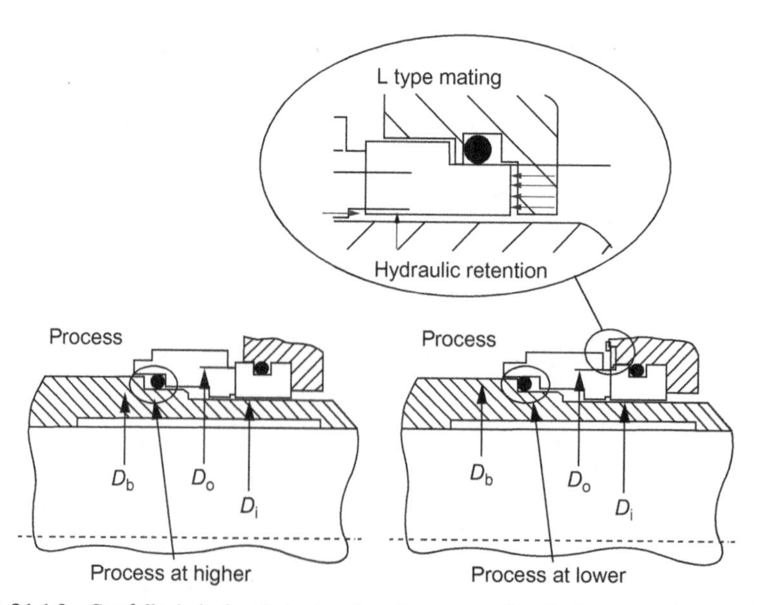

FIG. 21.1.3 Carefully designing O-ring location allows proper functioning even under conditions of pressure reversal.

DISADVANTAGES OF CONVENTIONAL FACE-TO-BACK CONFIGURATIONS

Of major concern in face-to-back designs is cooling of the inner seal. Seal designers have typically approached the issue by mounting component seals with adaptive hardware (sleeve and gland) to form a cartridge. However, the barrier fluid flow path to the inner seal is compromised by a region of low or zero flow. The temperature in these stagnant areas will be elevated by heat soak and face-generated heat (Fig. 21.1.4). The second edition of API-682 very eloquently provides a warning [8]:

> *Restricted seal chamber dimensions and the resulting cartridge hardware construction can affect the ability of the barrier fluid flush to adequately cool the inner seal. Inadequate cooling of the inner seal can result in reduced seal reliability. Selection of a back-to-back or face-to-face configuration may resolve an inner seal cooling problem.*

That would indicate we should look for a seal with optimized cooling that must, at the same time, provide resistance to "hang-up" and must tolerate pressure reversals [9].

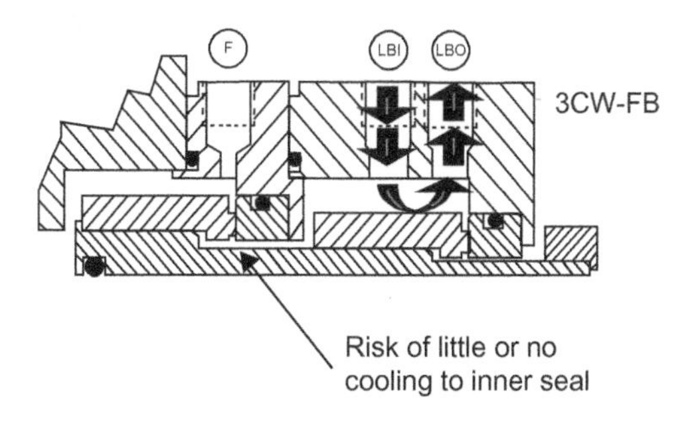

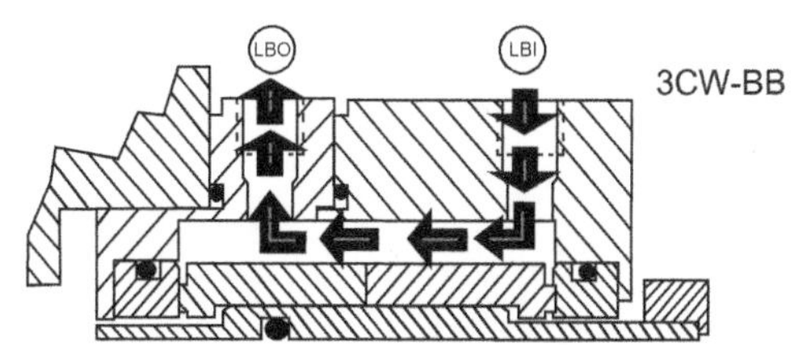

FIG. 21.1.4 Face-to-back (FB) technology (top) with ineffective inner seal cooling. Back-to-back arrangement (lower image) facilitates effective barrier fluid cooling.

Fortunately, superior cooling can now be achieved by including high-performance circulating devices and a flow deflector baffle in these seals. The deflector baffle diverts the barrier fluid flow to the inner seal; this provides cooling to the inner seal faces and represents an elegant solution to the dual-seal challenge [10].

DEFLECTOR BAFFLES AND SUPERIOR PUMPING RINGS

Deflector baffles have long been employed on single high-temperature seals connected to quench stream. Restrictions in seal chamber dimensions and the somewhat large cross section of conventional dual balanced pusher seals have generally inhibited the more widespread use of deflector baffles. Except for bellows seals with their traditionally smaller (Fig. 21.1.5) cross sections, incorporating deflectors has generally been limited to "engineered specials" rather than off-the-shelf designs.

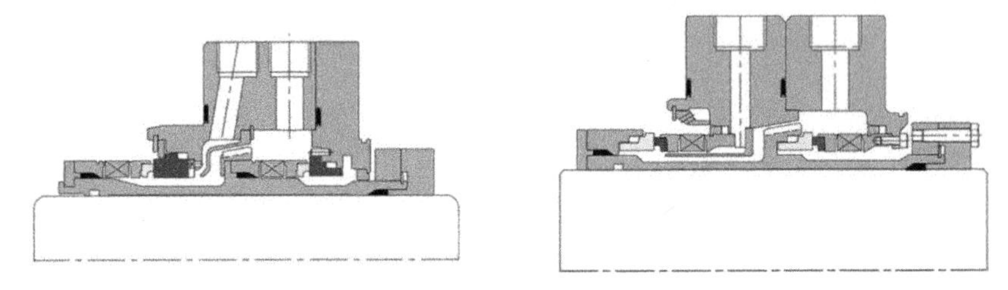

FIG. 21.1.5 Dual face-to-face seals. Rotating bellows are shown in the left image; the bellows in the right image are non-rotating.

Buffer fluid movement in Fig. 21.1.5 is substantially enhanced by the bidirectional tapered pumping ring shown in this dual seal. Fig. 21.1.6 depicts the pumping device [10].

FIG. 21.1.6 A highly innovative bidirectional tapered pumping device.

MORE DEVELOPMENTS IN PUSHER SEAL TECHNOLOGY

An alternative method toward achieving seal balance is to place O-rings on the outside of the seal faces. This is now common practice in the chemical process industries. Making the sleeve serve also as the face holder is made possible by modern CNC machining techniques. These design and manufacturing techniques facilitate a more compact design (Fig. 21.1.7); the techniques open up the inner seal envelope and provide an effective deflector baffle. Separation between barrier flow inlet and outlet concentrates the cooler barrier fluid at the inner seal faces; this can then become the key to greatly extended mechanical seal life.

FIG. 21.1.7 The bidirectional tapered pumping ring can be incorporated in many compact dual-seal types.

CFD FLOW PATH AND DEFLECTOR ANALYSIS

Barrier fluid circulation and flow regimes within a dual-seal cavity can be complex and will require the application of three-dimensional CFD to fully predict and understand.

As an example and confining ourselves to two-dimensional images, Fig. 21.1.8A illustrates the original velocity vector distribution (for the *X* and *Y* components) in the region adjacent to the deflector. It was expected that the flow would separate at the point where it makes a 180 degree turn in the deflector. However, with low flow velocities and the round profile, the flow remained

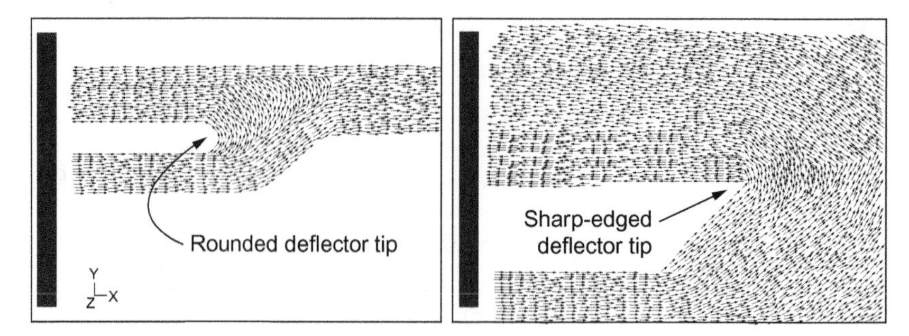

FIG. 21.1.8 Deflector end before (A) and after (B) flow optimization.

rather limited. Relatively small amounts traveled in the x-direction, that is, toward the seal faces.

Since it is, of course, desirable to maximize fluid flow in contact with the inboard seal faces, the end of the deflector was reprofiled to a triangular sharp edge. The follow-up analysis in Fig. 21.1.8B indicates there is now radial motion next to the extremity of the deflector. This recirculation reduces the flow path and prevents some fluid close to the deflector nose from escaping before even reaching the seal faces. Compared with the original round shape, the triangular sharp-edged shape promotes vortex motion and redirects additional coolant flow to the seal faces [10].

REFERENCES

[1] API Standard 682/3rd Edition, September 2004; also, ISO 21049: Shaft Sealing Systems for Centrifugal and Rotary Pumps, American Petroleum Institute, Washington, DC.

[2] API Standard 682/3rd Edition, Clause 7.3.4.2.1.

[3] API Standard 682/3rd Edition, Figs. 3 & 4.

[4] API Standard 682/3rd Edition, Clause 7.3.4.3.

[5] API Standard 682/3rd Edition, Clause 7.3.4.2.1 and NOTE therein.

[6] API Standard 682/3rd Edition, Clause 7.3.1.1 NOTE 1 therein.

[7] API Standard 682/3rd Edition, Clause 6.1.1.7.

[8] API 682/2nd Edition, July 2002; see note therein, regarding Clause 7.3.4.2.1.

[9] API Standard 682/3rd Edition, Annex G.

[10] C. Carmody, A. Roddis, J. Amaral Teixeira, D. Schurch, Integral pumping devices that improve mechanical seal longevity, Presented at 19th International Conference on Fluid Sealing, Poitiers, France, 25–26 September 2007.

Chapter 21.2

More About Pumping Ring Designs for Dual Mechanical Seals

INTRODUCTION

Dual mechanical seals are now widely accepted by pump users in many industries. In certain applications, they are the only choice suitable for safe and reliable containment. In other services, dual seals might be considered optional. Still, all major seal manufacturers can point out regulatory guidelines for using dual seals and assist in assessing their economics.

Safe and reliable fluid containment usually leads to pump maintenance cost avoidance in instances where the use of dual seals is mandatory. However, such benefits also accrue where using dual seals is discretionary. Figs. 21.2.1 and 21.2.2 are two of many different dual-seal styles available today.

As pump users then consider dual seals, they may often note differences in the proposed method of heat removal from both pairs of seals. It's in the nature of business that seal vendors are inclined to feature and favor their own design. Needless to say, not all vendors offer reliability and energy-optimized configurations. How the barrier fluid is pumped around deserves to be studied.

The clearance between the pumping rings or pumping screws and surrounding stationary parts is important. A close clearance makes the device relatively efficient because there is little backflow or slippage. Almost all the flow is from the device's inlet toward its discharge side. However, with close clearances, even a small out-of-concentric position of the rotor relative to the stationary parts is likely to result in contact. There can then be many potentially highly

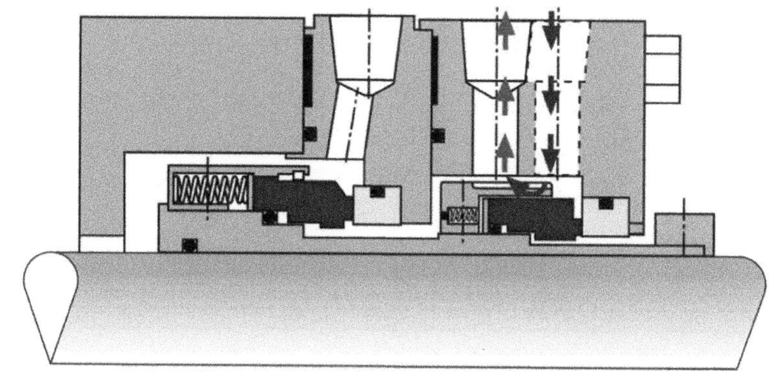

FIG. 21.2.1 Dual seal with straight-vane (cog-style) pumping rings are rather inefficient and not much cooling flow will reach the set of seal faces on the left.

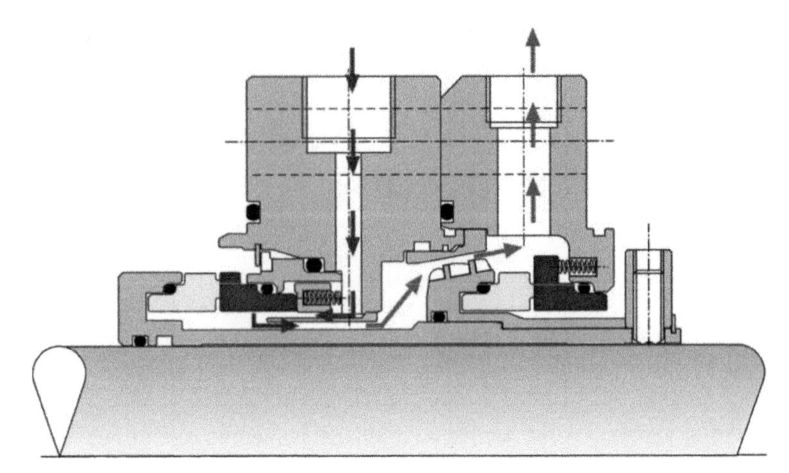

FIG. 21.2.2 Dual seal with tapered, bidirectional pumping ring. With a radial clearance gap of 0.060 in. (1.5 mm), this ring geometry is considered optimum. *(Courtesy AESSEAL Inc., Rotherham, UK and Rockford, TN.)*

undesirable consequences. The simple cog-style pumping ring in Fig. 21.2.1 allows visualization of inefficiency if the clearance is large and risk of making scraping contact if the clearance is small.

Although some pumping ring styles in dual mechanical seals are based on a straight-vane or paddle-type configuration, the head-flow (HQ) performance of other pumping ring styles will usually exceed that of straight-vane designs. Test results shown in Fig. 21.2.3 confirm that helical pumping screws and, in

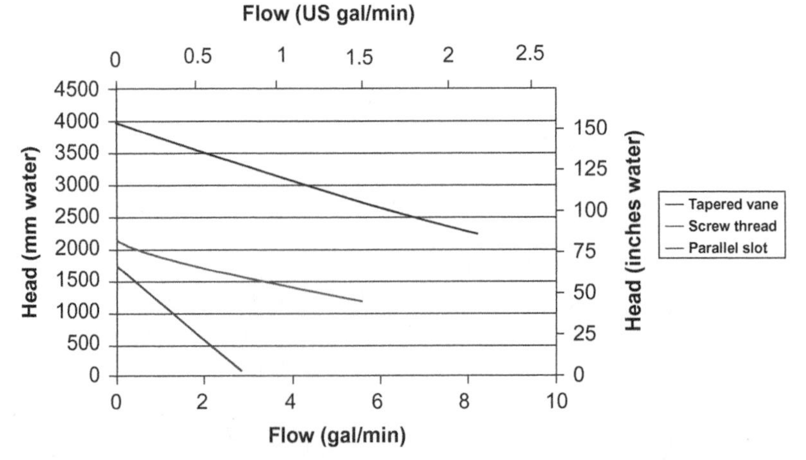

FIG. 21.2.3 Typical head-flow characteristics of a bidirectional "tapered-vane" pumping ring (*upper line*) compared to screw-thread (*middle line*) and parallel-slot (*lower line*) pumping devices used with older sealing technology. *(Courtesy AESSEAL Inc., Rotherham, UK and Rockford, TN.)*

particular, tapered bidirectional rings (Figs. 21.2.2 and 21.2.4) achieve better head-flow performance. Aside from that, straight-vane or paddle-type pumping rings will function only in the plane where the ports and the straight-vanes (paddles) are located. Therefore, tangential porting will be required, and even then, in many instances, little or no liquid is made to flow continually over the various seal faces.

While helical pumping screws are again more efficient than vane/paddle styles, high-efficiency helical screws rely on a dimensionally close clearance gap between screw periphery and housing bore.

Only the bidirectional tapered pumping device in Figs. 21.2.2 and 21.2.4 incorporates large clearances—low galling risk—and has excellent overall performance. The different head-flow characteristics in the comparison tests of Fig. 21.2.3 are of interest also. Bidirectional tapered-vane pumping rings incorporate the best of all approaches. They certainly deserve to be considered for reliability-focused installations.

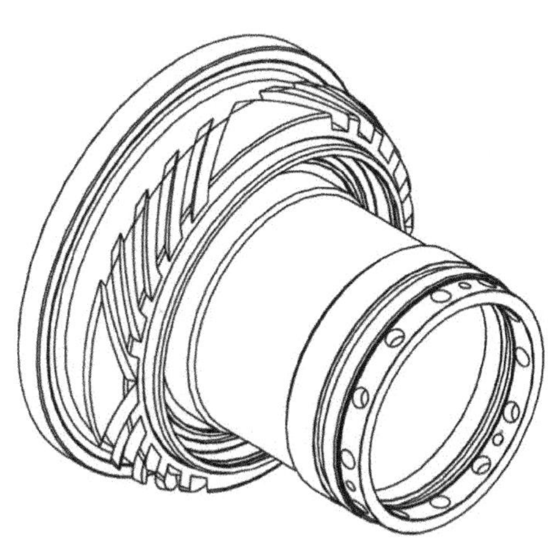

FIG. 21.2.4 Advanced-type bidirectional tapered pumping device used in a modern dual mechanical seal. *(Courtesy AESSEAL plc, Rotherham, UK and Rockford, TN.)*

Chapter 21.3

Unusual Flush Plans for Mechanical Seals

INTRODUCTION

Mechanical seals are used in millions of process pumps; the many available seal configurations are described in the standards of the American Petroleum Institute (API-682). These standards also describe the many flush plans (piping plans) used by modern industry. Unlike automotive, home appliance and similar applications where it is common for the pumpage to fully envelop the sealing components, a flush liquid stream and associated piping plans are used in process pumps to remove heat from the seal faces.

There are many manufacturers of mechanical seals, and their overall strategies appear similar: each desires to deliver safe products at reasonable cost. However, the business objectives of the very best mechanical seal manufacturers go beyond the obvious. Their objectives are expressed in marketing approaches that consistently represent value. Superior service and high customer satisfaction are among the discernibly beneficial aspects of good marketing. Additional benefits accrue if the seal's service and asset provider conveys educational or training updates to the ultimate seal user.

All of the objectives endorsed by client and provider can be summarized in "the three C's"—communication, cooperation, and consideration. The content of this chapter is rather typical of the "communications aspect" of a rewarding relationship between the parties. The manufacturer should make it their goal to alert user-clients to new opportunities. Such opportunities exist based on new flush plans found in the later editions of API-682; they are plans 03, 55, 65A, and B, also plans 66 and 99. Although these five flush plans and their derivatives are little known, they can be of great advantage in certain services.

Examining API Plan 03

The relatively recent API plan 03 (Fig. 21.3.1) is a great addition; it relates to a taper-bore seal chamber for an API-style pump. For decades, API pumps have been using closed (cylindrical) seal chambers and have relied on piping plans to maintain a chosen seal environment. However, because taper-bore stuffing boxes are now very well proved in American National Standards Institute (ANSI) pumps in contaminated services, we can now also specify tapered bores for API-compliant pumps.

In plan 03 (Fig. 21.3.1), the flush fluid flows into the pumpage. Circulation between the seal chamber and the pump is facilitated by the tapered geometry.

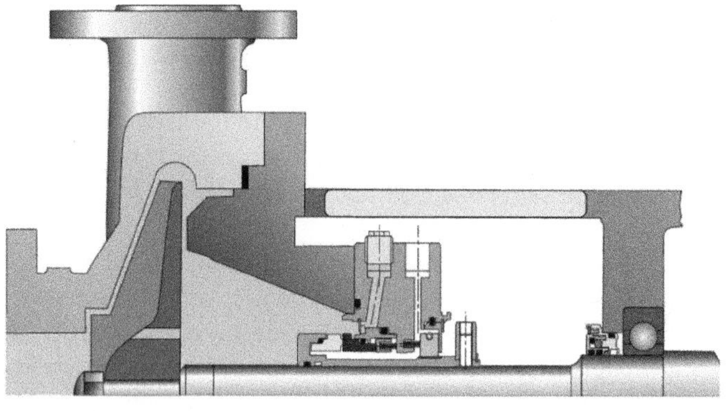

FIG. 21.3.1 API flush plan 03 recognizing tapered stuffing box environment (note also the bearing housing protector seal with large O-ring moving in axial direction). *(Courtesy AESSEAL, Inc., Rotherham, UK and Rockford, TN.)*

Solids accumulation risk is greatly reduced by the tapering, and the former stuffing box is now part of the back pullout cover of this pump. New pumps can accommodate the tapered design, as will preexisting pumps through a modification or upgrading process. It should be noted that the taper should be relatively steep; 30–45 degree inclination has worked well. Very shallow taper angles can be ineffective and should be avoided.

This seal chamber geometry promotes circulation that, in turn, provides cooling for the seal and vents air or vapors from the seal chamber. Flush plan 03 is most often used in applications where the seal faces generate relatively small amounts of heat. Plan 03 is also used in applications where the old-style cylindrical chamber would have allowed solids to collect. Occasionally, the tapered bore is fitted with antiswirl vanes (sometimes called "swirl-interrupting baffles") for even greater assurance against solids accumulation.

Note also the floating outboard throttle bushing in Fig. 21.3.1. This provision allows leakage monitoring, assuming the outlet port is located at the bottom (please note: as a drawing convenience, our illustrations may show certain porting arrangements above the centerline; on actual seals a particular port may be located near the bottom).

Standard Seal Flush Plan 55

In plan 55 (Fig. 21.3.2), there is an unpressurized external buffer fluid system supplying clean liquid to the buffer fluid seal chamber. Plan 55 is used with dual (double, tandem) liquid seal arrangements. The buffer liquid is typically maintained at a pressure less than seal chamber pressure and less than 0.28 MPa (2.8 bar/40 psi).

Plan 55 is similar to plan 54 except the buffer liquid is unpressurized. The plan 55 representation in Fig. 21.3.2 shows an efficient bidirectional tapered pumping ring. This particular ring greatly assists in moving the buffer fluid to and from an external reservoir and/or through an external heat exchanger (cooler). Also, the potential advantages of using a tapered pumping ring can be significant. One such model (Fig. 21.3.2) is offered with bidirectional functionality and a wide clearance between its vane tips and the opposing stationary parts. In the event of pump bearing distress, this wide clearance gap protects against scraping and extreme heat generation.

The outboard seal in Fig. 21.3.2 is a wet containment seal (API calls it configuration 2CW-CW—dual contact wet seal) and is normally used in services where process fluid leakage to atmosphere must be avoided, which is to say minimized and contained. Many users found plan 55 advantageous in applications where the process was prone to solidify in contact with atmosphere or in applications where additional heat removal from the inner seal was required.

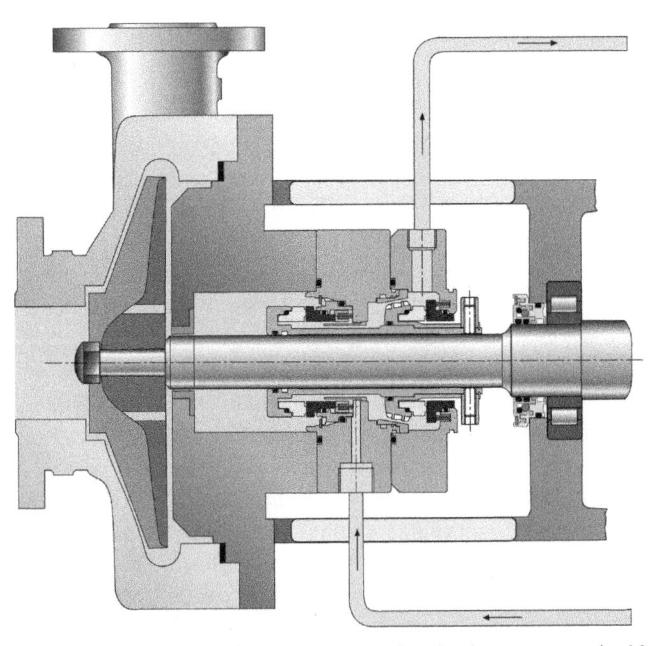

FIG. 21.3.2 Plan 55 and a bidirectional tapered pumping ring in an unpressurized buffer fluid loop. *(Courtesy AESSEAL, Inc., Rotherham, UK and Rockford, TN.)*

Examining major seal manufacturers' websites allows users to see how plan 55 differs from plan 52. In plan 52, the buffer liquid is not necessarily self-contained; with plan 52, buffer liquid circulation is created by an external pump or pressure system. If plan 55 is specified, carefully consider the reliability of the buffer liquid source and the possible contamination of the buffer flow with process liquid or vapor. However, suitable supervisory instrumentation may give ample warning of a compromised primary seal.

API Plans 65A and 65B Serve as Models for Leak Detection

In plan 65A/B, there is an atmospheric leakage collection and detection system for condensing leakage. Failure of the seal will be detected by an excessive rate of flow into the leakage collection system. Fig. 21.3.3 is intended to convey that many different *seal configurations* are allowed; the emphasis is largely on leakage *monitoring*. The central port is equipped with one of many feasible instruments. In any event, Fig. 21.3.3 depicts a standard setup when pumped fluid condenses at ambient temperatures.

Plan 65A/B differs only in that "A" is using a throttle bushing, whereas "B" uses an orifice arrangement in the leakage collection setup. Plan 65A/B is normally used with single seals in services where the anticipated seal leakage is mostly liquid, not gas. Piping is connected to the drain connection in the gland plate and directs any primary seal leakage to an exterior collecting volume or system.

The exterior collecting reservoir (the "volume") is not usually provided by the seal manufacturer; the "volume" could be an oily water sewer or some other environmentally acceptable liquid collection system in the plant. Within the seal, excessive flow rates would be restricted by the orifice located downstream of the reservoir and are redirected to it, causing the level transmitter to activate an alarm. The orifice is shown here with plan 65A (Fig. 21.3.4) that

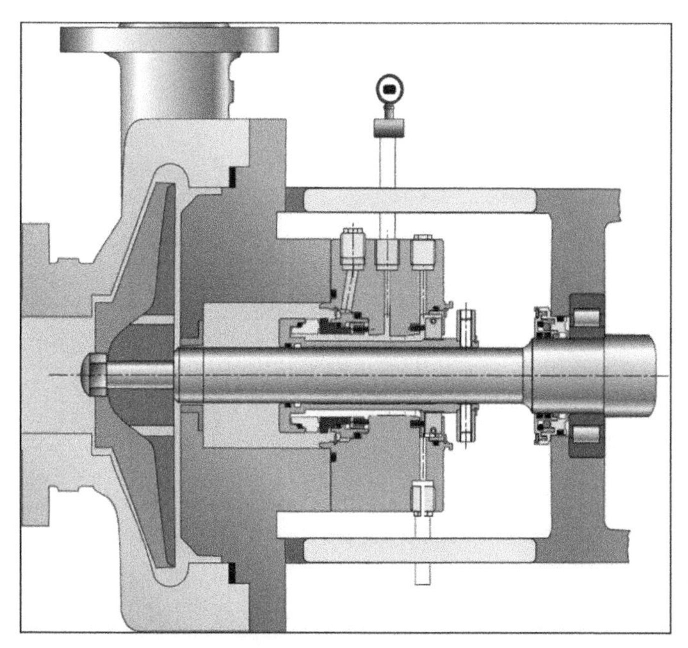

FIG. 21.3.3 Piping plan 65A/B is generally used for standard leak detection. *(Courtesy AESSEAL, Inc., Rotherham, UK and Rockford, TN.)*

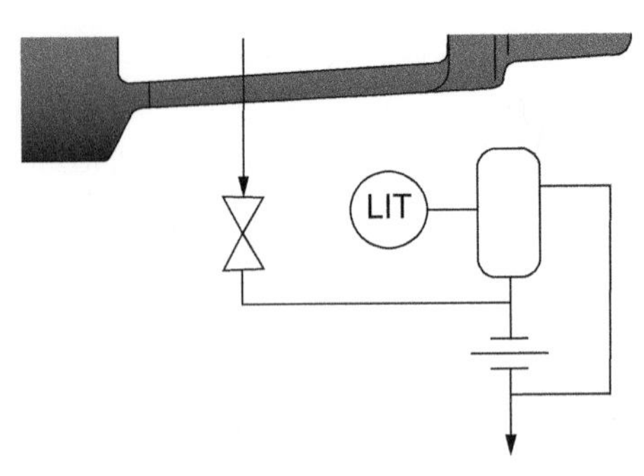

FIG. 21.3.4 In plan 65A, minor leakage flows away through an orifice; major leakage is noted by a level indicator-transmitter (LIT).

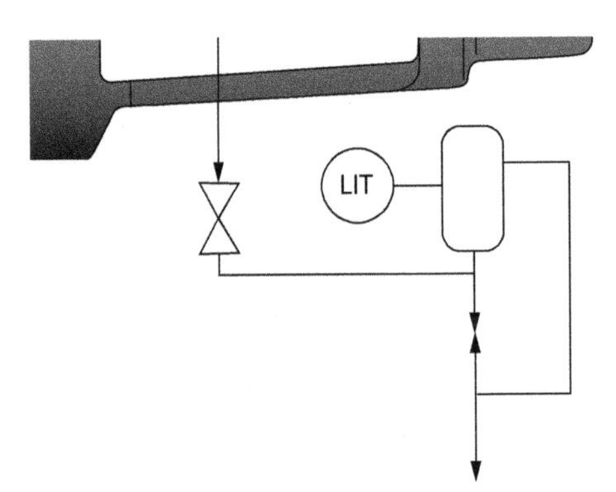

FIG. 21.3.5 In plan 65B, a needle valve can be finely adjusted to suit. Major leakage bypasses this valve and is led away. The rate of minor leakage can be tracked by the LIT device. *(Courtesy AESSEAL, Inc., Rotherham, UK and Rockford, TN.)*

is typically 5 mm (~0.25 in.); it should be located in a vertical piping leg to avoid accumulation of fluid in the drain piping. The piping allows bypassing the orifice so as to effectively self-drain excessive leakage amounts. A pressure transmitter can be provided as a monitoring alternative to the level indicator-transmitter (LIT) shown here.

Plan 65B is very similar (Fig. 21.3.5). A needle valve can be trimmed to suit the user's needs. Major leakage bypasses this valve and flows away. The rate of leakage can be safely tracked by the "LIT" device. The leakage collecting reservoir again has to be mounted below the seal gland to allow gravity flow from seal to reservoir. A valve is usually located between the seal and reservoir; it has to remain open during operation and should be closed during controlled maintenance events only.

How Plan 66 Fine-Tunes Leakage Detection

Plan 66 (Fig. 21.3.6) is a leakage detection plan often used by the pipeline industry sector for duty in remote applications. Here, high leakage flow is of prime interest. Note how a suitably orificed (or valve-equipped) pressure transmitter would be connected to the *central* port of this cartridge seal. Under conditions of high leakage flow, the resulting pressure rise would trigger an alarm. This approach will probably be similarly effective with more viscous fluids. Indeed, alternative versions have appeared in production areas with a closed valve on the outlet rather than the orifice. The valve will require periodic opening to drain off the "normal" or reasonably expected seal leakage. By trending the time interval between drain-downs, users obtain accurate data on the condition (or even failure trend) of a single seal.

Bearing protection takes on a special significance in remote pipeline pumping. Fig. 21.3.6 prompts the author to bring this to the reader's attention. An advanced bearing housing protector seal is illustrated, as in several of the preceding figures (see Subject Category 14 for more details).

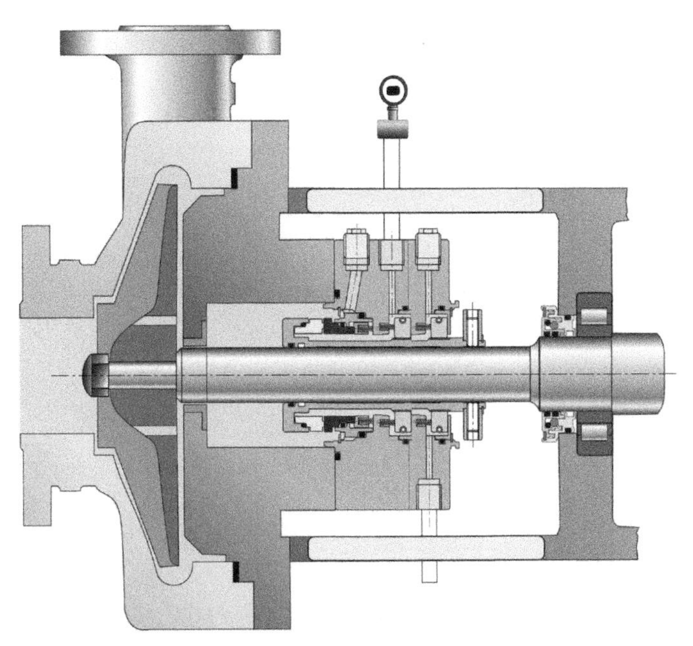

FIG. 21.3.6 API plan 66 is often used in remote pipeline services. *(Courtesy AESSEAL, Inc., Rotherham, UK and Rockford, TN.)*

API Plan 99—The Catch-All Plan

There could also be an engineered piping plan not covered by present API standards—a plan executed to the customer's orders (Fig. 21.3.7). A knowledgeable customer still wants to listen to manufacturer's advice and experience.

To recap and summarize our opening paragraphs, there are many manufacturers of mechanical seals, and their overall strategies seem similar. Special seals and special applications are of interest to reliability-focused users. Such users often seek out seal manufacturers whose overarching desire it is to go beyond delivering safe products at reasonable cost. These may be companies other than your traditional alliance partners; they will, by definition, be manufacturers whose marketing approaches consistently represent value. They must be able to point to superior service and high customer satisfaction. And they must have the desire to teach. We consider them seal service and asset providers who willingly convey educational and training updates to the ultimate seal user.

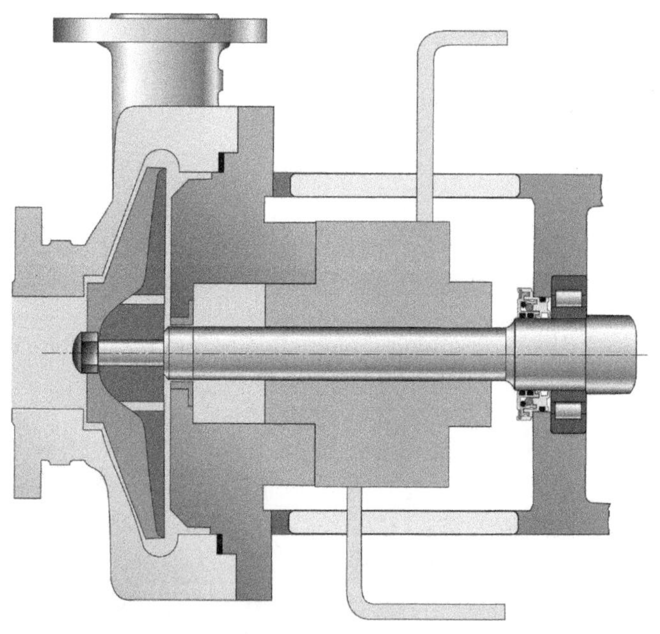

FIG. 21.3.7 API plan 99 allows the user to sketch in a desired configuration. *(Courtesy AESSEAL, Inc., Rotherham, UK and Rockford, TN.)*

Chapter 21.4

Cost Savings With Advantageous Mechanical Seals and API Seal Flush Plans

Many pump failure statistics point to mechanical seals as the components that fail most frequently. In 2008, a magazine feature article dealt with sealing issues of interest to industry.

The article explained the energy conservation and environmental benefits of certain seal-flush arrangements [1,2]. While not necessarily aimed at these issues, the current American Petroleum Institute (API) standards have enabled the hydrocarbon processing and other industries to take quantum leaps in reliability improvement and downtime avoidance. For decades, these API standards have contained seal-flush plans that allow users to specify and manufacturers to offer seal support systems that suit the specific requirements of a particular pumping service.

We will not even attempt here to deal with the many important piping plans described in the API literature and a number of websites such as http://www.aesseal.com/en/resources/api-plans/api-plan-54#sthash.rqsCGZC1.dpuf, which allow us to view these plans with great ease. Mechanical seal upgrading is another valuable pursuit that is best coordinated with a competent pump repair shop, or "CPRS." What a CPRS can do for users who wish to optimize fluid machinery will be further explained in Subject Category 22. Also, an earlier Chapter (21.2) illustrated a well-proved tapered pumping device that was first introduced in the 2006–08 time frame. It has been adopted by thousands of satisfied users. Why not make it your business to be informed about its application range and use in several different flush plans?

Only a brief summary can be given here. It highlights the importance of reliability engineers becoming familiar with the merits of different seal flush plans. In this instance, the value of changing from API plan 21 (Fig. 21.4.1) to plan 23 (Fig. 21.4.2) is briefly explained [3,4].

Visualize how, in fluid machinery operating with higher temperature media, single seals often require cooling for reliable long-term service. In many services, this cooling may be needed to improve the temperature margin to vapor formation or to meet the temperature limitations of certain secondary sealing elements (e.g., O-rings) that were chosen based on life cycle, polymerization, coke formation, or even chemical-resistance criteria. In services where the seal temperature environment must be controlled, both piping plans (API plans 21 and/or 23) are commonly used, but one may save energy, while the other will not.

Our example illustrates the efficiency difference between two "process side" flush piping plans. It involved a closely monitored boiler feedwater application in a modern combined heat and power (CHP) plant at a paper recycling mill. Operating

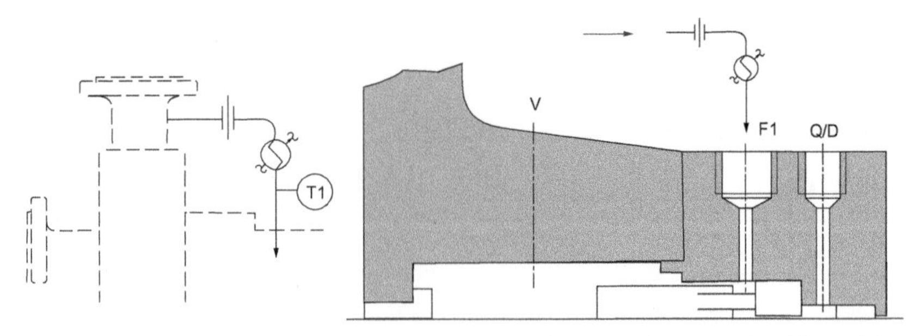

FIG. 21.4.1 Seal flush plan 21—product recirculation from discharge through orifice and heat exchanger.

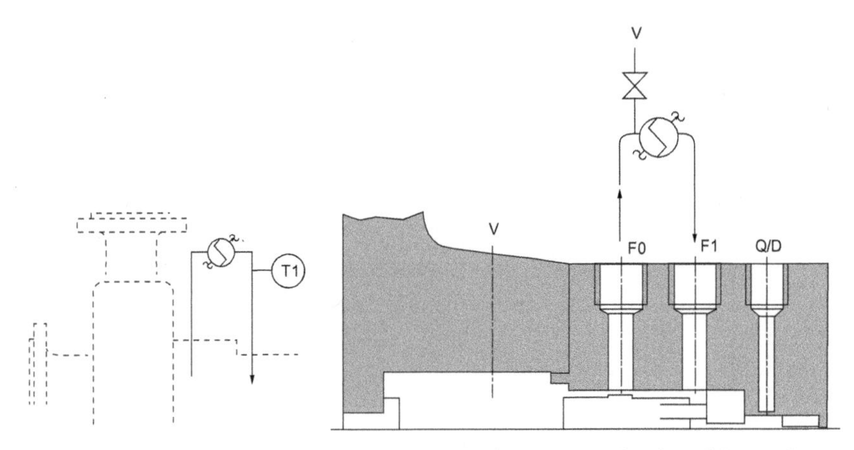

FIG. 21.4.2 Seal flush plan 23—product recirculation from seal chamber through heat exchanger and back to seal chamber.

at 160°C and a seal chamber pressure of 8 barg, the case history involved pumps originally fitted with a traditional 85 mm diameter seal. The seal faces received some cooling from the original API plan 21 configuration (Fig. 21.4.1). However, the service life of the seal was less than 12 months, and this (for water) disappointing seal life was attributed to heat exchanger fouling. Computer simulations indicated that the seal, with API flush plan 21, would be operating with a seal chamber temperature of 108°C (226°F) and with an exchanger heat load in excess of 14 kW.

Using the same basic operating parameters, a modern mechanical seal equipped with the bidirectional pumping device in Chapter 21.2 and piping arranged per API plan 23 (Fig. 21.4.2) yielded significant efficiency improvements. The seal chamber temperature was lowered from 108°C to now only 47°C (116°F). The heat exchanger load fell to 1.9 kW—less than 14% of the original plan 21 system. The yearly power savings proved sizable, and their value can be readily calculated. Moreover, switching to plan 23 and not having to contend with heat exchanger fouling did away with considerable and recurring maintenance expense.

REFERENCES

[1] H.P. Bloch, C.J. Rehmann, Mechanical Seals & Energy Efficiency, Presentation No. 60068 at ASME Power 2008, Orlando, Florida, July 2008.
[2] H.P. Bloch, Updating Your Sealing Technology, Machinery Lubrication, September/October 2008.
[3] D.W. Francis, M.T. Towers, T.C. Browne, Energy Cost Reduction in the Pulp and Paper Industry – An Energy Benchmarking Perspective, Pulp and Paper Research Institute of Canada, Office of Energy Efficiency of Natural Resources Canada, 2002.
[4] A. Gurría, Water: How to Manage a Vital Resource, Secretary-General, OECD FORUM 14–15 May, OECD, Paris, 2007, www.oecd.org.

Chapter 21.5

Consider Mutually Beneficial Strategic Mechanical Seal Partnerships Only

INTRODUCTION

Not all mechanical seal supplier agreements benefit both parties equally, which is to say that we, the user industry, may be offered some rather one-sided vendor alliance or sealing product partnership offers. At one industry conference, the suggestion that the "partner" should disclose pumping services where another manufacturer's seal would outperform the "partner's" product was met with derision by one and silenced by a second among the few world-scale mechanical seal manufacturers. Suffice it to say that partnership agreements along the one shown here exist where greed and adversarial relationships are not the first order of business.

BENEFIT SUMMARY

This is a basic proposal for an equitable agreement that serves both parties. While it may not be considered complete, it could easily form the cornerstone of your partnership agreement with a seal supplier. Work on these clauses, and start by reaffirming the many where there should be no dispute. They would be clauses where all parties are winners. After that, work on the ones that have been implemented elsewhere, and ask yourself why you cannot agree. Ask what you can do to break the pattern of adversarial relationships that so often gets in the way of plain common sense.

There is at least one well-known major mechanical seal manufacturer with lots of partnering or alliance experience. The manufacturer is perhaps in a position to offer us what we really need: sealing products that represent best life-cycle cost value. And of course, the application of sealing products that represent best long-term value is the only approach that makes sense, regardless of whether these products are

- already in your plant and originally supplied by his competition,
- supplied by your partner company and installed in your plant at the next downtime event,
- alternatively could be supplied by one or more of the alliance partner's competitors (!) and installed in your plant at the next downtime event.

Of equal or even greater importance is the fact that a knowledgeable partner supplier will undoubtedly point out substantial and valuable pump improvement measures. The implementation of these measures will not only extend seal life but also drive up equipment MTBR/MTBF as well. Valuable maintenance cost savings will result, and the partners must be apportioned their respective predefined share of benefits.

OBJECTIVES OF MUTUALLY BENEFICIAL AGREEMENTS

For an agreement to meet the test of mutual benefit, achieving at least 90% of the seal life obtained elsewhere for a given service should be one of the stipulated goals. Since energy efficiency, regulatory compliance, and effective supply chain management are part of the life-cycle cost objectives, the use of (dry) gas seals and perhaps even seals manufactured by the competition cannot be ruled out. Since the partners must share data on best available sealing technology, the supplier must compile and disclose this technology to the user. User and supplier will now have joint access to all available data and can compare their achievements with best available technology.

The supplier should benefit from the base fee paid and should, in addition, earn a success fee that gradually peaks at achieving the 90% mark mentioned above. Additional bonuses may be negotiated and predefined as the 90% mark is exceeded.

SCOPE OF SUPPLY DEFINED

The supplier should agree to furnish products and services including

- mechanical seal hardware (complete seals and components) and/or the repair of same for a given population of rotating equipment as defined in an appropriate tabulation that will have to be appended to the narrated agreement;
- detailed and specific instructions, assessments, life-cycle cost impact calculations (including energy efficiency), and other recommendations, as needed for enhancing the reliability of all sealing systems predefined and covered by the agreement;
- training of operators and maintenance workforces on a predefined time schedule (unless otherwise agreed to, quarterly training is to be conducted);
- an on-site technician (this technician should spend a specified number of man-hours per year at the user's facility);
- management of the supply chain of the inventory of complete seals, the rebuilding of seals (regardless of origin), without jeopardizing the availability of equipment within the population to meet the plant production requirements.

INVENTORY BUYBACK

The supplier should purchase the current and useable mechanical seal inventory from the user subject to the following conditions:

- The inventory must be utilized on equipment currently in use and expected to be in use over the term of the contract.
- The inventory must be of current design and suitable for providing reliability commensurate with the goals set forth in this agreement.

- The inventory meeting the above criteria should be acquired by the supplier at an agreed discount on its present book value.
- The inventory should be paid for in the form of a credit memo to be used to offset the base fees to be paid to the supplier. The credit memo should not exceed a defined share of the base fee in any given month over the life of the contract.
- Any inventory purchased by the supplier that has not been used in the fulfillment of this contract should be repurchased by the plant owner at the termination or completion of this contract. Repurchase should be at the same price as paid by the supplier, plus carrying costs of the unused inventory at 10% per year.

OBLIGATIONS AND RESPONSIBILITIES OF THE OWNER

The owner must provide the supplier with

- accurate data on the maintenance history and the parts consumption of the equipment population covered by this agreement (to the extent it is determined that insufficient history does not allow supplier to accurately estimate its cost of meeting obligations under the terms set forth in an agreement, the supplier should be allowed an adjustment of the fees to be paid to them or delete the equipment from the population covered by the agreement);
- timely implementation of recommended and mutually agreed-to improvements necessary to achieve the goals and/or objectives of the agreement (if the improvements are not made in a timely and as agreed-upon manner, the same remedies should be available to supplier as stated above);
- suitable access to plant and working space needed;
- access for training, at least annually, during normal working hours and at no cost to the supplier (the training should be extended to such operations and maintenance personnel as have a direct impact on the reliability and performance of the equipment covered by the agreement);
- adequately decontaminated equipment or the knowledge and facilities allowing supplier to adequately decontaminate any equipment or device that is to be handled by supplier;
- material safety data sheets and related documentation;
- payment, net 30 days and with surcharges of 1.5% per month of unpaid balance thereafter.

OBLIGATIONS AND RESPONSIBILITIES OF THE SUPPLIER

- The supplier should provide the products and services covered by this agreement, barring labor strife, catastrophes, shutdowns, etc., in which case equitable pro rata separation of responsibilities would have to be made.
- Nonexclusivity is required, that is, the partner is not the only supplier. No best-of-class company has ever secluded itself from access to others.

- The partner must become a technology resource to the owner; achievements beyond best-of-class MTBF reflect in repair cost avoidance. The aim is an 85% (partner) versus 15% (all others) split in rewards for such achievements.
- The partner must agree to develop comparison matrices that show API flush plans, materials, etc., used by its competitors for services.
- The order of priority of matrix development is to be determined by failure frequency (or seal repair frequency).

SUCCESS OVERSIGHT COMMITTEE AND DESIGNATED ARBITER

- Each party should appoint at least two but no more than three individuals to a committee charged with overseeing the timely and successful implementation of the agreement at each plant site. This committee should convene periodically, but no less frequent than every 60 days.
- In plants with two or more sites, an executive committee should monitor the overall progress. The committee should have at least one member from each side with sufficient authority to resolve disputes and direct corrective steps to resolve deficiencies in their company's performance under the terms of this agreement.
- Disputes that cannot be resolved by the committee(s) should be submitted to an independent third-party arbiter who will be designated by mutual agreement at the inception of this agreement and whose fees should be paid equally by both supplier and owner.

COMPENSATION FOR PRODUCTS AND SERVICES

- The base fee should be set by mutual agreement based on prior yearly seal consumption (equipment MTBR) and historic expenditures for such related outlays as seal failure induced collateral damage, production losses, etc.
- In addition, the supplier should earn a mutually agreed-to success fee equivalent to 1/3 of the overall savings generated by the enhancements and to the extent that these exceed the predetermined goals of the program. Savings are defined as direct and measurable savings in labor, parts, energy, incremental production, etc.

TERMS AND TERMINATION

A typical agreement should cover a 5-year period. Early termination of the contract without cause would cause monetary damage to both parties. Therefore, the party terminating without cause should agree to pay the other party damages equivalent to 12 months base fee and an annualized success (or, as the case may be, nonrealized success fee) based on the results achieved (or, as the case may be, forfeited) because of early termination.

Chapter 21.6

Repetitive Pump Seal Failures Can Cause Disasters

INTRODUCTION

Release no. 2004-08-I-NM, issued in Oct. 2005 by the US Chemical Safety and Hazard Investigation Board, addresses an incident at an oil refinery with a history of repeated pump failures. Located in New Mexico, this facility's total of three primary, electric, and steam-driven spare isostripper recirculation pumps had 23 work orders submitted (see Table 21.6.1) for repair of seal-related problems or pump seizures in the 1-year period prior to a fire and explosion. The catastrophic incident occurred during disassembly on Apr. 8, 2004 and caused over $13,000,000 in damage. At least six people were injured, and production at this alkylation unit was shut down for months.

Because of the serious nature of the incident, the US Chemical Safety and Hazard Investigation Board (http://www.csb.gov) produced a detailed write-up that describes the sudden release of flammable liquid and subsequent fire and explosion that occurred.

LESSONS FOR THOSE WILLING TO LEARN

For many decades, truly reliability-focused organizations have avoided repeat seal failures by insisting on understanding and eliminating failure causes. They realize that collecting only the generalized failure descriptions of Table 21.6.1 would be analogous to a trucking company cataloging repeated nonperformance as "engine problems." More detail would be needed to implement sound remedial action.

The reliability-focused also realize that when a seal failure combines with one or more other deviations from the norm, disasters result. More specifically, the reliability-focused make it their business to know what fit-for-service mechanical seals are available. But, of course, these components have to be properly installed and will usually require a pump-around circuit and dual seals, as depicted in a previous chapter of this text. Good dual seals will include a conservatively designed wide-clearance pumping ring.

We should remind ourselves of a key requirement of API 682/third edition: mechanical seals installed in refinery equipment should have a design life of 25,000h. Clearly, the mechanical seals in this particular installation fell far short of that requirement.

As to other lessons, there was an obvious breakdown in maintenance management at that refinery. It is hard to comprehend that an organization would

TABLE 21.6.1 Seal Failures Preceding Major Fire Event in a Refinery

Sequence	Date	Pump	Problem
01	Apr. 17, 2003	P-5A (elec.)	Seal leak
02	May 9, 2003	P-5B (stm.)	Pump spraying from seal
03	May 23, 2003	P-5A (elec.)	Repair seal
04	Jun. 9, 2003	P-5B (stm.)	Repair seal
05	Jun. 9, 2003	P-5A (elec.)	Repair seal
06	Jun. 18, 2003	P-5A (elec.)	Repair seal
07	Jun. 20, 2003	P-5A (elec.)	Replaced seal
08	Jul. 31, 2003	P-5A (elec.)	Replaced seal
09	Aug. 22, 2003	P-5B (stm.)	Seal leak
10	Aug. 25, 2003	P-5B (stm.)	Replaced seal
11	Sep. 26, 2003	P-5B (stm.)	Replaced seal
12	Sep. 26, 2003	P-5A (elec.)	Replaced seal
13	Oct. 14, 2003	P-5A (elec.)	Seal leak
14	Dec. 6, 2003	P-5A (elec.)	Replaced seal
15	Dec. 9, 2003	P-5B (stm.)	Seal leak
16	Dec. 9, 2003	P-5A (elec.)	Seal leak
17	Dec. 15, 2003	P-5B (stm.)	Replaced seal
18	Dec. 15, 2003	P-5A (elec.)	Seal leak
19	Jan. 28, 2004	P-5A (elec.)	Seal leak
20	Mar. 22, 2004	P-5A (elec.)	Seal leak
21	Apr. 1, 2004	P-5A (elec.)	Pump seal leaking
22	Apr. 3, 2004	P-5A (elec.)	Pump seal leaking
23	Apr. 7, 2004	P-5A (elec.)	Repair pump seal

tolerate 23 costly pump interventions without insisting on solid answers and remedies long before an inevitable catastrophic event occurs. Issues no doubt start with inadequate experience during the purchasing stage. After installation, it takes well-trained, motivated mechanical engineers to correctly diagnose seal failures and find the true root causes. Unless the root cause is determined, there will be repeat failures. Don't employ the nonteachable, and do develop an intense dislike for repeat failures. Be determined to reward managers and hourly employees who eradicate repeat pump failures.

For a certainty breakdown maintenance of machinery will lead to complete plant breakdowns, as occurred here. Identifying the risky parameters and process operating conditions and avoiding failure-prone components are important. Weak components will contribute to equipment failure risk. Avoiding them is perhaps the most cost-effective ingredients of a sound mechanical integrity program.

Finally, consider instituting a new regime by valuing, developing, and promoting competent leaders instead of favoring those who can fabricate a profit margin in the next quarter. Understand that house-of-cards and built-on-sand next quarter profits are crumbling at random, whereas house-of-brick and built-on-rock organizations have few surprises and will keep standing.

Chapter 21.7

Diamond-Coated Seal Face Technology

INTRODUCTION

The first of two essential mechanical seal requirements is that the contact faces must be flat and polished. This is normally achieved by a special machining process known as lapping. The adjacent primary seal surfaces are measured for flatness by viewing them through an optical flat; a monochromatic sodium light source supplies the needed illumination. Typically, silicon carbide (SiC) and tungsten carbide (WC) have been selected as long-lasting hard-face seal materials. With few exceptions, these components are manufactured as solid parts with faces highly polished and with near-perfect flatness. The second of the two essential requirements is that secondary sealing elements—usually O-rings—must survive the temperatures to which they might be exposed. A limiting temperature of 250°C (482°F) is assumed for elastomer O-rings in mechanical seals.

Although seal face separation and cooling are perhaps of greatest importance, mechanical seal performance is also dependent upon the tribology of the seal face materials. Over successive decades, many developments have been aimed at improving resistance to wear. However, while other low-friction coatings (among them tetrahedral amorphous carbon and graphene) were considered, realistic application research had to be restricted to best available technology at net economic cost, "BATNEC." With that in mind, some seal users have been steered toward a hard, low-friction diamond-like carbon (DLC) coating in the expectation that this coating will allow dry running of contacting seal surfaces. But many seal users wanted to know more about other promising alternatives, including polycrystalline diamond (PCD) and PA-CVD, an amorphous diamond material applied by using a patented plasma-assisted chemical vapor deposition (PA-CVD) method.

Published material is available claiming the advantages of PCD and its suitability for extended running under dry conditions. One such claim suggests that PCD-coated faces are immune to dry running. In order to investigate this claim, it was deemed appropriate to perform a series of rigorously controlled tests. The basic aim of these tests was to compare the performance of mechanical seal faces coated with a PCD coating against other faces with DLC coatings.

WHERE YOU CAN FIND THE DETAILS

The most recent findings on these advanced coatings can be found in Ref. [1]. Both its writer (seymourmd@aol.com) and the scientific journal in which his team-assisted findings were published [2] gave readers unrestricted access to the article. Our present aim is to condense and convey this peer-reviewed material—originally released in May 2016. One of the goals of Mr. Seymour's article was to describe a rigorous test program the intent of which was to give an authoritative and test-backed overview. He wanted to compare and quantify how effective one face surface treatment solution was relative to another. We, with the benefit of several decades of general pump seal user's experience, have anticipated the end users' "so what?" questions. Our bottom-line answer is directed at cost-conscious and reliability-focused readers. We wish to express it as: *the incremental or potential benefits derived from synthetic diamond coatings (SDCs, which could be DLC, PA-CVD, or PCD) may not be worth the additional cost of some aggressively marketed SDCs.*

First, however, it is worthy of note that the seals involved in rigorous testing at one of the two seal manufacturers had been designed and produced by two different manufacturers. Testing was carried out at the UK-based manufacturer's facilities; the responsible team benefited from access to the Materials and Engineering Research Institute (MERI) based at Sheffield Hallam University (the United Kingdom). The tests followed a rigorous protocol; the research team and its experienced leader had access to suitable apparatus and well-calibrated, often highly sophisticated means of measuring.

It was known that seemingly identical hard-face materials were applied to seal components by different coating methods; the research team led by Mr. Seymour wanted to study the impact of these different methods. Seal surface topography was accurately measured before and after testing and was used to evaluate the wear behavior of PLC and DLC coatings. From the test results, the research team concluded that neither of the two coatings is suitable for extended dry running use. However, it was evident that seal faces coated with *a new form* of DLC identified as PA-CVD lasted 18 times longer than PCD-coated seals before either of them reached a predetermined friction-induced temperature. This is explained in Fig. 21.7.1.

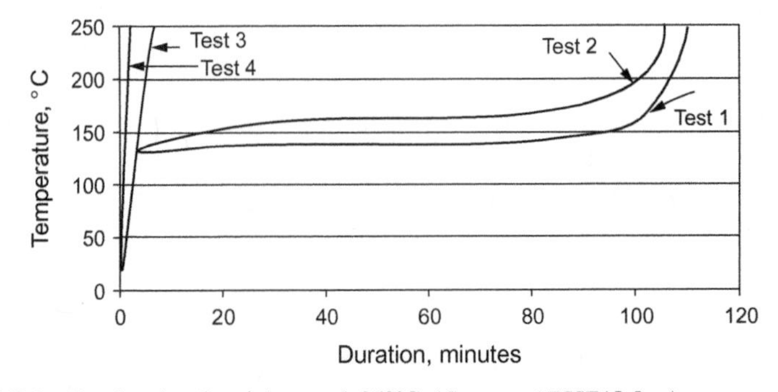

FIG. 21.7.1 Test duration (in min) to reach 250°C. *(Courtesy AESSEAL Inc.)*

Ref. [1] also cites published evidence that PCD-coated seal faces are capable of producing very high frictional temperatures that could, in a dry running situation, allow certain liquid fuels such as flashing hydrocarbons to reach their autoignition temperatures. In addition, it was revealed that the PCD-coated seal units are being sold at a higher cost than the equivalent DLC-coated ones by a factor of three.

INVESTIGATING DIAMOND-LIKE SEAL FACE COATINGS (DLCS)

The very name, DLC, alludes to and implies value imparted by such coatings. Perceiving value is not unfounded because DLCs indeed have the ability to provide some of the properties of diamond to surfaces of almost any material. The primary desirable qualities are hardness, wear resistance, chemical resistance, and low coefficient of friction. As of 2016, at least ten different DLC coatings have been developed; they can be defined as either nonplasma or plasma processes. Under operational conditions, some nonplasma process coatings can delayer (delaminate) due to poor adhesion to the substrate material, and this was seen in certain unrelated MERI tests on ion beam deposition.

INVESTIGATING PA-CVD PROCESS COATINGS

PA-CVD is another process used to form an amorphous diamond material coating. Although still referred to as a DLC, the bonds within this material are 99% diamond phase and have a hardness of 4000 Hv. The advantage of this process is that it can be applied to a variety of substrate materials at low temperature (120°C/248°F), which results in significantly extending the dry running lifetime of contacting surfaces (Fig. 21.7.1). Although it is not recommend for extended dry running conditions, the PA-CVD coating is ideal for short-term intermittent operation that could occur during a temporary process disruption. Also, while other processes limit coatings to 3–4 μm, the PA-CVD process can provide coatings in the order of 30–40 μm. Note, however, that in the interest of maintaining seal face flatness, 12 μm is considered the practical limit.

INVESTIGATING POLYCRYSTALLINE DIAMOND-LIKE (PCD) COATINGS

Diamond can be one single, continuous crystal, or it can be made up of many smaller crystals (polycrystal). Large, clear, and transparent single-crystal diamonds are typically used in gemstones. PCD refers specifically to diamond particles that have been sintered together into a coherent structure using a chemomechanical binder and high-pressure, high-temperature conditions. While similar to conditions encountered in single-crystal diamond synthesis, PCD structure and conditions are unsuitable for gems but find use in industrial applications such as cutting tools and rotary seal surfaces.

PCD is often described by the average size (or grain size) of the crystals that make it up. Grain sizes range from nanometers to hundreds of microns, usually referred to as "nanocrystalline" and "microcrystalline" diamond, respectively. These coatings have the extreme hardness of diamonds, in the order of 10,000 Hv, and are usually produced by various means of nonplasma chemical vapor deposition at or above 800°C. The various production techniques that have been developed over recent years differ by the grain size that is produced. There are several names given to these coatings such as "ultracrystalline diamonds," "nanocrystalline diamonds," and "polycrystalline diamonds," but all are essentially similar.

Normally, PCD coatings are 2–3 μm thick; increasing these coating thicknesses is feasible but will cost a significant premium. For example, the two PCD samples tested had coating thicknesses of 10 μm and were purchased "off the shelf" at a much higher price than the two PA-CVD-coated face seals. In contrast, the PA-CVD coating method can readily produce greater thickness coatings of up to 40 μm at considerably lower cost. Recall, please, that in order to maintain seal face flatness, 12 μm is considered the practical limit.

COMPELLING RESULTS AND WHAT THEY MEAN

The laboratory measurements recorded surface topographies before and after dry running tests and indicated the depth and location of wear across the faces at two diametrically opposite locations.

A certain depth of wear was expected because the tests were run dry with the PA-CVD coating acting as the only lubricant present. The tests revealed that PA-CVD-coated faces lasted as long as 1 h + 46 min and 1 h and 1 h + 47 min, respectively, although this coating was never designed to be used for extended dry running. Both tests were surprisingly consistent since they reached 250°C/482°F being within less than 2 min of each other.

It was also observed that wear rates, measured in microns/hour, of the two different surface coatings differed remarkably. In the two PA-CVD samples, the wear depth amounted to about 2.5 μm/h; the two PCD-coated samples showed considerably greater wear depths of 10 and 50 μm/h, respectively. These findings are restated in Table 21.7.1.

TABLE 21.7.1 Wear Observed on Dry Running (1.875 in. Shaft Diameter) Single Seals (8 m/s Sliding Speed)

Test 1, PA-CVD coating: 2.25 μm/h; face contact pressure 234 kPa/34 lbf/in.^{2}
Test 2, PA-CVD coating: 2.53 μm/h; face contact pressure 234 kPa/34 lbf/in.^{2}
Test 3, PCD coating: 10 μm/h; face contact pressure 110 kPa/16 lbf/in.^{2}
Test 4, PCD coating: 50 μm/h; face contact pressure 110 kPa/16 lbf/in.^{2}

An in-depth investigation into why the wear rates are so different would be the subject of further research into the structure, grain size, and application methods of both coatings. Recall, please, that this was not the aim of the test program—a program that was to *compare the effectiveness of one face surface treatment solution with another.* Interestingly, the international company producing the seal assemblies identified here as tests 3 and 4 quotes in their marketing literature that "wear, in pure dry running mode, is an outstanding 0.08–0.2 μm/h." Assuming these numbers are correct, one would surmise they were probably obtained at unspecified or unknown test pressures and velocities that favor obtaining low numbers. The point is that with the realistic test conditions described here, dry running wear rate measurements and calculations have shown the differences found in Table 21.7.1. It can be seen that these wear rate measurements and calculations favored PA-CVD coatings over the more expensive PCD coatings by multipliers ranging from 4:1 to 20:1.

One of several different observations obtained under identical test conditions compared the initial topographies of tests 1 and 2 with the initial topographies of tests 3 and 4. Although not of identical surface roughness, the topographies (profiles) were somewhat similar to each other. Accordingly, it was expected that the PCD-coated faces would last as long as or perhaps even longer than the ones coated with the PA-CVD coating. This run-length expectation was proved incorrect by these careful tests (note again that the thickness of PA-CVD is limited on the seal faces to 12 μm).

Again pointing to Fig. 21.7.1, tests 3 and 4 lasted only 109 and 351 s, respectively, before the coated faces reached 250°C/482°F. In test 3, a rougher profile was measured, and as expected, it took less time to reach this friction-induced designated cutoff temperature. Post-test measurements show that the wear scar on the PA-CVD-coated faces is deeper by a factor of between 4 and 5 than the PCD-coated faces. This result was not surprising as the greater depth of wear was possibly caused by a combination of a longer running period (18×) and a higher face contact pressure. Per Table 21.7.1, the test pressures were 234 kPa for tests 1 and 2, and 110 kPa for tests 3 and 4.

EXPECTATIONS VERSUS IN-DEPTH CONSIDERATIONS

Further promotional material in the form of a publically available "YouTube" video [3] shows a dry running test being performed to compare an untreated SiC face seal assembly with one that was PCD-coated. A remote temperature sensor showed that the PCD-coated seal was producing friction-induced temperatures of up to 360°C after 30 min but, nevertheless, was still intact, whereas the SiC seal unit only lasted 28 s before destruction. While all of this is visually impressive and no doubt technically correct, ask what would happen to the polymer O-rings that are only effective at lower temperatures. Ask also what risk would be incurred if seals were operating at this temperature when processing a fluid with low autoignition temperature (i.e., the minimum temperature required to ignite a gas or vapor in air without an ignition source being present). For example, automotive gasoline/petrol (247°C) or motor vehicle diesel (210°C) would ignite in the event of the pumping process being interrupted, which would likely allow the seals to run dry. In both of these high friction-induced temperature scenarios, a potentially hazardous state could occur. That is why it is strongly recommended for users not to operate these coated seals for extended periods in a dry running condition. In addition, one company claims, in a recently published release, that "the diamond coating (PCD) makes the seal immune to dry running and solids in the medium." The test team found this statement difficult to understand. The release at issue describes a bellows-type seal where polymer O-ring seals are not used. Yet, the same release contains a diagram referring to and clearly showing O-ring seals.

Fig. 21.7.1 graphically represents the results confirming the aim of the test program performed here. It demonstrates that the seal assemblies in tests 1 and 2 performed significantly better in terms of extended running times than the other two assemblies, tests 3 and 4.

CONCLUSIONS

(1) SDCs on solid SiC substrates of mechanical seals are not suitable for extended periods of dry running.

(2) PCD coating seal faces were found to be completely unsuited for extended dry running. They performed, in terms of dry running time, poorly against the equivalent PA-CVD-coated faces by a factor of 18. In short-term dry running conditions represented by processes being interrupted, the PA-CVD-coated seals would perform longer and safer than the PCD-coated equivalent.

(3) Users should be aware that dry running these seals for extended periods will create high temperatures. These temperatures will possibly render the O-rings ineffective. In addition, high frictional temperatures could exceed the autoignition limits of some process fluids and cause a fire.

(4) A high wear rate was measured on the PCD-coated faces; it provided evidence that under extreme dry running conditions, the PA-CVD coating performed better by factors ranging from 4 to 20 times.

(5) Although a precise difference in seal unit costs is difficult to obtain, it is generally recognized that processes to produce PCD faced coatings can cost up to three times the amount of an equivalent PA-CVD process.

(6) Finally, users and buyers should be aware that diamond coatings are not immune to dry running, as this rigorous test program has revealed in 2015/2016. The testing of two different manufacturers' equivalent mechanical seal assemblies showed that both design solutions were effective for short periods of time. However, the PA-CVD coating lasted considerably longer than the more costly PCD-coated seal faces. Providing test-backed proof is an incredibly valuable contribution to the concept and principles of "BATNEC"—using best available technology at net economic cost.

REFERENCES

[1] M.D. Seymour, Are diamond surface coatings immune to dry running? J. Coating Sci. Technol. 3 (1) (2016) 1–8.
[2] http://www.lifescienceglobal.com/journals/journal-of-coating-science-and-technology.
[3] YouTube.com/watch?v=mndO_uC3bXA.

Chapter 21.8

Converting From Gland Packing in a Boiler Feed Pump: A Case Study

INTRODUCTION

As of 2016, every rotating (dynamic) process pump in oil refineries and petrochemical plants is either equipped or could be equipped with mechanical seals. The fact that someone chooses to use labor-intensive and/or leakage-prone packing has nothing to do with the accuracy of the preceding statement. Automobiles burst on the transportation scene 100 years ago, and for good reason. Although the reasons vary, some people opt for more elementary means of locomotion.

For now, please assume that uncertainty over conversion costs has been a barrier to technological progress. Perhaps, an actual case history will be of value; it shows why thorough consideration should be given to the full picture.

CONVERSION COST AND SAVINGS

This case history involved converting eight multistage boiler feed pumps from the traditional soft packing gland to a plan 23 mechanical seal arrangement (plan 23 was illustrated in Chapter 21.4). As the reader no doubt knows, whenever packing is used, one must reduce rubbing friction and damage to shaft and packing. The packing gland follower must be adjusted, and some leakage must be allowed. One aims for a drip rate of one drop per second to assist in cooling and lubricating the shaft contact region. As packing and shaft wear, periodic adjustment will be needed, and a competent workforce is needed to strike a balance between excessive tightness (too many repairs) and too much gland leakage. Ask yourself what this labor component would cost.

In any event, the boiler feed pumps incorporated packing glands at both drive and nondrive ends. Although installed at a power generation facility, this arrangement is typical of many applications around the world at chemical works and refineries. Most of the feedwater pumps installed in this time frame had been fitted with packed glands, and many continue to run today using this outdated approach. Leakage from the packed glands will be a pure loss to the operation. In this example, the boiler feed was at 121°C, and losses through the packed gland had to be made up with water from the treatment plant. The calculation of the energy loss is based on the energy required to take the makeup water from 10°C to a feedwater temperature of 121°C. The steam generators were gas-fired, and the heat energy requirement could be translated into a net CO_2 contribution. Because plant manpower had been reduced, gland follower adjustments were only made when the leakage was severe. As a result, the average leakage rate from the pumps was about 1 L/min per gland.

With eight boiler feedwater pumps and 16 glands leaking on average one quart (roughly 1 L) per minute per gland, energy loss was calculated at 124 kW. The plant operates 24 h per day, 365 days per year, causing an annual energy loss of 1,086,240 kWh. These energy savings are purely based on heating requirements and do not include energy costs for water treatment, deaeration, and pumping. It is worth noting that these energy savings do not include possible pump power reductions. The stated energy savings refer to the combustion process and boiler operation costs only.

Site personnel determined that the combustion process emitted 0.0282 kg of CO_2 per liter of water heated. With losses of 1 L per minute per gland, the calculated savings amounted to 237 tons of CO_2 per year. Just to compare, an average European high-efficiency diesel-fueled vehicle covering 20,000 km (12,500 miles) per year would emit 3.2 tons per year. The savings equate to removing about 80 automobiles from the roads. In some parts of the world, very substantial carbon tax payments are eliminated by intelligent sealing pursuits.

Aside from any tax issues, the curtailment of gland leakage would require employing trained craftspeople who do the gland adjusting day in, day out. Training, rewarding, and retaining these crafts has become an insurmountable challenge for most facilities. Reliability-focused engineers will make a compelling case for using mechanical seals with flushing in accordance with plan 23.

Reliability-focused engineers and managers will make it their obligation to understand water management systems that keep pace with 21st century thinking. Some such systems go well beyond API flush plan 23, and special opportunities abound in the power generation, mining, and pulp and paper industries. Substantial profitability and reliability gains are possible by making a competent seal manufacturer your technology resource. Some old and unprofitable sealing arrangements (read packing) may benefit from redesigned seal housings; see Fig. 21.8.1.

CONCLUSION

Thoughtful selection of mechanical seals and their support systems deserves the time and attention of plant engineers. Thoughtful selection must reach into issues that have to do with the kind of quality control and leakage testing performed by mechanical seal manufacturers. Some of these manufacturers give a "passing grade" to mechanical seals that leak many times the amount allowed by safety-focused and highly principled seal manufacturers. Ask questions and become familiar with the leakage and testing matter. Stand for buying wisely and reward the best performing manufacturers.

The energy indirectly consumed with traditional, nonoptimized systems should not be ignored, and sole emphasis on minimum capital cost rarely results in best operational efficiency. Sizable maintenance cost savings are possible when traditional gland packing arrangements are upgraded to modern

FIG. 21.8.1 Designing and reconfiguring a new mechanical seal environment is being discussed here. *(Courtesy Hydro Inc., Chicago, IL, http://www.hydroinc.com.)*

mechanical sealing strategies. Labor, downtime, operating costs, environmental load, and plant safety can be affected in the positive sense by these conversions.

The mechanical seal industry has developed and documented well-engineered solutions that provide owner-operators with higher equipment reliability and lower life cycle costs. Sound solutions add profits to the bottom line and can be shown to reduce environmental impact as well. Based on experience in oil refineries around the world, these are win-win propositions that many other industries will find worth investigating.

Subject Category 22

Non-OEM Repairs Combine Effective Upgrade Strategies

Chapter 22.1

Competent Repair Shops and Some Upgrade Options that Extend Pump Uptime

Consider a large power plant with two 10,000 kW boiler feed pumps that operate in parallel. These two pumps are scheduled for repair during the next scheduled shutdown. There may be a tendency to interpret this as traditional maintenance work that could be accomplished within the scheduled time and would cost $200,000. But what if a CPRS which fittingly stands for both competent pump repair shop and competent pump rebuild shop could be found and it could be ascertained that this CPRS could perform a combined maintenance and upgrade job? Suppose the upgrade would result in an efficiency gain of 2% and power is worth $0.07/kWh. The savings would amount to $245,280 per year. Suppose further that the CPRS would charge $300,000 to do the work; that's an incremental cost of $100,000 that would return handsomely, regardless of the precise calculation method employed.

A well-informed pump user will have captured much pertinent information on his pumps and, especially, will have failure frequency data relating to his equipment. These data and an understanding of what caused a given pump deficiency will enable both user and competent pump rebuild shop to point out and explain, specify, or recommend a number of appropriate options. Once cost-effective options are selected, the competent pump rebuild shop should be asked to implement such measures as upgrading of sensitive components, avoidance of vulnerable lubricant application methods, and so forth. Some of this work is design and engineering ahead of the actual opportunity to put upgraded parts into the machine. At all times, preplanning is very desirable.

Consider involving the CPRS in failure analysis; make the CPRS your technology resource. Based on understanding what failed and why it failed, a

Petrochemical Machinery Insights. http://dx.doi.org/10.1016/B978-0-12-809272-9.00022-0

reliability-focused user will take steps to authorize and implement routine shop upgrades. We define *routine* upgrades as those done on every important pump that enters the CPRS facility. Routine shop upgrading measures are rarely pursued by the OEM, whereas the CPRS will be eager to explain and advocate them. Most, if not all, of these routine upgrade measures are embedded in this text. As just one of numerous routine upgrade examples, we might mention pressure-balanced constant level lubricators where previously pressure-unbalanced constant level lubricators had been used. The user's reliability professional should discuss these with the CPRS. All are measures or options that are further explained in Refs. [1–3].

Of course, pump repair and rebuilding efforts very often go beyond just the routines that we have described above and in the captions to Figs. 22.1.1–22.1.3.

FIG. 22.1.1 Elementary defect assessment is part of incoming inspection. It ranges from the obvious in this image to the much more elusive. *(Courtesy Hydro Inc., Chicago, IL, http://www.hydroinc.com.)*

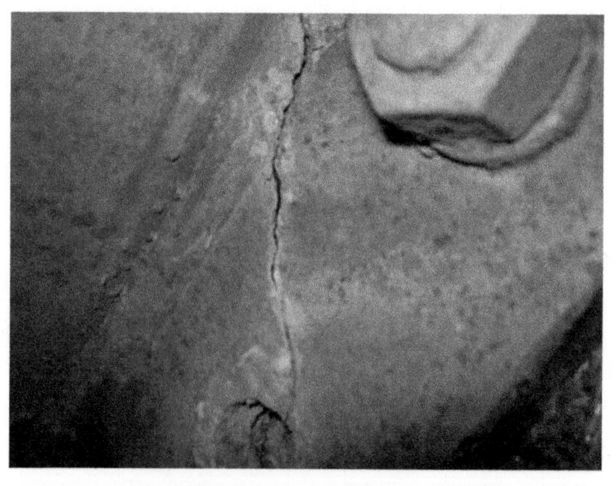

FIG. 22.1.2 Still at the elementary defect assessment; cause examination is usually initiated at this point. *(Courtesy Hydro Inc., Chicago, IL, http://www.hydroinc.com.)*

Repair scopes differ from pump to pump and must be defined if the highly desirable goals of uptime extension and failure risk reduction are to be explored and achieved. Think also of combining repairs with achieving greater energy efficiency (Fig. 22.1.4). [4,5]

FIG. 22.1.3 After elementary defect assessment, time-intensive nondestructive testing and mapping will commence. *(Courtesy Hydro Inc., Chicago, IL, http://www.hydroinc.com.)*

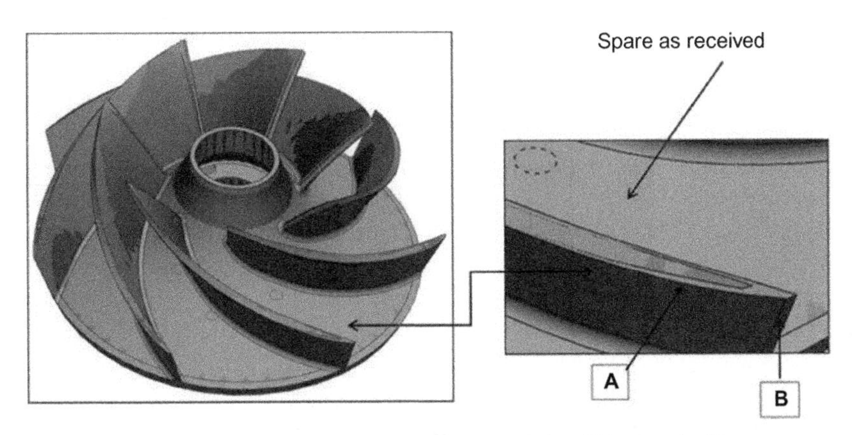

FIG. 22.1.4 Moving toward the more elusive defect assessment. Such work now includes more precise measurements and comparisons. *(Courtesy Hydro Inc., Chicago, IL, http://www.hydroinc.com.)*

FINDING A CPRS

Let us assume you represent a facility plant seeking a capable alliance partner willing to maintain, repair, and upgrade your pumps. The first order of business would be to engage in a well-focused assessment. You would not just pick such a potential partner facility on the basis of cost and delivery promises. Instead, your choice would be based on vendor/partner competence.

Consider working with a CPRS that has testing facilities that rightly deserve to be called a testing laboratory. This allows users to ascertain the before-versus-after performance of impeller upgrades and similar modifications, Figs. 22.1.5 and 22.1.6.

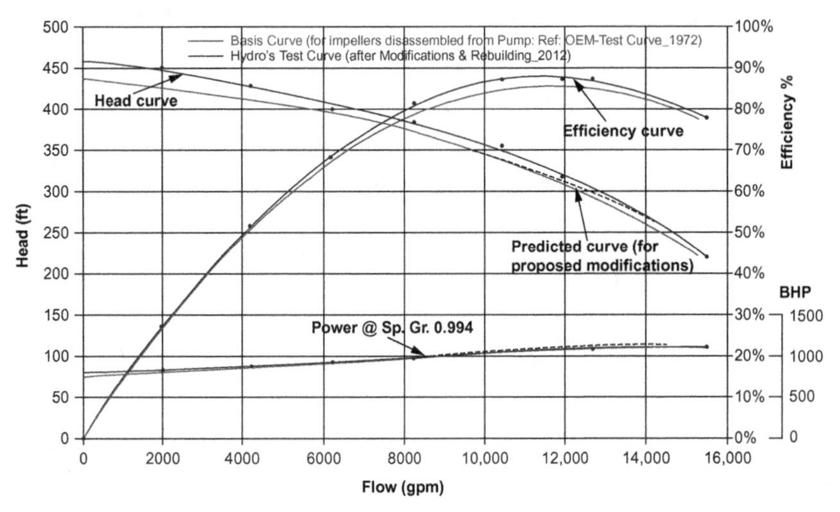

FIG. 22.1.5 Performance prediction checked against performance achieved. *(Courtesy Hydro Inc., Chicago, IL, http://www.hydroinc.com.)*

COMPETENT PUMP REPAIR SHOP FACILITY ASSESSMENT

There are many assessment formats, matrices, or templates that allow pump user-owners to determine, review, or assess the competence of a given repair facility. Indeed, true reliability engineering includes making an assessment of potential bidders for both the procurement of new and refurbishment old equipment [1–4]. One such format starts with a general listing of items, names, and similar logistical and general information. It progresses to specialization reviews and 60 or more additional questions that will be of real interest. It investigates a large number of attributes and factors (see Appendix I). These data or findings must be captured somehow. Pictures and descriptions of work scope, quality, and work processes that can be expected from a particular CPRS are helpful. They are highlighted in the next few pages.

FIG. 22.1.6 Extent of vendor experience can often be determined by subjective examinations of quality and complexity of work. A number of observations relate to the quality of pump impellers. *(Courtesy Hydro Inc., Chicago, IL, http://www.hydroinc.com.)*

These data entries might represent the obvious, but there's much more to it. For instance, to make a relevant assessment, the experience background of the CPRS employees needs to be reviewed. Some of this can only be explored in personal interviews that take a certain amount of your valuable time. It will be worth the effort!

The degree of automation access or actual utilization of computer-based machining and fabrication techniques (including welding, Fig. 22.1.7) is of value. Data collection, ease of retrieval, and data security are important to a CPRS and the clients (Fig. 22.1.8). In fact, computerized access to a wide customer

FIG. 22.1.7 Extent of vendor capability can be determined by observing details such as automated welding in progress. *(Courtesy Hydro Inc., Chicago, IL, http://www.hydroinc.com.)*

FIG. 22.1.8 Demonstrated capability in cataloging major and minor machine components should be determined as part of your assessment of a potential partner CPRS. *(Courtesy Hydro Inc., Chicago, IL, http://www.hydroinc.com.)*

base may place the CPRS in the position of matching customer "A" with spare parts owner "B" half way around the globe. These are opportunities worth pondering—opportunities that exist if you select your CPRS with great care and diligence.

Establish if the CPRS has access to or owns a modern testing laboratory (Fig. 22.1.9). Performance testing can be of great value to the pump user.

HOW THE CPRS MUST ASSIST WITH PUMP REPAIR SCOPE DEFINITION

The competent pump rebuild shop (CPRS) must have the tools and experience needed to define a work scope beyond the routine upgrading we spelled out above. Brain matter counts as tools. The CPRS takes a lead role in defining the repair scope, and all parties realize that reasonably accurate definitions will be possible only after first making a thorough "incoming inspection." On a written form or document, on both paper and in the computer memory, the owner-customer, manufacturer, pump type, model designation, plant location, service,

FIG. 22.1.9 Non-OEM test stand/testing laboratory.

direction of rotation, and other data of interest are logged in, together with operating and performance data. The main effort goes into describing the general condition of a pump, and this effort might be followed by a more detailed description of the work. Either way, it constitutes the condition review.

Condition reviews include photos of the as-received equipment and close-up photos of parts and components of special interest.

End floats, lifts, and other detailed measurements are taken and recorded on a computerized dimensional record (Figs. 22.1.8 and 22.1.10. See also index words on "incoming inspection"). Accurate observations and measurements are the key to effective root cause identification.

Certain measurements must be recorded before and others after total dismantling. Existing components are marked or labeled, and hardware is counted and cataloged. Bearings, bushings, and impellers are removed or dismantled. Bead blasting, steam, or other cleaning methods are listed as initial and routine deliverables, and a completion date for these preliminary steps is agreed upon. It should be noted that only now would a competent shop be ready to move to the next phase, which is to define more closely the scope of its repair and upgrading efforts. An example of combining repairs and upgrading might deal with wear materials.

FIG. 22.1.10 Dimensional mapping "before-versus-after" is important whenever restoration and/or upgrading are involved. *(Courtesy Hydro Inc., Chicago, IL, http://www.hydroinc.com.)*

AN UPGRADE EXAMPLE: WEAR MATERIALS FOR IMPROVED ENERGY EFFICIENCY OF PUMPS

Fluid processing industries and CPRS facilities have embraced the use of current-generation composite materials in centrifugal pumps to increase efficiency, improve mean time between repairs (MTBR), and reduce repair costs. You should be able to depend on your CPRS to use these materials in upgrades. The CPRS must be familiar with one such material, a proprietary reinforced carbon fiber fluoropolymer resin (RCFFR) that has been used successfully by major refineries. RCFFR is a composite material with uniquely low coefficient of expansion and superior temperature stability. As described in Refs. [1,6–9], this one particular high-performance polymer composite has replaced traditional metal and previous-generation composite materials in pump wear rings, throat bushings, line shaft bearings, interstage bushings, and pressure-reducing bushings. The properties of this particular RCFFR reduce the risk of pump seizures. Using RCFFRs allows internal rotating-to-stationary part clearances to be reduced by approximately 50%. The resulting efficiency gain for a typical process pump is in the vicinity of 3–4%. Composite wear materials are included in the 9th (2003) and later editions of the American Petroleum Institute's Centrifugal Pump Standard, API-610.

Only low-expanding, high-temperature-capability, proprietary RCFFR materials have proved to virtually eliminate pump seizures. When properly dimensioned, these materials provide dry-running capability and greatly reduce the severity of damage from wear-ring contact at application points seen earlier in Fig. 20.2.1. Users report freedom from pump seizures during temporary periods of suction loss, off-design operation, slow-rolling, or start-up conditions. When the upset condition has been corrected, the pump continues to operate with no damage or loss of performance. Conversely, when metallic wear components make contact during operation, they generate heat, the materials gall (friction weld), and the pump seizes. This creates high-energy, dangerous failure modes, which can result in extensive equipment damage and even the potential release of process fluid to atmosphere.

CPRS engineers know that correctly chosen proprietary RCFFR wear parts undergo less than the thermal expansion of some other high-performance polymers. This is a very important distinction that contributes to the success of this engineering material. To restate, properly applied and configured pump components made from RCFFR material will reduce the risk of damaging expensive parts. The components must be carefully engineered, but once fits and clearances are properly defined and the parts properly machined, RCFFR materials will reduce repair costs and mitigate safety and environmental incidents.

Minimized wear-ring clearance also increases the hydraulic damping of the rotor, which can reduce vibration severity and shaft deflection during off-design operation. The lower vibration and reduced shaft deflection increase seal and bearing life and help users achieve reliable emissions compliance. This reduction in clearance also reduces the NPSHr on the order of 2–3 ft (~0.6–0.9 m), which can eliminate cavitation in marginal installations.

Field experience shows remarkable success when installing proprietary RCFFRs to achieve all of these benefits. One facility installed such wear rings and line shaft bearings to eliminate frequent seizures in 180°F condensate return service. The condensate return pumps have subsequently been in service for many years without failure. Another user improved the efficiency and reliability of two product shipping pumps by installing proprietary RCFFR wear rings, interstage bushings, and throat bushings. These shipping pumps also have been in service for many years without failure or loss of performance. Hundreds of other services and applications have benefited from properly selected composite wear components; they include pumps in light hydrocarbons, boiler feedwater, ammonia, sour water, and sulfuric acid.

As usual, there are many ways to investigate the cost justification for upgrading to high-performance proprietary RCFFRs in pumps, and a CPRS—by definition—will have no difficulty assisting the user/owner with written return-on-investment calculations. The user/owner should also consider obtaining advice from the manufacturers or suppliers of the advanced RCFFR composites at issue here. Include these suppliers on your list of technology resources and ask to be placed on their list of newsletter recipients.

RECOMMENDED RADIAL GAP GUIDELINES FOR LARGE PROCESS PUMPS

It can be said that not every OEM delivers pumps with internal clearances designed and manufactured for optimized hydraulic and mechanical performance. Adhering to the radial gap criteria listed in Fig. 22.1.11 is very important.

		Gap "B" as Percentage of Impeller Radius		
Type	Gap "A"	Minimum	Preferred	Maximum
Diffuser	0.0 50 in	4%	6%	2%
Volute	0.050 in	6%	10%	2%

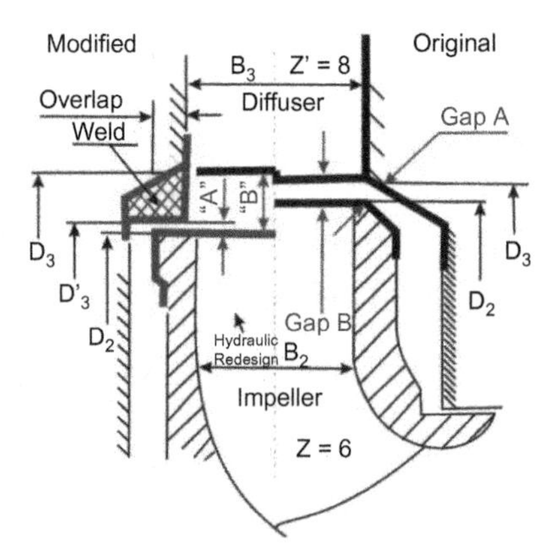

FIG. 22.1.11 Eliminating impeller gap flaws often improves both hydraulic and mechanical performance. *(Courtesy Dr. Elemer Mackay, drawing courtesy of Hydro Inc., Chicago, IL, http://www.hydroinc.com.)*

It should be noted if the number of impeller vanes and the number of diffuser/volute vanes are both even, the radial gap must be considerably larger—10% minimum. We include questions in our assessment; we wish to explore how well the CPRS understands these matters.

So then, we use this as an example of verifying that a repair facility must maximize impeller performance. Best-of-class CPRS facilities know the details that are of consequence:

1. Bring the impeller middle shroud plate out to the impeller OD to reinforce the impeller structure.
2. Stagger the right and left side of the vane to reduce hydraulic shocks and alter the vane-passing frequency.
3. Reduce clearances to optimum between shrouds and casing (Gap "A").
4. Avoid even number of impeller vanes for double volute; or, if the diffuser vanes are even numbered, increase the impeller sidewall thickness.

5. Impellers manufactured with blunt vane tips can also cause trouble by generating hydraulic "hammer" even when the impeller OD is the correct distance from the cutwater. The blunt tips cause disturbance in the volute. This effect may be partly or entirely eliminated by tapering the vanes by "overfiling" or "underfiling" the trailing edge, as described in many texts; among them are Refs. [4,5].

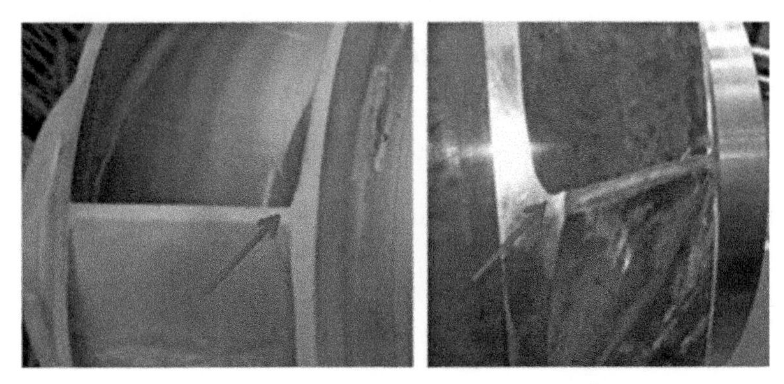

FIG. 22.1.12 Close examination of critical components by a CPRS often uncovers meaningful differences between the "as-is" and "should-be" dimensions. *(Courtesy Hydro Inc., Chicago, IL, http://www.hydroinc.com.)*

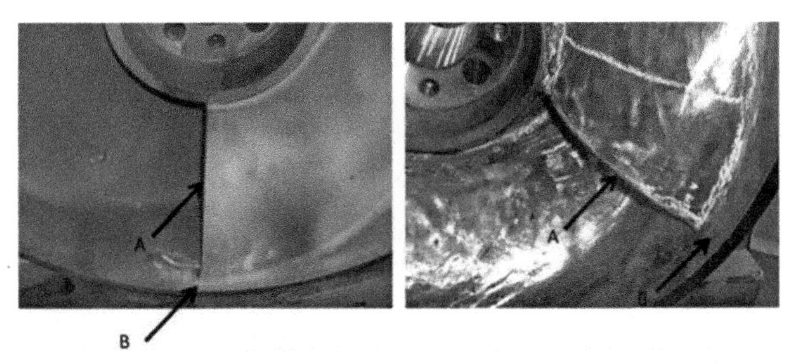

FIG. 22.1.13 Nondestructive testing (NDT) done by a CPRS is a mandatory inspection activity. *(Courtesy Hydro Inc., Chicago, IL, http://www.hydroinc.com.)*

Figs. 22.1.12 and 22.1.13 give us guidance. We expect a CPRS to have well-documented photos, illustrations, and case histories. We want him to establish convincingly how his involvement as our potential non-OEM partner can add value to our enterprise.

CASE HISTORY DEALING WITH AN INGERSOLL RAND MODEL 6CHT 9-STAGE BOILER FEEDWATER PUMP

The importance of a CPRS knowing these guidelines is again best explained by an example that involved a 9-stage boiler feedwater pump that had been previously repaired and was subsequently shipped and reinstalled at the owner's site. When the field installation crew was ready to hand the pump over to operations, the rotor would not turn, and there was zero axial float. The owners lost little time shipping the critically important pump to the CPRS.

When the pump reached the CPRS's workshop, a work scope had to be developed on an expedited basis. The CPRS first disassembled and inspected the components. It was immediately realized that the channel rings had to be investigated and mapped. Typical mapping data were captured in computer-generated sketches.

Here is the action sequence that followed:

1. Set up each channel ring and perform a total indicator reading (TIR) inspection.
2. Remove the old alignment ring from stages 7, 8 and the last stage. Manufacture three new alignment rings.
3. Install the new alignment rings.
4. Set up and finish machining the interstage fit diameters to obtain the proper fit-up.
5. Deburr the channel rings.
6. Make note that, on the last-stage channel ring, the 8th stage vane tips were protruding into the impeller diameter area.

The CPRS knew that protruding vane tips are unacceptable and must have caused interference contact. Indeed, that correlated well with the owner-operators complaint after the pump had been repaired by "others" and it was realized that the rotor could not be moved after the initial, defective repair.

The "as-found" and "as-repaired" conditions were recorded. The vane contour was cut back and ground to the clearance dimensions mentioned in the Gap "A" and Gap "B" table given earlier (Fig. 22.1.11). The element was assembled, using all the existing components except for the gaskets. On the discharge head, the CPRS set up and performed a total indicator reading (TIR) inspection of the critical diameters and faces. New gaskets were supplied for the element assembly and preparations made for quick shipment to the owner-operator.

SHOP REPAIR PROCEDURES AND RESTORATION GUIDELINES ARE NEEDED EVEN BY A CPRS

It should be standard policy for a CPRS to warrantee refurbished equipment and parts of its own manufacture against defects in material and workmanship under normal use. A typical warrantee period for parts and services is 1 year from the date of initial start-up, but not exceeding 5 years from date of shipment, provided that final alignment, lifts, and floats are witnessed by a CPRS service technician. It is only fitting that any deviation from the stated policy must be authorized in writing by the CPRS. All of this should be part of a user's assessment of a competent pump repair shop or CPRS. The user industry must be fully aware that repair work performed by unqualified or careless bidders will represent unacceptable risks.

Pump manufacturers usually supply pump maintenance manuals with detailed assembly and disassembly instructions that are either generic or specific to a particular pump style and model. A number of important checks should be performed by the CPRS for users whose serious goal is to systematically eradicate failure risk. Both CPRS and user have responsibilities in ascertaining that all quality checks are performed with due diligence.

Experience shows that after years of repairs, many pumps are ready for a series of comprehensive dimensional and assembly-related checks. Even a spare part may not meet expectations (Fig. 22.1.14). As a minimum, every critical pump deserves much scrutiny. After the well-known dial indicator checks are complete, the dimensional "before-versus-after" findings should be recorded in electronic and paper format. Unless paper is part of the documentation process, certain glitches or readability oversights will not be discovered until much later. At that time, much finger-pointing will take place, and nothing of value will be the end result. Documentation and data recording includes photo images and close-up details. These will be the key to an equipment owner's desired failure reduction objectives.

SPARE PARTS AVAILABILITY AND PROCUREMENT DECISIONS

Shop repair and spare parts issues are intertwined. There are geographic locations where pump users prefer to make their own replacement parts. Some do so effectively and save money in the process; other locations do so ineffectively and fare poorly in terms of pump MTBF and plant reliability. Since parts need to be in hand before shop repair can be completed, we will deal first with general spare parts availability and procurement issues. That said, we don't want to lose sight of situations where an independent or non-OEM expert repair shop has the capability of making spares and replacement parts. In those important instances, the CPRS is taking the lead and will, of course, assume the requisite responsibilities.

In recent years, some major equipment OEMs have streamlined order entry and production efficiencies, have placed consigned repair part stocks at key user locations, have offered to manage user pump maintenance, and have become more competitive on pricing. But, there have also emerged a small handful of expert repair and upgrade shops that offer both reasonable pricing and highly

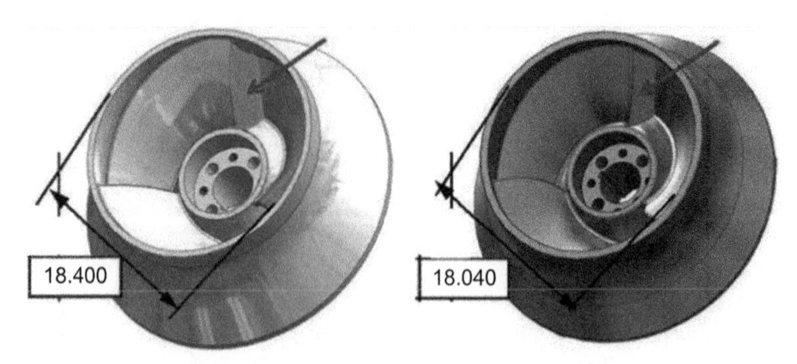

FIG. 22.1.14 Spare parts adequacy must be verified. At one owner plant, and if used in their as-found condition, 14% of the spares in stock would not have been performing as intended. *(Image contributed by Hydro Inc., Chicago, IL, http://www.hydroinc.com.)*

efficient component upgrades. They have been able to match or go beyond the OEM's performance in both regards. Very often, then, the pump user has a choice of buying from any of three parties: the OEM, the low-cost replicator, and the competent pump rebuild shop. It is often the CPRS that can impart both efficiency upgrading and mechanical reliability improvements to the user's pumps.

Buying spare parts on the basis of cost and delivery alone is not the best approach for reliability-focused pump users. There are critical *functional* factors to consider, other than price and delivery, when making the decision of whether or not to purchase OEM or non-OEM parts. To the writer, an unbiased approach requires listing some of these factors in point and counterpoint style:

- The use of non-OEM parts will affect the manufacturer's warranty.

  Note that the CPRS (and remember what these letters stand for) warranties his work just as the OEM would.
- Non-OEM parts may not meet OEM performance specifications.

  Note that the CPRS may be in a position to improve this original performance and will give proof of such claims.
- Since non-OEM parts are often copied from worn OEM parts, critical fits and tolerances on copied parts do not meet OEM specifications, which can lead to premature failures. For example, if a shaft fillet radius is too small, it will increase the shaft stress concentration and could lead to a fatigue failure. If a shaft fillet radius is too large, it may interfere with proper assembly stack-up. Machining marks under a shaft lip seal may cause it to leak in operation. Oversized ball bearing fits on the shaft can cause excessive preloads, which in turn drastically shorten bearing life.

  Remember the meaning of CPRS. Therefore, a CPRS is aware of these risks and will use its pump design and manufacturing experience to avoid these issues.
- Non-OEMs do not have R&D facilities, so they cannot keep up with the latest part changes (mechanical and metallurgical improvements).

Realistically, the CPRS does not tread into R&D territory and makes no claims to provide warranties on unproved solutions.

- Most pump OEMs meet ISO 9001 quality control, while non-OEMs cannot meet these requirements without OEM drawings.

  The same is not true for a top-notch CPRS, whose design experience and response time can be superior to those of the OEM.
- Metallurgies and material mechanical properties may not meet OEM specifications, possibly increasing corrosion rates or reducing pressure ratings.

  While this may have been the case with low-cost replicators, it is not true for experienced CPRS.
- Many part-oriented non-OEMs only offer the most popular repair parts, so that users will still have to deal with OEMs for low volume parts. Sometimes, the user misses out on OEM quantity discounts.

  Again, that should not be an issue with a good CPRS; he's not "part-oriented."

CONSULT REPAIR AND RESTORATION GUIDELINES BEFORE DECIDING WHERE TO PURCHASE

Using a CPRS assessment scoring matrix (Appendix I), a CPRS will have been preselected. Still, when the repair-and-upgrade opportunity presents itself, it is incumbent upon both parties to agree on work scope and critical repair part sourcing decisions. In other words, more definitive repair and restoration guidelines should be consulted in a meeting of user/owner and CPRS representatives at this time. Some of the data provided earlier will have been rolled into a good CPRS operating mode or reflect in the manner whereby the CPRS is conducting his business.

Jointly with the CPRS, the user makes critically important repair part sourcing decisions. The right choice will lower overall maintenance costs by improving equipment MTBF. The ultimate effect will be reduced life cycle costs for the pump.

CONCLUSION

Rebuilding a vintage process pump to original OEM specifications makes no sense given current pump rebuilding technology and changes to the system performance that occur over time. A highly qualified independent rebuild shop with guidelines, checklists, procedures, and a willingness to cooperate with the owner/user is called a CPRS. Its competent and experienced personnel can verifiably offer high-quality upgrades that improve both uptime and efficiency consistent with current system performance requirements. With the considerable consolidations in the pump industry, the distinct possibility exists that the OEM is not able to offer the same engineering competence he previously had. If one simply makes these statements, the issues can debated for a long time. However, assessing vendor competence and making comparisons using the various points made in this chapter tend to become more objective. To the pump user, such assessments may be worth a small fortune in maintenance cost avoidance and run-length extension. Predefine your CPRS and agree on deliverables.

REFERENCES

[1] H.P. Bloch, A.R. Budris, Pump User's Handbook: Life Extension, second ed., Fairmont Press Inc., Lilburn, GA, ISBN: 0-88173-517-5, 2006.

[2] H.P. Bloch, Practical Machinery Management for Process Plants. Vol. 1: Improving Machinery Reliability, second ed., Gulf Publishing Company Houston, Houston, TX, ISBN: 0-87201-455-X, 1988.

[3] H.P. Bloch, F.K. Geitner, Practical Machinery Management for Process Plants. Vol. 4: Major Process Equipment Maintenance and Repair, Gulf Publishing Company, Houston, TX, ISBN: 0-88415-663-X, 1985.

[4] W.Ed. Nelson, J. Dufour, Centrifugal Pump Sourcebook, McGraw-Hill Publishing Company, New York, NY, ISBN: 0-07-018033-4, 1993. pp. 186–188.

[5] V.S. Lobanoff, R.R. Ross, Centrifugal Pumps: Design and Application, second ed., Gulf Publishing Company, Houston, TX, ISBN: 0-87201-200-X, 1992.

[6] M.A. Corbo, R.A. Leishear, D.B. Stefanko, Practical use of rotordynamic analysis to correct a vertical long shaft pump's whirl problem, in: Proceedings of the 19th International Pump Users Symposium, Turbomachinery Laboratory, Texas A&M University, College Station, TX, 2002, pp. 107–120.

[7] M.A. Corbo, S.B. Malanoski, Pump rotordynamics made simple, in: Proceedings of the 15th International Pump Users Symposium, Turbomachinery Laboratory, Texas A&M University, College Station, TX, 1998, pp. 167–204.

[8] R.P. Komin, Improving pump reliability in light hydrocarbon and condensate service with metal-filled graphite wear parts, in: Proceedings of the Seventh International Pump Users Symposium, Turbomachinery Laboratory, Texas A&M University, College Station, TX, 1990, pp. 49–54.

[9] J.P. Pledger, Improving Pump Performance and Efficiency With Composite Wear Components, World Pumps, 2001. Number 420.

Chapter 22.2

Upgrading as an Aftermarket Parts Alternative for Process Pumps

INTRODUCTION

Over the years, many fluid machinery owners have accepted the now somewhat out-of-date practice of separating spare parts by criticality and by statistical probability to fail after so many hours of operation. The practice probably made considerable sense when it originated many decades ago. At that time, engineering, procurement, and construction (EPC) contractors used criticality parameters for primary guidance. Additional subjective factors taken into account by owners and EPCs were geographic location, proximity to experienced repair facilities, and related concerns.

FIG. 22.2.1 Parts quality control takes on many forms. Impeller vane tips are under scrutiny in this semiopen impeller. *(Courtesy Hydro Inc., Chicago, IL, http://www.hydroinc.com)*

As you read this text and perhaps more than ever before, parts availability and quality control are of great importance (Fig. 22.2.1). Downtime costs are probably very high. Qualified maintenance labor is probably in short supply—just as it was in 2010 and 2015.

Today, in 2016, advanced or best-of-class (BoC) owners interweave reliability engineering concepts and upgrade studies in their spare parts decision-making processes. All options are being explored by the truly reliability-focused. As a reliability professional observed in 2014, over the last 20 years, his world-scale fertilizer plant had used a piecemeal approach that was frequently found wanting. His plant inevitably stayed down for 2 days when it should only have been down for only 18–24h after an unplanned shutdown. He was now further challenged by being asked to set up the spares for a new "greenfield" world-scale methanol plant. He wrote:

Surely the spares that we stock for a particular rotating machine should be somewhat generic. The fact that, for turbines alone, we have 3 different manufacturers simply means getting the relevant part numbers/serial numbers to the warehousing people. They would complete an administrative exercise which looks at other factors, i.e. risk, production loss, and so forth. In any event, I would appreciate your thoughts and guidance on the matter.

Our guidance was straightforward and direct. We recommended that this fertilizer plant should abandon the way it—and others—had made spare parts decisions in decades past. Accordingly, the focus of this chapter is on how modern spare parts programs should be handled in 2016.

INVOLVING THE CPRS IN SPARE PARTS DECISIONS

Each plant differs from the next one in certain respects. Although two refineries or hydrocarbon-based fertilizer plants may have been designed identically, their respective staffers neither may have received identical training nor are they equally motivated. One plant will often take greater risks in areas where the second plant would practice greater operating prudence. Some plants allow adequate time for steam turbine warm-up, while others use the dangerous "full speed ahead on lukewarm" approach. Or, although professing to perform failure analysis, a subpar plant will replace failed parts before understanding why the part failed in the first place. We will simply allude to the very straightforward examples found under the index words, constant level lubricators, oil ring issues, and many others in this text. By using risky or failure-prone components, some user facilities will unknowingly set themselves up for repeat failures.

Meet with your CPRS and agree on routine upgrades that avoid the risks associated with less-than-best available technology. Refuse to depend on the spare parts you presently have in hand. They may not fit, may not be sufficiently strong, or may not be as efficient as upgraded parts. Preengineer the upgraded spares that you plan to install during the very next downtime opportunity. Or simply use the next downtime event to take all requisite measurements to engineer, cast, or fabricate these parts. Appoint an "owner" of this strategy and make full use of every one of the methods, techniques, and work procedures in Subject Category 22, of which the chapter you're presently reading is part.

Chapter 22.3

Non-OEM "Billet-to-Final-Test" Capabilities

INTRODUCTION

At least one leading CPRS has the capability of fabricating, replicating, and duplicating multistage high-pressure pumps from billet to final testing (Fig. 22.3.1). The CPRS owns the rights to producing certain legacy pump models or has full reverse engineering capabilities. Working in collaboration with the CPRS, a user can narrow the focus on the best of a number of options and producing from the ground on up is certainly an option to be considered.

FIG. 22.3.1 A welder at work in a CPRS facility. The CPRS duplicates the design of a multistage feed pump, and the job starts with either a stainless steel casting or a forged billet. *(Courtesy Hydro Inc., Chicago, IL, http://www.hydroinc.com.)*

For vertical pumps with their traditional multibowl design, widely differing steel or cast iron engineered start-from-scratch options present themselves (Figs. 22.3.2–22.3.4). For vertical pumps, fabricated steel impellers and bowls (Fig. 22.3.5) are not to be ruled out.

SAND PRINTING OPENS NEW POSSIBILITIES

Three-dimensional sand printing of the cores used in gray cast iron and malleable castings (Fig. 22.3.2) is now available as a significant time-saver in producing cast parts. Best-of-class competent pump rebuild shops (CPRS) have capabilities in that regard.

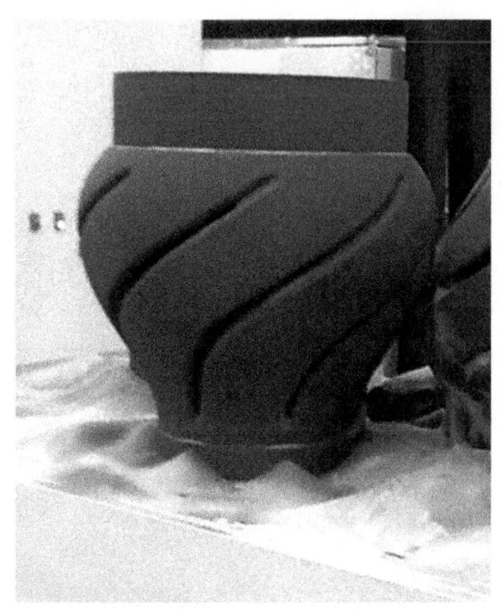

FIG. 22.3.2 Three-dimensional sand printing of a core speeds up the delivery of cast parts such as this vertical pump bowl. *(Courtesy Hydro Inc., Chicago, IL, http://www.hydroinc.com; also Hydro Middle East, Technopark, Dubai, UAE.)*

Fig. 22.3.3 gives an indication of what is involved in engineering gray cast and malleable cast iron parts. Modeling the cooling rate is a predictor of casting quality and a precursor activity which works toward on-schedule delivery of cast parts. The same is true for Fig. 22.3.4: only a thoroughly engineered design and manufacturing process will give the best possible assurance of on-schedule delivery and as-desired performance.

There is also the fabricated steel option (Fig. 22.3.5). Let a CPRS assist you in considering all available alternatives. There are pros and cons to each; there have also been cases where parallel pursuits served as a contingency or backup solution in case the other option encountered a snag.

The feasibility of entering into special arrangements is best discussed early in the discourse or conversation with a CPRS. With worldwide facilities or service teams, their involvement in field installation, recommissioning, and training may be advantageous (Fig. 22.3.6).

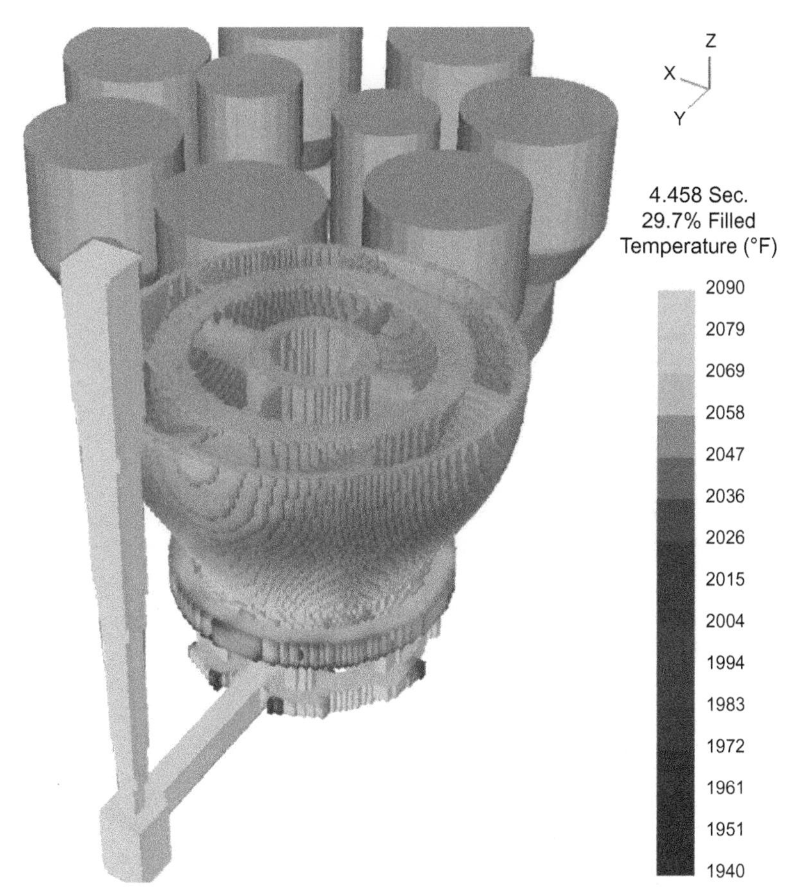

FIG. 22.3.3 Modeling cooling rates is a predictor of casting quality. *(Courtesy Hydro Inc., Chicago, IL, http://www.hydroinc.com.)*

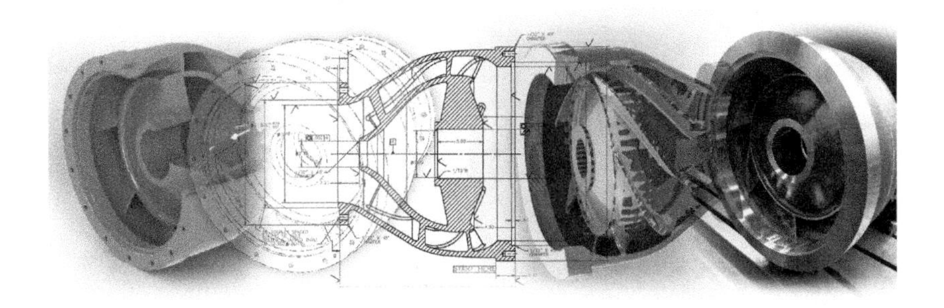

FIG. 22.3.4 Replacement orders often move through concept-model-upgrade design study casting to finish machining of vertical pump parts. *(Courtesy Hydro Inc., Chicago, IL, http://www.hydroinc.com.)*

FIG. 22.3.5 Fabricated steel impellers and bowls are among suitable options that can be pursued by a cooperative CPRS and user team. *(Courtesy Hydro Inc., Chicago, IL, http://www.hydroinc.com.)*

FIG. 22.3.6 A large fabricated steel impeller during field assembly in an emerging technology country. The CPRS and user team cooperate in doing the work. *(Courtesy Hydro Inc., Chicago, IL, http://www.hydroinc.com.)*

Chapter 22.4

Repurposing Major Process Pumps

INTRODUCTION

At times, it makes economic sense to acquire a used or surplus machine and to adapt or modify it rather than placing an order for such a machine and accept very long delivery times. At other times, the original equipment manufacturer (OEM) no longer exists. In that case, the choices include making a duplicate of the OEM's equipment and buying a near-identical machine. The near-identical machine can be reengineered and transformed into the needed machine.

It would be quite correct to point to any of numerous examples that time and again demonstrate the value of employing a well-trained work force. Some plants practice structured and well-guided maintenance efforts; good supervisors often do double duty as experienced teachers and mentors. Their work execution and follow-up inspections are well planned and executed. Nevertheless, other plants often neglect allocating time, brain power, and monetary resources to these essential pursuits. Also, one plant may be located in a geographic area with an abundance of competent repair shops, while another plant is not.

When the implications of different repair strategies were pointed out to groups of maintenance and reliability technicians in four different countries, it was clear that issues and opportunities were not viewed in the same manner at different sites. It was also clear that organizational initiatives such as asset management and operational excellence had gained prominence, whereas attention to detail and the systematic acquisition of highly relevant in-depth knowledge remained vastly underappreciated to this day. When it's all said and done, understanding these details is the absolute bottom line, one of the true keys to superior asset performance. The other true keys to asset performance include design knowledge, maintenance knowledge, and operating knowledge. Each of these branches of knowledge needs to be implemented responsibly. There must be accountability, evenhandedness, and equitability. An ethical reward system needs to be in place for knowledge to be practiced with unerring continuity and wisdom.

Repurposing of fluid machinery (Figs. 22.4.1–22.4.3) is an important adjunct to the spare parts question. The best reliability professionals make use of the opportunities and the solid input available from CPRS entities with worldwide presence. Fig. 22.4.4 is again just a reminder of destaging and its opposite, increasing the number of stages on a pump rotor. The image simply reemphasizes the upgrading aspects of well-designed rotor modification options and verification capabilities made possible by exploiting the testing capabilities of a CPRS.

FIG. 22.4.1 Liquid inlet side ready for fit-up of a transition piece that will join it to pump discharge side. The reconfigured high-pressure multistage pump is ~30 days away from shipment to the user site. *(Courtesy Hydro Inc., Chicago, IL, http://www.hydroinc.com.)*

FIG. 22.4.2 Process pumps can be modified by either removing or adding stages. The casing is lengthened or shortened, and inlet and outlet sides rejoined by welding. *(Courtesy Hydro Inc., Chicago, IL, http://www.hydroinc.com.)*

FIG. 22.4.3 Mounting hole locations verified and baseplate ready for paint application. This reconfigured high-pressure multistage pump is just days away from shipment to the user site. *(Courtesy Hydro Inc., Chicago, IL, http://www.hydroinc.com.)*

FIG. 22.4.4 A five-stage rotor being slow-rolled prior to balancing. For at-speed balancing, the safety gate (left) will be lowered. *(Courtesy Hydro Inc., Chicago, IL, http://www.hydroinc.com.)*

Chapter 22.5

How to Capture and Apply More Pump Expertise

INTRODUCTION

We do not want to lose the focus of "Non-OEM Repairs and Effective Upgrade Strategies"—the overall heading of our Subject Category 22. Recall that strong indications point to fewer spare parts being needed by facilities that implement and follow up with the necessary pursuits in well-structured ways. Plants and facilities staffed by highly competent maintenance and reliability professionals make it their primary goal to avoid repeat failures. They view every single maintenance intervention as a challenge to their desire to upgrade or to design out the need for excessively frequent preventive maintenance actions. Highly competent maintenance and reliability professionals loathe unplanned downtime events and repeat failures. They agree without equivocation that all repeat failures must be ascribed either to the fact that the root cause of a failure has not been determined or to the fact that the root cause is indeed known but remedial action is not being pursued. Highly competent practitioners of maintenance and reliability improvement skills will never be at ease with either of these two possibilities, that is, not knowing what led to a failure or knowing why an asset failed and doing nothing about it. Competent practitioners view these challenges as personal affronts—a bit like a medical doctor losing a patient to the common cold.

SPARE PARTS "POOLING ARRANGEMENTS" CAN BE BENEFICIAL

Smart plants do a considerable amount of pooling of major turbomachinery spares, that is, several plants or units have access to a common spare. Moreover, some plants have found it prudent to specify and procure certain internal (stationary) volute and/or stage components made from readily repairable steel rather than difficult-to-repair cast iron. Some users will only purchase pump impellers that represent prior art, while others will buy custom-designed vane contours that promise (and usually deliver) a higher energy efficiency. Certain free-standing blades in axial pumps falling into the prototype category may have been subjected to high operational stresses and are thus prone to fail prematurely. A leading CPRS could participate in failure analysis; the CPRS would zero in on these possibilities. Of course, even well-designed blades could be at risk if the flow rate is unsteady or deficient in some other ways. Needless to say, the list could go on.

Any reasonable determination of recommended spare parts must include not only the above but also an analysis of prior parts consumption trends and an assessment of storage practices, to name but a few key items. It's no secret that most users are reluctant to publish and thus share their field experience and related pertinent information. Broadcasting past mistakes, existing shortcomings, and underperformance threatens the job security of plant management. Conversely, educating others as to the details that had ensured past successful operations is not popular. Such information sharing, while relatively common in 1960, has progressively declined. Today, in 2016, such sharing is largely frowned upon. The legal departments of some companies consider information sharing tantamount to giving away competitive advantage. Perhaps, reliability professionals should stand up and explain when and why we differ with these legal departments in a number of instances.

ASSESSING AND ASSISTING IN DEFINING SPARE PARTS QUANTITY

Experience shows that many plants or facilities would benefit from periodically engaging competent consultants with years of practical field experience. These consultants should conduct (and teach) periodic audits of oil refineries and major chemical plants. In such an audit, spare parts consumption and maintenance intensity of a facility's process pumps would be properly assessed. However, such an assessment should only be viewed as a first step at best. To be of deeper and lasting value, the assessment would have to include identification of the root causes of repeat failures.

Notice how we have now returned to a multifaceted spare parts assessment strategy. As part of any spare parts audit, the details of tangible remedial steps and the monetary value of upgrade options explained to management would be explored. That is the only logical answer to the question of spare parts stocking in a highly competitive environment. There is certainly no magic computer program that can manipulate the almost endless number of variables that must be weighed and taken into consideration to determine how many spares are needed in different petrochemical plants. Solid experience and an open mind are of greatest value.

While value adding for new plants starts with pump vendor selection, finding the right CPRS is part of the strategy. Parts replication and using a combined rebuild-and-upgrade approach have great merit. So there is no misunderstanding: vendor selection for grass roots construction and design includes up-front identification of experienced repair facilities. These could be original equipment manufacturers' shops (OEM ownership); they could also be non-OEM facilities. A relevant audit would be needed to make this selection. As was mentioned earlier in this chapter of our text, vendor personnel experience and overall capabilities of the OEM and non-OEM would be thoroughly assessed.

YOU STILL WORK WITH PUMP MANUFACTURERS IN NEW CONSTRUCTION

Selecting from among the many pump *manufacturers* may not be easy. Indeed, picking the right *bidders* may well be an important prerequisite for choosing the best pump. Vendor qualifications change over time, but three principal characteristics will always float to the top and impress us as we try to identify capable, experienced vendors:

- They are in a position to provide extensive experience listings for equipment offered and will submit this information without much hesitation.
- Their centrifugal pumps enjoy a reputation for sound design and infrequent maintenance requirements.
- Their marketing personnel are thoroughly supported by engineering departments. Also, both groups are willing to provide technical data beyond those that are customarily submitted with routine proposals.

Vendor competence and willingness to cooperate are shown in a number of ways, but data submittal is the first test vendors must pass. When offering pumps that are required to comply with the standards of the American Petroleum Institute (i.e., the latest edition of API-610), a capable vendor will make diligent efforts to fill in all of the data requirements listed on the API specification sheet. A manufacturer's real depth of technical know-how will show in the way a vendor-manufacturer explains exceptions taken to API-610 or supplementary user's specifications. Most users are willing to waive some specification requirements if the vendor is able to offer sound engineering reasons, but only the best-qualified centrifugal pump vendors can state their reasons convincingly [1].

As a modern user, you must become familiar with machinery quality assessment (MQA). MQA is an up-front activity that uses the input of experienced subject matter experts (SMEs) to determine which equipment best serves the owner/purchaser's long-term asset reliability goals.

Early involvement of a CPRS makes considerable sense in view of the fact that many of the CPRS's key employees have experience with legacy manufacturers.

Part of a combined MQA-SME collaborative effort relates to the review of drawings and the acceptance or upgrading of initial proposals. Drawing reviews and the scrutiny of bills of materials are alluded to; we submit Fig. 22.5.1 as a fitting reminder.

Potential design weaknesses can be discovered in the course of reviewing dimensionally accurate cross-sectional drawings. In such reviews, the diligent and experienced user-purchaser will possibly uncover design weaknesses or even design flaws. What are the case wear ring and bearing materials? Would closed impellers be more appropriate in this service? Make the CRPS your technology resource and hear him out on the matter.

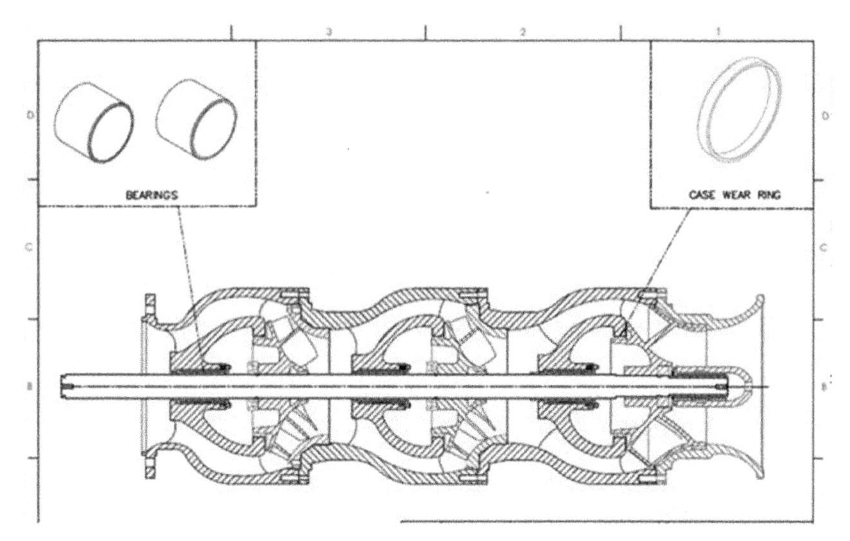

FIG. 22.5.1 Examination of assembly drawings may uncover design weaknesses. *(Courtesy Vendor-neutral drawing courtesy of Hydro Inc., Chicago, IL, http://www.hydroinc.com.)*

Examination and review of the references (to this chapter of our text) will disclose dozens of areas to be reviewed or questioned. There are two compelling reasons to conduct this drawing review during the bid evaluation phase of a project: first, some pump vendors may be unable (or unwilling) to respond to user requests for accurate drawings after the order is placed; second, the design weakness could be significant enough to require extensive redesign. In the latter case, the purchaser may be better off working with a different pump manufacturer or inquiring about a different pump style or model [2]. Again, input from a CPRS with legacy pump experience will be of value.

NEITHER CHEAP PUMPS NOR UNCOOPERATIVE REPAIR SHOPS ARE WISE CHOICES

It is intuitively evident that purchasing the least expensive pump will rarely be the wisest choice for users wishing to achieve long run times and minimized future maintenance outlays. Although a new pump manufacturing company may, occasionally, be able to design and manufacture a better pump, it is not likely that such newcomers will suddenly produce a superior product. It would thus be more reasonable to choose from among the most respected *existing* manufacturers, that is, from the list of manufacturers that *currently* enjoy a proved track record.

The first step should involve selecting and inviting only those bidders that meet a number of predefined criteria. Here's the process by which to determine acceptable vendors for situations demanding high reliability:

- Acceptable vendors must have experience with size, pressure, temperature, flow, and service conditions specified.

- Vendors must have proved capability in manufacturing with the chosen metallurgy, surface conversion treatment, and fabrication method, for example, sand casting, weld overlay, etc.
- Vendor's "shop loading" must be able to accommodate your order within the required time frame (time to delivery of product).
- Vendors must have implemented satisfactory quality control and must be able to demonstrate a satisfactory on-time delivery history over the past several (usually two) years.
- If unionized, vendors must show that there is virtually no risk of labor strife (strikes or work stoppages), while manufacturing of your pumps is in progress.

USE SUPPLEMENTAL SPECIFICATIONS

Imparting reliability to a process pump is important. Without an adequate specification, many pump suppliers "default" their offers and supplies to least expensive initial cost and relatively uncertain delivery times. Neither of these default situations will be in the user's best interest. Accordingly, the second step would be for the owner/purchaser to the following:

- Specify for low maintenance. As a reliability-focused purchaser, you should realize that selective upgrading of certain components will result in rapid payback. Components that are upgrade candidates have been described in the references at the end of this chapter. Be sure you specify upgraded component whenever these are available. Here's a simple example: review failure statistics for principal failure causes. If bearings are prone to fail, realize that the failure cause may be incorrect lube application, or lube contamination. Address these failure causes in your specification and insist on the best available bearing protector seals (Subject Category 14).
- Evaluate vendor response. Allow exceptions to the specification if the exceptions are well explained and valid. The potential reliability impact of waivers should be quantified.
- Future failure analysis and troubleshooting efforts will be greatly impeded unless the pump design is clearly documented. For a certainty, your plant will require pump cross-sectional views (Fig. 22.5.1) and other documents for future repair and troubleshooting work. Do not allow the vendor to claim that these documents are proprietary and that you, the purchaser, are not entitled to see them.

  Therefore, place the vendor under contractual obligation to supply all agreed-upon documents in a predetermined time frame and make it clear that you will withhold 10% or 15% of the total purchase price until all contractual data transmittal requirements have been met.
- On critical orders, *contractually* arrange for access to a factory contact. Alternatively, insist on the nomination of a "management sponsor." The

local salesman is not your management sponsor. A management sponsor is a vice president or director of manufacturing (or a person holding a similar job function) residing at the manufacturer's factory or head offices. You will communicate with this person for redress on issues that could impair pump quality or could seriously delay delivery.

Following these guidelines will give best assurance of meeting the expectations of reliability-focused owner/purchasers.

DON'T JUST REPAIR: UPGRADE!

Recall our earlier comments with regard to legacy pump manufacturers. With time, pump manufacturers have gone through cycles of consolidation and employee attrition. In the process, they may have lost valuable employees to early retirement or to non-OEM pump rebuilders. The rebuilder's focus is competence and customer satisfaction, although valuing the customer's needs should be of great importance to both OEMs and non-OEM pump upgrade providers. We single out competent pump repair shops (CPRS) and know from experience that the CPRS will spot the omissions described in other chapters of this text.

A CPRS with the ability to use three-dimensional mapping of existing pump parts can catalog these for future use. The facility and/or its worldwide associates can rapidly produce three-dimensional casting patterns and molds for needed parts. Suppose your plant has units with two process pumps per service. In that case, the CPRS can rapidly produce the replacement parts for the pump that is out of service, while your process unit stays on line. Even more exciting is the prospect of your process plant entrusting the CPRS with manufacturing these replacement parts with whatever predefined contours and dimensions that will yield higher throughput, or greater efficiency, or extended safe operating time. Engineering and modeling are generally involved. Testing in a modern laboratory environment has great merit (Fig. 22.5.2). Again, it is a possibility that can be explored during the spare parts definition activity for new facilities or during a spare parts audit at an existing plant.

If you were to visit the pump shop of a repair-focused facility, you would expect to see the mechanics and machinists busily working on repeat repairs. The facility's average pump operating life would probably not exceed 2 or 3 years, although the stated design life of many components is more likely in the range of five to as many as ten actual operating years. Ten operating years would mean 20 years of field installation for redundant (spared or parallel-but-standby) pumps—an average seldom reached by even the very best refineries in the United States. To their credit, the very best will not allow themselves get locked into an unproductive cycle of frequent failure and repair. Best-of-class pump user facilities are certainly not excessive consumers of pump spare parts.

FIG. 22.5.2 A modern fluid machinery test stand is thoroughly instrumented and includes variable speed drive provisions. *(Courtesy Hydro Inc., Chicago, IL, http://www.hydroinc.com.)*

Because the best refineries and petrochemical plants are reliability-focused, they view every repair event as an opportunity to upgrade [2]. Reliability professionals report to a manager who insists on fully understanding the exact component selection methods, installation procedures, and work processes that his facility is presently pursuing. This manager then commissions comparisons against best-of-class practices and asks for systematic upgrading at your facility. And so, regarding Figs. 22.5.1 and 22.5.3, your facility would realize that seemingly minor issues exist and seemingly minor upgrades can be immensely helpful in reaching the stipulated higher reliability and desired maintenance cost avoidance goals.

In all you do, remember that some repair-focused refineries and process plants have already fallen by the wayside. Others are sure to close down unless they become reliability-focused. Do your utmost to understand the root causes of equipment failures and accept the fact that there is much low-hanging, ripe fruit on the tree, so to speak. Harvest the "ripe, long hanging fruit" first. And realize that even the most glitzy maintenance cost reduction program will fail unless you step in to do a professional's job. That all translates that you or your designated professional:

- Understand that equipment vendors base on their judgment on operation of machinery under near-perfect conditions and that your field conditions are far from ideal.
- Accept the fact that cost-effective upgrades are often possible and that your plant must become familiar with these.

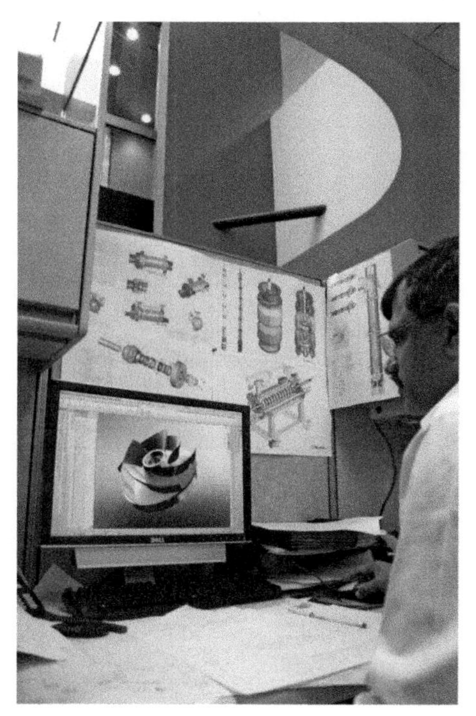

FIG. 22.5.3 A CPRS engineer investigates pump upgrading opportunities. *(Courtesy Hydro Middle East, Technopark, Dubai, UAE.)*

- Implement such basics as have been explained here.
- Insist on root cause failure identification.
- Refuse to tolerate repeat failures.

Most importantly, enable and empower someone at your plant to answer two all-important questions:

1. Is an upgrade possible, and if affirmative, what is the benefit-to-cost ratio of such upgrading?
2. Which CPRS should be prequalified and preselected as our pump spare parts and upgrading provider?

As to spare parts, this author's recommendation is not only to be informed on the supply capabilities of OEMs but to also become thoroughly familiar with the competence and response times offered by leading non-OEM vendors. The best of these non-OEMs both repair and systematically upgrade machines during a maintenance or downtime event. Being on their mailing list for relevant bulletins and inviting them to explain their reverse-engineering and pump upgrading capabilities will prove even more helpful in cases where upgraded spare parts cannot be quickly obtained from the OEM.

ASSESSING AND ASSISTING IN DEFINING SPARE PARTS QUANTITY

Experience shows that many plants or facilities would benefit from periodically engaging competent consultants with years of practical field experience. These consultants should conduct (and teach) periodic audits of oil refineries and major chemical plants. In such an audit, spare parts consumption and maintenance intensity of a facility's process pumps would be properly assessed. However, such an assessment should only be viewed as a first step at best. To be of deeper and lasting value, the assessment would have to include identification of the root causes of repeat failures.

As part of the audit, the details of tangible remedial steps and the monetary value of upgrade options explained to management would be explored. That is the only logical answer to the question of spare parts stocking in a highly competitive environment. There is certainly no magic computer program that can manipulate the almost endless number of variables that must be weighed and taken into consideration to determine how many spares are needed in different petrochemical plants. Solid experience and an open mind are of greatest value.

Value adding starts with vendor selection [3]. But it is certainly implied that vendor selection includes our choice and up-front selection of experienced repair facilities. These could be original equipment manufacturers' shops (OEM ownership); they could also be non-OEM facilities. A relevant audit would be needed to make this selection. Vendor personnel experience and overall capabilities of the OEM and non-OEM would be thoroughly assessed.

Our final comment is again directed at reliability professionals: intelligent spare parts definition involves many parties. As regards managers, please explain to them that what you've just read that represents best practice [4,5]. Be prepared to annoy the "vested interest folks" by asking them to read and to acquire factual knowledge. Expect strong pushback when trying to inculcate zero tolerance for deviations from best practice at every level in an organization. As you then aim to substitute accountability for today's all-pervasive indifference, you will really ruffle many feathers. Then again, ruffling feathers is a very small price to pay for avoiding a great number of unexpected and potentially unpleasant outcomes.

REFERENCES

[1] S. Bradshaw, et al., in: Proceedings of the 17th TAMU International Pump, 2000.
[2] H.P. Bloch, Pump Wisdom: Problem Solving for Operators and Specialists, John Wiley & Sons, Hoboken, NJ, 2011.
[3] H.P. Bloch, Machinery Reliability Improvement, third ed., Gulf Publishing Company, Houston, TX, 1998.
[4] H.P. Bloch, A. Budris, Pump User's Handbook: Life Extension, fourth ed., Fairmont Publishing Company, Lilburn, GA, 2013.
[5] H.P. Bloch, Updating your Sealing Knowledge, Hydrocarbon Processing, 2008. March.

Chapter 22.6

Spare Parts Availability and the Need for Non-OEM Options

INTRODUCTION

Reliability engineers and their associated job functions often try hard to make their responsible supply chain staff or purchasing departments understand that it is impossible for the maintenance department to plan or accurately predict spare parts demand. The reason is that a major proportion of equipment and systems fail randomly. A good example would be rolling element bearings in all industries worldwide.

BEARING REPLACEMENT STATISTICS

According to bearing manufacturers' statistics, 91% fail *before* they reach the end of their conservatively estimated design lives, and only 9% reach the end of design life. The millions that prematurely fail every year can all be slotted in one or more of the seven basic failure categories: (1) operator error, (2) design oversights, (3) maintenance mistake, (4) fabrication/processing defect, (5) assembly/installation error, (6) material defects, and (7) operation under unintended conditions.

Experience clearly shows a preponderance of failures that could be avoided but will continue to persist because of human error. That is also the reason for the randomness with which they occur. There is generally neither the investment in education, personnel, tools, time, nor systematic grooming of a culture of knowledge and loyalty. The latter two would be needed to reduce failure frequencies and improve the predictability of failures.

Large-scale investments have been made in predictive maintenance devices and tools that are then sitting idle or cannot be used because plant staff members have not been trained, mentored, or taught to interpret reams of data. A good example would be the many computerized maintenance management systems (CMMS) that to this day are being fed with useless information such as "bearing replaced," "bearing failed," and "bearing repaired," instead of "bearing failed because loose flinger ring abraded and brass chips contaminated the lube oil. Corrections made by upgrading to a clamped-on steel flinger disk." Of course, the prerequisite is for the facility to practice failure analysis and take remedial action instead of merely replacing parts.

People in authority sometimes make frightfully wrong decisions. As an example, they legislate to standardize on just one type of lubricant, or to buy critically important machinery and components from the lowest bidder, or to procure replacement parts without linking them to a well thought-out specification, or to allow parts to be stocked without first thoroughly inspecting them for dimensional and material-related accuracy or specification compliance.

RANDOM FAILURES DEMAND READY AVAILABILITY OF GOOD OPTIONS

Because of the randomness of failures in HP plants that are not adhering to true best-of-class concepts, it is necessary to have certain parts in stock. We know that some theoreticians extrapolate widely from data that apply to predictable wear-out failures. However, what transpires in hydrocarbon processing facilities is a function of numerous variables, most of which have lots to do with human error. And so, spare parts predictions determined from mathematical models are generally too far-off to merit intelligent discourse. Instead, reasonable specifics are a function of equipment type, geographic location, skill levels of workforce members, etc. For decades, it is has been understood that relevant answers require auditing or reviewing a particular local situation.

After reflecting on the issue, we believe some of today's probable statistics are still close to the ones published 15 years ago. However, some definitions or updated numbers may be helpful:

(a) 25% of all failures are preventable but not prevented because of an arbitrary decision that is simply not rooted in knowledge and experience. For example, "use the cheap oil" may overlook the fact that the cheap oil lacked demulsifiers, antifoaming agents, etc.

(b) 15% of all failures are predictable but not predicted. For example, the random appearance of "black oil" is attributable to O-ring degradation of a certain style of bearing protector seal. The bearings will soon fail but nobody has read the books and articles that describe the occurrence (the occurrence should be linked to a certain risky design feature on a widely used product [1]).

(c) 20% of all failures are predicted, but the machine is not shut down to allow making repairs. Chances are someone in authority overruled an expert who asked for a shutdown when vibration increased beyond a safe level. The risk takers prevailed.

(d) 25% of all failures are predicted, and equipment is shut down for repair. That's good; everybody is happy. Unfortunately, the organization's energy is funneled into restorative maintenance efforts instead of proactive upgrade efforts that would prevent failures in the first place. We believe prevention is better than spending money for restoration.

(e) Only 1% of all failures are neither preventable nor predictable. As of 2010, we changed our mind and no longer believe the old 15% figure was correct. Human beings make the decision to build cities in earthquake zones—either *with* suitable building codes or by *disregarding* such codes. Strong

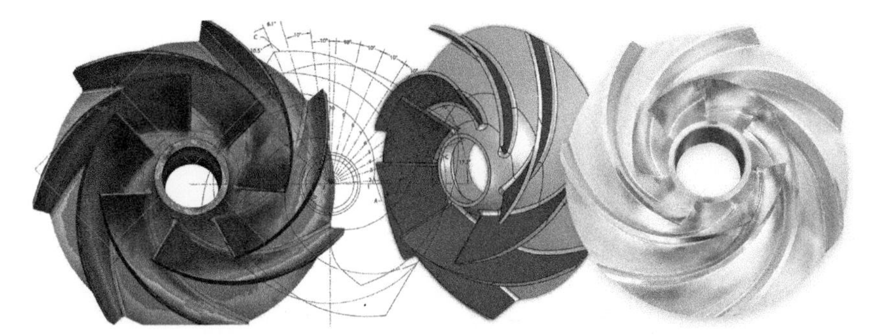

FIG. 22.6.1 Informative bulletins inform us of a non-OEM's capabilities for combining rapid repairs with fabrication of superior components. The narrative sent with this image highlighted the progression from computer simulation to finished part. *(Courtesy Hydro Middle East, Technopark, Dubai, UAE; http://www.hydroinc.com.)*

levies *can be* built or *not* built and can be maintained or *not* maintained. But yes, some machines might fail because a neighboring pressure vessel exploded or because a fire in another unit spreads. These may indeed fit the "neither preventable nor predictable" 1% category. The remaining 14% should simply be distributed among categories (a) through (d).

As to spare parts, our recommendation is not only to be informed on the supply capabilities of OEMs, but to also become familiar with the competence and response times offered by non-OEM vendors. The best of these both repair and systematically upgrade machines during a maintenance event. Being on their mailing list for relevant bulletins (Fig. 22.6.1) and understanding their reverse engineering capabilities will prove even more helpful in cases where spare parts cannot be quickly obtained from the OEM.

REFERENCE

[1] H.P. Bloch, Getting all the Facts is More Important Than Ever, Hydrocarbon Processing, 2009. May.

Subject Category 23

Oil Mist Lubrication and Preservation

Chapter 23.1

Oil-Mist Systems for Plant-Wide Lubrication of General Purpose Machinery

INTRODUCTION

Plant-wide oil-mist systems have been in use in numerous reliability-focused petrochemical and other large plants since the mid-1960s. The latest edition of the American Petroleum Institute (API-610) standard for centrifugal pumps describes oil mist as one of the advantageous lube application methods. In the United States, Canada, South America, the Middle East, and Pacific Rim countries, oil-mist lubrication technology has matured to the point where plant-wide systems are now being specified by many major design contractors. In the United States alone, an estimated 75,000 pieces of machinery were thought to be lubricated with oil mist in early 2016. The worldwide number probably exceeds 150,000.

OIL MIST IS EASILY CONTROLLED AND APPLIED

Modern plants use oil mist as the lube application of choice. Plant-wide piping distributes the mist to a wide variety of users (Fig. 23.1.1). Oil mist is easily produced, and its flow to bearings is easily controlled. Flow, of course, is a function of application orifice ("reclassifier") size and piping ("header") pressure. Unless plugged by an unsuitable, for example, elevated pour point or "waxy" paraffinic lubricant, reclassifiers have a fixed flow area. They require no maintenance whatsoever.

Petrochemical Machinery Insights. http://dx.doi.org/10.1016/B978-0-12-809272-9.00023-2

FIG. 23.1.1 Oil-mist console and associated header system in a plant-wide application. *(Courtesy Colfax Lubrication Management, Lube Systems Division, Houston, TX.)*

Depending on maker and systems provider, header pressures range from 20 to 35 in. (500–890 mm) of H_2O. Modern units are provided with controls and instrumentation that will maintain these settings without much difficulty. However, mixing ratios, typically 160,000–200,000 volumes of air per volume of oil, are frequently incorrect on old-style mist generators that incorporate gaskets and O-rings in the mixing head, unless these elastomers have been periodically replaced or properly serviced. Newer-style vortex-type mixing heads cannot plug. They do not incorporate O-rings and require no maintenance; they are self-cleaning.

Since about 2013, modern electronics have been incorporated in the supervisory instrumentation modules located inside the oil-mist cabinet in Fig. 23.1.1. The signal light on top of the console remains green unless—exceedingly rare—deviations occur. In case of deviation from normal operation, a red flashing signal light will alert process operators. A failure was reported (in 2015) on a precursor electronic module that had been in service for 10 years. Typical reliability statistics confirm that computer boards fail after about 10 years. On critical equipment, we must either use precautionary board replacement (preventive maintenance) or install suitable redundant elements.

SYSTEM AND MIST FLOW DESCRIPTION

In plant-wide systems, oil mist flows into a 2-in. header; top takeoff lines are typically 1-in. nominal diameter. After turning into a vertical line often called a "drop," the drop terminates in a small manifold. Usually, the restriction orifices (also called "reclassifiers" because the mist is partially reformed into a thin stream of oil) are threaded into the manifold. However, there is one very important exception: for shaft surface velocities greater than 610 m/min

(~2000 ft/min), reclassifiers must be located approximately 3/8 in. to ½ in. (~10 to 12 mm) from the bearing rolling elements. We call them "directed" oil-mist reclassifiers. Oil mist exits from a directed reclassifier at a high enough velocity (or with "momentum") to overcome the windage effect (identical to a fan effect) generated by the inclined cage (the ball separator) of an angular contact bearing (refer back to the subject category "Bearings" for more detail).

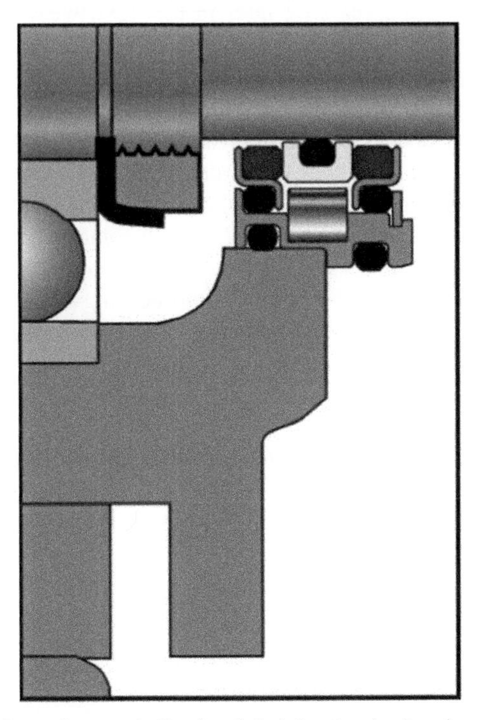

FIG. 23.1.2 An advanced magnetically closed dual-face bearing housing protector seal. It incorporates a face geometry essential for leak-free use on equipment connected to closed oil-mist systems. *(Courtesy AESSEAL Inc., Rotherham, UK, and Rockford, TN.)*

One of our later chapters (23.3) gives considerable detail comparing plants with nonoptimized versus optimized mist entry into equipment bearing housings. But visualize, first, that in order for oil mist to flow, there have to be an entry and an exit port. In other words, oil-mist supplies cannot be dead-ended. Suppose the exit port is at the bottom of the bearing housing and the oil-mist entry is at the top of the bearing housing shown in Fig. 23.1.2 and suppose both of these ports are located to the left of the bearing shown here.

Had oil mist been introduced to the left of the bearing shown in Fig. 23.1.2, we would call it a nonoptimized oil-mist route. It would be considered nonoptimized because not every globule of oil will have gone through the bearing. In contrast, bringing oil mist into the space between the bearing and the magnetically tight-closing-closed bearing protector seal shown in Fig. 23.1.2 would be called a fully optimized application method. On its way to an oil-mist exit port somewhere near the bottom of the bearing housing and to the left of the bearing shown in Fig. 23.1.2, every globule of oil mist will have traveled through the bearing. With optimized entry and exit, oil-mist consumption is about 40% less than at plants that allow mist to enter to the left of the bearing and to then exit at the bottom of a bearing housing. In Chapter 23.3 later in this text, we will describe function and configuration in considerable detail and will again emphasize the advantage of introducing the oil mist between the bearing and its adjacent bearing protector seal in Fig. 23.1.2. The technically and environmentally superior mist entry and vent locations were reported in the October, 1990 issue of *Hydrocarbon Processing* and can also be found in API Standard 610 (eighth and later editions, from 2000). As regards the face-type bearing housing protector seal shown in Fig. 23.1.2, a rotating labyrinth seal with axial microgap can also be used but will allow a somewhat greater volume of oil mist to escape. For an average pump bearing housing, however, the amount of oil escaping as an oil mist into the atmosphere is well below a teaspoon per day. It will not exceed reasonable environmental and pollution guidelines (see index words "Bearing Protector Seals" or Subject Category 14 of this text).

The point must be made again: forward-looking plants have actually used the API recommended oil-mist application route since the mid-1970s. It was then already recognized that mist entering at locations too far from the bearings could have difficulty overcoming bearing windage effects. If, as a result of the reclassifier location being too far from a high velocity bearing, the mist is kept away from the bearing, then insufficient amounts of oil mist would reach the bearing's rolling elements. The bearing would be starved of lubrication and would fail prematurely.

WHERE AND HOW OIL MIST IS WIDELY USED

Oil-mist lubrication and oil-mist preservation differ in purpose. Oil mist exists in the form of mist (an oil fog) similar to hair spray in appearance. It will remain in the misted or atomized form until the very small globules come into contact with each other. At the point of destination or at the point of reclassification, many small oil globules collide. They will become large droplets; they will plate out and coat (or "wet") the bearing elements. A light coating of oil stays on the bearing, and excess oil mist leaves through a bearing housing drain or exit port. Because there is no liquid oil level maintained in the bearing housing, we call this application mode "dry-sump" or "pure" oil mist.

In bearing environments with liquid oil purposely kept in the sump, application of oil mist to fill the space above the oil level is called "wet sump." In wet-sump applications, the mist is simply serving the purpose of keeping out external contaminants, including water vapor. Filling the bearing housing with oil mist at a slightly higher than atmospheric pressure keeps external contaminants away from the bearing housing interior. Wet-sump systems depend on some means of transporting the liquid oil into the bearing. The oil mist in wet-sump applications does little, if anything, to provide lubrication.

Because dry-sump oil mist virtually eliminates preventive maintenance (no oil changes, no constant level lubricators, no oil rings, no abrasion product from oil rings, no oil flinger disks, and no intrusion of atmospheric contaminants), dry sump is used far more extensively than wet sump. Dry-sump oil-mist lubrication is always possible on any conceivable type of rolling element bearing; there are no exceptions to this universal rule or its applicability for different machines. Many users find it easy to cost-justify plant-wide oil mist because they include electric motor bearings in their calculation. If electric motors are not part of one's oil-mist lubrication strategy, the payback time is twice what it would otherwise be.

It should be noted that three or four different styles of reclassifiers are in successful use by modern industry. The reliability professional must select the correct one of these. Alternatively, engage an experienced oil-mist service company and listen to their advice.

A general overview of the different machinery applications is provided in Figs. 23.1.3–23.1.13.

FIG. 23.1.3 A large fluid machine equipped with oil-mist lubrication via the small modular unit in the lower left. This is a "per machine" module sized to provide enough oil mist for one very large or as many of six small pumps. It would not, however, be called "plant-wide." *(Courtesy Colfax Lubrication Management, Lube Systems Division, Houston, TX.)*

FIG. 23.1.4 Vertical pump (motor not yet installed) in storage and equipped with temporary oil-mist supply tubing. Once installed at site, permanent stainless steel tubing would replace the temporary plastic tubing. *(Courtesy Colfax Lubrication Management, Lube Systems Division, Houston, TX.)*

FIG. 23.1.5 The governor end of a small steam turbine equipped with wet-sump oil-mist lubrication. *(Courtesy Colfax Lubrication Management, Lube Systems Division, Houston, TX.)*

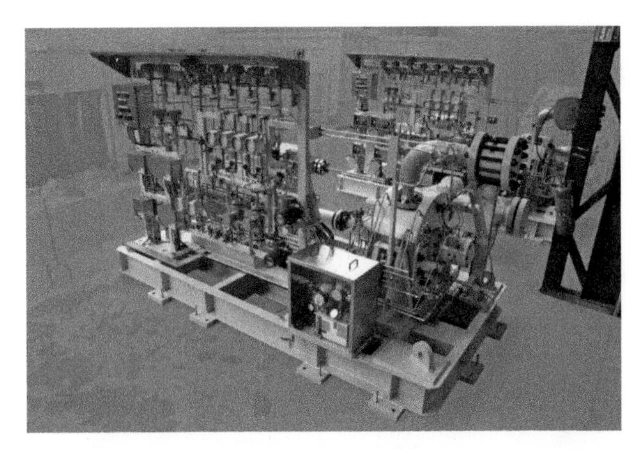

FIG. 23.1.6 A "packaged" vacuum pump. The small oil-mist supply module is located on the I-beam in the foreground. *(Courtesy Colfax Lubrication Management, Lube Systems Division, Houston, TX.)*

FIG. 23.1.7 Electric motor in a storage (preservation) location. After field installation, the oil mist will float above a liquid oil sump, making this a "wet-sump" oil-mist lubricated motor. *(Courtesy Colfax Lubrication Management, Lube Systems Division, Houston, TX.)*

FIG. 23.1.8 A between-bearing fan (or blower) with wet-sump oil mist on the bearing pedestal. *(Courtesy Colfax Lubrication Management, Lube Systems Division, Houston, TX.)*

FIG. 23.1.9 The two pillow block bearings of this forced-draft (overhung) fan are lubricated by oil mist. Excess mist coalesces and reverts to oil in liquid form, collected in the container at the lower right. A small hand pump allows an operator/technician to pump this oil back into the 2-in. header. A small globe valve allows direct draining of coalesced oil. *(Courtesy Colfax Lubrication Management, Lube Systems Division, Houston, TX.)*

FIG. 23.1.10 A combination of wet-sump oil mist and dry-sump oil mist is used on this positive displacement gas moving machine. As in Fig. 23.1.6, a self-contained oil-mist reservoir/generator unit is seen in the foreground. *(Courtesy Colfax Lubrication Management, Lube Systems Division, Houston, TX.)*

FIG. 23.1.11 Multistage moderate pressure blower with wet-sump oil mist. This user collects coalesced oil but will drain it through a small globe valve instead of using the pump-back method in Fig. 23.1.9. *(Courtesy Colfax Lubrication Management, Lube Systems Division, Houston, TX.)*

FIG. 23.1.12 Medium-size pump and driver equipped with dry-sump oil mist. Both drain valve and pump-back provisions exist on the collecting container. *(Courtesy Colfax Lubrication Management, Lube Systems Division, Houston, TX.)*

FIG. 23.1.13 Fin-fan bearings are often lubricated with pure oil mist. Trouble-free operation in excess of two decades has been achieved. *(Courtesy Colfax Lubrication Management, Lube Systems Division, Houston, TX.)*

Header Temperature and Header Size

Ambient temperature has never been an issue for properly designed systems. Once a mist or aerosol of suitably low particle size has been produced, it will travel in header systems regardless of whether the system is located in Alaska or the Middle East. Particle size is influenced by the temperature constancy of both air and oil in the mixing head. We had earlier called it the oil-mist generator; it is located inside the oil-mist cabinet in Fig. 23.1.1. Clean, dry, compressed air meets the lubricating oil and forms a mist that will move along inside noninsulated headers toward the many possible destinations shown in Figs. 23.1.3–23.1.13.

To restate, ambient temperature has very little influence on mist quality and effectiveness. Mist temperatures in headers have ranged from well below freezing in North America to well over 122 °F (50 °C) in the Middle East. Regardless of geographic location, responsibly engineered systems will incorporate both oil and air heaters since these are helpful in maintaining constant and optimized air/oil mixing ratios in the oil-mist generator. The heaters must have low watt density (low power input per square inch of surface area) in order to prevent overheating of the oil. Users that try to save money by omitting heaters or using undersized headers will not be able to achieve optimum life cycle cost of their assets. Undersized headers will sag unless supported at many locations. In undersized headers, oil will perhaps travel at relatively high velocity. There would then be many collisions of small oil globules. These would combine into liquid oil before even reaching their intended destinations. In that case, an excessively lean mist would arrive at the point to be lubricated.

Recall our earlier point: the application of dry-sump oil mist is highly advantageous for a number of reasons. Among these, we found lower bearing temperatures, the presence of nothing but uncontaminated oil mist, and the exclusion of external contaminants. However, two reasons that are often overlooked involve oil rings (see Subject Category 24) and bearing cooling concerns. Neither oil rings nor cooling is used with the dry-sump oil-mist application method.

Oil rings represent outdated 18th century technology; they were initially developed for slow-speed machinery during the industrial revolution. Deletion of oil rings is one of the many keys to improved reliability of virtually *any* type or style of bearing. Oil rings are known to have journal surface velocity limitations, sometimes as low as 2000 fpm or 10 m/s. So as not to "run downhill," which might cause the rings to make frictional contact and slow down, ring-lubricated shaft systems must be installed with near-perfect horizontal orientation. That can become a time-consuming maintenance task.

If rings run downhill and contact any part of the bearing housing interior, there is a high risk of frictional wear. Abrasive wear products then cause contamination of the oil. Also, oil rings will malfunction unless they are machined concentric within relatively close tolerances. They suffer from limitations in allowable depth of immersion and, to operate as intended, need narrowly defined and closely controlled oil viscosity.

ENVIRONMENTAL AND HEALTH CONCERNS FULLY ADDRESSED

For decades, these concerns have been addressed by using oil formulations that are neither toxic nor carcinogenic. Such formulations are available to responsible users. Appropriate lubricants have also been formulated for minimum stray mist emissions. These, too, are readily available to responsible users. Stray mist emissions can be kept to very low values by installing suitable face-type bearing housing seals. Face-type magnetically closed seals were shown in an earlier chapter (also Subject Category 14) of this text and are further explained in the many references. Unlike old-style labyrinth or other housing seals that allow highly undesirable communication between housing interior and ambient air, face-type devices seal off this contamination route.

Closed oil-mist systems have been available since being first applied in the Swiss textile industry in the late 1950s. Today, closed systems are in use at many US petrochemical plants and other industries. Closed oil-mist systems allow an estimated 99% of the lube oil to be recovered and reused. Closed systems emit no oil mist into the environment and are available for use by concerned users.

EXPERIENCE WITH MODERN OIL-MIST SYSTEMS

Actual statistics from a world-scale facility convey an accurate picture of the value of properly applied oil-mist technology. This petrochemical plant went onstream in 1978 with 17 oil-mist systems providing dry-sump oil mist to over 400 pumps and electric motors in the facility. With dry sump, the oil mist is introduced at a location that guarantees its flow through the bearings and to an appropriate vent location. There are neither oil rings nor any other provisions for the introduction of liquid oil on pumps and motors with rolling element bearings at the plant.

Over a period of 14 years, one qualified contract worker serviced these systems by visiting the plant one day each month. In this 14-year time period, there was only one single malfunction; it involved a defective float switch in one of the 17 systems. The incident caused a string of pumps to operate (and operate without inducing bearing failure!) for eight hours. In 1992, the combination of availability and reliability at this US Gulf Coast plant was calculated to be 99.99962%.

In 2016, feedback was obtained from a highly respected pump expert with consulting assignments at a different refinery from the one just mentioned. He wrote: "The choice of properly designed and installed oil mist is not an issue with this refinery. The latest unit built in this refinery, circa 2001, has all of its pumps running on oil mist for 14 years and has not had a pump out for maintenance. This is, so far, an infinite MTBR. The process is relatively clean gasoline products. The pumps were properly sized for the services. Sister units using oil mist have had similar long MTBRs. Operations teams and reliability groups believe in the results achieved with oil mist along with proper sizing and proper installation. The statistical evidence is very supportive of the choices made since the early 1970s when oil mist was first tried in this facility."

The pump expert also mentioned that no disagreements existed within the refinery: oil mist contributed to their high reliability and profitability. The issue had to do with new project funding being curtailed. A rather uninformed subject matter expert (SME) was supporting a "go-cheap" high-risk corporate philosophy. From our own experience, we believe this SME's mind is made up; he will never accept the findings of some of his corporation's best refinery units. Why aim to be informed if remaining uninformed is well rewarded?

EMPHASIS ON PURE (DRY-SUMP) OIL MIST

Being aware of the relative unreliability of conventional lubricant application methods involving oil rings and constant level lubricators, knowledgeable reliability professionals can attest to the utility and overall advantages of properly engineered dry-sump oil-mist systems. Certainly, the known advantages of properly engineered oil-mist systems far outweigh the actual or perceived disadvantages. It is unfortunate that much information to the contrary either is anecdotal or pertains to systems that were not correctly designed, installed, maintained, or upgraded as new technology became available.

Only dry-sump applications will lubricate, preserve, and protect both operating and stand-by rolling element bearings. At all times, only clean fresh oil will reach the bearings. If wet-sump lubrication is used, any debris produced by oil ring inadequacies will very likely reduce bearing life. In many instances, bearing operating temperatures with dry-sump oil-mist lubrication are 10 or even 20°F (6 to 12°C) lower than with wet-sump lubrication. Industry experience with dry-sump oil-mist systems is well documented, and its superiority over both conventionally applied liquid oil and wet-sump oil-mist lube applications has been solidly established.

Regrettably, there are still entire plants that try to get by on wet-sump oil mist. They either misdiagnosed the root causes of a prior failure or have elected to stay uninformed for reasons we will never fully understand. Wet-sump lubrication makes economic sense on sleeve bearings only. With wet sumps, the only function of oil mist is the exclusion of atmospheric contaminants. This exclusion is achieved due to the oil mist existing at a pressure slightly above that of the surrounding ambient air.

Chapter 23.2

Oil Mist as a Preservation Strategy and What to Do if No Oil-Mist Preservation Was in Place

INTRODUCTION

Suppose a major oil refinery had been planning for a substantial modernization around 2008. Energy efficiency gains and the ability to process a more readily available crude slate were key motivating factors. But that's when an economic downturn hit and field erection was stopped for the usual financial considerations.

But in late 2008, much of the equipment either was on its way or had just arrived at the field staging area. There was no budget for the type of state-of-art equipment protection and preservation successfully practiced in the 1960s and often described in books and articles [1–3]. Fig. 23.2.1 gives a glimpse of protective oil-mist lines connected to a fluid machine. Storing unprotected equipment in the open air always invites trouble. Without protection, expect adversity.

FIG. 23.2.1 A fluid machine with oil mist connected to seal and bearing regions. The oil mist will prevent atmospheric contaminants from entering. *(Courtesy Colfax Lubrication Management, Lube Systems Division, Houston, TX.)*

ISSUES WITHOUT PROTECTION

Without oil mist as a preservation method, standby equipment is exposed to the risk of oil being "wiped off" due to vibration transmitted from the adjacent running equipment. The corrosion risk of the bearings in nonrunning equipment is greatly reduced by the use of oil mist. The rate of corrosion is a function of ingress and egress of air (the "breathing" action) of the affected bearing housings. This breathing action takes place only in unprotected bearing housings and is not possible in bearing housings filled with oil mist at about 1 psi over atmospheric pressure. Industry now has in excess of four decades of experience with this protection method on many thousands of pumps and motors. It's one of the important topics found in "Pump Wisdom," [3]. Both indoor and outdoor protection methods are often used in modern industry (Fig. 23.2.2).

FIG. 23.2.2 Indoor storage protection under a "blanket" of oil mist. *(Courtesy Colfax Lubrication Management, Lube Systems Division, Houston, TX.)*

Whenever the conversation turns to oil mist, we need to highlight the applicability of oil mist to electric motor bearings. All motors with rolling element bearings will benefit from dry-sump oil-mist application. Thousands of motors have been so lubricated since the early 1970s [2]. Many books and dozens of papers have been written about the subject. Inferior grease lubrication has been superseded by superior dry-sump oil mist for decades. Oil mist allows plants to eliminate slinger rings, constant level lubricators, desiccant breathers, bull's-eye sight glasses, overgreased motor bearings, and mountains of failed bearings. Closed oil-mist systems are proved technology; they excel by further reducing lubricant consumption and protecting the environment.

OIL-MIST PROTECTION STATISTICS

Filling or "blanketing" all internal equipment spaces with oil mist was proved effective and economical since the mid-1960s. Both indoor and outdoor storage were applied on hundreds of process pumps in many plants beginning about 1965 [2]. Staging yards similar to the one shown Fig. 23.2.3 have successfully set up the same oil-mist consoles that would later serve hundreds of pumps in their designated process units. In other words, the oil-mist cabinet and oil-mist generator initially used for the protection of many pieces of equipment are later used as the day-in-day-out lubrication service for process pumps and their electric motor drivers. Combining project management and maintenance/reliability wisdom, this style of planning for success has proved eminently successful for over 40 years. At one facility, in 1979, equipment infant mortality of machines kept in an oil-mist staging yard was kept well below 3%. In stark contrast, the probable failure rate upon starting up after 18 months of unprotected outdoor storage at a facility in 2013 was estimated at about 30%, which is not a pleasant prospect.

FIG. 23.2.3 Outdoor storage under a "blanket" of oil mist. More typically, open-to-atmosphere storage yards serve 100-600 pieces of equipment and no shipping crates are involved. *(Courtesy Colfax Lubrication Management, Lube Systems Division, Houston, TX.)*

WHAT IF THERE WAS NO SUCH PROTECTION?

Proceeding toward starting up such a refinery unit without first implementing proactive remedial steps could prove very costly. To have 30% of 200 or more process pumps fail upon startup is deemed unacceptable. At an (optimistically) estimated $20,000 per pump repair, well over a million dollars would be spent on repairs. The cost of even a single unit downtime day may be a multiple of that amount. So, evasive action and risk reduction will be needed on equipment that has not been optimally preserved. Fluid machinery left in the open without full protection for more than six months should not be expected to run flawlessly for very long. Bearings and mechanical seals are likely to fail before they will have reached their respective design lives; accordingly, five reasonable risk reduction steps should be pursued:

- Equipment dismantling and cleaning (or part replacements) should be prioritized by criticality.
- Dismantling and reassembly of critical machines should be entrusted to highly competent individuals. Deliverables and accountabilities should be defined in writing. With these definitions spelled out and looking at vendor responses, thorough bid evaluations should be next. They will point to qualified vendors. It should come as no surprise that the lowest bidders are rarely the most qualified vendors.
- Designate an "owner" for each piece of equipment that is being dismantled. Use retirees or other conscientious, experienced individuals to act as such "owners" during the equipment inspection and rebuilding process.
- Give the "owner" a checklist of points or items to ascertain and which equipment vendors have often overlooked. Then, have the "owner" certify that elusive failure causes have been addressed and will be avoided.
- Have the "owner" stand next to an operator when the equipment is first started up. On all machines that have been stored with inadequate protection and irrespective of criticality, let a designated "owner" take vibration and thermal imaging readings during the start-up process and daily thereafter. An "owner" can handle 6–8 machines, and his/her involvement may be discontinued once the operating staff can take over their routine monitoring duties.

HOW OIL MIST WORKS IN PRESERVATION SYSTEMS

Oil mist-protected equipment was shown in Figs. 23.2.1–23.2.3. So, how does it work? As a general rule, one could proceed with the following:

(a) Oil mist will "plate out" (actually, coalesce) when equipment operates. Operation causes a measure of turbulence; the atomized particles get knocked together and become large and heavy. Turbulence is obviously not created in equipment that's standing still. Therefore, the conversion of oil mist to liquid oil proceeds at a rather slow rate in nonrunning equipment.
(b) A low-point drain location should be identified and left open at all times. This low point is useful because it prevents oil accumulation and ensures through-flow of a small amount of mist. That's really your makeup mist.
(c) With rare exceptions, a 3 mm drain orifice is all that's needed.
(d) A machine in storage and ready for preservation can be slanted or skewed to create or locate a low-point drain.
(e) A rotor should be manually rotated two-and-a-half turns every 6 months.

(f) Prior to machine commissioning, one should blow steam through the equipment. This will remove the thin coating of oil that will have accumulated on the surfaces. Use oil with formulations that will not promote future stress-corrosion cracking

(g) Collect some of the exiting blow-through steam and observe if it contains oil.

As a *not so general rule*, one could proceed with (h), below:

(h) One could calculate the total interior surface areas that are being wetted on a particular machine and then assume an oil film of a thickness of 0.0005 in. coating these surfaces. Knowing area and oil film thickness, one could then calculate the total volume of oil that will exist in the equipment at any particular time period and convert this to milliliters of oil. Using conversion formulas from the Internet, this milliliter volume can be readily converted to gallons. One could then assume the oil costs $50 (or whatever) per gallon and figure out what it all costs, after subtracting oil lost. One would have to account for oil lost through the 3 mm orifice, based on an equipment-internal pressure of 0.3 psi above ambient.

Belatedly implementing these failure-reducing action steps in 2016 will not be cheap. The monetary outlay ranks somewhere between oil-mist preservation (~$200,000 in 2007) and having 50 or 60 near-catastrophic failure events if nothing is done. From an organizational point of view, reliability thinking and equipment preservation issues should always be laid out for management at an early stage.

We do know that inadequately stored equipment will be risky to run and that money spent on suitable up-front preservation will pay dividends for a long time to come. Numerous documented accounts are available where oil mist has provided reliable protection for idle and standby equipment at many refineries over the years. There is no reason not to provide that same protection for current projects and its new equipment during storage.

NITROGEN FOR STORAGE PROTECTION

Questions regarding nitrogen as a protective blanket for equipment storage are occasionally asked, and our answer is worth considering. Although most of the atmosphere (78.08%) consists of nitrogen, nitrogen is an inert gas, and breathing this gas without an admixture of oxygen can be deadly. In fact, dozens of people have suffocated when they inadvertently or without breathing apparatus entered vessels filled with nitrogen. However, nobody has ever suffocated in oil mist applied to stored equipment. And so, it's important to say this again: safety is an ethical imperative. To some, safety is a concept that is mentioned in passing and brushed aside all too easily.

Regrettably, we cannot share details of a consulting experience that dates back to about 2013. Relating details is often thwarted by clients who first make a competent advisor sign a nondisclosure and (occasionally) a nondefamation clause. However, years ago, I spoke with a client whose facility had initially "saved" a few dollars by not using the right equipment protection. They later paid the price.

Back to the client refinery that had no money for storage protection when the economy took a downturn a few years ago, they did not like the bulleted risk reduction strategies (above) that caused them to spend millions undoing the (potential) damage. I simply believe that reliability professionals must aim for the implementation of best available technology. If they don't see this as their role, they have no business calling themselves reliability professionals. We found that competent advisors are often brought in *after* the damage is done. On average, uninformed opinion voicers are often involved *initially*. Of course, they are held in much more esteem because they always support the boss—but only until the proverbial chickens come home to roost!

REFERENCES

[1] H.P. Bloch, A. Shamim, Oil Mist Lubrication: Practical Applications, The Fairmont Press, Lilburn, GA, 1998. pp. 143–154.

[2] A. Budris, Pump User's Handbook: Life Extension, third ed., The Fairmont Press, Lilburn, GA, 2010. pp. 279–304.

[3] H.P. Bloch, Pump Wisdom, John Wiley & Sons, Hoboken, NJ, 2011. p. 16.

Chapter 23.3

Closed Oil-Mist Systems and an Update on Maintenance Cost Avoidance

INTRODUCTION

Oil mist successfully lubricates operating machinery, protects and preserves standby equipment, and provides superior lubrication to electric motors driving process pumps. Oil mist also protects or preserves standby equipment and has been cost-justified in all categories of equipment with rolling element bearings. First applied to the onstream (continuous) lubrication of pneumatic tools in 1937, oil-mist lubrication had migrated to high-speed spindles in textile machinery by the 1940s. Two decades later, many plant-wide oil-mist systems were commissioned in the oil refining and petrochemical industries. As of 2016, an estimated 3000–3500 plant-wide systems are in use worldwide.

FIG. 23.3.1 Small oil-mist console for 20 pumps (on left), and return mist coalescer-collector vessel (on right). *(Courtesy Colfax Lubrication Management, Lubrication Systems Division, Houston, TX.)*

OIL SAMPLING RECOMMENDATIONS

Beginning with the mid-1980s, many facilities have discontinued using open oil-mist systems. In open systems, the mist was allowed to escape into the environment after first passing through the rolling elements of bearings. In closed systems, the spent oil mist is piped to a central collecting point where coalescing (turning it back into liquid oil) takes place in a suitable device.

Knowledgeable lubrication engineers or practitioners of predictive maintenance (PdM) would occasionally sample the oil that has passed through the equipment bearings. The coalesced bulk oil in the coalescer-collector vessel of Fig. 23.3.1

should be sampled twice per year. Sampling of the coalesced mist drawn from the bottom of a particular bearing housing is optional; it would pinpoint where a defect (or contaminant intrusion) originates, whereas monitoring bulk oil and finding contaminants would merely determine if the oil should be filtered before being reused. Bulk oil sampling can be done either by portable analyzers or by a generally more precise laboratory analysis. Bearing condition monitoring is made feasible through the use of any of a number of widely available PdM devices. Ultrasonic monitors can indicate bearing defects; so can accelerometers and stress wave monitors, the latter sometimes called incipient bearing flaw detectors.

Some PdM devices monitor vibration amplitude and frequency spectra. Still, with superior synthetic lubricants made cost-effective by virtue of very low oil consumption and cooler running bearings, a 5-year oil life should be reached without difficulty. It's only when standard mineral oils are used that it might be wise to schedule more frequent oil replacement intervals. Because mineral oils tend to oxidize more readily than synthetics when hot, it may become necessary to send the (oxidized) oil to a waste oil recycling firm or to dispose of the oil by mixing it with the fuel used in boilers or furnaces.

The difference between wet-sump ("purge mist") and dry-sump ("pure mist") applications was explained in more detail earlier. In the wet-sump method, a liquid oil level is maintained near the bottom of the bearings, and the mist fills the housing space above the liquid oil. Dry-sump oil mist describes the application method whereby no liquid oil level is maintained in the bearing housing (Figs. 23.3.2 and 23.3.3). Lubrication is provided entirely by oil mist migrating through the bearing [1].

"OLD-STYLE OPEN" AND "NEW-STYLE CLOSED" OIL-MIST APPLICATION

In old-style "open" oil-mist systems (Fig. 23.3.3), the air/oil mixture fills the bearing housing, but not all of the mist passes through the bearings. A portion of the coalesced droplets takes a straight top-to-bottom path through the bearing housing. Only this portion of the coalesced oil and stray mist then exit near the bottom of the housing and can be collected for disposal at the drain location. For bearings to be properly lubricated, the oil mist will have to pass through the bearings and then escape at the two unsealed regions where shafts protrude through the bearing housing.

Efforts to simply provide highly effective bearing housing seals at the ends labeled "oil mist out" in Fig. 23.3.3 had unexpected consequences for inexperienced users. Oil mist works by causing small globules of lubricating oil to provide an oily coating on the bearing components. However, so as to provide continuous oil replenishment on bearing surfaces, the oil mist must flow and cannot be stagnant, or dead-ended. In "old-style" configurations, tight-sealing bearing protector seals placed at the "oil mist out" locations in Fig. 23.3.3 would very often result in dead-ending. Without oil-mist flow, there was then neither cooling nor lubricant replenishing on bearing surfaces.

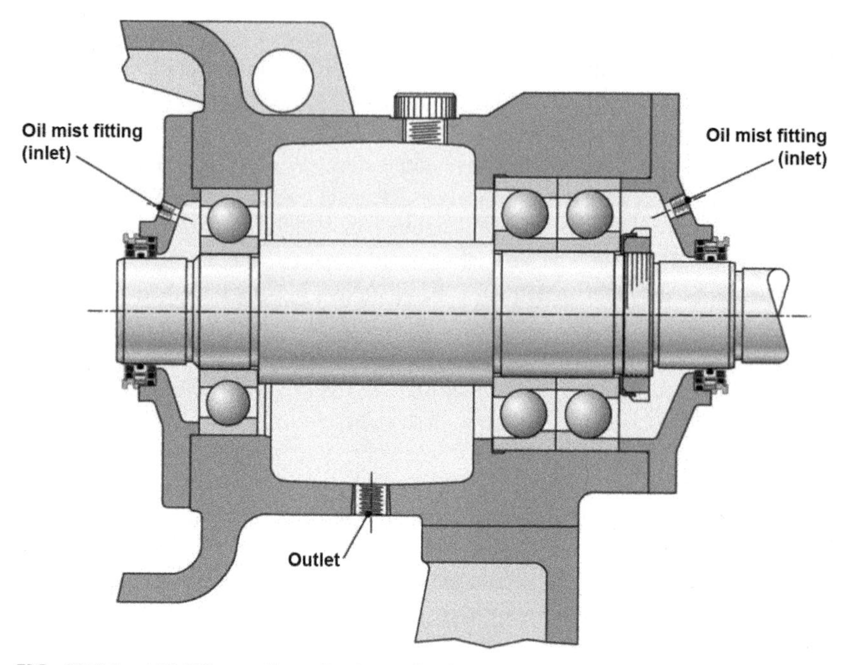

FIG. 23.3.2 API 610-compliant oil-mist application at locations between the rolling element bearings and the magnetic dual-face bearing isolators. *(Courtesy AESSEAL Inc., Rotherham, UK, and Rockford, TN.)*

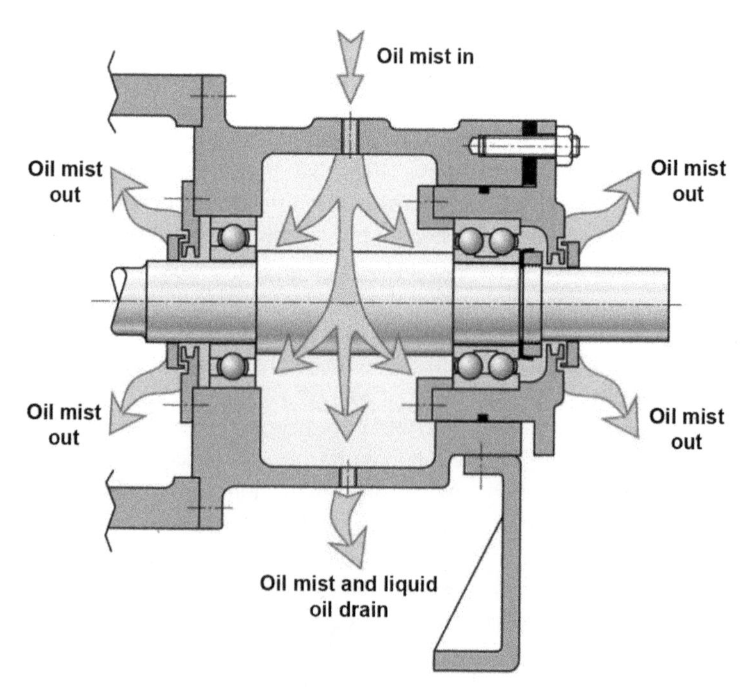

FIG. 23.3.3 Old-style (non-API-type) oil-mist introduction at midpoint of bearing housing.

Again, a simple reminder, oil mist applied per Fig. 23.3.2 is eminently successful in pumps and electric motors [2]. Either a dual magnetic face-type bearing protector seal or an axial O-ring-style bearing protector seal (Subject Category 14) in a "closed" oil-mist system represents state-of-art installations.

REFERENCES

[1] H.P. Bloch, Pump Wisdom: Problem Solving for Operators and Specialists, Wiley & Sons, Hoboken, NJ, ISBN: 9-781118-04123-9, 2011.
[2] H.P. Bloch, Hydrocarbon Processing, March 1977.

Chapter 23.4

Oil-Mist Lubrication Reliability Statistics

INTRODUCTION

Pure ("dry-sump") oil-mist systems contain no moving parts. With pure oil mist, service-intensive oil rings, constant-level lubricators, and the traditional oil refill labor requirements are eliminated. The labor requirement for pumps with oil mist has been estimated as one-tenth that of traditional lubrication. In the past decade, the earlier practice of allowing excess oil mist to escape into the atmosphere has been superseded by widespread use of closed systems. Closed systems avoid polluting the environment. Some closed systems have been in highly successful service since the mid-1980s and represent best available technology in all respects (questions on electric motor wiring in an oil-mist environment are addressed in Subject Category 11) [1].

WHY OIL MIST EXCELS AS A LUBE APPLICATION METHOD

While the purpose of this chapter is to provide the reader with a relevant experience update on closed oil-mist systems, we want to again summarize the primary advantages of oil mist:

- Plant-wide systems are almost completely maintenance-free and fully self-checking. Users do not have to rely on operators or maintenance workers to check and fill housings with oil.
- Better lubrication conditions exist because the oil coating on the bearings is always new [2].
- Lower bearing operating temperatures are routinely obtained. Reductions typically range from 10 to 20 °F (6 to 13 °C).
- Power requirements are reduced by typically at least 1% and sometimes even 3% since bearings operate on a thin oil film instead of plowing through a drag-inducing pool of oil.
- Oil mist is applied without using oil rings (oil rings are subject to abrasive wear and/or slowing down if the shaft system is not absolutely parallel).

In Chapter 23.3 of this text, we had shown a closed oil-mist system supplying continuous lubrication to pumps and drivers. We had explained that reliability leaders at best-of-class (BoC) companies soon implemented the most ideal routing of oil mist through bearings and how this routing harmonizes with the best practices conveyed in API-610 [3]. This through-the-bearing routing (whereby oil mist is introduced into the space between a modern bearing housing protector seal and the bearing) became an industry standard almost 20 years ago. Any oil reaching the bottom center of the bearing housing will have first lubricated and cooled the bearing. Also, having a single centrally located exit hole makes it easy to collect the coalesced oil or residual oil mist. We explained that the system is now "closed," and excess oil is ready for filtration and reuse.

WORKERS' HEALTH GUIDELINES

The workers' health guidelines of many industrialized nations permit the relatively small amount of oil mist released from open systems. We had estimated the oil discharged through bearing protector seals to be less than a teaspoon full per day. These estimates are easy to verify by a simple mass balance: suppose we start with 10.5 l of oil in our oil reservoir at "time zero" and find 5.7 l of oil in the reservoir after 109 h of operation. We also retrieve 4.1 l of oil from the drain ports of 47 bearing housings. We have discharged into the atmosphere 4.8 − 4.1 = 0.7 l = 700 ml of oil in 109 h. That's 155 ml per day or 3.3 ml per bearing housing per day. A teaspoon holds 5 ml: you do your own numbers [4].

Using a closed system is clearly best available technology. Irrespective of prevailing or mandated clean-air requirements, an environmentally conscious user would not allow avoidable oil releases of oil mist into the atmosphere. Moreover, from a housekeeping viewpoint, it is clearly advantageous not to have smudges of oil on the ground near pumps and other equipment. Oil smudges or rain runoff will reach a waste oil pit, and it takes money to extract that oil before the water can be discharged responsibly. We are advocates of proved closed oil-mist systems technology [1].

The details of closed oil-mist system technology may differ, but Fig. 23.3.1 in Chapter 23.3 incorporates a collecting tank (shown at the right) to which a return header system is connected. A small blower is provided at the top of the collecting tank, and the suction intake effect of this small blower causes excess or "stray" oil mist to be pulled into the tank. Fig. 23.1.1 in Chapter 23.1 incorporated a similar collecting tank shown on the left. In each case, the design includes a coalesce maze inserted in the blower. Coalesced oil droplets fall out, and the oil can be reused. Only virtually oil-free air is vented to the atmosphere.

The extended mean-time-between-failure (MTBF) benefits of oil mist over traditional liquid-oil-in-sump lubrication have been well-documented, and oil mist was included in API-610 pump standards for many years. In describing the basic oil misting process, we had noted that the bearing housing in dry-sump (pure) oil-mist lubricated equipment contains no liquid oil [2]. Instead, an oil-mist generator (OMG) with no moving parts creates the mist in a central console; a typical console and its OMG serve all of a facility's process pumps within about a 600 foot (160 m) radius. The mist is a mixture of microscopic (<3 micron) oil droplets combined with clean air at a volume ratio of about 1:200,000. The mist is conveyed in 2" pipe headers to virtually any equipment incorporating rolling element-style bearings headers at a low pressure and velocity (<7 ft/s to reduce the effects of globules becoming too large for suspension in the carrier air). Near each process pump or electric motor, the oil mist passes through a nozzle or reclassifier—essentially a metering orifice. The mist velocity is thereby greatly increased; also, a bearing in motion further promotes atomized droplets to collide and coalesce into larger liquid drops of oil. Oil mist is supplied to standby equipment as well. In standby or shutdown equipment, oil mist serves as a protective blanket. Because its pressure is marginally higher than the surrounding atmosphere, oil mist prevents the entry of airborne dirt and moisture [3].

An entire system was depicted in Chapter 23.1 of our text. The oil-mist generator (OMG, where oil meets air) is located inside an oil-mist cabinet. Piping consists of a delivery header in the foreground and a return oil collection header in the background of Fig. 23.1.1. Oil-mist takeoffs to and from process pump and motor bearing housings are connected to the top of their respective headers [3].

FAVORABLE EXPERIENCE ASCERTAINED

Books and articles also describe the unqualified success of oil-mist lubrication for electric motors [1,4]. Contrary to unfounded (and fully refuted) opinions about oil mist attacking electric motor windings or oil mist being a fire hazard, this lubrication method is fully proved to be superior to alternative lube application methods. Closed-loop oil mist best protects industry's physical plant and our environmental assets, whether in spare equipment (standstill) or full operating mode. In best technology closed oil-mist systems, from 97% to 99% of the lube oil is recovered and reused. These systems emit no oil mist into the surrounding atmosphere and, for years, have been praised by refineries and petrochemical plants concerned about the environment [4,5].

RELIABILITY REAFFIRMED

In meetings with the primary providers of plant-wide oil-mist systems (in 2012 and 2013), the author had ascertained that mere three shutdown incidents were known to have occurred on oil-mist systems in three decades of highly successful operation. An estimated number of 1800 plant-wide systems had been in service during that 20-year period tracked by these primary providers. As of 2016, an estimated 3000 plant-wide systems are in use.

Only three of the 1800 systems tracked from 1982 until 2012 experienced an interruption of service. The first systems interruption involving a modern plant-wide oil-mist system occurred at a US Gulf Coast facility in about 1982. At that time, a thorough analysis traced the failure to pipe shavings in the 5 gallon capacity misting chamber reservoir. This is where ferrous debris became attached to a magnetic level switch; it prevented the switch from activating a solenoid, which would have allowed oil from a bulk oil-holding tank to replenish the much smaller chamber reservoir. When the small reservoir was depleted, none of the connected equipment received oil mist. The bearings ran dry but did so without incident or bearing failure. It was known that bearings coated with oil in horizontally installed shaft systems can operate for a few hours after discontinuing oil mist had been reported and explained by Allen Clapp (with Dow Chemical Company, Freeport, TX) in 1973 [1]. These findings were again corroborated in academic research by Shamim and others [4]. Allen Clapp made the additional point that a small pool of oil will exist in the five-to-seven O'clock segment of the contoured raceway in a bearing's outer ring. Oil-mist lubricated pumps can safely stay in service for approximately eight hours before this small pool of oil is depleted. Given that there is ample supervisory instrumentation to annunciate deviations, no modern oil-mist system has ever encountered unavailability in excess of 8 h.

The second unavailability event developed at an oil refinery in Enid, Oklahoma, where a single oil-mist console was serving two adjacent process units. When the process unit where the oil-mist generator (OMG) was located had to be shut down in preparation for scheduled maintenance and repair downtime, the OMG was inadvertently valved off. A day or so later, the adjoining process unit experienced a pump failure. It was then realized that there had been no oil mist supplied for at least 24 h. The oil-mist supply was restored, and there were no other bearing failures on any of the connected pumps. The cause of failure was clearly human error; it could have been averted with a simple advisory note posted at the appropriate switch or valve.

A third incident report relates to a Texas Gulf Coast oil refinery where the OMG did not have an automatic fill option. A reservoir refill line connected a bulk storage tank to the small oil reservoir located inside the main oil-mist console. An operator ("when in doubt, always blame the operator….") decided to crack open the needle valve in the refill line and let it slowly maintain the oil level in the small reservoir. After a period of time, the entire piping distribution system was full of liquid oil; oil had, in fact, displaced the oil mist. About 10 or 12 pump bearing housings and their respective motor driver bearings were filled with oil. A lot of oil was wasted, but there were again no equipment failures in this incident.

Because modern oil-mist systems are provided with suitable supervisory instrumentation, no system had ever been disrupted for more than two hours from 2000 until early 2014. But then, a fourth incident happened later in 2014. A console with an aging programmable logic controller (PLC) was blamed on a card failure after slightly over 10 years of uneventful operation—or so it seemed. However, it was also reasoned that the owner had not paid attention to a warning sent by the same PLC. In any event, at least one of the involved parties thought the PLC's electronics should have been upgraded because others had sold and marketed different PLCs for several years. Some analyses simply favor certain parties, and things get fuzzy after a few years. Suffice it to say that in 1998 the overall reliability and availability of a cross section of oil-mist consoles were calculated as 99.99962% and is now (in 2016) still estimated to be somewhere near 99.999987%. That is a number that has never been approached by any other lubrication method [5].

WHAT—IF ANYTHING—CAN SHUT DOWN AN OIL-MIST SYSTEM?

The most profitable and reliable plants have found it neither cost- nor risk-justified to install oil-mist systems with fully connected, ready-to-go spare backup. Nevertheless, some plants ask oil-mist suppliers to propose and provide 100% redundancy. In those facilities, a backup or auxiliary oil-mist system can be placed in operation on a moment's notice. Switching from the main unit to a full backup takes 30–60 s to complete. The switchover procedure calls for one-quarter turn of the handle of a ball valve. A two- or three-sentence procedure sheet is posted on the inside of the cabinet door. The sheet explains what to do in the highly unlikely event of such switchovers ever becoming necessary.

In 2002, one of the authors visited eight petroleum refineries located in the US Gulf Coast region. It was quickly discovered during these visits that oil-mist lubrication was the predominant method of lubricating pumps throughout the refining industry in the USA. Also in 2002, an equipment sales specialist with over 20 years experience as a refinery reliability engineer estimated that oil mist was being used by 24 of the 30 refineries in the Beaumont-Port Arthur region of Southeast Texas. He believed that about 80% of the pumps in each facility were lubricated by oil-mist systems. One US West Coast consulting engineer with considerable background as a refinery engineer estimated that about 50% of all US refineries were using oil mist in 2001.

As of 2015, several of these refineries have been employing closed systems oil-mist technology for over four decades and are calling this application method an unqualified success. The refineries consider closed oil-mist systems a competitive advantage and have fully endorsed the application routines illustrated in this chapter of our text. Moreover, these users are doing their part toward achieving a cleaner environment while imparting reliability to their rotating equipment assets.

What else could cause an oil-mist unit or system to shut down? Well, one manager added to the four events mentioned above. He said running a forklift into the 2-in. oil-mist header would shut the system down. That may be true but certainly has never happened on any of the estimated 3000 plant-wide oil-mist systems now in service all over the world. And if it did happen, it would take considerably less than eight hours to repair the low-pressure nonflammable oil-mist header.

ACKNOWLEDGMENTS

Oil-mist systems illustrations and updated statistics were provided by Don Ehlert, Total Lubrication Management, a Colfax Fluid Handling Company, Houston, TX. Additional illustrations and narrative explaining bearing housing protector seals were contributed by Chris Rehmann, AESSEAL, Inc., Rockford, TN.

REFERENCES

[1] H.P. Bloch, Dry Sump Oil Mist Lubrication for Electric Motors, Hydrocarbon Processing, March 1977.

[2] H.P. Bloch, Practical Lubrication for Industrial Facilities, second ed., Fairmont Publishing Company, Lilburn, GA, ISBN: 0-88173-579-5, 2009.

[3] H.P. Bloch, A. Budris, Pump User's Handbook: Life Extension, fourth ed., Fairmont Publishing Company, Lilburn, GA, 2014. ISBN 0-88173-720-8; 978-1-4822-2864-9.

[4] H.P. Bloch, A. Shamim, Oil Mist Lubrication: Practical Applications, Fairmont Publishing Company, Lilburn, GA, ISBN: 0-88173-256-7, 1998.

[5] H.P. Bloch, Pump Wisdom: Problem Solving for Operators and Specialists, Wiley & Sons, Hoboken, NJ, ISBN: 9-781118-04123-9, 2011.

Subject Category 24

Oil Ring Issues

Chapter 24.1

Grooved and Other Oil Rings: Claims and Counterclaims

INTRODUCTION

Some pump manufacturers advocate using grooved oil rings instead of flat rings; others have recommended the opposite. A plain flat oil ring, often also called an "oil pickup ring," is shown in Fig. 24.1.1. Some vendors supply these with concentric

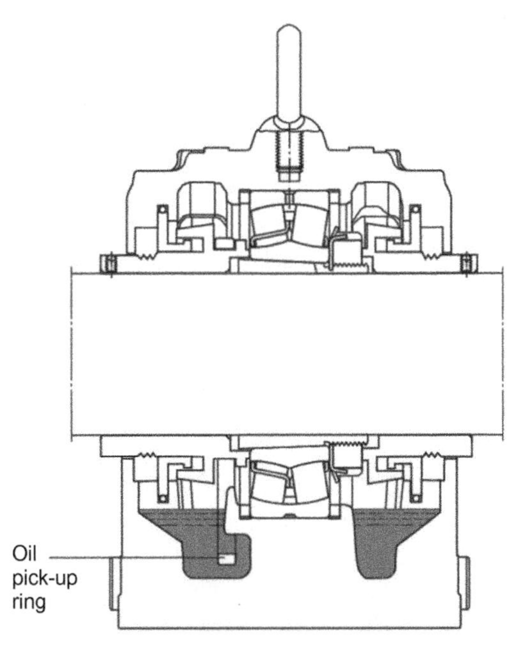

FIG. 24.1.1 Oil pickup ring (oil ring) in a double spherical roller bearing housing. *(Courtesy SKF America, Kulpsville, Pennsylvania.)*

Petrochemical Machinery Insights. http://dx.doi.org/10.1016/B978-0-12-809272-9.00024-4

grooves machined in the bore of the ring. Others do just the opposite and send out service bulletins asking to "replace grooved oil rings by the newer, and more effective, flat oil rings." A third option, contoured oil rings, was discussed at the Texas A&M International Pump User's Symposium in 2014.

WHY CLAIMS CAN BE CONTRADICTORY

It has been said that all of the above recommendations are relevant to a particular shaft speed, or a particular bearing housing configuration, or oil viscosity, etc. Efforts to standardize are involved, as are attempts to save initial cost. But the advice given by certain manufacturers and vendors also typifies a general loss of basic knowledge, and this short chapter of our text brings these concerns to the reader's attention. Thoughtful users must be on guard and may even have to challenge the erroneous advice sometimes dispensed by the harried, stressed, underinformed, or occasional "shallow thinkers."

GROOVED OIL RINGS GENERALLY OUTPERFORM FLAT OIL RINGS

Our example dates back to a "go with the new flat oil rings" letter that was sent to an electric power producer in late 2000 or early 2001. Claiming superior performance for flat oil rings contradicts the comprehensive paper by Baudry and Tichvinsky published in a 1937 vintage *ASME Journal*. While employed by Westinghouse Electric in the late 1920s, these two researchers experimented with different oil ring configurations. They demonstrated and graphically explained how and why, in fact, grooved oil rings, as shown in Fig. 24.1.2 [1], deliver *more* oil than conventional ones. Similarly, authoritative findings can be found in a very rigorous 1985 study published by Heshmat [2], not to mention later articles based on prior art, such as the ones that appeared in other trade journals a few years ago [3].

While flat rings are undoubtedly cheaper than the grooved version in Fig. 24.1.2, it is important to note that in almost every case, the capacity of flat rings to feed oil proved inferior to the slightly more expensive grooved rings. The pump manufacturer's claim is therefore in error and the user might be at risk, were he to favor flat over grooved oil rings in some of his equipment.

CONTOURED OIL RINGS

Consider what steps to take if your pumps incorporate the "flat" oil rings shown later in Chapter 24.2, Figs. 24.2.1 and 24.2.2, and in many earlier illustrations throughout this text. If you're happy with them, stay the course. If not, it's time to write better specifications and insist on specification compliance. Meanwhile, consider putting relevant details into your CMMS (computerized maintenance management system). Measure and record the as-installed oil ring out-of-roundness

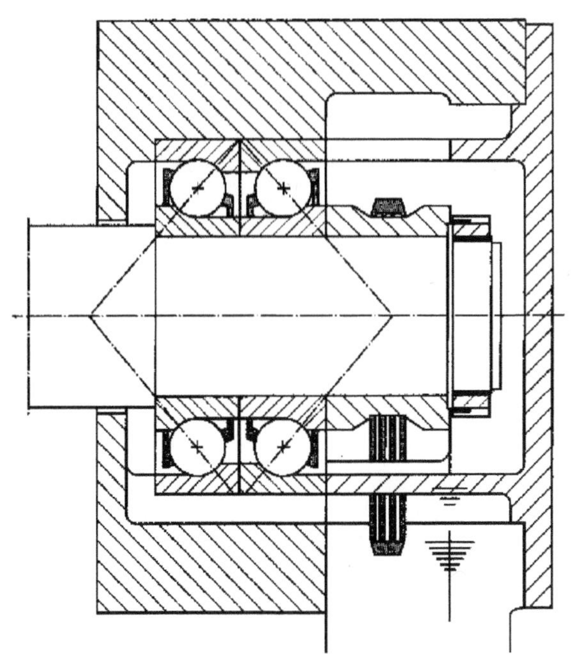

FIG. 24.1.2 Multigrooved oil ring riding on a carrier ring dated back to the 1930s. This thrust bearing set is back-to-back mounted; note the generously sized oil return slot in this bearing housing end cap. *(Courtesy SKF Americas.)*

when doing repairs. In the time period from 2010 to 2015, we measured oil ring eccentricities from 0.017 to 0.060 in., although 0.002 in. is usually quoted as the maximum permissible eccentricity.

Measure oil ring width when, in the future, dismantling a pump in the shop. The difference between the measurements is evident to the naked eye; in our later Chapter 24.2, Fig. 24.2.1, we notice how the chamfers are worn off. Measure the difference in width—before versus after, new versus used. That difference will have been converted to abrasion product; it will cause premature failure of the bearings.

Remedies are being sought, and, in 2014, the results of research and testing involving the oil ring in Fig. 24.1.3 were published [4]. In a number of short-term tests, a pump manufacturer established that oil rings with cross sections resembling the mirror-image trapezoids in Fig. 24.1.3 may tolerate wave motion on seaborne vessels. The trapezoidal oil rings were retained on contoured ring carriers and oil was flung into oil galleries. But if your pump bearing housings do not have oil galleries shown earlier as two upward-pointed lines in Fig. 4.5.8, you may have to deal with a risk-prone oil application design.

Suppose a vendor offers new oil ring geometries; does this mean the old ones left something to be desired? If a vendor tests new oil rings for a few hours and finds no oil ring abrasion, does that mean the test results can be extrapolated

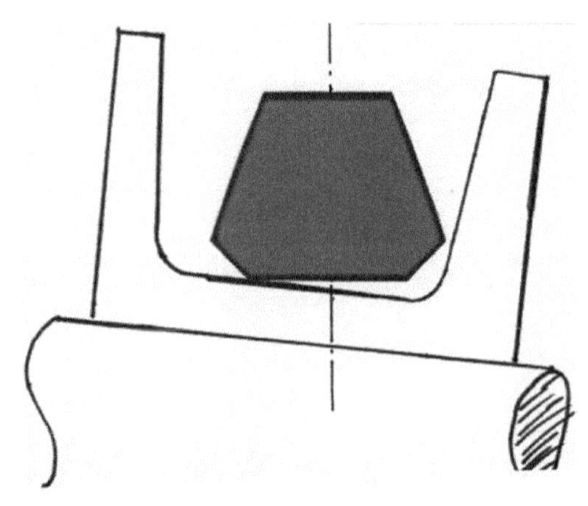

FIG. 24.1.3 Mirror-image trapezoidal oil ring on a contoured carrier sleeve [4].

to a three-year run? Pumps on the vendor's test stand always start with fresh oil, but seldom will your plant be on the same oil replacement schedule as the vendor's test stand [5].

These are among the questions for which competent reliability engineers seek answers. If no answers are discovered or developed, the facility may find that repeat failures continue in random intervals.

In a series of rigorous tests, a pump manufacturer established that oil rings with cross sections resembling the trapezoid cross-sectional ring in Fig. 24.1.3 may be preferable to "flat" oil rings. The vendor wanted to demonstrate better tolerance to wave motion on seaborne vessels [4]. The trapezoidal oil rings were retained on contoured ring carriers and oil is flung into oil galleries. (Note that a so-called oil gallery is indicated by the "inverted VEE" in Chapter 4.5, Fig. 4.5.8.) Oil galleries are often cast into the inside of a bearing housing. Their design intent is to collect oil and direct it toward the bearings. Shaft alignment is achieved by putting shims under equipment feet that, as a logical consequence, tends to cause shaft systems to be at a slight angle relative to the true horizon. On shipboard, pumps pitch and roll. Equipment surveillance and precautionary oil changes differ on shipboard from what we find at many land-based installations.

The author does not believe that a reasonable person needs to show data to prove that driving automobiles with worn tires is a greater risk than driving on tires with tread. Also, what looks worn to "A" looks normal to "B." It is no different with oil rings in pumps at location "X" versus location "Y." Gravity being gravity, the sketch in Fig. 24.1.3 is a scientific fact. The ultimate ramifications of a trapezoidal oil ring operating in this manner can be foreseen: a trapezoidal oil ring has two "pointed" or "circumferential" ridges—one on the left and one on the right side. Pointed ridges have a very small total surface area. As the oil ring slews from side to side in its carrier sleeve, it will touch the side of the carrier sleeve. That means the force per unit area, the "pounds per square inch" (commonly known as "pressure"), will be rather high. When that happens, the "pointed ridge" will break through the oil film and abrasion will occur.

Whenever a pointed ridge breaks through the oil film, there will be increased friction and the oil ring will slow down. Long-term satisfactory operation will be at risk, unless the pump owners invest heavily in preventive maintenance action. But preventive maintenance costs money, and that is simply an additional reason why oil rings rank very low on our scale. And so, assume that your pumps may incorporate the "flat" oil rings shown in Chapter 24.2, Fig. 24.2.1, and several other illustrations throughout this text. If you're happy with them, stay the course. If not, it's time to write better specifications and insist on specification compliance. Be aware that oil rings that were fabricated or processed without proper heat treatment will very often deform in operation. Therefore, be sure to purchase or produce only oil rings that have been properly stress-relieved (annealed). Realize also that the use of plastic or composite materials in oil ring is often of marginal value at best. Give thought to lube application methods that avoid oil rings [5]. While we readily acknowledge that oil rings work quite well in the majority of cases, it is equally true that they are among the most overlooked contributors to seemingly elusive or random repeat failures. Those are the failures which can cause anything from minor annoyances to serious disasters in a process plant.

WHY SLINGER RINGS MAY NOT BE YOUR BEST CHOICE

Even the most advantageous laser optic shaft alignment system will not usually ensure that the shaft centerlines are absolutely horizontal. Visualize, therefore, how slinger rings installed on shaft systems that are not totally parallel with the true horizon will run downhill. In so doing, the slinger ring will encounter frictional contact with either a groove machined in the shaft or some stationary surfaces associated with the bearing housing. The slinger ring now tends to slow down, feeding less oil into the bearing. We have also seen slinger rings that showed evidence of edge wear and metal loss. Needless to say, the lost metal shavings ended up contaminating the lubricant—not a desirable condition.

Slinger ring movement or circumferential speed is also affected by the degree of immersion in the lubricant and by lubricant viscosity. Typical immersions were shown in Figs. 24.1.1 and 24.1.2. An immersion depth of 10–12 mm (roughly 3/8 in.) is the norm, but recommendations may vary for different types of equipment. Quite obviously, a more deeply immersed oil ring, or oil rings in contact with an excessively viscous lubricant, will not perform as intended.

ALTERNATIVES TO SLINGER RINGS

Fixed slinger disks [2] avoid many of the problems encountered by slinger rings. Users may also have an incentive to upgrade ring-lubricated bearings to cost-effective and technically superior circulating lube retrofit using an inductive pump [5,6]. There are also alternatives such as adding a stand-alone pump-around system similar to Chapter 27.1, Fig. 27.1.1. Although originally designed for dual mechanical seals requiring a barrier fluid, these systems can be used to pressurize lubricant and aim this oil through small nozzles directly at the bearing cage (ball separator). After flowing through the bearings, the oil would be routed back to the reservoir portion of the pump-around unit. Numerous other alternatives would exist if pump manufacturers offered designed-in, shaft-driven oil pumps. Markets respond to user demands; it follows that purchasing from willing and able pump manufacturers would be reliability drivers.

REFERENCES

[1] "Bearings in Centrifugal Pumps", SKF Publication 100–955, Page 21, Figure 2.8, second ed., 1995.

[2] H.P. Bloch, Pump Wisdom: Problem Solving for Operators and Specialists, John Wiley and Sons, Hoboken, NJ, 2011.

[3] H.P. Bloch, Going for Better Sleeve Bearings, Hydrocarbon Processing, February 1996.

[4] S. Bradshaw, J. Hawa, Factors affecting oil ring and slinger lubrication delivery and stability, in: Proceedings of the 30th Texas A&M University International Pump Users Symposium, Houston, TX, 2014.

[5] Technical Literature, Inductive Pumps, Inc., www.inductivepump.com.

[6] H.P. Bloch, Lubrication delivery advances for pumps and electric motor drivers, in: Proceedings of the 31st Texas A&M University International Pump Users Symposium, Houston, TX, 2015.

Chapter 24.2

When Not to Use Oil Rings

INTRODUCTION

For process pumps in certain size and duty categories, oil rings are rarely (if ever) the most reliable means of lubricant application. They tend to skip around and even abrade (Fig. 24.2.1). The skipping and abrading can be avoided if five principal attributes or requirements fall into place:

- The shaft system must be truly horizontal [1]. Unfortunately, horizontality is very rarely obtained when shims are used to obtain shaft centerline alignment.
- Ring immersion in the lubricant is just right. Consistent depth of immersion can be difficult to maintain over time.
- Ring eccentricity is closely controlled. It rarely is, unless stress-relief annealing and careful final machining are included in the various manufacturing steps.

FIG. 24.2.1 Oil rings in as-new ("wide and chamfered") condition on the left and abraded (worn narrow and now without chamfer) condition on the right side [1,2].

- Ring bore RMS (root mean square) finish and oil viscosity are maintained within close limits. (Experience shows that viscosities sometimes change; they depend on oil temperature and contamination-related oil quality.)
- Shaft surface velocities are in the acceptable range.

The various parameters are probably within limits on the pump manufacturer's test stand, and the manufacturer feels exonerated. However, taken together, these parameters are rarely within suitably close limits in actual operating plants [1]. Other vulnerabilities exist and cause repeat pump failures [2].

To avoid premature equipment failure or, alternatively, avoid the need for frequent precautionary replacement of rings and lube oil changes, serious reliability-focused owner-purchasers often specify and select pumps with flinger disks (see Chapter 4.5, Fig. 4.5.2). Reliability-focused plants often follow the advice of a then prominent pump manufacturer whose 1970s vintage brochures asked buyers to opt for its superior—and we quote here from an advertisement—"anti-friction oil thrower [oil ring] ensuring positive lubrication to eliminate the problems associated with oil rings" [1]. A flinger disk (oil thrower) can, in fact, serve as an efficient oil spray producer. Of course, the proper flinger disk diameter must be chosen, and solid stainless steel flinger materials are generally preferred over plastics. Insufficient lubrication will result if the disk diameter is too small to dip into the lubricant. Conversely, high friction-induced operating temperatures can result if the disk diameter is too large or if disk contours are left to chance.

Often, and so as to accommodate the correct flinger disk diameter, the pump manufacturer will have to mount the thrust bearing set in a cartridge (our earlier Fig. 4.5.2). Cartridge-mounted designs will increase pump price slightly, but avoided failures easily make up for the higher incremental cost.

INTERESTED IN BETTER PUMPS

The term "better pumps" describes fluid movers that are designed beyond just soundly engineered hydraulic efficiency and modern metallurgy. Better pumps are ones that avoid risk-inducing components or geometries in the mechanical portion commonly called the drive end. Oil ring improvement research was described in a 2014 release [3].

An overemphasis on (initial) cost cutting by some pump manufacturers and many pump purchasers has negatively affected the drive ends of many thousands of process pumps. Flawed drive end components contribute to elusive repeat failures that often plague these simple machines. Pump drive end failures represent an issue that has not been addressed with the urgency it deserves. Keep in mind that repeat failures can only happen if the true root cause of failure has remained hidden or, if the true root cause is known, someone decided not to do anything about it. Either of these two possibilities defeats asset preservation and operational excellence goals.

Finally, tracking metal wear [4] by oil analysis, or by observing the shaft (Fig. 24.2.2), or via micrometer measurement of the oil ring width [2] should be of interest to reliability specialists who want to move their respective plants into the best-of-class category. Best practices companies measure and record the ring's original as-installed and also its after-removal widths. These measurements are entered in their CMMS (computerized maintenance management systems). Lost material widths are an abrasion product that causes serious oil contamination and premature bearing failures [5].

FIG. 24.2.2 Wear tracking on both equipment shaft and oil ring should be of interest. Note how this oil ring makes contact with housing internal surfaces. *(Courtesy Bob Matthews.)*

REFERENCES

[1] H.P. Bloch, A. Budris, Pump User's Handbook—Life Extension, fourth ed., Fairmont Press, Lilburn, GA, ISBN: 0-88173-627-9, 2014.

[2] H.P. Bloch, Pump Wisdom: Problem Solving for Operators and Specialists, John Wiley and Sons, New York, ISBN: 9-781118-041239, 2011.

[3] Bradshaw, et al., Proceedings of the 30th TAMU International Pump User's Symposium, 2014.

[4] S. Bradshaw, Investigations into the contamination of lubricating oils in rolling element pump bearing assemblies, in: Proceedings of the 17th International Pump User's Symposium, A&M University, Houston, TX, 2000.

[5] H.P. Bloch, Practical Lubrication for Industrial Facilities, second ed., Fairmont Press, Lilburn, GA, ISBN: 088173-579-5, 2009.

Subject Category 25

Operation and the Operator Function

Chapter 25.1

The Operator's Role in Achieving Equipment Reliability

INTRODUCTION

There can be no reliability without operator involvement. Just as the most well-designed and best-maintained automobile will fail in the hands of a thoughtless or inexperienced driver, the best and most reliable machine will not perform optimally if the operator lacks training, care, or motivation. We accept the responsibility of viewing the dashboard instruments of a modern automobile; similarly, the operator in a modern process plant must accept equipment surveillance as his prime responsibility.

SURVEILLANCE IS AN OPERATOR TASK

The purpose of surveillance is to spot deviations from normal operation. Surveillance is done by the operators assigned to a process unit. Although the surveillance task includes collecting vibration data (Fig. 25.1.1), the operator is primarily searching for deviations from normal vibratory activity. Analyzing and finding the root cause of vibration are not the operator's assignment. Deviations are reported to maintenance or reliability personnel who then determine significance, root cause, and remedial action.

It is of extreme importance to realize that process pumps in US oil refineries typically have a mean time between failures (MTBF) of 9 years, while a few report an MTBF of only about 2 or 3 years. Each group purchases its pumps from the same cluster of manufacturers; in each group, we find old and new pumps. Both unionized and nonunionized plants are represented in each group. It may take a good and well-executed audit to pinpoint the many small and generally overlooked

Petrochemical Machinery Insights. http://dx.doi.org/10.1016/B978-0-12-809272-9.00025-6

reasons for drastically different reliability performance of essentially identical pumps. The audit results have to be explained to management and shared with operators and maintenance technicians. Targeted training would be the next step.

"BEST-PRACTICES" SURVEILLANCE AND OTHER MISUNDERSTANDINGS CLEARED UP

Each approach to reliability achievement has its place, but a pacesetter "best-practices" or best-of-class (BoC) plant will carefully select and (a) explain when, why, and where electronic or console surveillance makes economic sense and (b) where only the operator's eyes and ears will do the best job. In that case, patrolling the actual field unit and seeing the asset will be required.

FIG. 25.1.1 Vibration monitoring is one of many operator functions. Once a certain vibration threshold is exceeded, the findings are given to a maintenance/reliability worker who will analyze the root cause. *(Courtesy Ludeca, Inc., Miami, FL; www.ludeca.com.)*

Process plants with reliability focus make it a routine to thoroughly explain many seemingly small issues to their process operators. Among them are

- conditions that could cause bearings to be deprived of lubrication even though constant level lubricators are full of oil;
- operating conditions wherein plenty of cooling water is applied, but bearings fail due to overheating;
- situations where compressor suction drums are free of liquid, but liquid is still found in the compressor and causes serious damage;
- unacceptable flow conditions that will drastically reduce pump life;
- pump switchover and warm-up issues that have reliability impact; and
- the operators that are the eyes and ears of the facility are indispensable (there can be no sustained reliability without dedicated operator involvement).

And while the aforementioned can often be remedied by informed operators, other misunderstandings are best addressed by informed engineers and managers. More detail is given in subject categories dealing with organizational structure and training; you may find them as you carefully examine this experience-based text.

MISUNDERSTANDINGS INVOLVING ENGINEERING AND SPECIFICATIONS

Plants that subscribe to the logical premise that operations, maintenance, and technical groups must cooperate will encourage their engineers to participate in operator training sessions. At one location with highly successful training, engineers and maintenance technicians benefit from functionally overlapping learning experiences. The training approach includes informing people of what they are about to learn and why they should listen and absorb why, where, and how:

- Vented bearing housings are being phased out.
- Lubrication by *oil rings* often has serious shortcomings over other methods.
- Oil-mist lube: why so successful at some locations and seemingly deemed of little value elsewhere?
- Mechanical modifications can extend bearing life.
- In certain easily definable pumping services, mechanical seals do not require external flush fluids.
- Catastrophic failure and loss of life occurred because there was water in the lube oil.

WHERE ONLY A COMPETENT OPERATOR CAN AVOID MACHINERY FAILURE

There are many areas where only a well-informed, competent, and conscientious operator can avoid equipment failures. In each of these areas, success is linked to "operator buy-in." In other words, things are more likely to be done in a timely and procedurally correct fashion if the operator understands not only how but also why a reliability-focused plant engages in the following:

- Slow rolling (warm-up) of turbines.
- Measures that prevent water slugging of turbines.
- Understanding contaminated lube oil: how bad is bad?
- Exercising steam turbine trip-throttle (T/T) valves.

Without exception, a well-trained workforce is essential to a plant's safety and reliability performance. Process plants and utilities can reverse unfavorable trends by paying attention to training topics that limit exposure to philosophies and, instead, emphasize the tangible details to which operators and mechanical workforce members can respond. This would be the essence of operator-driven reliability (ODR).

Chapter 25.2

Consider Procedure Cards, Technical Books, and Maintenance Conferences

INTRODUCTION

Memories are imperfect, and we don't want to risk making mistakes. Therefore, we learned long ago to jot down brief reminder notes on pieces of paper. People called it a "cheat sheet," or a checklist. An airline pilot has a set of checklists in the cockpit. Even after hundreds of successful takeoffs and landings, he will use the checklist. Chances are that consistently using such a checklist is a condition of employment for airline crews. But, because errors are costly and can even be dangerous, best-of-class petrochemical companies have also made it a practice to use pertinent checklists or procedure cards, Fig. 25.2.1.

WHY USE CHECKLISTS

At one company that uses checklists, whenever operators patrol their process units, they have to have the checklists in one of their shirt pockets. These checklists became confidence builders. Because it was in their culture, so to speak, that company's successor affiliates still believe that an informed workforce is a happy, responsible, and productive partner.

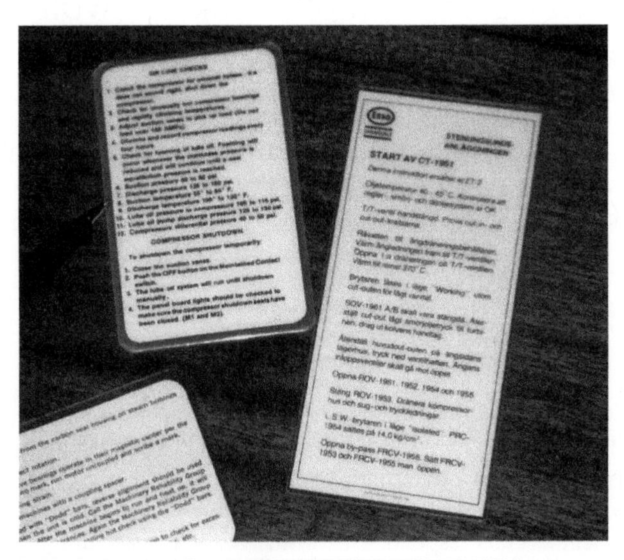

FIG. 25.2.1 Shirt-pocket checklists used at a best-of-class plant in 1985.

FIG. 25.2.2 Important technical books within reach of a maintenance technician's work station in 1985.

Using checklists sends a positive message and, at best-of-class companies, process unit supervisors lead by example. Accordingly, they, too, have these checklists with them and will look at them every time a machine is being started, stopped, or, in the case of important valves, when making adjustments. At best-of-class companies, technical books continue to be displayed within reach of operators and technical staff (Fig. 25.2.2). The Internet is not an adequate substitute for good books.

Good books on reliability improvements and failure avoidance can be found in Appendix II. The list gives the names of over 20 different authors; it will help you to reconnect with good experience-based technical books.

Chapter 25.3

Reassess Your Pump Start-Up and Shutdown Procedures

INTRODUCTION

It may sound far-fetched, but the ramifications of incorrect pump SU and SD procedures have recently been debated in reliability-related litigation in US courts. At issue was the performance of multistage centrifugal pumps in pipeline service at a location that had to limit motor inrush currents. The pumps operated above their first rotor critical speed and operation at the critical speeds had to be limited to two or three seconds in duration. Should the discharge block valves be fully or partially open, or should they be closed at the moment of pushing the start button?

NO ONE RULE FITS ALL

As can be seen from our four-part illustration, Fig. 25.3.1, the power demand of impellers with different specific speeds can increase or decrease as the flow increases. Radial flow impellers (part A with specific speeds up to about 120 metric or 6200 US) have power curves (P) that increase with flow. This would favor starting against a fully closed discharge valve. The same would be true for the mixed flow pumps of parts B and C, with their specific speeds from 40 to about 200 metric, or 2000–10,000 US. Obviously, their power slopes are also increasing with flow. It is different with the axial flow pumps of part D. Their specific speeds typically range from 160 to about 350 metric, or 8000–18,000 US. These axial flow pumps are best started with the discharge valve in the fully open position because they exhibit lower power demands at lower flows.

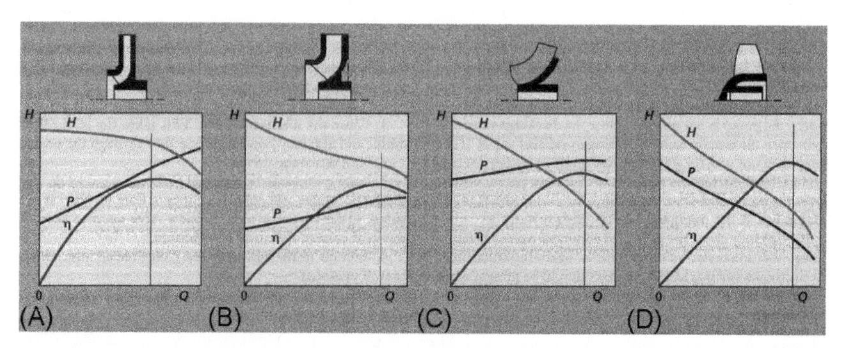

FIG. 25.3.1 Impeller shape determines head-versus-flow characteristics and the slope of a power curve. Per common practice, axial flow pumps ("D") are best started with the discharge valve in the fully open position.

It should be realized, however, that pumped fluid pushing against a fully closed discharge valve will often make it near impossible to open the valve. The recommended procedure would thus call for starting against a partially opened discharge valve. Partial opening in this context means a valve gate that has just been lifted off its seat. The pumped medium thus wets both upstream and downstream sides of the sliding gate.

REVIEW AND UPDATE YOUR PROCEDURES

Operator knowledge and execution are critically important ingredients to the achievement of reliability. We call it operator-driven reliability (ODR). Reliability-focused pump users consider using the following procedures for pump start-up and shutdown.

Starting Centrifugal Pumps

1. Arrange for an electrician and a machinist/specialist to be present when a pump is initially commissioned. Ascertain that large motors have been checked out.
2. Close the discharge valve and open the suction valve. Except for axial flow pumps, the almost closed discharge valve creates a minimum load on the driver when the pump is started. Assuming that the motor inrush current allows and that the motor will not kick off, the discharge valve may be just "cracked"—about 1/8 open—before the pump is started.
3. Be sure the pump is primed. Opening all valves between the product source and the pump suction should get product to the suction but does not always ensure that the pump is primed.

Only after ascertaining that fluid emissions are not hazardous or are routed to a safe area, open the bleeder valve from the pump casing until all vapor is exhausted and a steady stream of product flows from the bleeder. It may be necessary to open the bleeder again when the pump is started or even to shut down and again bleed off vapor if pump discharge pressure is erratic.

Note: Priming of a cold service pump may have to be preceded by "chill-down." A cold service pump is one that handles a liquid that vaporizes at ambient temperatures when under operating pressures. Chilling down of a pump is similar to priming in that a casing bleeder or vent valve is opened with the suction line open. There are three additional factors to be considered for cold service pumps:

- Chilling a pump requires time for the pump case to reach the temperature of the suction fluid.
- The chill-down vents are *always* tied into a closed system.
- On pumps with vents on the pump case and on the discharge line, open the vent on the discharge line first for chill-down, and then open the pump case vent to ensure that the pump is primed.

Should it be necessary to have a cold service pump chilled down and ready for a quick start, for example, refrigerant transfer pumps during unit start-up, then the chill-down line can be left cracked open to get circulation of the suction fluid.

- If a minimum flow bypass line is provided, open the bypass. Be sure the minimum flow bypass is also open on spare pump if it starts automatically.
- Never operate a centrifugal pump without liquid in it.
- Check lube oil and seal pot level (assumes dual seals).
- Start the pump. Confirm that the pump is operating by observing the discharge pressure gauge. If the discharge pressure does not build up, stop the pump immediately and determine the cause.
- Open the discharge valve slowly, watching the pressure gauge. The discharge pressure will probably drop somewhat, level off, and remain steady. If it does not drop at all, there is probably a valve closed somewhere in the discharge line. In that case, close the discharge valve.
- Do not continue operation for any length of time with discharge valve or line blocked.
- If the discharge pressure drops to zero or fluctuates widely, the pump is not primed. Close the discharge valve, and again open the bleeder from the casing to exhaust vapor. If the pump does not pick up at once, as shown by a steady stream of product from the bleeder and steady discharge pressure, shut down the pump and driver, and check for closed valves in the suction line. A dry pump will rapidly destroy itself.
- Carefully check the pump for abnormal noise, vibration (using vibration meter), or other unusual operating conditions. An electrician and machinery engineer should be present when pumps are started up for the very first time, that is, upon being initially commissioned.
- Be careful not to allow the bearings to overheat. Recheck all lube-oil levels.
- Observe whether or not the pump seal or stuffing box is leaking.
- Check the pump nozzle connections and piping for leaks.
- When steady pumping has been established, close the start-up bypass and chill-down line (if provided), and check that block valves in minimum flow bypass line are open.

Watching Pump Operation

1. Not only during commencement of pumping but also on periodic checks, note any abnormal noises and vibration. If excessive, shut down.
2. Note any unusual drop or rise in discharge pressure. Some discharge pressure drops may be considered normal. When a line contains heavy, cold product, and the tank being pumped contains a lighter or warmer stock, the discharge pressure will drop when the line has been displaced. Also, discharge pressure will drop slowly and steadily to a certain point as the tank

level is lowered. Any other changes in discharge pressure while pumping should be investigated. If not explainable under good operating conditions, shut down and investigate thoroughly. Do not start up again until the trouble has been found and remedied.

3. If lubricated by oil mist, periodically check oil-mist bottom drain sight glass for water and drain, if necessary.
4. On open oil-mist systems, regularly check both oil level and oil-mist vapor flow from vents or labyrinths.
5. Seal-oil pots need to be checked regularly for correct level. Refill with fresh sealing liquid (oil, propylene glycol, or methanol, as specifically required and approved for the particular application).
6. Periodically check for excessive packing leaks, mechanical seal leaks, or other abnormal losses. Also, check for overheating of packing or bearings. Do not trust your hand; use a surface pyrometer or infrared temperature gun instead. Note that excessive heat will cause rapid failure of equipment and may result in costly and hazardous fires.
7. At a minimum, operate all spare pumps at least once a week to prevent the bearings from seizing and to ensure that the pump will be operable when needed.

   Realize, however, that "best-practices plants" have determined monthly switching of the "A" and "B" pumps to be the most appropriate and cost-effective long-term operating mode.
8. When a pump has been repaired, place it in service as soon as possible to check its operation. Arrange for a machinist to be present when the pump is started up.

Shutting Down Centrifugal Pumps

1. Close the discharge valve. This takes the load off the motor and also may prevent reverse flow through the pump.
2. Shut down the driver.
3. If pump is to be removed for mechanical work, close the suction valve, and open vent lines to flare or drain as provided. Otherwise leave suction valve open to keep pump at correct operating temperature.
4. Shut off steam tracing, if any. Continue oil-mist lubrication, if provided.
5. Shut off cooling water, sealing oil, etc., if pump is to be removed for mechanical work.
6. At times, an emergency shutdown may be necessary. If you cannot reach the regular starter station (e.g., in case of fire), stop the pump from the starter box, which is located some distance away and is usually accessible. If neither the starter station nor the starter box can be reached, call the electricians. Do not, as a part of regular operations, stop pumps from the starter box. Use the regular starter station instead.

Please note that these procedures are of a general nature and may have to be modified for nonroutine services. Review pertinent process data, as applicable. Rewrite in concise sentences if instructions are to become part of checklists that operators are asked to have on their person while on duty. And remember that best-in-class organizations are best-of-class because they insist on using checklists and procedures.

Organizations without written procedures and work processes will likely encounter operators who make it a habit to tweak process instruments and control valves. At times, the "control valve tweakers" and general mischief makers have made it a fun sport to test the brilliance of the next shift of operators. Organizations that close an eye to such events will sooner or later pay huge penalties for allowing such behavior.

In any event, process plants without equipment operating checklists may have a difficult time maintaining consistency and cost-effectively training the next generation of operators. The same comment is true for maintenance technicians or mechanical workforce members. Before a worker can "go by the book," a proper plan must be in place. Written checklists are used by airline pilots, and it should be no different in the process industries.

Chapter 25.4

Pump Switching: What's Best Practice?

INTRODUCTION

Because this text reflects decades of experience, it often speaks to different generations of reliability professionals. As expected, old questions are occasionally raised again. Likewise, new questions are asked and deserve to be answered. Many of these are intriguing; especially those that seem to indicate previous generations of reliability professionals have, for whatever reasons, not passed on their knowledge.

RULINGS OFTEN DESERVE UPDATING

From a sequence of correspondence, we were able to discern that certain rulings were being made years ago, although the rationale employed to make these rulings has never been explained. This example involves a reliability engineer in Australia. He works for a polyethylene manufacturer that formerly belonged to "MNXC," a leading multinational corporation. The plant is now owned by a Chinese company that we will call "PEYP," and PEYP inherited operating and maintenance manuals from MNXC.

The reliability engineer was aware that MNXC in its maintenance practices manual had documented "proven best practices" for spare pump operation. MNXC was said to have claimed that the more frequently a pump is started, the lower its mean time between repairs (MTBR). This, MNXC explained, was attributable to pump start-up being the most severe or stressful transient operation to which pump seal(s) and bearings are being subjected.

The Australian reader informed us that MNXC's maintenance practices manual advocated a four-category criticality ranking of all pumps and then to swap the pumps as follows:

Emergency: Every 4 weeks
Vital: Every 12 weeks
Normal: Every 52 weeks
Low: Never

Our Australian reader noted—quite correctly—that the MNXC maintenance practices manual recommendation does not seem to follow logic. It was apparent to this reliability professional that, for its most critical equipment, MNXC applied a frequent start strategy that, as they correctly inferred, would yield the shortest MTBR. The Australian engineer summarized his correspondence by recalling that, because of prior consulting assignments in Australia, we were familiar with MNXC's practices and asked for our opinion on the matter.

Well, the manual didn't exist at the time we visited the company years ago. In the intervening years, we had ample opportunity to research the issue and to observe prevailing practices in many countries. We subsequently alluded to pump "swap-over" topics in books and articles. As to the issue raised by the reader, there are several possibilities. Some, we will admit, are alluded to with a twist of irony but deserve to be mentioned as (a) through (d):

(a) There is confusion with the terminology "swap" versus "test run." The old manual might advocate a "swap" once a year but (elsewhere) probably asks the operators to test-run the second pump for at least 4 h every month.

(b) The old owner company MNXC and the person(s) compiling the manual have overlooked the fact that letting a pump sit idle for 52 weeks and to then expect it to run flawlessly is stunningly naive and will—ultimately—cost the owner dearly. Experienced reliability professionals well know that after a year the bearings of the standstill "spare" pump will have degraded due to two actions:

- One action is microvibration transmitted from adjacent running equipment; such vibratory motion causes the oil film to be wiped off. The result is metal-to-metal contact and false brinelling, an indentation without a surrounding "mound" of displaced metal.

  The most likely location of an indentation called false brinelling would probably be in the outer race spot closest to the lowermost bearing balls. Good images are found in the bearing failure identification literature issued by the major bearing manufacturers.

  False brinelling sets itself apart from true brinelling. In true brinelling, the indentation would be surrounded by a "mound" of displaced metal.
- The second typical action results from corrosive damage unless, of course, dry-sump oil mist is used for both lubrication and standstill protection.

(c) Whoever made the rules is inexperienced and/or bases his/her advice on the premise that "one has to do something in order to justify being on the company's payroll."

(d) The owner company is planning to save money and will later sell the asset to some unsuspecting buyer. The enterprise will look good—low operating cost, high profits. Two years later, the facility will be well on its way to becoming maintenance-intensive and unprofitable.

Personally, I believe what a leading producer of petrochemicals (LPC) advocated and practiced in 1986 is still correct: swapping every 4 weeks will protect the bearings and will keep the stagnant liquid product in the piping and seal regions from partially vaporizing. Periodically swapping pumps will also serve as a training exercise to keep operators knowledgeable and alert.

An LPC engineer told us, at a conference in 2008, that LPC had started to change their pump swapping routine and was doing the switch every 6 weeks. That's sufficiently close to our old 4-week rule, and we can live with that. The "new rule" the reader mentioned in his note makes no sense, but I'm past arguing. The wheel is being reinvented even as we speak, and interestingly, pump MTBFs at some plants have declined since 2002. The "new rule" undoubtedly contributes to the MTBF decline. In any event, the important answer and technical explanation are found in (b), above. We might as well disregard the rest.

Chapter 25.5

More Than Procedures are Needed

INTRODUCTION

The subject "procedures" fits squarely between operator issues and organizational issues. More than procedures are needed to run a profitable plant. Regrettably, old habits seem to persist in the typical disaster scenario: An explosion rips through a refinery (and when an explosion happens, the heat is on). All too often, the organization reacts swiftly and predictably. The customary investigative task force is assembled, and the search for data is started. A few thousand dollars are allocated to solve a 10-million-dollar problem, and many thousands of e-mails are generated. Some of these only recite the obvious; others will caution not to put faith in whatever assertions are made "because it really depends on this and that." A person willing to risk his or her career will be asked to find the true root cause, but without stepping on anyone's toes. Hopefully, they will uncover facts and essential input from the operators. However, after a while and as is customary, the investigative task force is disbanded; the organization goes back to its old ways of protecting predominant interests. They zero-in on the usual suspects, the very same process operators, who now regret having cooperated with the failure cause investigator. And the cycle repeats itself.

Admittedly, over the past years, a disturbing number of accidents have occurred because operators and maintenance technicians did not follow existing and generally well-documented procedures. Adhering to procedures is a mandatory requirement for safe and reliable operation of modern facilities. However, the problem with many accidents goes much deeper and involves management flaws. This contention is implied in our heading, which makes the point that more than procedures are needed. Management's attention needs to move to a higher plain.

DISTINCTION BETWEEN PLANTS

Off and on, we have explained that *all* job functions in a process plant, oil refinery, or petrochemical plant influence equipment reliability. Reliability contributions thus include process and managerial personnel. Yet, we have observed an interesting distinction between facilities that seem to have lots of operator-associated incidents and facilities that find problem causes concentrated in flawed maintenance, faulty design, material defects, and so forth. We believe that true best-of-class ("BoC") facilities, the ones with fewer operator-associated incidents, are fortunate in having enlightened management. BoCs consistently apply best practices, and these include training and nurturing professional talent. The training methods of true top performers lead to high profits and superior safety performance. Top performers enjoy high employee morale and "peace at the workplace." There are few knee-jerk reactions; there is balance between euphoria and dejection at these plants. We choose to call them the tranquil facilities.

Tranquil facilities have a culture that includes two important core beliefs:

1. You cannot teach what you yourself don't understand or consider important.
2. A good teacher explains why he is giving a particular instruction.

At tranquil facilities, these core beliefs are translated into tangible action. In one of these plants, the top manager many years ago told the heads of his maintenance and operations departments that he would, someday in the future, ask them to switch their respective positions and responsibilities. He correctly reasoned that this occasional switching back-and-forth would force them to learn each other's jobs. It precluded the posturing, bickering, and finger-pointing we have seen elsewhere. This top manager taught his midlevel managers to interact, communicate, cooperate, and show due consideration for both human and physical assets. Throughout the workweek, the two department heads met in the control center early in the morning to practice the "three Cs": communication, cooperation, and consideration. The results were remarkable and have remained consistently remarkable over several decades.

At this particular facility, a very similar approach was also being practiced at the next lower supervisory level. Contact engineers, sometimes called superintendents, learned the operators' jobs. They became intimately familiar with the procedures, work processes, and hardware systems that made up the plant. To this day, many tranquil facilities use these approaches. They seem to enjoy sustained prosperity and low rates of unexpected failures. Their contact engineers review reams of data to spot spikes and deviations. They correct the course *before* incidents happen. And individuals with any involvement in the root cause failure analysis (RCFA) process are taught the RCFA process and have bought into it.

In sharp contrast, facilities that do not use this approach seem to deal with the unexpected more often. Some have asserted that unexplained catastrophic failures are just part of the cost of doing business. Talking proactive language but *reacting* to events, they are unwilling to insist on shared operator-engineering responsibility and accountability. Their engineers and the designated failure team coordinator go over reams of records to spot deviations only *after* an incident has occurred. In some isolated cases, they seem to deliberately avoid drilling down to the true failure cause because doing so would embarrass people. And they neither understand nor support the structured RCFA process. We believe these facilities are condemned to sooner or later experience a worse disaster.

At plants that have not yet gotten the message, we have occasionally advised the failure team coordinator to add an unusual set of action items to their report:

- Strongly recommend making it a condition of employment that contact engineers *know* whether or not procedures are being followed. Permanently locate these employees in the control center, if necessary.
- Ascertain that contact engineers *fully understand* the potential consequences of procedures not being followed.
- Insist that contact engineers *explain* to operators *why* and what deviations will sooner or later lead to definable, foreseeable events.
- Insist that the contact engineers write those down and periodically elevate the organization's awareness.
- Teach the structured RCFA process to every professional employee, and insist on its application.

We believe that some recent near-misses and industrial events are wake-up calls that must be taken very seriously. These events should trigger a big change in what top managers must expect from the engineering and management staff at their various refineries and petrochemical plants. Recent events and the overall incident record dating back several years point to hidden and potentially catastrophic dangers if the risk-prone approach practiced by many is not modified.

To become a best-of-class "tranquil" plant, the facility must employ process superintendents and/or contact engineers who, early in their careers, get to learn their units from end to end and from top to bottom. At many facilities, the systematic upgrading of personnel cannot be limited to just the operator or equipment maintenance functions. Upgrading must include existing engineering and management staff. Reliability professionals learn all about certain process units. In case of labor strife, these salaried personnel can step in after a few days of refresher training and operate the process unit just as effectively as did the people who are temporarily absent from their usual assigned positions. There are major opportunities here. We should refuse to let them slip past us!

Checklists and procedures are of immense value. However, more than mere procedures are needed. The procedures must be correct, and they must be followed. At BoCs, management is taking well-defined proactive steps in the direction of having the right mind-sets in place at all times.

Subject Category 26

Organizational Issues

Chapter 26.1

Organizing to be Reliability-Focused

INTRODUCTION

Companies are either repair-focused or reliability-focused. We believe that a best practices (BPs) facility will be reliability-focused. Just as quality must be built into high-quality machines so is reliability. Quality and reliability are forethoughts, not afterthoughts. Key observations and comparisons are listed here in brief, bulleted format:

- The reliability function at *repair-focused* facilities is not generally separated from the plant maintenance function. At repair-focused plants, traditional maintenance priorities and "fix it the way we've always done it" mentality win out more often than warranted.

  In contrast, BP facilities know precisely when upgrading is warranted and cost-justified. They view every maintenance event as an opportunity to upgrade and are organized to respond quickly to proved opportunities.
- The reward system at the RFI plant is often largely production-oriented and is not geared toward consistently optimizing the bottom-line life-cycle cost (LCC) impact. At RFI facilities, the LCC concept is not applied to upgrade options.

  That differs from BP facilities that are driven by consistent pursuit of longer-term LCC considerations. At BPs, life-cycle costing is applied on both new and existing (upgrade) equipment options.
- Competent reliability professionals have insufficient awareness of the details of successful reliability implementations elsewhere.

  BP facilities provide easy access to mentors and utilize effective modes of self-teaching via mandatory exposure to

  (a) trade journals and related publications,
  (b) frequent and periodic "shirt-sleeve seminars" (briefing sessions that give visibility to the reliability technicians' work effort, disseminate

Petrochemical Machinery Insights. http://dx.doi.org/10.1016/B978-0-12-809272-9.00026-8

technical information in single-sheet laminated format, and serve to upgrade the entire workforce by slowly changing the prevailing culture).

- The lack of continuity of leadership is found at many RFI plants. These organizations do not seem to retain their attention span long enough to effect a needed change from the present repair focus to the urgently needed reliability focus. The influence of both mechanical and I&E equipment reliability on justifiably coveted process reliability does not always seem to be appreciated at RFI plants.

We know of no BP organization (top quartile company) that is repair-focused. Experts generally agree that successful players must be reliability-focused to survive in the coming decades.

Some of the most successful BP organizations have seen huge advantages in randomly requiring maintenance superintendents and operations superintendents switching jobs back and forth. There is no better way to impart appropriate knowledge and "sensitivity" to both functions.

- At RFIs, failure analysis and effective data logging are often insufficient and generally lagging behind industry practices.

  BP organizations involve operations, maintenance, and project/reliability personnel in joint failure analysis and logging of failure cause activities. A structured and repeatable approach is being used. Accountabilities are understood.
- At the typical RFI plant, there are gaps in planning functions and process-mechanical coordinator (PMC) assignments. There is also an apparent emphasis on cost and schedule that allows nonoptimized equipment and process configurations to be installed and, sometimes, replicated. Reliability-focused installation standards are rarely invoked, and responsible owner follow-up on contractor or vendor work is practiced infrequently.

  BP organizations actively involve their maintenance and reliability functions. LCC considerations are given strong weight. Also, leading BP organizations have contingency budgets that can be tapped in the event that debugging is required. They do not tolerate the notion that operations departments must learn to live with a constraint.
- A reliability-focused (BP) organization will be diligent in providing feedback to its professional workforce. The typical RFI does not use this information route.

ROLE STATEMENTS AND TRAINING PLANS ESSENTIAL FOR PROGRESS

- A lack of role statements can lead to inefficiency and encourages being trapped in a cycle of "firefighting." Not having written role statements deprives the entire organization of a uniform understanding of roles and expectations for reliability professionals.

BP organizations use role statements as a clear roadmap to achieving mutually agreed-upon goals. Among other things, this allows meaningful performance appraisals.

- In addition to structured self-training and participation in periodic "shirt-sleeve" seminars, the proficiency of reliability personnel must be enhanced by well-thought-out training plans. Key learnings must be consistently employed. Site management must verify continuity of application.

  BP organizations encourage salaried professionals to submit their projected training plans, both long-term and short-term, in writing. These plans are then critically reviewed, and employer requirements reconciled with an employee's developmental needs. Input from competent consultants is often enlisted. BP organizations make active and consistent use of what they have learned.

MENTORING, RESOURCES, AND NETWORKING

- Occasionally, even a repair-focused organization has both business improvement and reliability improvement teams in place. As it plans to move toward BP status, the RFI plant must make an honest appraisal of the effectiveness of these teams. Their value obviously hinges on the technical strength and breadth of experience of the various team members.
- At the typical RFI location, maintenance-technical personnel are often unfamiliar with helpful written material that could easily point them in the right direction.
- RFIs often use only one mechanical seal supplier. Moreover, access to the manufacturer is sometimes funneled entirely through a distributor.

  BP organizations have full access to the design offices of several major OEM firms. They have acquired and actively maintained a full awareness of competing products. They will select whichever meets their profitability objectives. This is reflected in their contract with a seal alliance partner.
- At RFIs, a single asset may require costly maintenance work effort every year, while another, seemingly identical asset, lasts several years between shutdowns.

  BP organizations provide mentor access, which will result in true root cause failure analysis and the authoritative and immensely cost-effective definition of what's in the best interest of the company. This could be repeat repair, upgrading, or total replacement.
- Repair-focused RFI plants seem to "reinvent the wheel" or use ineffective and often risky trial-and-error approaches.

Finally, BP organizations make extensive use of networking. Relatively informal, very low-cost network newsletters are issued in paper or electronic form. They use input from grassroots contributors who gain "visibility" and "name recognition" by being eager to communicate their successes to other affiliates. A network chairperson is being used to communicate with plant counterparts. This job function is assigned to an in-plant specialist on a rotational basis.

Chapter 26.2

Ten Sure Steps to Substandard Reliability Performance

INTRODUCTION

Equipment reliability can be enhanced or jeopardized by many different actions, inactions, commissions, and omissions. We're happy to report that entire corporations have prospered because of a consistent and highly productive reliability focus. But some companies have disappeared because they only paid lip service to reliability or because company executives decided to put reliability priorities at the very bottom of their concerns. Because there can be no reliability performance without organizational leadership, we are placing this chapter in the Subject Category Organizational Issues. And we know full well that what one can read in this chapter is causing discomfort to some. Then again, discomfort is part of change.

WHY SOME FALL SHORT OF REACHING THEIR GOALS

We recognize that reliability is a hollow term to many. Still, first and foremost, a reliable plant is a safe plant. Safety and reliability cannot be separated. And, while certain of these actions, inactions, commissions, and omissions may take a company close to the brink, we believe that when five or more combine, they are almost certain to push a company over the brink. Here, then, are *10 sure steps to substandard reliability performance* listed in alphabetical order and written with a touch of irony:

1. *Benchmarking fix*

   Statistics? What statistics? Besides, why call it a failure in the first place? Why not just call it a routine repair? All machines need repairs sooner or later, don't they? We have no failures. We just do lots of maintenance.
2. *Buying cheap notions*

   We always buy cheap. We can always fix it up if it doesn't work. Or we can always sell the whole plant to investors with lots of money. All we need is good lawyers and dumb buyers.
3. *Confirmation bias*

   We listen to those who have figured out what we want to hear. We reward those who have gotten away with guessing the way we would want them to guess. Alternatively, we support those who find ways to blame "issues" on things other than our guesses. And we can always blame it on the manufacturer.

4. *Imperfection argument*
   Nobody's perfect and everything is relative. That's why we don't even have to try. We must be OK doing what we've always done; if it was good enough for our CEOs in the previous decades, it sure is good enough for us today.
5. *Opinion support*
   We prefer to listen to opinions instead of facts. We allow our staff to comingle opinions and facts and see no need to separate the two.
6. *Overconfidence*
   Our failure record might indicate that we are not nearly as smart as we think, but nobody's data collection effort is perfect. Therefore, we are comforted by the thought that our failure record could be flawed, which is the reason why we lose no sleep over the matter.
7. *Repeat failure tolerance*
   My guys report to the operations department. Operations are interested in production. As long as we produce, all of us are happy. That's why we have spare pumps—get it? One pump out for repair has no production impact.
8. *Status quo preference*
   My boss was promoted into the skybox by not making any waves. Because I also want to get to the skybox, I will not make waves either. I say the failure record doesn't bother me—it's always been that way.
9. *Team player demands*
   The guys who don't jump on my bandwagon just aren't team players. At a minimum, I'll see to it that their careers will be redirected into dead ends.
10. *Training permissiveness*
   I let my team players choose what training they need. The fact that they prefer Honolulu over Detroit is not one of my concerns and travel is one of their fringe benefits.

Reading what we called "sure steps to substandard reliability performance" may raise eyebrows. We tend to believe we're smarter than we actually are. But if we use introspective and see any of the above traits in our organization, we might consider listing them in the order of importance. We might then proceed to map out an appropriate improvement strategy and, hopefully, stick to following the map to recovery before it's too late.

Chapter 26.3

How to Turn ODR-OPPM Into a Worthwhile Initiative

INTRODUCTION

Consultants often invent new initiatives; ODR, which stands for operator-driven reliability, represents one of them. Around 2010, ODR has morphed into a new arrangement of letters, OPPM. These four letters stand for one-page project management. OPPM's original premise was that any project or work execution can be reduced to a simple, single-sheet instruction template. An OPPM initiative would thus involve creating and issuing simple, one-page instruction and/or status documents. In its initial phases, the structure of each document was to perfectly express essential details, communicate those details to upper management, and track project progress. If, as you read this book, OPPM no longer exists, it will have joined the many short-lived consultant-conceived initiatives that preceded it. If OPPM is still with us, you may benefit from the guidance found worthwhile in 2016.

FORMALIZING ODR

The new twist or evolving application of OPPM seems to formalize what used to be labeled ODR. In general, the various possible "evolved" OPPM templates are aimed at enabling an operator or reliability technician to correctly and consistently carry out straightforward reliability-enhancing maintenance actions on a variety of physical assets found in industrial facilities. The operator would become a multiskilled employee.

WHO IMPLEMENTS ODR

The old ODR concept has been around for many decades. Its initial intent was to train operators in minor maintenance tasks. ODR has sometimes worked well and added value to an enterprise. But ODR has also, at times, been judged a failure. That is because all work process implementers are people, be they operators, multiskilled technicians, or maintenance craftspeople. Some of us are highly motivated and responsive to targeted schooling, but many of us can be indifferent to training and can't get away from whatever it is we have always done.

At present, there are three ODR implementation routes:

(1) The user company does its own ODR without outside assistance.
(2) The user company selects an ODR/OPPM contractor-implementer to develop the needed documentation.

(3) The user company hires an ODR or ODR/OPPM provider and defines its scope of deliverables. The user also engages a highly experienced detail-oriented third party to supplement the input obtained from commercial consulting companies. This detail-oriented third party ensures that upgrade (and update) opportunities are fully reflected in the ODR/OPPM directives.

We will not know which of the two approaches will prove more valuable in the long run. For now, we base our report on feedback obtained. This feedback on the success or failure of ODR is often informal; it covers a rather wide range of observations and opinions. We have feedback on (1), above, indicating that user companies have not retained the ODR approach over periods of several years. Regarding (2), some user companies report their expectations were not met by the company selling ODR/OPPM services. We know of users that are presently debating if they should go with route (3). Clearly, these services have the potential of strengthening ODR, but time will tell if they live up to the promise.

CONSIDERING FIXES

We believe companies that reported flaws in self-started ODR execution should engage competent third parties—parties that authoritatively recommend improvements. We give the same advice to users whose experience with ODR/OPPM service providers was not favorable. These too should communicate with competent third parties. Truly competent third parties are helpful since the user (or the companies marketing OPPM consulting services) may not be staffed by subject-matter experts.

A user company reporting flaws in ODR execution might wish to establish if, for example, their chosen ODR/OPPM implementer (the second party) is unfamiliar with the fact that different electric motor bearings require different grease application strategies. If the ODR/OPPM service provider's employees are unfamiliar with such details, there will be risks in hiring them. They will probably pile up lots of consultant-conceived generalities and may simply be unable to implement an optimized grease application strategy. In other words, OPPM template writers can only contribute the things they know. To add tangible value to an enterprise, a template writer would have to be a gray-haired subject-matter expert.

Our experience shows that people respond far better to sets of instruction that include the reasons why we should do something. Without these explanations and because they are given encouragement toward becoming innovative self-starters, operating technicians tend to simplify, circumvent, or even abandon the execution detail spelled out on a single-sheet OPPM template. Writing down explanations of why we do something will be helpful in most situations and, quite frankly, may engage the one issuing a template to better understand what the whole thing is about in the first place.

Let's again use electric motor bearing grease replenishment as but one of many hundreds of possible examples. The template writer may ask for grease drain plugs (Chapter 17.1 and Fig. 17.1.1) to be temporarily removed while regreasing electric motor bearings. Without an explanation as to why a plug must be removed, the operator or maintenance worker will assume it's just extra work—"we have never had to do this before." That kind of thinking will probably earn his company a grade of "D" in the maintenance effectiveness rankings of reliability-focused plants. It would be much better for the template to state that not removing drain plugs causes the pressurized grease to push bearing shields into contact with rolling elements. This then causes frictional heat, which causes the grease to rapidly oxidize. Incorporating this explanatory note, the improved OPPM template might get the plant moved up to a "C" ranking.

But suppose we had the experience, insight, or knowledge that smart user companies (in this instance, ARCO Alaska) had 40 years ago. They replaced the grease drain plug with a length of pipe and an elbow (Chapter 17.1 and Fig. 17.1.4). Spent grease would reside inside the pipe and form a "plug." The plug would be pushed out whenever new grease was applied at point 1. With no drain plug requiring removal, excessive pressurization of the grease cavity is impossible. Another problem is solved, and the facility would perhaps earn a grade of "B" among reliability achievers.

Of course, oil mist could be applied to the electric motor bearings. That would allow the plant to earn an "A" ranking among reliability achievers. Such a facility would probably experience the lowest electric motor maintenance cost and the fewest downtime events.

ADVICE TO THOSE CONTEMPLATING ODR/OPPM

Unless an asset manager or ODR/OPPM implementer knows the requisite details and incorporates the "why we do it this way" detail in his approach, the entire initiative is likely to fall short of its true potential. If the ODR/OPPM effort is unable to point out best available technology (BAT) and cannot give guidance on the use and cost-justification for BAT, lost opportunities will abound. A client company must carefully examine the service provider's detailed knowledge. The value potentially contributed by the service provider is wholly dependent on detailed knowledge. Determine diligently if the service provider's other clients have moved ahead of the pack in terms of failure avoidance. If a service provider's clients experience many repeat failures of assets, something isn't right. As said earlier, competent third parties may have to be enlisted for help. In essence,

(1) ODR requires a close partnership between operations and maintenance to ensure and ascertain all tasks are carried out in accordance with best practice; maintenance is notified immediately regarding conditions that require actions beyond the scope of ODR;

(2) ODR and presumably OPPM require leadership and coordination beyond what is expected for a single-function program like condition monitoring (CM) or condition-based monitoring (CBM) or even lubrication;

(3) Close-in supervision, continuing training, and follow-up are essential ingredients; analogous to safety, it is imperative to identify deficiencies, conduct training, and follow-up to assure the needed training has been implemented [1].

You can turn ODR and/or ODR/OPPM into worthwhile initiatives if you explain to your operators what these concepts are all about. You should consider obtaining buy-in from the operators before you go the buyout route. Then, you should hire or, better yet, groom the right subject-matter experts. These SMEs need to have intimate knowledge of the work processes and procedures in place at best-of-class companies. Also, they must have the desire and motivation to pass on their detailed expertise to your operators. You can then empower your operators and mold them into multiskilled employees by teaming them up with the right subject-matter experts. Then, and only then, will you be able to look back on ODR-OPPM as a worthwhile initiative.

REFERENCE

[1] H.P. Bloch, A.R. Budris, Pump User's Handbook: Life Extension, in: fourth ed., Fairmont Publishing, Lilburn, GA, 2013.

Chapter 26.4

Reliability Implementation Is an Organizational Issue

INTRODUCTION

We cannot think of a better example to highlight the urgency of having managers understand just how far behind one can get relative to one's competition. There was an article on "Eliminating Cooling Water from General Purpose Pumps and Drivers" in the Jan. 1977 issue of *Hydrocarbon Processing*. The article documented why, in the early and mid-1970s, many of the world's largest and most profitable "best-of-class" oil refineries had discontinued using cooling water on pump bearing housings. In the decades since the 1970s, many write-ups in books and articles have dealt with cooling water elimination from process pumps. However, three or even four decades after the 1970s and in spite of numerous exhortations and admonitions to conserve precious water resources, there still are thousands of process pumps needlessly and wastefully connected to bearing cooling water lines. We believe indifference and the lack of motivation are, ultimately, an organizational shortcoming. We find it at facilities that talk preservation but simply do not practice it. Not keeping up with best practices can injure the environment and the profitability of companies and corporate entities.

CASE 1: WHY ARE THEY STILL USING COOLING WATER ON PUMPS?

The headlined question was prompted by a visit to a fairly large refinery in one of the US Mountain States. Note that the visit took place over three decades after the "Cooling Water Not Needed" article was published [1]. The facility continued to use cooling water on hundreds of pumps that would have had greatly extended bearing lives if only the cooling water had been deleted years ago. In case you're wondering, the basis for deletion of cooling water from pump bearing housings equipped with rolling element bearings is found in the immutable laws of physics. Placing a cooling water jacket around the bearing's outer ring will prevent it from thermally expanding. As the hot inner ring "grows" and the internal clearances vanish, the bearing is excessively preloaded and fails prematurely. Or, if cooling coils are provided in the oil sump and the cooling water is indeed cold, some of the water vapor in the moist air floating above the oil will condense. The oil is now contaminated with water (water has a considerably lower film strength than oil), and the bearings fail prematurely.

Over the past three-plus decades, this refinery probably spent a small fortune on avoidable pump repairs. It undoubtedly consumed hundreds of thousands of gallons of cooling water unnecessarily and, for a certainty, used more energy than was really necessary. That's food for thought, especially since the facility's pump repair frequencies were inferior to those typically seen at best-of-class companies. The refinery is the loser here, and we'll leave it to the reader to decide where the problem originates. (Hint: dig deep and peel off the many layers of indifference.)

CASE 2: UPGRADING IS THE USER-CUSTOMER'S JOB

We know of a major overseas manufacturer of very thoughtfully engineered upgrade components for process pumps. This manufacturer sent two of its managers to explain their superior upgrade products to a well-known pump manufacturer. After carefully laying out why reliability-focused users have implemented pump upgrades using the components at issue, the managers concluded that

> *....pursuing the OEM approach, i.e., convincing the pump manufacturers of the merits of upgrading their pumps, is not possible. For some time now the technical sales approach does not seem to have worked with U.S. pump manufacturers. By any measure they only want to cut the cost, not improve quality. We determined that the only way many U.S. pump manufacturers will listen is if their end-customers demand our superior upgrade components on new equipment. We are convinced that we must begin with selling to the end-customers first.*

The attitude displayed by original equipment manufacturers (OEMs) should not surprise us at all. This experience is not new, and an article in Ref. [2] by Joe Askew ("Tales of an Engineering Consultant," *World Pumps*, Jun. 2006) is among many publications that fully support the same findings. The article prompted the comment, below, from a technical editor in the United Kingdom:

> *It's all about cost—and wide-spread ignorance of how inefficient some systems are. What makes it so tragic is the terrible waste of energy and the consequences of that waste—Europe has experienced one of its hottest, and most uncomfortable summers to date. It's only when plants appreciate the missed opportunities and cost savings inherent in energy conservation that they will be motivated to take real steps to install and run their systems efficiently.*

Again, it's up to the user to demand better performance. Surely, a solid and well-detailed user specification that includes upgrade measures implemented by best practices refiners would do wonders here. And unless a corporation or facility has such a specification, the kind of narrow focus and repair-intensive response one gets from the pump manufacturers will be perpetuated.

Finally, we'll relate Case History Three. It was excerpted from an e-mail by an observant engineer. He concludes the following:

CASE 3: OEMS WANT YOUR SPARE PARTS BUSINESS

> *One of our designers was recently meeting with a major client, his engineering company and a major manufacturer of lobe blowers in Europe. Both the owner and the engineering company wanted pure oil mist on the bearing end and purge*

mist on the timing gear end of the blower. The manufacturer refused to approve pure mist and warrantee it. What is more interesting is that he told everybody the reason why. He said that he would not be able to sell as many bearings and parts if he allowed the installation of pure oil mist.

That too is not surprising. When the bid invitations for a major petrochemical facility were sent to a number of pump manufacturers, three potential pump vendors replied they could not accept warranty responsibilities for the several hundred pumps they proposed to furnish for the project. These pumps were to be provided with pure oil mist. No constant level lubricators and no troublesome oil rings were involved.

The user-customer stood up to these manufacturers. Not to be fooled, the owner-purchaser responded by making an intelligent and entirely logical reply to the three pump manufacturers. Each was notified that they would be released from all responsibilities for *bearing performance and bearing life*. However, if they would not accept all customary responsibilities for the *hydraulic performance* of their products, they would be disqualified from supplying pumps for this project and all future jobs requiring oil mist lubrication.

The three vendors immediately agreed to warrantee the hydraulic ends of their pumps. Every one of their pumps was still in service decades later, although none of the three pump manufacturers still exist today.

Here's the issue: Are your reliability engineers willing to do what others did 40 and 50 years ago, that is, understand the immense value of intelligent lubrication technology and take a stand for it? If the answer is "yes," we commend you. If it's "no," you might ask a few questions and include the most important one: What are you planning to do about it? Best-of-class companies became what they are because they, the user-customers, understood that they had to educate themselves. Unless the user-customers insist on quality, they will receive maintenance or repair-intensive products. Relying on design contractors and equipment manufacturers is simply not enough. In the final analysis, the customer gets what he deserves, and it's either more downtime risk and bloated maintenance budgets or higher equipment reliability and more profits. The choice is clearly up to the buyer. The buyer's management staff must accept these facts and take the lead in reversing costly mistakes. It is indeed an organizational issue.

REFERENCES

[1] H.P. Bloch, Eliminating Cooling Water from General Purpose Pumps and Drivers, Hydrocarbon Processing, 1977. January.

[2] J. Askew, Tales of an Engineering Consultant, World Pumps, 2006. June.

Chapter 26.5

Counting Interventions Instead of MTBF: An Organizational Issue

INTRODUCTION

Keeping track of one's mean time between failures is a means of benchmarking, that is, comparing oneself to the competition. But there are pitfalls when people start chasing after the numbers rather than establishing only meaningful facts. Also, successful managers strive to understand relevant principles in combination with details of interest.

MTBF TRACKING CAN BE SUBVERTED

A very experienced machinery reliability professional at a refinery in the United Kingdom alerted his managers to an interesting way by which mid-level staffers had subverted the aims of management's reliability improvement efforts. Among the machinery reliability professional's many tasks was the tracking of mean time between failures (MTBF) of centrifugal pumps at his place of work. The refinery was divided into several sections or grouped process units. It was evident from the tracking charts that some sections had better pump MTBF than others.

But this professional was not too fond of MTBF tracking and offered an important reason: to achieve high MTBF, some unit managers scheduled and performed frequent interventions. They thereby expended considerable money and resources for precautionary oil changes and the like. He offered the very valuable suggestion of tracking interventions of any kind. He had come to realize that tracking interventions is smarter than striving for an artificially inflated MTBF as a performance indicator. Of course, we agreed.

THINKING "OUT OF THE BOX"

We should value this reliability engineer's example of "thinking out of the box" and not accepting business as usual. There are still many opportunities in the hydrocarbon processing industry (HPI) and other industries to excel and save precious resources, be they environmental or financial or both. Here are a few examples worth pondering:

1. From 1971 until 1975, one of the unquestionable multinational leaders in the HPI dismantled cooling water piping on all centrifugal pumps equipped with rolling element bearings. Many of these pumps were in hot services, up to 740°F. Using a higher viscosity grade lubricant and, later, switching to

superior PAO-/dibasic ester-based synthetic lubes caused these bearings to run well within acceptable operating temperature ranges and their average life improved over the "cooled bearing" versions. For over 30 years, this success has been documented in articles and books. Yet, as of this writing, there still are refineries that use cooling water on process pumps with rolling element bearings. One of these changes bearing oil four times per year.

2. Ever since the early 1990s, synthesized hydrocarbon lubricants have become the number one lubrication problem solver. Properly formulated, these high-performance products will form a tenacious oil film in rolling element bearings. Frictional resistance and heat generation are reduced. They are easily cost-justified for virtually any of the more problematic rolling element bearings found in roughly 7% of a typical refinery's pumps. We call these 7% our "bad actor population."
3. Think about a widely advertised 1980s version of a rotating labyrinth seal. You may think that external contaminants cannot enter into the bearing housing through an assembly fitted with the dynamic O-ring shown in one of the chapters in this text. Yet, a dynamic O-ring operating directly opposite the sharp-cornered O-ring groove in Fig. 14.2.1 will be severely damaged if there is axial movement of the shaft or while the shaft is being slow-rolled or whenever the O-ring is moving from standstill to liftoff speed. In contrast, after 2006, technically superior top-performing rotating labyrinth seals became available. Superior bearing protector seals are designed with an axially moving O-ring contacting a generously dimensioned area (see Subject Category 14). These products achieve exceedingly high lifetime averages.
4. In 1978, a major olefin facility was commissioned in the Houston, Texas, area. Many hundreds of centrifugal pumps and their drive motors that had been purchased without oil slinger rings and without the maintenance-intensive constant level lubricators were started up with pure oil mist lubrication (also called "dry sump") supplied to the bearings. Hundreds of pumps and motors have since been added in expanding this highly profitable facility. The pumps are still oil mist lubricated today, many decades later. Papers and books have been written explaining why now approximately 170,000 pump sets are operating flawlessly on oil mist in some of the most advanced facilities in different parts of the world. If you're not in a position to use plant-wide oil mist systems, take heart. Small oil mist units serving two to four pump sets have been in service for many years. Small oil mist lubrication units have been added to the market in about 2000 (see index words on oil mist). Whenever they are retrofitted to "bad actor" equipment and eliminate costly repeat failures, the payback is measured in months, not years.

Any of the above measures or implementation steps will likely reduce the number of maintenance interventions—both planned and unplanned. Managers working in close cooperation with well-informed reliability engineers start by budgeting for these and other betterments. Keep the opportunity in your focus.

Chapter 26.6

Thinking Independently: An Organizational Issue

INTRODUCTION

Another chapter of this text touches on the subject of tracking mean time between failures (MTBF) of general-purpose machinery. It was noted that, to achieve high MTBF, some unit managers used frequent interventions and thereby expended considerable money and resources for precautionary oil changes and the like. We made the point that tracking interventions of any kind are smarter than striving for an artificially inflated MTBF as a performance indicator. It follows that "doing things the way they have always been done" may not be the path to higher reliability and profitability. In other words, employ teachable and highly motivated staffers. New ways of thinking and implementing better approaches may be worth a lot (Fig. 26.6.1).

FIG. 26.6.1 Should process operators be thinking or just wait until being told and do what they're being told?

But here's the rub: Does your organizational setup encourage independent thinking? Do you have accountability? Are you enabling before empowering?

HOW THINKING INDEPENDENTLY CAN ADD MUCH VALUE

Over the years, attentive readers have been alerted to numerous value-adding or cost-reducing opportunities. None of them were speculative or untried or untested, although many of them had not been widely publicized. Here then are just five more seal or lubrication-related examples that smart reliability professionals have been made aware of and which you might consider implementing:

1. The shortcomings of traditional pumping rings in dual mechanical seals were fully understood in 1978. Seal manufacturers then provided pumping screws that were more efficient—but only as long as there were rather small clearances between rotating and stationary seal components. However, relevant clauses in the venerable API pump standard, API 610, anticipated shaft deflection and, therefore, recommend relatively liberal wear ring clearances. Following that line of reasoning, one should avoid "tightening" the clearances in the close-fitting regions of certain mechanical seals. With close-clearance pumping screws, one increases the risk of galling or even seizing contact. Fortunately, bidirectional pumping rings have been available from at least one innovative seal manufacturer, and thousands are in successful use today. Could bidirectional pumping rings be an opportunity for you?
2. In 1985, we copresented, at the ASLE 40th Annual Meeting in Las Vegas, Nevada, a rather comprehensive paper relating our experience with "Mechanical Seals in Medium-Pressure Steam Turbines." Our logic that high-temperature seals would be well suited to survive medium temperature steam was proved correct, and bellows seals were easily cost-justified in small steam turbines. The Bloch/Elliott paper was also published in the Nov. 1985 issue of *Lubrication Engineering*. Although never acknowledging our work in the intervening decades, a few mechanical seal manufacturers have since picked up on the idea. How much steam are you losing through conventional carbon ring seals used in your small steam turbines? Another opportunity perhaps?
3. Aerospace-type magnetic seals solved a major problem with oil contamination at a Texas plant in 1979. From 1979, that plant's reliability-focused users have been advocating the selective and often highly cost-justified use of magnetic bearing protector seals. A number of companies now produce industrial versions; a dual magnetic bearing housing protector seal is being marketed by the world's fourth largest manufacturer of mechanical seals (see Subject Category 14). Would not these products be highly effective for protecting, say, your cooling tower fan gearboxes? Or the pillow block bearings on a forced draft fan? If this fan is mounted high up in a hot process unit, will it really be properly maintained? What is the cost of properly maintaining

it? What will be the true cost of having it fail randomly? What would be the bottom-line value of extending its repair interval by factors of four or five?

4. The shortcomings of millions of pressure-unbalanced constant level lubricators were recognized in the 1980s. Many problems can be overcome by balanced constant level lubricator models. Better yet, have you considered properly engineered bearing protector seals together with unvented (!) bearing housings for containment and protection of the oil in equipment bearing housings? You could then dispense with these often not so perfect constant level lubricators and observe oil levels through a suitably selected sight glass. For a certainty, you wouldn't be the first user to do so! Without fanfare, many users have done just that.
5. Pure oil mist and application of oil in the "jet oil" mode are the most effective, most durable, and most reliable lubrication methods for rolling element bearings. While initially more expensive than a set of generally more failure-prone brass or bronze oil rings and a constant level lubricator, the high reliability and extremely low-maintenance requirements of jet-oil-lubricated installations have been published and should be readily acknowledged. Some oil mist application modules applied to "bad actor" bearings have resulted in greatly reduced failure frequencies and had paybacks of 4 and 5 months. (See index words "oil mist" and "oil rings." Also, Subject Category 24 expands on the issue.)

REACH OUT FOR THE OPPORTUNITY

As reliability professionals, we should question sales talk and should examine things with full objectivity. We must look at the full picture, which inevitably takes into account life-cycle cost, reduced maintenance effort, and even opportunity cost. Serious reliability improvements require stepping away from the "business as usual" approach. Just as sooner or later, common sense and the laws of physics will converge, so will the overall understanding of competent reliability professionals. Reaching out for true opportunities will add great value.

Understanding Oil Mist: Learning by Reading

On another recent occasion, a reliability professional expressed annoyance that the $4,000,000 cost proposal he had just received for a plant-wide oil mist system was widely off the numbers I had published in an *Oil Mist Lubrication Handbook* in 1987. Why was the man surprised? Hello? He seemed even more baffled when I reminded him of page 25 of the Oct. 1990 issue of HP. There we had noted, decades ago, "together with an appropriate amount of a suitable state-of-the-art synthetic lubricant, this low-cost retrofit (referring to a modern magnetic seal and a plugged vent instead of the customarily open-to-atmosphere bearing housing vent port) may extend bearing life to the point where oil mist lubrication is no longer economically attractive."

Regrettably, many people who so often proclaim expertise in modern technologies (such as oil mist) have opinions, but have not read very much. They tend to take comments out of context and deal with very old and often highly biased data. Some have no idea of the statistics of oil mist *preservation* and its highly beneficial effect on the nonrunning standby pump in areas with high humidity or blowing sand. They seem to be unaware that electric motors at best-of-class plants are included in the equipment served by modern oil mist systems. There is thus great value in the concurrent life extension of electric motor bearings. Of course, one cannot overlook that best-of-class plants and facilities have used through-flow ("directed") mist exclusively in certain bearings as early as 1969 and applied the mist through-flow method per API 610 eighth and later editions (see Chapter 23.3 and Fig. 23.3.2), a full 23 years before the API finally adopted it.

Chapter 26.7

More About Unreliability, Global Procurement, and You

INTRODUCTION

We quite obviously believe that reading is the key to professional growth and that Mark Twain was right in stating that the man who refuses to read is no different from the illiterate person who cannot read. We also know that, while technical texts can be pricey, a book costing $100 will often alert its reader to the solution to a million-dollar problem. In that case, the return on the investment would be 10,000:1. And so, it makes much sense to put reading in your training plans and to have either a budgetary item or subsidy for books at your facility. Reading and reliability improvements are related to each other and cannot be separated. In another chapter, this text alluded to the reliability professional's job of identifying critical parts and to write detailed specifications, which then facilitate global procurement of high-quality parts. Without such details, one simply gets the cheapest part that generally meets a typical description.

COPING WITH GLOBAL PROCUREMENT AND HEADING IN THE RIGHT DIRECTION

Unless proved otherwise, you should assume that the lowest bidder utilizes neither quality control nor exacting specifications. Perhaps, this explains why an entity is the lowest bidder. Again, you must provide and sometimes personally write a specification for these critical parts. Once critical spare parts (even the ones originating from vendors accepting your specifications and professing to have quality control) are delivered to your facility, the job is far from finished. You must add value by personally verifying the full specification compliance of these parts. Alternatively, take responsibility by arranging for competent inspectors that verify specification compliance of the critical spare parts received. These parts should be accepted by the storeroom clerk only after compliance has been verified. The clerk can then proceed to tag and preserve the parts for future use.

As to the misguided direction where some oil refining and related process industries are headed, we recall a very strong message from a well-known asset management consultant. After visiting a major refinery, he considered its management system completely broken. He expressed the view that, due to past failures, decision makers now seem afraid to make any decision that carries even a whiff of risk—so they do nothing! He met a young reliability engineer who had poured heart and soul into a project, submitted it to the plant manager, and heard nothing—not even the simple courtesy of an acknowledgement! The consultant was struck by this refinery's bewilderment why so many of the young men and women who should be its lifeblood and future were quitting their jobs. Those who remained seemed to have the attitude, "just tell me what you want, manager, and I will get myself involved somehow." Of course, the manager doesn't have a clue, so nothing of substance gets done at that location.

Then, there are the many recurring accountability issues mentioned by the asset management consultant. In one review meeting, a reliability engineer was asked why he thought he had to spend so much time in the plant during turnarounds, watching things like gasket replacements. His answer? "Because I'm held directly responsible even though the fault may be solely attributable to the carelessness of a mechanic. The mechanic will not be held accountable, but I will be." Word spreads, and we heard rumors that, a few years later, not a single graduating engineer accepted the offer of employment made by one particular major oil company.

SHUNNING CHEAP TEMPORARY FIXES

A huge problem at one refinery seems to be its constant pursuit of cheap temporary fixes. Managers at this location have no discernible concept of the bigger overall picture and have enunciated neither sound strategy nor anything resembling long-term improvement. At one location, a highly experienced management consultant judged as totally inoperable the functional asset hierarchy on which all cost and reliability data are based. Upon being briefed about the issue, the refinery managers considered corrective action "too difficult" and elected to again do nothing but maintain a very precarious status quo.

That then gets us back to the original point and a conversation as to where global procurement involves all kinds of service providers. Once we identify the most successful service providers, we must ascertain that they will continue to add value every step of the way. They will join us in viewing every maintenance event as an opportunity to upgrade. Upgrading means strengthening the weakest link in the component chain whenever cost-justified. It will make the operator's life easier and will open wide the (presently very narrow) door to operator-driven reliability (ODR). Conscientious upgrading will have merit beyond that of traditional maintenance.

Whatever your job function, you can make a big difference and be an effective change agent. Start by understanding or personally defining critical spare parts and take it upon yourself to describe them in an appropriate purchase specification. Read what others have done in this regard and how they persevered and excelled not just recently, but decades ago. Other facilities became best practices companies by having professional employees totally involved—these employees took the lead in advancing the reliability improvement process. They were among the first to view every maintenance event as an opportunity to upgrade components and machines and initiated action where it was both feasible and cost-justified.

If you are *not* a manager, write down what you have found in the various books and articles and discuss it with your manager. If he doesn't take action, find someone who will. And if you *are* a manager, do something about the critical situation we have accurately described in this chapter of our text. A bit of introspection will let you know who you really are. Make adjustments in your course, which is another way of asking you to either lead, follow, or move out of the way.

Subject Category 27

Packaged Machinery

Chapter 27.1

Upgrading Fan Bearings

INTRODUCTION

There are never enough hours in a day or in a workweek or even in a full year for any one of us to become an expert at anything under the sun. Nevertheless, we are being asked by management personnel to solve problems or to avoid problems altogether. As mechanical engineers, we are perhaps trained in calculation methods, machine design, thermodynamics, vibration technology, and so forth. However, we have, hopefully, concentrated on certain areas of expertise. This is where I want to briefly mention a highly relevant component, rolling element bearings, which are decidedly *not* my area of expertise. To understand much about rolling element bearings, I determined decades ago that I had to make a renowned bearing manufacturer my technology resource. I picked a manufacturer with an application engineering department ("AED") and did the same with mechanical seal manufacturers, vendors of packaged machinery, service providers, and others. They had to grant me access to people who knew far more than I knew [1,2].

They had to have an application engineering department and I explained the value of such departments to my managers. I wanted these managers to understand that the initial per-item cost of components from manufacturers with AEDs was bound to exceed those of the lowest bidder and that the incremental cost was often paid back in a single week. Failure avoidance was the motivator, and failure avoidance was achieved with the help of these manufacturers. It's a sustained lesson that I wanted to pass on. It is embedded in several of the chapters and subtopics you are reading in this book.

STATUS OF RELIABILITY THINKING

Communications originating in different corners of the world gave us an overview of the status of reliability thinking. We often picked the best questions,

Petrochemical Machinery Insights. http://dx.doi.org/10.1016/B978-0-12-809272-9.00027-X

and the following discourse with a gentleman in the Caribbean might serve as an example. Here was his dilemma, in his own words:

I have not been in communication with you for quite some time. Once again, I need you to point me in the right direction. It concerns the fan train for packaged boilers.

As you know on ammonia/methanol plants, the steam/methane reformer includes relatively large forced draft and induced draft fans. These are typically supplied by reasonably well-known fan manufacturers; the general purpose drive turbines are in the 1000 kW range. You may recall I had a project a few years back that I had to fight for and won; it improved production by 100 MT/day of methanol when completed.

Anyway, over the years, the reliability has been pretty good as these trains sit on large piled foundations with excellent damping effect. They typically have sleeve bearings, and the trains' major weakness has been a rolling element thrust bearing in the turbine. The bearings are supported by force-fed oil lubrication systems, with the main oil pump drive coming from the gearbox and an auxiliary waiting to auto-start in the event of failure. No worries.

However, the same cannot be said for the packaged boilers. These boilers produce anywhere from 50 to 150 tons of steam per hour and are critical to plant needs, especially during start-up and shutdown. No steam, no methanol or ammonia.

The fan trains we inherited sit on welded plate frames with rubber vibration isolators and have grease-lubricated bearings. Resonance is normally a problem. During the design stage of the methanol project, I pointed out this shortcoming to the designers and stakeholders but unfortunately was unsuccessful in getting fans with sleeve bearings fed by a force-fed lubrication system sitting on top of a nice big concrete foundation. The result is three bearing failures to date; all are attributed to excessive load and poor lubrication.

These draft fans are normally governed by either the design and engineering company's standards or the fan vendor's standard. I have never seen a standard for fans similar to API 611/612 for turbines. If there is one, please enlighten me as I want to review the minimum requirements to see how best to make my case for an upgraded design for my existing plant and also the new methanol plant currently under construction by my company in the Middle East. Naturally, they want to give me the same fan design and are using the same vendors again.

We asked the reliability engineer to use rigorous specifications and to explain to his management that reliability cannot be an afterthought. Reliability must be a designed-in quality and quality comes at a cost. Quality will produce attractive payback.

UPGRADING STARTS WITH API STANDARDS

We replied that while there's an API standard for centrifugal fans and blowers (API 673), the term "packaged" often implies bare bones or minimum acceptable. Replicating an entire package saves time and money and changes made in the interest of upgrading upset the manufacturer's routine. This is why package suppliers often tend to charge significant extra amounts for compliance with API 673 and other reliability-focused standards.

Also, what seems to be conveniently overlooked by some purchasers is that stand-alone, nonspared, packages are intended for users that generate (or need) steam intermittently and for users that generally continue on the path of being repair-focused. That, now, contrasts with certain management circles erroneously claiming the desire to be reliability-focused. If there has ever been a truism: "desiring" and "having" are rarely the same.

Therefore, when buying packaged boilers, one should consider buying them with standby spares, especially if one is repair-focused and geared up for frequent repairs. Some people like to do repairs because it keeps the workforce happy with overtime income. More often, if the project people have been allowed to make their cost estimates on the basis of lowest available equipment pricing, the plant will usually end up with repair-intensive equipment. In that case, achieving best-of-class status will almost certainly elude the plant. It is no joke to anticipate that the project will probably come in slightly under budget and likely will go onstream 2 days ahead of expectation. The project team will (almost certainly) be commended and its members promoted.

At this point then, the burden is being shifted to the facility's machinery engineer. The machine now becomes his responsibility and that brings us back to the writer from the Caribbean. He must somehow selectively upgrade the fans and blowers at his plant. Kingsbury, Waukesha, and other manufacturers offer retrofit circulating systems that integrate suitable sleeve-type radial and tilt-pad thrust bearings in a flange-on subassembly. Such a flange-on upgrade would cost money, but it would also solve his problem.

The engineering manager making the inquiry could also do his own upgrading by using an inductive pump or a prepackaged pump-around (a fluid distribution unit, "FDU") system of the type normally used with double mechanical seals (Fig. 27.1.1). Optimized application of the lubricating oil is important. Synthetic oils should be used and the bearing housings should be retrofitted with modern bearing protector seals. In other words, "standard equipment packages" can often be upgraded by the purchaser; look into it under the index word "MQA" [1].

Finally, the facility could certainly continue to use rolling element bearings and use a small oil mist system. Oil mist has been highly successful not only on process pumps and their motor drivers but also on boiler fan and blower bearings (Fig. 27.1.2). Some installations date back to 1975 and small self-contained oil mist supply units are also shown in Subject Category 23.

The last and also least desirable alternative would be to do nothing. Doing nothing would be demonstrating the opposite of reliability thinking. In the final analysis, it would certainly be the most costly option because it infers high downtime risk and costly maintenance intensity.

FIG. 27.1.1 A fluid distribution system ("FDU") could also be adapted for use as a bearing lube circulation unit. *(Courtesy AESSEAL Inc., Rotherham, the United Kingdom, and Rockford, TN.)*

FIG. 27.1.2 Fan bearings lubricated with oil mist. *(Courtesy Colfax Fluid Handling Company, Lubrication Systems Division, Houston, Texas.)*

REFERENCES

[1] H.P. Bloch, A. Budris, Pump User's Handbook: Life Extension, fourth ed., Fairmont Press, Lilburn, GA, ISBN: 0-88173-720-8, 2013.

[2] H.P. Bloch, Pump Wisdom: Problem Solving for Operators and Specialists, Wiley & Sons, Hoboken, NJ, ISBN: 978-1-118-04123-9, 2011.

Subject Category 28

Piping and Related Components

Chapter 28.1

Configure Pump Suction Piping Carefully

FUNCTION OF SUCTION PIPING

The function of suction piping is to supply an evenly distributed flow of liquid to the pump suction, with sufficient pressure to the pump to avoid cavitation (the formation and implosion of vapor bubbles) and related damage in the pump impeller. An uneven flow distribution is characterized by strong local currents, swirls, and/or an excessive amount of entrained air. The ideal approach is a straight pipe, coming directly to the pump, with no turns or flow-disturbing fittings close to the pump. Further, the suction piping should be at least as large as the pump suction nozzle and be sized to ensure that the maximum liquid velocity at any point in the inlet piping does not exceed 8 ft/s (~2.5 m/s).

If the suction piping fails to deliver the liquid to the pump in this condition, a number of pump problems can result. More often than not, these include one or more of the following:

- Noisy operation
- Random axial load oscillations
- Premature bearing and/or seal failure
- Cavitation damage to the impeller and inlet portions of the casing
- Occasional damage from liquid separation on the discharge side

HYDRAULIC CONSIDERATIONS FOR SUCTION PIPING

In general, pumps should have an uninterrupted and nonthrottled flow into the inlet (suction) nozzle. Flow-disturbing fittings should not be present for some minimum length. Flow disturbances (Fig. 28.1.1A) on the inlet side of the pump can lead to the following:

Petrochemical Machinery Insights. http://dx.doi.org/10.1016/B978-0-12-809272-9.00028-1

- Deterioration in performance
- Damage to the impeller and shortened impeller life (especially with high suction energy pumps)
- Shortened mechanical seal life
- Shortened bearing life

Isolation valves, strainers, and other devices used on the inlet (suction) side of the pump should be sized and located to minimize disturbance of the flow into the pump. Eccentric reducers should (usually) be installed with the top horizontal and the bottom sloped, as noted in Figs. 28.1.1 and 28.1.2 [1]. By understanding that the design intent is to eliminate vapor pockets, we will understand when and where and why there are installations when the flat (nonsloped) part of an eccentric reducer needs to be at the bottom (Fig. 28.1.2B).

The most disturbing flow patterns to a pump are those that result from swirling liquid that has traversed several changes of direction in various planes. Liquid in the inlet pipe should, therefore, approach the pump in a state of straight, steady flow. When fittings such as "T" fittings and elbows (especially two elbows at right angles) are located too close to the pump inlet, a spinning action or "swirl" is induced. This swirl could adversely affect pump performance by reducing efficiency, head, and net positive suction head available (NPSHa). It also could generate noise, vibration, and damage in high suction energy pumps.

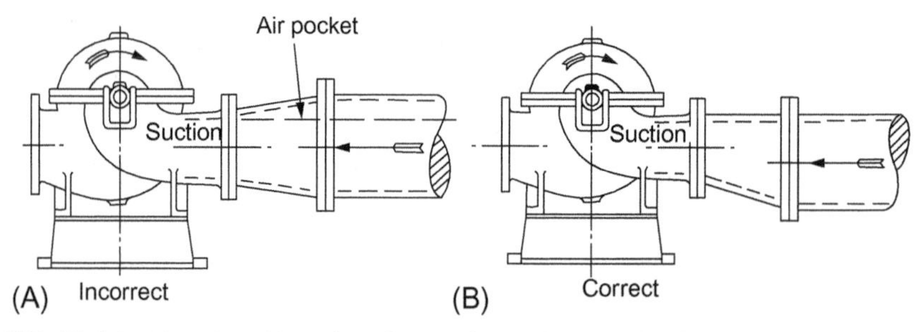

FIG. 28.1.1 (A and B) Illustration of eccentric reducer mounting from Hydraulic Institute standard [1].

Only vertically oriented elbows should be allowed directly at a pump (Fig. 28.1.2B). In all other instances, it is always recommended that a straight, uninterrupted section of pipe be installed between the pump and the nearest fitting. This should follow the minimum straight pipe length guidelines listed in Ref. [2], which range from the following:

- One to eight pipe diameters (for low suction energy/low specific speeds)
- Three to 16 pipe diameters (for high suction energy/high specific speeds)

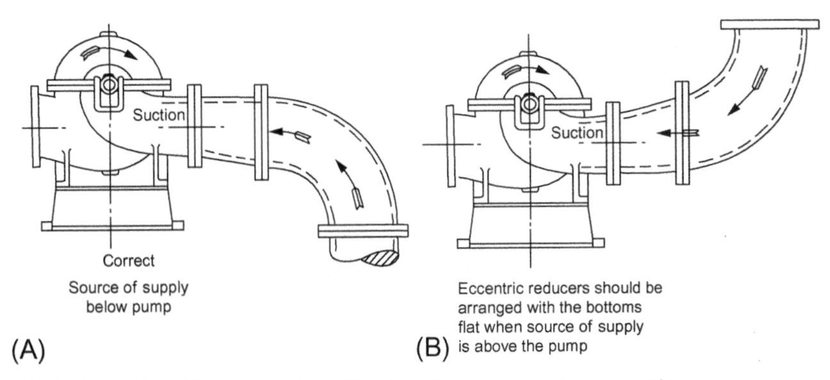

FIG. 28.1.2 (A and B) Suggested modifications for eccentric reducer mounting.

The specific straight pipe length recommendation depends on the type of fitting(s), pump type, suction energy level, and pump specific speed. Generally, high suction energy pumps have suction nozzle sizes larger than 10 in. and 1800 or more revolutions per minute (rpm) and larger than 6 in. at 3600 rpm. High specific speed starts above a value of 3500. If it should prove impossible to provide the minimum recommended pipe lengths, contoured vane flow-straightening devices should be considered. These contoured, twisted-angle inserts are preferred over the occasional straight vane welded-in-place insert.

Questions relating to the proper application of reducers in centrifugal pump suction lines date back many decades. Until his death (at age 84, in 1995), world-renowned pump expert Igor Karassik frequently commented on such issues [3]. In a sequence of letters to Karassik, pump users referred to Figs. 28.1.1A/B and 28.1.2A/B and noted that the four images were quite typical of illustrations found in many textbooks. In essence, Fig. 28.1.1A indicates that, with a suction line entering the pump in the horizontal plane, a concentric reducer would allow formation of a vapor pocket. Ingesting the compressible vapor at random, unpredictable, intervals can upset both hydraulic and mechanical pump performance. Figs. 28.1.1B and 28.1.2A show an eccentric reducer placed with the flat at the top.

Some available texts, however, give no indication as to whether the pumpage comes from above or below the pump. If the source of supply was from above the pump, the eccentric reducer should be installed with the flat at the bottom, as shown in Fig. 28.1.2B. Entrained vapor bubbles could then migrate back into the source instead of staying near the pump suction. If the pump suction piping entered after a long horizontal run or from below the pump, the flat of the eccentric reducer should be at the top, per Figs. 28.1.1B and 28.1.2A.

The aforementioned is amplified by comments found in early 20th century books. Many older texts assume that the source of the pumpage originates at a level below the pump suction nozzle. Also, older Hydraulic Institute standards commented on the slope of the suction pipe:

...Any high point in the suction pipe will become filled with air and thus prevent proper operation of the pump. A straight taper reducer should not be used in a horizontal suction line as an air pocket is formed in the top of the reducer and the pipe. An eccentric reducer should be used instead.

This instruction applies regardless of where the pumpage originates. Again, note that, depending on the particulars of an installation, trapped vapors can reduce the effective cross-sectional area of a suction line. Should that be the case, flow velocities would tend to be higher than anticipated. Higher friction losses would occur and pump performance would be adversely affected.

In the case of a liquid source above the pump suction and particularly where the suction line consists of an eccentric reducer followed by an elbow turned vertically upward and a vertical length of pipe—all assembled in that sequence from the pump suction flange upstream—it will be mandatory for the flat side of the eccentric reducer to be at the bottom. That said, Figs. 28.1.1 and 28.1.2 should clarify what reliability-focused users need to implement.

Also, whenever vapors must be vented against the direction of flow, the size of the line upstream of any low point must be governed by an important criterion. The line must be of a diameter that will limit the velocity of pumpage to values below those where bubbles will rise through the liquid.

LENGTHS OF HORIZONTAL PIPE RUNS

In general, it could be stated that wherever a low point exists in a suction line, the horizontal run of piping at that point should be kept as short as possible. In a proper installation, the reducer flange will thus be located at the pump suction nozzle, and there is usually no straight piping between reducer *outlet* and pump nozzle. Straight lengths of pipe are, however, connected to the *inlet* flange of the eccentric reducer. On most pumps, one usually gets away with 5 diameters of straight length next to the reducer. In the case of certain unspecified velocities and other interacting variables (e.g., viscosity, NPSH margin, and style of pump), it might be wise to install as many as 10 diameters of straight length pipe next to the reducer inlet flange. The two different rules of thumb explain seeming inconsistencies in the literature, where both the 5D and 10D rules can be found.

REFERENCES

[1] Piping for Rotodynamic Pumps (Centrifugal Pumps), HII 9.6.6 Draft Document, Hydraulic Institute Pump Piping Working Group, Parsippany, NJ, 2003.

[2] American National Standard, Centrifugal and Vertical Pump NPSH Margins, ANSI/HII 9.6.1 Hydraulic Institute, Parsippany, NJ, 1998.

[3] I.J. Karassik, Centrifugal Pump Clinic, second ed., Marcel Dekker, New York and Basel, 1989.

Chapter 28.2

Pump Suction Strainer Issues

INTRODUCTION

Does your process piping have strainers located just upstream of the pump suction nozzles? Perhaps you are using them to protect pumps from unintended ingestion of tower packing, nuts, bolts, and other debris. While it would be smart to investigate why this stuff shakes loose, we will confine our brief comments to common misunderstandings about strainers.

Whenever strainers are used because the upstream equipment is flawed, be sure to understand the important requirements imposed on strainers by reliability-focused engineers. These engineers recognize, first and foremost, that a distinction is to be made between temporary and permanent strainers.

Temporary strainers are generally installed with the tip pointed in the upstream direction, which places the material in compression instead of tension. These temporary strainers must be removed about one week after commissioning the piping loop. They are often fabricated on-site, using the general configuration shown later in Fig. 40.1.1.

In contrast, permanent strainers are designed to be left in place and must be cleanable without shutting down the pump. They are typically available from a variety of commercial sources, must be made of high-grade corrosion-resistant materials, and can be expensive.

Chapter 28.3

Flexible Connectors for Pumps: Points and Counterpoints

INTRODUCTION

As is always the case, there are two sides to even the thinnest of pancakes. Yet reliability-focused engineers take the position that safety is of paramount importance. Reliability engineering is the profession that identifies and eliminates vulnerabilities. This is the reason why a very large number of responsible and reliability-focused companies have, for many decades, refused to allow flexible connectors on pumps. Points and counterpoints are voiced in this write-up.

WHY NOT TO USE FLEXIBLE PIPING

Reliability focus and safety are principles that should never be undermined. That said, "Petrochem Machinery Insights" takes the firm position that expansion joints have no place in a risk-averse plant. Quite aside from the fact that API specifications require "hard piping" for refinery pumps, it is of interest that

1. the typical refinery experiences one costly pump fire per (roughly) 1000 pump failure events (since fires inevitably risk engulfing also the piping of other equipment (including, of course, other pumps), and because flexible pump connectors are the weakest link in the piping, it would be unthinkable to consider employing these connectors in refineries or petrochemical plants);
2. some of the very worst industrial disasters, including Flixborough (on Jun. 1, 1974, and many others before and after) involved expansion joints, bellows, or similar nonrigid connections [1];
3. responsible engineering effort will always accomplish pump installations that avoid flexible connectors in services other than, perhaps, water (in other words, while possibly cost-justified on water pumps and low-pressure steam, proper piping configuration and geometry are always attainable without the use of flexible pump connectors. That the upstream equipment has to be properly located relative to the pump is a rather obvious prerequisite).

But there are sides that contest or modify the three points made above. And so, when given the opportunity to comment, a vendor-manufacturer of flexible connectors addressed each of the earlier three points by stating the following:

1. I think the point being made here is that the connectors are susceptible to fire damage and are therefore a risk. In cases where this is a major concern,

I would suggest using an all-metal connector. Either a braided connector or a bellows connector would be appropriate.

2. Although it is possible to apply expansion joints and flexible connectors incorrectly at pumps (see below), their proper use is completely safe. On more critical applications, additional safeguards can be used, such as limit rods.
3. My third point and my point 2 as well are both directed more at expansion joints for thermal expansion rather than flexible connectors used at pumps for reasons stated in the chapter. But to address the concern, expansion joints for thermal expansion and other uses are part of responsible engineering efforts. The suitable applications often cover cases other than water and low-pressure steam. The use of hard pipe loops and similar alternative measures is simply not practical, or even possible, in many cases.

Says the vendor-manufacturer:

I would like to make one further point. It has been the experience of those in our business that there is one misapplication that dominates those failures that have been seen in the field: improper understanding of the thrust load that is generated with the use of expansion joints. While sometimes mentioned in articles, it may not always be explored fully due to the limited scope of a particular write-up.

Thrust loads are anchor loads that only exist when an expansion joint is in use under pressure; its magnitude equals the pressure in the system times the effective area of the joint. These loads can become quite large, and if the pipe system anchoring cannot withstand the load, it may fail. The result of the anchor failing is that the joint will stretch apart, which can make the joint appear at fault. The remedy is to either (a) ensure that the anchors are designed for the load or (b) ensure the joint has control rods or another way to absorb this load.

The other misapplication with expansion joints, not so with pump connectors, is not using pipe guides appropriately. The proper number and location of guides are very important whenever expansion joints are used. An engineer should not design-in expansion joints without understanding these issues. Once understood, it is a simple matter to design solid systems.

Well, we gave the writer of this pro-flexible connector feedback the full benefit of the doubt. As trained professionals, vendor-manufacturers undoubtedly understand the limitations and guidelines that apply to flexible connectors and would not make some of the mistakes made by their average customers. But even on water pumps, the use or advocacy favoring flexible connectors can often be traced to sloppy engineering. There might be flawed expectations that the flexible device will tolerate misalignment or unknown loads. Imagine, for the sake of illustration, a facility that could not vouch for the sound installation of conventional fixed piping by its work crews. In those (many) instances, it would be quite a leap of faith to entrust the crews with installing and servicing certain types of flexible connectors. And that's just one more additional reason why reliability-focused plants stick to the position we've outlined earlier.

REFERENCE

[1] T. Kletz, What Went Wrong?—Case Histories of Process Plant Disasters, third ed., Gulf Publishing, Houston, TX, ISBN: 0-88415-027-5, 1995.

Chapter 28.4

Pipe Supports and Spring Hangers

INTRODUCTION

Hanger and spring support-related installation oversights are responsible for many excessive pipe stress incidents. Piping-induced loads will very often cause pump casings to deflect; internal parts, especially bearings and mechanical seals, become misaligned and fail prematurely. Pump reliability improvement efforts must include verification of acceptable pipe stress values. But have you ever looked at a concrete sidewalk or driveway? Are not the different segments misaligned relative to each other? Soil instabilities may be at fault or the concrete piers and foundations in your plant may be too small. In any event, piping that looked good when the plant was built and commissioned is no longer well fitting. What does it mean to you?

INSTALLATION

Start at the beginning. It is always wise to consult the plant's record system for listings of line numbers and system spring identification numbers, types and sizes of springs, calculated cold setting values, and calculated travel. In new construction projects, such lists are generated for field follow-up and future checking of springs in their respective hot positions. Observe the following and compile a work procedure with mandatory "do this" items:

- Spring support/hanger movement arresting pins (locking pins) that were inserted by the contractor's or by the plant's mechanical work forces shall not be removed prior to completion of steps (a) and (b) below.
  (a) Line hydrotesting, flushing, and cleaning and, if the line requires insulation, with the pipe insulation either fully installed or replicated by a dummy load such as sandbags.
  (b) Flange alignment has been verified to be correct and free of excessive stress.
- Special precautions are required for lines carrying liquids. Standard practice calls for spring hangers/supports to be calibrated such that when the piping is at its hot position, the supporting force of the hanger/support is equal to the calculated operational pipe load. Note that the maximum variation in supporting force will therefore occur with the piping in its cold position, when stresses added to piping and/or equipment nozzles are less critical.
- Weight calculations and spring calibration must, therefore, include pipe, insulation, and liquid totals.
- After the locking pins have been removed, the construction contractor or plant maintenance technician shall adjust all springs to their required cold load marking values.

- Substantial forces could develop when machinery flanges have been aligned/bolted up on liquid-filled lines, the pins have been removed, and the spring travel indicator has been adjusted to the cold marking. Most of these forces would act directly on the pump nozzles, possibly exceeding allowable load limits. It is therefore recommended to provide sandbags or similar counterweights on 8″ or larger liquid-carrying lines. This will account for the weight of the liquid *after* the lines have been connected to the pumps. (Alternatively, springs in those particular systems could be reviewed by responsible engineering groups and different load values calculated and provided for the simulation of start-up conditions.)
- If the earlier step shows a particular spring load or travel indicator settling outside of the middle third of the total working range of the spring, the contractor's engineering staff and/or the plant mechanical workforces must determine appropriate remedial steps.
- At this time, the proper cold setting of hangers/supports will have been achieved. Equipment is typically turned over to plant operations personnel and piping is free of liquid. When flange bolts are *now* removed, a certain amount of pump misalignment may be introduced.
- Before a flange is opened up for any reason, all preloads on springs resulting from previous cold setting adjustments have to be eliminated first. Preferably, the spring settings should first be restored to the position they were in before the pins were pulled.

A few final thoughts: Following the earlier steps will go a long way toward ensuring that piping forces and moments will not be excessive. Excessive pipe stress will shorten machinery MTBF (mean time between failures). Neither centrifugal pumps nor pump support elements such as base plates and pedestals are designed for zero deflection or warpage when subjected to excessive piping loads [1].

When you see piping being pulled into place, something is wrong. Although today may not be the time to fix it all, make this the last time you tolerate a "wrong" connection. Designate a staffer who will design a spool piece, or design a foundation fix, or exchange a spring support or does whatever it takes to implement a solution. The next time the machine undergoes maintenance, implement the fix.

Never tolerate pipe stress in excess of what you can push into place with two hands. This two-hand rule will work every time. Teach the work crews that allowing misaligned piping will cause bearings to be edge-loaded. When bearings are edge-loaded, there will be no oil film; instead, there will be metal-to-metal contact that will cause premature bearing failure. And pipe stress will negatively affect mechanical seal and coupling life. Tolerating pipe stress will inevitably result in high maintenance expenditures and reduced equipment reliability.

REFERENCE

[1] H.P. Bloch, F. Geitner, Maximizing Machinery Uptime, Gulf Publishing Company, Houston, TX, ISBN: 0-7506-7725-2, 2006.

Chapter 28.5

Remembering Flixborough

INTRODUCTION

If there is a common thread to all true reliability engineering pursuits, it is for reliability professionals to understand why equipment fails. Generalizations sound nice, but details matter. Learn from the incident when one of six cyclohexane oxidation reactors at the Flixborough (the United Kingdom) plant had developed a crack. The reactor was bypassed by constructing a 20in. bellows line. The bypass failed immediately after being put into operation in 1974. Many tons of inventory were released and a vapor cloud was formed. More than two dozen lives were lost and close to 100 others were injured. So, how can it be that one still encounters an uninformed person occasionally lobbying for flexible connections even as you read this text?

FEASIBILITY VERSUS DESIRABILITY

While sometimes feasible, using flexible piping on centrifugal pumps in a process plant environment is generally a bad idea. We leave it to others to decide if it is sloppy workmanship or lack of engineering that impels some to consider flexible piping in the HPI. In any event, most reliability professionals would not tolerate the ensuing risk. Since there can be no doubt that best-practices plants are using only hard-piped connections in similar pumping services, users should understand and research facts before allowing anything other than diligently configured conventional piping.

You might imagine yourself as a plant manager dealing with a technician who came to you advocating anything other than best practices. A good reaction would be to instruct that technician to look into the matter by googling "Flixborough," and to later report back to you what they had discovered. But why Flixborough?

LEARNING FROM THE FLIXBOROUGH DISASTER

In early 1974, one of the six reactors at a cyclohexane oxidation plant in Flixborough (the United Kingdom) had developed a crack. It was considered appropriate then to bypass the reactor by constructing a 20in. bellows bypass. With operating conditions of 8bar and 150°C, the bypass failed immediately after being put into operation on Jun. 1, 1974. An estimated 40tons of inventory was released and a vapor cloud formed. The ensuing explosion was equivalent to an estimated 15tons of TNT; it totally destroyed the plant and many nearby houses. The control room collapsed; 28 operators were killed and 89 operators and neighbors were injured. A vicious fire lasted 3 days and as many as 250 firemen were involved in fighting the conflagration. The last fires in some storage vessels were allowed to burn under controlled conditions and were finally extinguished 13 days later [1].

While it is true that the modification at Flixborough was constructed by people who simply did not know the vulnerabilities of their 20 in. bellows configuration, I would insist on answers to questions of failure risk in the event of a fire in the vicinity of any flexible line. Unless the flexible connection is designed with a suitably large margin of safety and is both flawlessly installed and maintained, its use around flammable and toxic fluids will be risky and should never be allowed.

It would be good to remember this and the many other devastating incidents that happened since then. Learning from disasters is the best way to avoid future disasters. To an extent, this should make us reluctant to deviate from proven ways of doing things. But we must also find ways to progress and both identify and accept better procedures, work processes, and components as these become available.

Knowing how parts fail is the key to understanding when to insist on code compliance in one instance versus allowing changes to the code in another case. A perfect example would be hard-piping the liquid flush and similar connections on process pumps even in cases where hydraulic tubing might be feasible and might allow higher pressures to be accommodated. Understanding the rationale of hard-piping will shed light on the issue. Hard pipe is strong enough for a careless worker to stand on, whereas hydraulic tubing might deflect under the load and risk leaking at the connector fitting. So, one generally stays with hard pipe and utilizes tubing only when protected from the indifferent or unknowing folks.

This is just one of hundreds of examples where the secret to understanding a certain engineering code or rule is buried somewhere in the literature. Rediscovering the rationale for a code or ruling should be a training task assigned to new engineers. Understanding how and why disasters happen is the key to avoiding the next disaster.

READ, AND TRY TO GET THE FULL PICTURE

Understanding how parts function and malfunction is of great importance and cannot be overestimated. As a simple, albeit not life-affecting, example, hundreds of rotating labyrinth-style bearing protector seals are being sold each day. Some styles come with one half of an O-ring surrounded by a stationary part and the other half of the O-ring surrounded by a rotating part. The purchasers of this style of bearing housing protector seal do not seem to have the full picture. They need to think about what will happen to the O-ring under slow-roll conditions (it drags and degrades) or under conditions of axial movement (it gets sheared). And some people get hugely annoyed (usually at the messenger) when it finally dawns on them that they have been asleep at the switch, instead of having acted as searching, highly principled professionals. Along the same lines, let it be known that better products are available from "others," but to really absorb the full story on lubrication, sealing, coupling selection, and literally a hundred other things, disciplined reading is needed. Plants that arrive at decisions by guessing will never reach best-of-class status.

If one allows people to guess, the future scenarios should scare us all. Flawed purchasing strategies and horrendous failure events show that the direction given by (and to) a profession that no longer reads utterly lacks focus. Where there is no focus, bad things happen. Where there is no focus, reliability professionals are no longer nurtured; they have no access to technical libraries. To be fair, there are still a few good examples where a company listens to the need to encourage reading and to disseminate relevant material in a structured manner. These best-of-class companies insist on their professional staff collecting useful information. They give reliability professionals authority to purchase relevant texts while holding them accountable for what they learn from the texts that they have purchased.

Unfortunately, though, many HPI facilities with billion-dollar yearly profits have no budgets for books. Nevertheless, they have generous budgets that pay for repeat failures and repairing the same equipment year in, year out. They are the ones we wanted to reach with this book.

REFERENCE

[1] T. Kletz, What Went Wrong—Case Histories of Process Plant Disasters, third ed., Gulf Publishing, Houston, TX, ISBN: 0-88415-027-5, 1985.

Subject Category 29

Professional Growth

Chapter 29.1

Reliability Professionals and How They Are Perceived by Others

INTRODUCTION

It should be clear that John Bachner's editorial (see Main Focus, above) forcefully made the point that communication is inextricably tied to leadership. Professions that cannot communicate are flawed and engineers must learn to communicate. Those that refuse to learn are being outflanked by often ruthless advertisers, shameless self-promoters, and shrewd marketers.

It was certainly no coincidence that a very similar point was made by Dr. Hans-Juergen Kiesow, Vice President of Gas Turbine Engineering at Siemens Energy. Although himself a professor, he chastised the academic community in an address at the ASME Turbo Expo in Glasgow, Scotland (June 6–10, 2010). Mr. Kiesow pointed out that "we need less nano hype and more stress on basic physics."

TOO MUCH NOISE, NOT ENOUGH ENGINEERING

To be fair, in Turbomachinery International's July/August 2010 issue, Siemens executive Dr. Kiesow tried to balance his criticism by commenting on engineering students and their performance in the business world. "The smart engineers do not learn how to sell what they invent," he said. "MBAs (individuals with Master's Degrees in Business Administration) learn to successfully sell mediocre ideas while engineers struggle even to get heard."

Dr. Kiesow was not alone. In an editorial in Design News (July 2010), Dr. Geoffrey Orsak, the Dean of Engineering of the SMU (Southern Methodist University) Lyle School of Engineering, expressed the view that "companies extracting value from our earth have a responsibility to invest some of this value into increasing the reliability of these complex systems. And because no

Petrochemical Machinery Insights. http://dx.doi.org/10.1016/B978-0-12-809272-9.00029-3

engineering system is ever foolproof, we better have a good back-up plan when oil is released into the environment."

Regarding the 2009 BP/Transocean oil rig explosion and spill tragedy in the Gulf, we have seen editorials ranging from a basic "accidents happen, so let's just move on" to wholesale condemnations heaped on an entire industry. Must we always take an adversarial stance? Is everything stark black and white?

As to other voices heard on the matter of professionalism, a perceptive few pointed out that scenarios calling for plugging an undersea leak with golf balls and rubber tire shards seemed concocted by executives, not engineers. We also noted that BP made a presentation at the NPRA Reliability & Maintenance Conference in San Antonio (May 26, 2010) focusing on "the key foundational elements of a world class reliability program that were established for BP's largest and most technical refinery in 8 months vs. the typical industry practice of 3 years." Two months later, in early August 2010, BP agreed to pay a record $50.6 million fine for failing to correct safety hazards at its Texas City oil refinery after a 2005 explosion there killed 15 workers. In all of this, where were the engineers' voices? Why this obvious disconnect?

BUYING CHEAP WILL IMPOVERISH MANY

For decades, there has been a trend to buy commodities from the lowest bidder. Lowest bidders are often the shrewdest marketers. It's fair to say that the lowest bidders are rarely the providers of highest quality products. PINOs blame the purchasing department; the purchasing department blames PINOs for not taking the time to develop solid specifications. At times, it is indeed the fault of unresponsive managers who will not listen to true professionals. As is so often the case, there are two sides to such stories.

At one of its annual Technical Press Days in Philadelphia, top bearing manufacturer SKF briefed their audience on a topic with which we had been acquainted for a number of decades. In fact, we were thoroughly familiar with the topic because we had identified best-of-class (BoC) companies as ones that, among other things, purchased bearings from highly respected manufacturers only. Years ago already, these BoCs recognized that manufacturers with competent application engineering departments were providing far more than mere commodity products. They passed on priceless expertise in failure avoidance and became mentors to the relatively few *true* reliability professionals. True reliability pros are defined as the men and women who made it their life's ambition to add value to an enterprise.

On that occasion, SKF asked its audience to join forces in the important global fight against industrial counterfeiting. Counterfeit products continue to flood the marketplace worldwide. Fig. 29.1.1 is counterfeit, and a reliability expert must learn to distinguish it from the real product. There is only one thing that is obvious in cases of industrial counterfeiting: more than a brand will be

FIG. 29.1.1 Counterfeit double-row spherical roller bearing. *(Courtesy SKF America.)*

at risk. Many components, including bearings and mechanical seals, are safety-critical in most applications. Their "knockoffs" can pose very serious hazards in addition to performance issues.

The SKF presentation profiled why and how this company is striving to protect both its brand and customers from the illegal fakes that can cause real damage in service. But the standing of thousands of reliability professionals is affected by this issue. They are often perceived as ineffective in preventing downward slides in equipment quality or stopping the drift toward nonoptimized allocation of funds.

Reliability professionals must agree with the validity of concerns raised by SKF and actively do their part in not ever using any kind of fakes in their plants. Realize that the purchasing department is being rewarded on the basis of cost and schedule. The manager manages people, not bearings. Recognizing the difference and realistically assessing the performance of counterfeit versus genuine is not their areas of expertise. That leaves it to reliability professionals to make it their business to know the difference and to be the strongest possible advocates of what's in the best interest of shareholders, employers, and employees.

Subject Category 30

Pumps and Associated Issues

Chapter 30.1

Suction Specific Speed Choices have Consequences

INTRODUCTION

A refinery engineer was in a quandary over requests from a project group. High-volume/pressure/temperature pumps in hydrocarbon service were involved, and the refinery's standards had been changed so as to avoid purchasing pumps that might not operate well in the lower flow region. The engineer asked if it was really practical to insist on accepting only pumps with an Nsss (meaning "suction specific speed") below 9000, although decades ago, his company had allowed pumps with Nsss values up to 12,000. But first, a greatly simplified introduction to Nsss and its importance.

DEFINING NSSS

Note that pump suction specific speed (Nss or Nsss) differs from the pump specific speed parameter Ns. Suction specific speed is calculated by the straightforward mathematical expression:

$$\mathrm{Nss} = \frac{(r/\mathrm{min})\left[(\mathrm{gal}/\mathrm{min})/\mathrm{eye}\right]^{1/2}}{(\mathrm{NPSHr})^{3/4}}$$

In this equation, both the flow rate and NPSHr pertain to conditions observed at 100% of design flow (at the best efficiency point or BEP) on the maximum available impeller diameter for that particular pump.

The higher the design suction specific speed or **Nsss**, the closer the point of troublesome internal flow recirculation to BEP. Similarly, the closer the internal recirculation capacity is to BEP, the higher the hydraulic efficiency. Pump system designers are tempted to aim for the highest possible efficiency,

Petrochemical Machinery Insights. http://dx.doi.org/10.1016/B978-0-12-809272-9.00030-X

hence, high suction specific speed. However, such designs might result in systems with restricted pump operating range; if operated inside the restricted (high recirculation) range, disappointing reliability and frequent failures will be experienced.

TREND CURVES OF INTEREST

Although more precise calculations are available, illustrative trend curves of probable **NPSHr** for minimum recirculation and zero cavitation erosion in water (Fig. 30.1.1) are sufficiently accurate to warrant our attention [2]. The **NPSHr** needed for *zero* damage to impellers and other pump components may be many times that published in the manufacturer's literature. The manufacturers' **NPSHr** plot (lowermost curve in Fig. 30.1.1) is based on observing a 3% drop in discharge head or pressure; at $Q = 100\%$, we note **NPSHr** = 100% of the manufacturer's claims. Unfortunately, whenever this 3% fluctuation occurs, a measure of damage may already be in progress. Assume the true **NPSHr** is as shown in Fig. 30.1.1 and aim to provide a net positive suction head available (NPSH**a**) in excess of this true NPSH**r**.

In Ref. [2], Irving Taylor compiled his general observations and alerted us to this fact. He cautioned against considering his curves totally accurate and mentioned the demarcation line between low and high suction specific speeds somewhere between 8000 and 12,000. Field experience came into focus after 1980 and some investigators pointed to 8500 or 9000 as numbers of concern. Empirical data for pumps with **Nss** numbers higher than 9000 were found to fail more often than pumps with lower **Nss** numbers. If operated at flows much higher or lower than BEP, their life expectancy or repair-free operating time was reduced.

Trend of probable NPSHr for zero cavitation-erosion

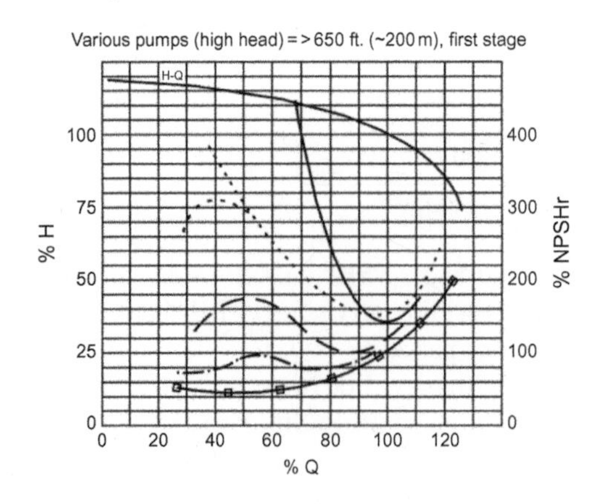

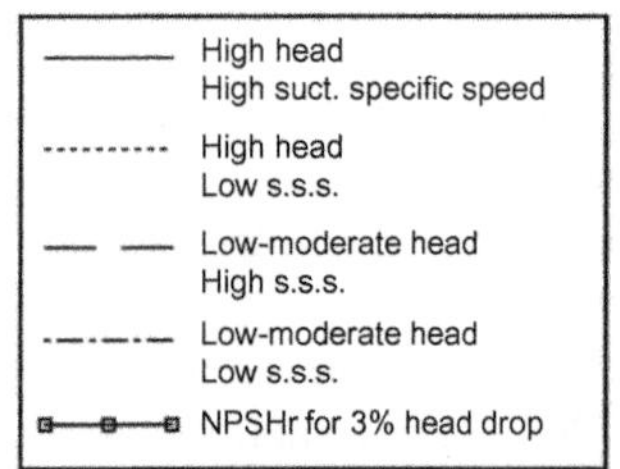

FIG. 30.1.1 NPSHr trends based on experience [1].

In the decades after Taylor's presentation, controlled testing has been done in many industrialized countries. The various findings have been reduced to relatively accurate calculations that were later published by HI, the Hydraulic Institute [3]. Relevant summaries can also be found in Ref. [1]. Calculations based on Refs. [3] and [1] determine the minimum allowable flow as a percentage of BEP.

Note, again, that recirculation differs from cavitation, a term that essentially describes vapor bubbles that collapse. Cavitation damage is often caused by low NPSHa. Such cavitation-related damage starts on the low-pressure side and proceeds to the high-pressure side. An impeller requires a certain net positive suction head (NPSHr); this NPSHr is simply the pressure needed at the impeller inlet (or eye) for relatively vapor-free flow.

WHAT GUIDELINE TO USE

As a general rule, adhering to the lowered Nsss requirement (not exceeding 9000) will probably force the industry to require the purchase of 1800 rpm multistage pumps instead of the obviously less expensive (albeit somewhat higher efficiency) 3600 rpm single-stage pumps that have made inroads in the past few decades.

Sometimes, it will be acceptable to purchase and operate 3600 rpm and higher speed/higher Nsss pumps. However, doing so should require a *written waiver* by the parties accepting the procurement of high Nsss pumps. Issuing such a waiver should (hopefully) force the various "parties to such agreements" into understanding, describing, and clearly communicating the associated risks. The limited operating range of such pumps would have to be stated.

There is no easy choice. However, at low flows, the less expensive pumps cannot operate safely for long periods. In the end, these are judgment issues and reliability professionals must advise the rest of the organization relative to risks incurred. Again, I believe reliability professionals will have to caution and properly advise operating and management personnel of the obvious risk of allowing high Nsss pumps to operate at low flows. Or, in some instances, the purchaser must choose between low-speed between-bearing pumps (Fig. 30.1.2) and less expensive overhung impeller offers.

The ultimate decision as to what to buy and how safe a plant they really want is management's. They, and operating personnel, must be informed of the need to stay away from certain flows. Alternatively, flow control devices would have to be purchased for these services.

We always caution against the easy alternative to just cautioning others verbally. Verbal operating instructions are soon overlooked or forgotten. The right course of action would be to budget reliable instrumentation that would have to be procured and installed. Such instrumentation might mitigate the risk of operating at low flow by automated means. But automated means may shift the life cycle cost calculation into numbers that were not previously anticipated. It might affect other elements in the unit's chain of operation.

FIG. 30.1.2 Between-bearing process pump train. It comprises a full ready-to-ship "package." *(Courtesy Dengyosha Machine Works (DMW), Tokyo, Japan; www.dengyosha.com.)*

A second alternative would be to have only experienced and thoroughly trained operating personnel—people who will simply not allow these (high Nsss) pumps to operate at low flows.

A third alternative would be to accept the need of more frequent monitoring and for performing repairs more often on high Nsss pumps. What was saved in capital will soon be spent as maintenance-related money.

REFERENCES

[1] H.P. Bloch, A. Budris, Pump User's Handbook, fourth ed., Fairmont Press, Lilburn, GA, ISBN: 0-88173-720-8, 2014.

[2] I. Taylor, The most persistent pump-application problems for petroleum and power engineers, in: Energy Technology Conference and Exhibit, Houston, Texas, September 18–22, 1977. ASME Publication 77-Pet-5.

[3] ANSI/HI9.6.3, Allowable Operating Region, Hydraulic Institute, Parsippany, NJ, 1997.

Chapter 30.2

One Pump Fire per 1000 Pump Repairs

INTRODUCTION

While performing reliability audits decades ago, pump failure statistics were made available or could be recovered with relative ease. But even then, the sources were usually kept confidential because fire incidents are stressful, to say the least (Fig. 30.2.1). It was known (in 1974) that for every 1000 pump repairs, there was a pump-related fire incident. Decades later, at a facility with approximately 2000 installed pumps, the acknowledged MTBR (mean time between repairs) was 6 years. This would allow us to calculate that approximately 333 pumps underwent repair at that refinery each year. The plant experienced five pump-related near-disasters in the span of 14 years, and performing a simple calculation tells us that this plant's rate of major pump issues tracked the 1000 per 1 rule within 6% accuracy.

FIG. 30.2.1 Assets and human lives are at stake when there are pump fires in refineries.

FAILURE STATISTICS TELL THE STORY

An airplane has about 4,000,000 parts, an automobile has approximately 10,000 parts, and a centrifugal process pump has only about 200 parts. It's fair to say that if a machine is made up of a large number of parts, more parts *could* malfunction. However, this does not mean that more parts *will*, in fact, malfunction during an operational cycle. So, what's the point of this reminder?

As we think about the reasons why the average process pump requires a repair after approximately 6 years, we realize that not all of its components are designed, fabricated, assembled, maintained, operated, or perhaps installed with the same diligence as aircraft components. The most appropriate mechanical seal support systems are not always selected.

It doesn't have to be that way. Alloys can be upgraded and better components are sold to owner-purchasers that insist on such upgrades. Advanced computer-based and reasonably priced design tools are available for the pump hydraulic assembly. It has been shown that computational fluid dynamics (CFD) can be used to define the improvement potential of impellers and stationary passages within pumps. Fig. 30.2.2 highlights the desirable progression from computer imaging to simulation, fabrication, and testing of a semiopen pump impeller. Thorough engineering is involved at every phase of this progression.

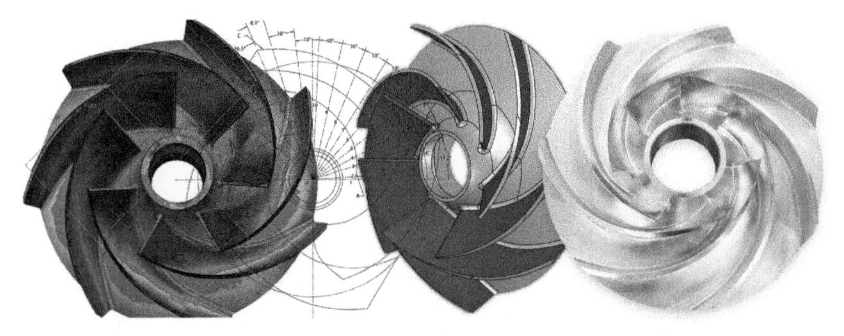

FIG. 30.2.2 Engineered solutions for 2016 and beyond. *(Courtesy Hydro, Inc., Chicago, IL, http://www.hydroinc.com.)*

Component strength and vibratory response can be studied and improved; Fig. 30.2.3 alludes to both in the context of fluid machinery upgrading. Upgrading means failure avoidance and failure avoidance reduces fire incidents.

Fire incidents are reduced in frequency by better sealing systems. Superior dual seals and environment-friendly sealing systems (Figs. 30.2.4 and 30.2.5) are available and routinely selected by reliability-focused owner-purchasers. The cost justification for superior sealing solutions is often quite relevant. Greater safety and reduced stress on the environment are associated with virtually every one of the many seal improvement solutions offered today. Better lubricant application and sealing methods reduce fire risks.

The mechanical assembly (drive end) of many pumps deserves our attention, especially since this portion of the pump has been neglected in some brands or models. Fortunately, expert advice is available for the specification and selection of better drive-end geometries for process pumps [1].

Thoughtful specification and selection used to be par for the course at best-of-class companies, and there is really no reason why this thinking should have

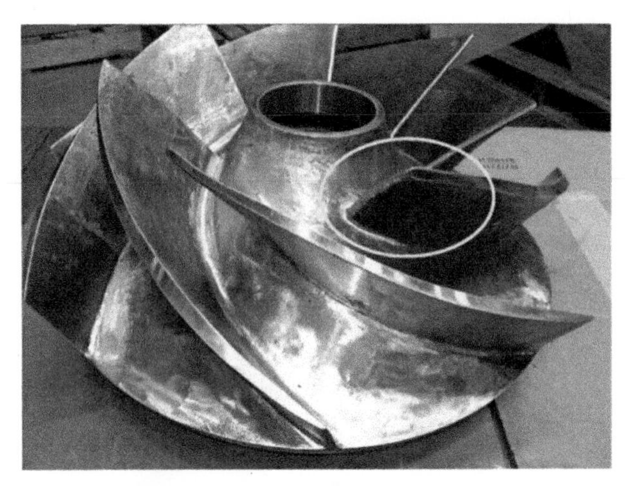

FIG. 30.2.3 Vane strengthening and vane contour modification are among the most frequent upgrade tasks. *(Courtesy Hydro, Inc., Chicago, IL, http://www.hydroinc.com.)*

undergone change. However, what we see lacking in 2016 is an awareness of the precise steps that are needed for such specifying and selecting. Management has fallen prey to consultant-conceived generalities, including "lean and mean" and similar catchy slogans. These slogans are often interpreted as a license to buying cheap. "Cheap is permissible" then gets into the management mind-set and the mental pendulum reverses direction only after an incident happens.

In contrast, we believe that good managers must lead and must reinforce an unwavering failure avoidance culture. That said, the user industry should

FIG. 30.2.4 A dual mechanical seal. The space between the sleeve and the inside diameter of the two sets of seal faces is taken up by a pressurized barrier fluid. *(Courtesy AESSEAL Inc., Rockford, TN, and Rotherham, UK.)*

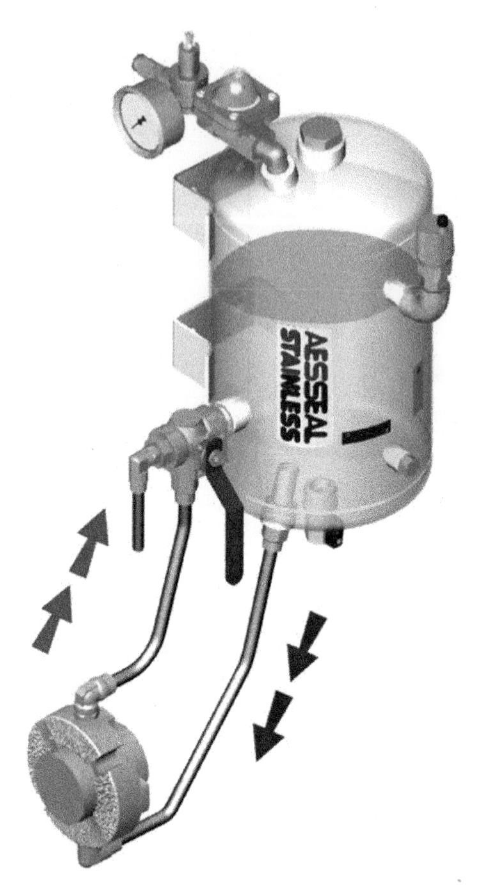

FIG. 30.2.5 Pre-engineered self-contained water management system found highly effective for dual mechanical seals in slurry services. Some water management system retrofits have had a payback (return on investment (ROI)) of as little as 3.5 months. *(Courtesy AESSEAL Inc., Rotherham, UK, and Rockford, TN.)*

get away from the obviously flawed approach whereby people are instructed to run equipment to failure and to then engage in costly attempts to resurrect the equipment to a better life than ever before. It also makes no sense to wait for things to go wrong and to then heap praise on someone's superhuman overtime-consuming efforts to rebuild the physical asset as quickly as possible. Upon closer examination, the old adage of an ounce of prevention being worth a pound of cure is as valid as ever. So, consider this a call to do something about your repeat failures and start asking questions long before machines are in distress.

REFERENCE

[1] H.P. Bloch, Pump Wisdom—Problem Solving for Operators and Specialists, John Wiley & Sons, Hoboken, NJ, ISBN: 978-1-118-04123-9, 2011.

Chapter 30.3

Not Ready for 29-Year-Old Pump Packing

INTRODUCTION

In 2009, there was a nuclear power plant that was never commissioned after its original construction. The facility had been sitting idle for 29 years and was only then (in 2009) in the process of being started up. The original request from an engineer was a bit lengthy but could be reworded in the short paragraph below:

> *Many pumps and valves have packing installed and the plant is wondering if they should repack. I think the answer is yes, but the facility is looking for some supporting documentation before advocating repacking all of the Nuke's equipment. Please ask HP's equipment editor if he has—or knows of—any books or literature relating to long-term storage of equipment in regards to packing in the stuffing boxes.*

Although someone seemed sufficiently embarrassed to later claim "the question was a bit off the cuff," our first reaction was to see this as an example of what happens when the industry is no longer staffed with experienced workers and people willing to make wise decisions. Clearly, many employees and their managers have neither the time nor the inclination to acquire true in-depth knowledge. Anyway, we thought the people who asked could perhaps do their own research in an old machinist's handbook or in a vintage pump catalog. The one I found in my files was silent on the matter of packing life. Its authors probably assumed (in the 1950s) that commonsense answers were obvious and that writing about them would bore the readers. Smart people consider it naive and inappropriate to put faith in packing on pumps or valves that stood idle for 29 years. So, replace the packing!

But what about packed pumps and associated maintenance in general? As an example, certain inexpensive or old-style firewater pumps have limited packing box space. On those, conceivably, one's only choice might be to continue using braided packing. One text gives a nine-step procedure for packing maintenance and repair. While interested parties might wish to read the whole story, a short overview sequence includes the following:

- Distinguish between different packing styles and materials.
- Preferably use a mandrel to properly wrap and cut new packing.
- Use the right tool to remove old packing.
- Remove corrosion products from shaft or sleeve surface.
- Coat ring surfaces with an approved lubricant.
- Insert one new ring at a time and then seat each with a proper tool (Fig. 30.3.1).

- Install packing gland, initially only finger-tight.
- Operate for run-in, initially allowing a stream of sealing water leakage.
- Tighten sequentially to allow a leakage flow of 40–60 drops per minute.

For more details on storage protection, refer to the same text [1]. Here's the short version: fill the pump, including its stuffing box or seal housing, with a mineral oil containing 5% rust-preventive concentrate. Be sure to turn the shaft four times a year.

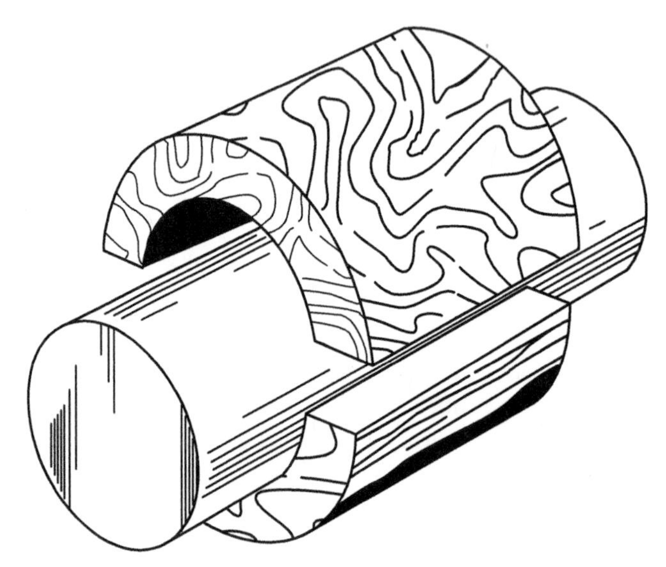

FIG. 30.3.1 Split plastic, brass, or aluminum bushing used for proper sequential seating of packing rings in pumps.

Reverting to the question relating to the recommissioning of a nuclear power facility, old data on firewater pump sealing surfaced as well. The claim is that firewater pumps in buildings must comply with applicable codes that require packing and disallow mechanical seals. We have reason to believe that old codes deal with technology prevailing at their time of issuance. With few exceptions, plant safety and reliability are the ultimate intent of codes and standards. That intent is best met by using solid experience and up-to-date technology.

Adhering to procedures is probably the main reason why nuclear plants are highly reliable. It follows that the answer is probably at the original questioner's fingertips—if only he took the time to read. If there really is no official guideline for the nuclear facility, then a sound recommendation would be to treat the packed pumps as if they were in full-time service. A rational person might simply ask how much time would elapse before a complete overhaul is due based on normal operation. If the answer is less than 29 years—which it better be—then the original question is answered. Look at your own normal packing replacement intervals and be governed by these. Hopefully, they reflect best practices!

PACKING NOT BEST PRACTICE FOR FIREWATER PUMPS

It might surprise some that packing is not the first choice of knowledgeable reliability professionals. Best practice in modern firewater pumps has been (since about 1968) to use single-spring mechanical seals. These mechanical seals are to be backed up by a floating throttle bushing and a deflector guard. If, for some unrealistic and hard-to-understand reason, a facility still uses packing, please think again. Do research what happened at a major refinery in 2015 and about that refinery's total economic losses when a firewater pump equipped with packing was out of service because a packing leak had destroyed the pump's bearings. The firewater pump was unavailable just when urgently needed to fight a major fire. "Designing out" maintenance, serious professionals would also check into the feasibility of securely installing an advanced rotating labyrinth bearing housing seal instead of the deflector guard used decades ago. But first, here's what led up to using single-spring mechanical seals instead of packing in the firewater pumps at modern refineries as early as 1965.

In the 1960s, accurate statistics were kept (for insurance purposes) by a major multinational oil company (MMOC). The statistics for firewater pumps showed that leaking packing tended to ruin bearings. As of the late 1960s, well-designed mechanical seals were selected by MMOC because these seals generally leak much less than packing and are considerably less likely to allow water spray to enter an adjacent bearing housing. Of course, we know that brittle mechanical seal faces might shatter when abused. However, seals that are properly designed, selected, and installed are highly unlikely to shatter. Moreover, floating throttle bushings represent a "second line of defense" in firewater pumps.

Informed engineers would *not* advocate the use of packing in modern firewater pumps. The contention that packing no longer represents best practice is amplified by the frequent lack of training of maintenance personnel in some plants. Included in best practice is periodic testing of all standby equipment. The question has occasionally been if switching the "A" and "B" pumps and running each for 1 month, or if turning on the standby pump once a month and then running it for 4–6 h, is a preferred choice. When people argued—many decades ago—that plants might get away by testing only twice a year, responsible reliability professionals took the position that testing only twice a year would *not* be acceptable and monthly testing was needed. Depending on lubricant selection and lube application method, switching "A" and "B" every 2 or 3 months is now considered best practice. This 2- to 3-month switchover keeps the bearings lubricated and prevents seal faces from sticking.

REFERENCE

[1] H.P. Bloch, F.K. Geitner, Major Process Equipment Maintenance and Repair, second ed., Gulf Publishing Company, Houston, TX, ISBN: 0-88415-663-X, 1998.

Chapter 30.4

Understanding Filter Beta Ratio and Beta Efficiency

INTRODUCTION

Reliability engineers are occasionally puzzled by the terms "beta ratio" and "beta efficiency." The two are related and we can calculate the filter efficiency after ascertaining the beta ratio.

By convention, beta ratio is a measure of a filter's efficiency. The term was introduced to give both filter manufacturer and user an accurate and representative comparison of available filter media. It is determined by a multipass test that establishes the number of particles upstream from the filter that are larger than the micron rating of the filter and then dividing that number by the number of particles downstream from the filter larger than the micron rating of the filter. The example illustrated here demonstrates the concept of beta ratio quite clearly.

AN EXAMPLE ILLUSTRATES THE CONCEPT

It is reasonable to see twin filters installed in oil or other hydraulic systems. Suppose that's the case in Fig. 30.4.1, and upstream conditions for both filters are 400 particles larger than 3 μm. We are not showing the shutoff valves allowing isolation and filter changes one at a time.

If, as an example, we find the downstream condition of the upper filter 100 particles larger than 3 μm, then the beta ratio = 400/100 = 4.

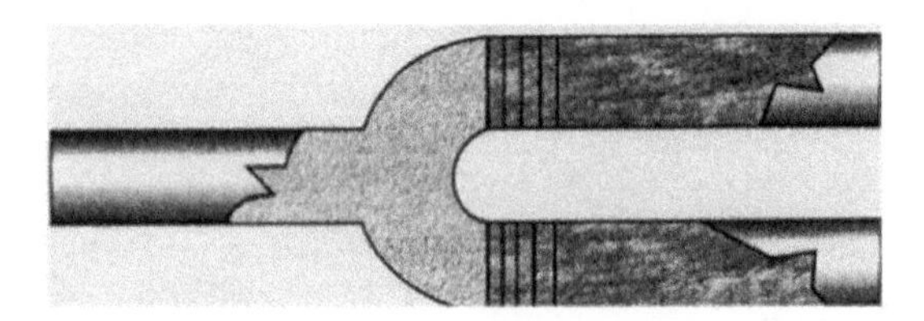

FIG. 30.4.1 Twin filters in a hydraulic system. Debris collects upstream of each filter.

Say we find the downstream condition of the lower filter one (1) particle larger than 3 μm. In that case, the beta ratio = 400/1 = 400.

For this example, there are 400 particles upstream from the filter larger than 3 μm in size. A filter having a lower beta ratio is less efficient because it allows more particles through it. Again, referring to the example above, it can be seen that while the filter at the top allows 100 particles to pass through, only one particle is allowed to pass through the filter at the bottom.

Accordingly, the beta ratio for the filter at the top is given by beta = 400/100 = 4, which is a less efficient value, whereas the beta ratio for the filter at the bottom is given by beta = 400/1 = 400, which is a more efficient value.

However, the following equation is used to determine the efficiency value of a filter, known as beta efficiency:

Beta efficiency = (number of upstream particles) minus (number of downstream particles), all divided by (number of upstream particles) where the particle size is greater than a specified value of *N* microns.

The relationship between beta ratio and beta efficiency can thus be represented as beta efficiency = 1 − (1/beta ratio).

For example, a filter with a beta ratio of 40 would have an efficiency of 1 − (1/40) = 97.5%. The higher the beta ratio, the higher will be the beta efficiency.

Chapter 30.5

Why Breaking the Cycle of Process Pump Repairs is Important

INTRODUCTION

We may have spent decades living with process pump failures and many chapters of this text have alluded to this fact. Breaking the cycle of process pump repairs is important and a few reminders are in order. This is why several keywords reappear in the narrative of this particular Chapter 30.5. Except for one or two illustrations, we had to save space and redundancy. This may mean that you should look for (and will find) certain relevant illustrations elsewhere in this text. If needed, please do so even if it will take you 5 min. If taking these 5 min prevents another $12,000 pump repair, it will have been worth the effort.

COST OF FAILURES

We start by exploring the value of extending pump mean time between failures (MTBF). One way to do this quickly and convincingly is to examine the likely savings if we could improve the MTBF from presently 4.5 years to a projected 5.5 years. Say, a facility has 1000 pumps; that's 1000/4.5 = 222 repairs before and 1000/5.5 = 182 repairs after understanding and solving the problem.

Avoiding 40 repairs at $6000 each is actually a very low estimate, but would be worth $240,000. Avoiding repairs frees up manpower for other tasks: at 20 man-hours times 40 incidents times $100 per hour; it follows that reassigning these professionals to certain prioritized reliability improvement and repair avoidance tasks would be worth at least $80,000.

There is also one ~$3,000,000 fire per 1000 pump failures. Data obtained in 2012 at a refinery in the Great Lakes area considered the number low; their estimate was $6,000,000. A facility in one of the Plains states (in 2009) thought $3 million was about as accurate as it gets. Engineered parts make economic sense, and the reliability professional must take the lead in explaining ROI (return on investment) or time elapse to achieve payback to managers (Fig. 30.5.1).

PAYBACK (ROI) IS USUALLY VERY RAPID

Data and contributory details of catastrophic incidents are often closely guarded secrets. Virtually all consulting done from 1990 on by qualified independent professional engineers is linked to a legally binding nondisclosure agreement. The client is often compelled to file reports with local and federal regulatory agencies. These reports might differ from the findings of consulting engineers

FIG. 30.5.1 Explain to management why engineered parts make economic sense. *(Courtesy Hydro Inc., Chicago, IL, http://www.hydroinc.com.)*

who understand the true root causes of failures or whose sense of priorities is tuned to higher standards. Diverging statements or findings might feed a bureaucratic machine that will busy itself with issues of this type. Meanwhile, the above estimate for the value of fire damage restoration cost seems quite reasonable: avoiding 40 repairs would be worth 40/1000 × \$3,000,000 = \$120,000. Together, the three items (\$240k, \$80k, and \$120k) add up to \$440,000.

Although \$6000 was used for repair cost avoidance calculations earlier, an average API (American Petroleum Institute) pump repair at a Texas refinery costs slightly over \$10,200; a refinery in Mississippi reported \$11,000. If the incremental cost of upgrading during the next repair adds \$2000 to the repair bill and avoids even a single failure every 3 or 4 years over the 30-year total life of a pump, the payback will have been quite substantial. It would be reasonable to assume 8 avoided repairs at \$6000 give payback of 48k/2k = 24:1.

The quoted repair cost numbers of close to \$11,000 reflect what needs to be considered in a pump repair cost calculation: direct labor, direct materials, employee benefits at roughly 50% of direct labor, refinery administration and service costs at close to 10% of direct labor, mechanical-technical service personnel overhead costs amounting to ~115% of direct labor, and materials procurement costs from 7% to 8% of materials outlay [1]. Disregarding the true cost of failures or repairs is likely to deprive some users of seeing the true benefit-to-cost ratio associated with pump upgrades.

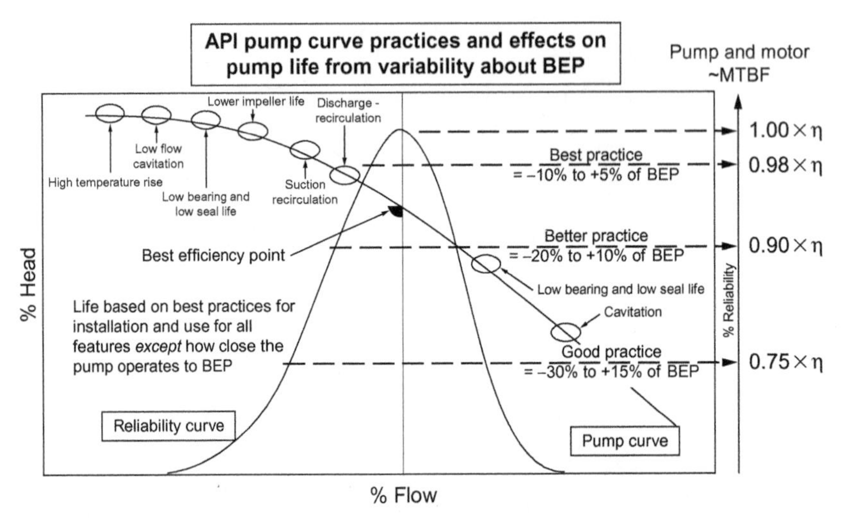

FIG. 30.5.2 Barringer-Nelson curves show reliability impact of operation away from BEP. *(Courtesy of Paul Barringer, http://www.barringer1.com.)*

We could examine other ways to calculate as well. It would be reasonable to assume that implementing a component upgrade (generally the elimination of a weak link) extends pump uptime by 10%. Implementing five upgrade items yields $1.1^5 = 1.61$—a 61% mean time between repair (MTBR) increase. Or, say, we gave up 10% each by not implementing six reasonable improvement items. In that instance, $0.9^6 = 0.53$, meaning that the MTBR is only 53% of what it might otherwise be. That might explain the industry's widely diverging MTBR. The MTBR gap is quite conservatively assumed to range from 3.6 years to 9.0 years in US oil refineries, and, as of 2013, no well-informed pump professional has disagreed with this range of MTBR numbers.

STAYING WITHIN DEFINED PUMP OPERATING RANGE

The onset of pump problems is not the same for different pumps or different services. Attempts to identify best practices are to be commended. Both Paul Barringer and Ed Nelson contributed to Fig. 30.5.2, the somewhat typical HQ curve. They plotted eight traditional non-BEP problem areas on that curve. The plot supports the conclusion that pump reliability can approach zero as one operates farther away from the best efficiency point or BEP.

The implications of Fig. 30.5.2 are important. Just because pumps are able to run at lower than BEP flows does not mean that it's good to operate there. Compare it with a vehicle able to go 12 mph in sixth gear or 47 mph in first gear. It can be done, but will likely prove costly if done for very long. Pioneering efforts to define minimum allowable flows can be traced back decades, and attention belongs to work originally published by Taylor [2] (also Fig. 30.1.1 in the previous Chapter 30.1). Taylor's work is worth mentioning because he approximated in a single illustration what others have tried to convey in complex words and mathematical formulas. Taylor kept the average user in mind.

We should keep in mind that the actual NPSHr needed for zero damage to impellers and other pump components may be many times the number published in the manufacturer's literature. The manufacturers' NPSHr plots are commonly based on observing a 3% drop in discharge head or pressure. Unfortunately, whenever this 3% fluctuation occurs, a measure of damage may already be in progress. It is prudent to assume a more realistic NPSHr and to provide an NPSHa in excess of this likely NPSHr. Doing so builds a certain margin of safety into the pump and reduces the risk of catastrophic failure events.

There are hydrocarbon services where an NPSHa surplus of just 1 ft (~0.3 m) over NPSHr will be sufficient to avoid cavitation. However, there are services, such as carbamate, where a 25 ft surplus is not nearly enough. Reliability-focused users should enlist the help of competent pump manufacturers and experienced design contractors to agree on NPSH multipliers or bracket the right NPSH margins for a particular liquid or pumping service.

While no rigorous Nss value exists, cautious reliability professionals observe safe margins. Many users choose Nss = 9000 as the limit for flows away from BEP. There are, however, some pumps (including certain high-speed Sundyne designs) that will operate quite well with Nss values higher than 9000. But these are special cases, and a close pump user-to-pump CPRS (competent pump repair shop) or user versus manufacturer relationship is needed to shed light on long-term experience. As is so often the case, marketers get ahead of users; it will take a good reliability engineer to separate exuberant claims from factual experience. Roughly 50% of all claims are bogus and 50% are within the realm of facts. Getting to know which 50% are bogus is the tricky part. Gullibility is never an ingredient of sound reliability engineering.

Repeat failures and mechanical issues. The vulnerability of operating process pumps in parallel is not always appreciated by pump purchasers, although API-610 clearly advises against parallel operation for pumps with relatively flat performance curves. Reasonable, yet general, specifications require a 10% minimum head rise from BEP to shutoff.

There are problems with short elbows near the suction nozzle of certain pumps, and flow stratification and friction losses are sometimes overlooked. Some sources advocate a minimum straight pipe run at the pump suction equal to five pipe diameters, but others recommend a 10-diameter straight-run equivalent. The multipoint trouble illustration in Fig. 30.5.2 is of interest here. Suffice it to say that tight-radius elbows and incorrect pipe reducer orientation can quickly wreck certain pump configurations [3]. Neglecting piping issues can be a costly mistake.

Pulling piping into place at a pump nozzle can cause edge loading of the pump's bearings, which will lead to premature bearing failures. Did soil settlement under pipe supports play a role in misalignment? Just as concrete driveway and sidewalk sections near residences often misalign a few scant years after construction, many pipe support columns are no longer truly vertical decades after they were first installed. The trouble is that few facilities regularly check into the matter, and few are the ones who budget time and money for corrective work on these supports.

Flow separation issues must be considered. The flow velocity at the small-radius wall of an elbow will differ from that at the large-radius wall. And again, because these facts are generally well known and many symposia have been devoted to them, we limit the topic to pump mechanical or drive end (i.e., power end) issues. Failure avoidance in the pump's drive end must concentrate on elusive reasons why many pumps fail repeatedly.

DISALLOW DEVIATIONS FROM BEST AVAILABLE TECHNOLOGY

User plants will usually get away with one or two small deviations from the best available technology. But when three or more deviations occur, failure risks usually increase exponentially. That said, there are a number of reasons why a few well-versed reliability engineers are reluctant to accept pumps that incorporate the drive end shown in Fig. 30.5.3 [4]. The short overview of reasons is that reliability-focused pros take seriously their obligation to consider the actual, lifetime-related, and not just short-term, cost of ownership. Specialists realize that the bearing housing in Fig. 30.5.3 (which was also highlighted in an earlier chapter of this text) will work initially and yet will fail prematurely. The housing is shown here exactly as originally provided, including its several risk-increasing features. Allowing these features to exist will sooner or later hurt the profitability of users and vendors alike. All are related to lube application, and process pump users should pay very close attention to these and other lube application matters.

As they examine Fig. 30.5.3, careful viewers can be certain of five facts:

- In Fig. 30.5.3, oil rings are used to lift oil from the sump into the bearings. These oil rings tend to skip and jump at progressively higher shaft surface speeds, or if not perfectly concentric, or if not operating in perfectly horizontal shaft systems.
- As the pump is transported from shop to field, an oil ring can become dislodged and get caught between the shaft periphery and the tip of the long limiter screw
- The back-to-back oriented thrust bearings of Fig. 30.5.3 are not located in a cartridge. This limits flinger disk dimensions (if they were to be retrofitted) to no more than the housing bore diameter.

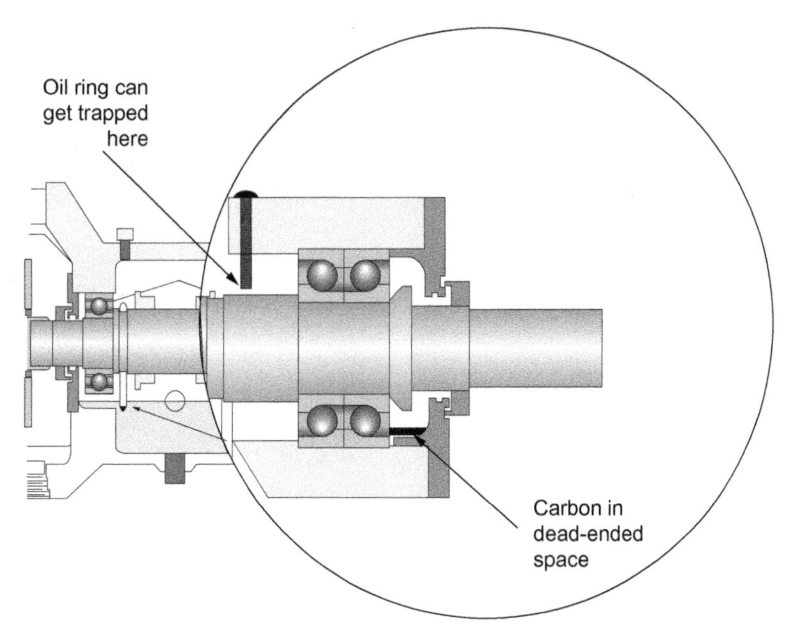

FIG. 30.5.3 A bearing housing with several potentially costly vulnerabilities.

- Bearing housing protector seals are missing from the picture in Fig. 30.5.3. Once added, bearing protector seals in this housing will change the flow of venting air.
- Although the bottom of the housing bore (at the radial bearing) shows the needed oil return passage, the same type of oil return or pressure-equalizing passage seems to have been left out near the 6 o'clock position of the thrust bearing. A small pool of oil can accumulate behind the thrust bearing and this oil will probably overheat. Carbon debris will form.
- No particular constant-level lubricator is shown in Fig. 30.5.3, and there is uncertainty as to the type or style of constant-level lubricator that will be provided. Unless specified, OEMs rarely supply the best available constant-level lubricator.

It should be noted that the angular contact thrust bearings in Fig. 30.5.3 will usually incorporate cages (ball separators) that are angularly inclined, which means they are arranged at a slant. These cages often act as small impellers [5], and impellers promote flow from the smaller toward the larger of the two diameters. This is more readily evident from Fig. 30.5.4, and particular attention should be given to windage created by the impeller-like air flow action of an inclined bearing cage. In many cases, the pump is designed with an oil ring to the left of this bearing. While the design intent is for oil to flow from left to right, windage from an inclined cage will act in the opposite direction.

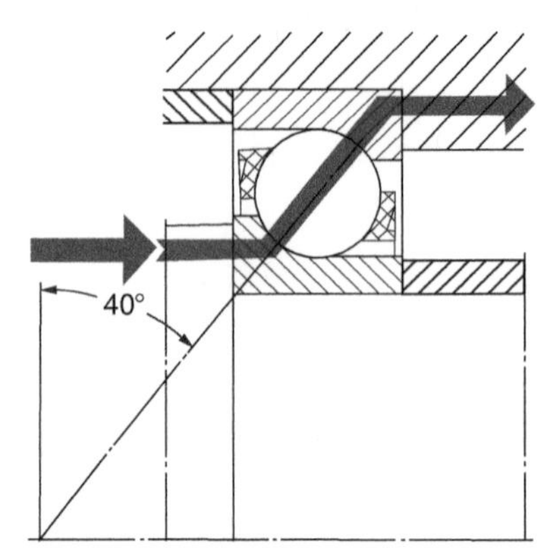

FIG. 30.5.4 Attempts to apply lubricant in the direction of the *arrow* (oil flow from left to right) meet with windage (air flow right to left) from an inclined cage. The two directions often oppose each other [5].

One might ask: how does one alleviate windage and/or its effects? The fact that windage may be generated by some of these bearings and is more likely found in particular bearing housing configurations requires thoughtful—and sometimes purely precautionary—abatement of unequal pressures inside a bearing housing.

PRESSURE EQUALIZATION

Decades ago, it was well recognized that the pressures surrounding bearings inside a pump bearing housing had to be equalized for best possible lubrication. Balance holes are shown on each side of the radial bearing and on each side of the thrust bearing set in Fig. 30.5.5. Why such balance provisions are shown only on the radial bearing in Fig. 30.5.3 or are altogether omitted from thousands of pumps marketed over the past 2 or 3 decades is not known. All we know is that not having a balanced pressure profile inside a pump bearing housing presents an increased bearing distress or failure risk. Catch it during machinery quality assessment (MQA) or while troubleshooting a process pump with unsatisfactory mechanical performance.

Before we leave the subject or progress further into the topic, note how carefully the now defunct Worthington Pump Company ascertained that pressures on each side of a bearing were equalized. Worthington's design went through the trouble of drilling balance holes right above the bearings (Fig. 30.5.5). Do you have balance holes (or pressure equalization ports) in your pump bearing housings? If not, then why not? Perhaps you don't need them, but then again—maybe you do. It's all about risk reduction [4,6].

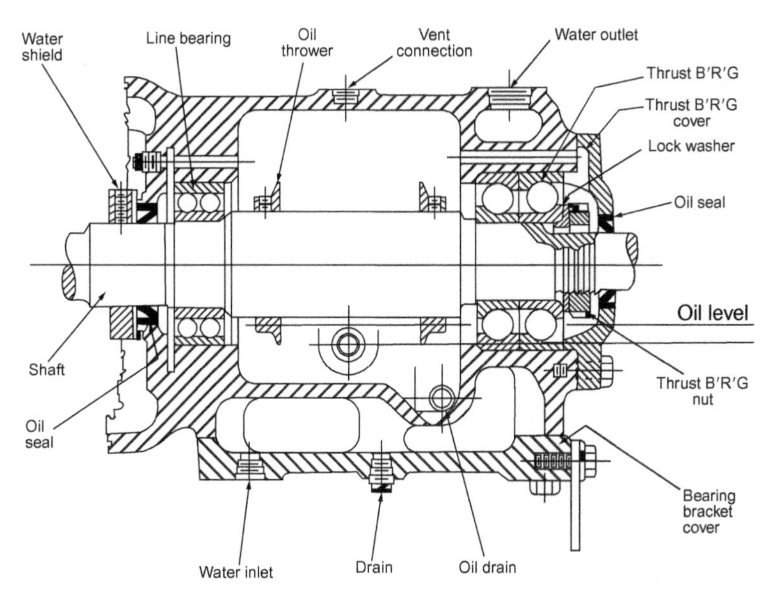

FIG. 30.5.5 Worthington Pump Company made sure that pressure-balancing holes surrounded bearings.

DISALLOW UNNEEDED COOLING

Note that the bearing housing in Fig. 30.5.5 incorporated a cooling water jacket. However, lubricant technology has made considerable progress since the Worthington Pump Company issued the drawing in about 1960. Bearing housing cooling is *not* needed on process pumps that incorporate rolling element bearings. Cooling is harmful if it promotes moisture condensation (water cooling coils) or restricts thermal expansion of the bearing outer ring (water cooling jacket). In 1967, these concerns were seen to influence pump reliability. The jacketed cooling water passages in Fig. 30.5.5 were from then on left open to the ambient air environment. The decision to delete cooling water from pumps with rolling element bearings was first implemented in 1967 at an oil refinery in Sicily. The owner's engineers had recorded bearing lube oil in four identical pumps reaching an average of 176°F with cooling water in the jacketed passages. Without cooling water, the lube oil averaged 158°F, which is 18°F cooler. The bearings now lasted much longer. These findings and experiences were first shared in 1983 and have since been described in many articles and texts. Years ago, API-610 updated its pump standard to require a manufacturer to disclose if they still planned using cooling water on process pumps with rolling element bearings. An experienced MQA engineer would suspect bad bearing choices or inadequately sized oil sumps if the need for bearing cooling were claimed by a pump manufacturer in 2016 [7].

Still, as of 2016 and quite obviously many decades after 1967, not everybody has acted on the message. The message is clear: there is no need to waste water, an increasingly more precious resource. Basic physics and an elementary understanding of temperature-related growth and shrinkage of steel make us realize that cooling water is very often responsible for actually reducing the life of rolling element bearings in process pumps [1,4].

MORE ON LUBRICATION AND BEARING DISTRESS

Only 9% of all bearings actually reach their as-designed life, and lubrication-related issues are often at fault in the estimated 50–60% of pump failures that involve bearing distress. Obviously, having the correct oil level should be a consideration in bearing housings with rolling element bearings. Oil level settings are part of our progressive investigation of elusive failure causes in process pumps.

The traditional oil sump was depicted, in Fig. 30.5.5, with the lubricant reaching to about the center of the lowermost bearing elements. This arrangement works well at low shaft surface velocities. To gain reliability advantages, synthetic lubricants, oil mist application (called "oil fog" in some languages), and liquid oil jets (also known as "oil spray") are often used. Oil jet lubrication existed long before the development of plant-wide oil mist systems [8].

Circulating systems also merit consideration in certain high-load or very large pumping services. Generally speaking, circulating systems include well-engineered systems such as Fig. 27.1.1. Although originally designed for barrier fluid circulation in dual mechanical seal services, these pump-around systems are easily adapted for bearing lubrication in large process pumps that utilize sleeve bearings. If needed, the oil can be passed through a heat exchanger in these systems before being returned to the bearing. However, and to again make an earlier point: regardless of lube application method—on rolling element bearings, cooling will not be needed as long as high-grade mineral or synthetic lubricants are utilized [1,4,7]. High-performance mineral oils developed after 2010 are deliberately mentioned here.

Irrespective of base stock and oil formulation, the required lubricant viscosity is a function of bearing diameter and shaft speed. Technical reasons are described in numerous books and articles, among them Refs. [4,9,10]. Most process pump bearings will reach long operating lives if the oil viscosity (at a particular operating temperature) is maintained in a range from 13 to 20 cSt [11]. It should be noted that whenever oil rings are used to "lift" the oil from sump to bearings, the need to maintain a narrow range of viscosities takes on added importance [12]. In the special case of the same bearing housing containing both rolling element and sliding bearings, it will be prudent to address the implications of (some) oil rings not being able to function optimally in the higher viscosity (ISO grade 68) lubricant that's often chosen for rolling element bearings. The oil ring may have been designed to cater to sleeve bearings, which normally need a lower viscosity lubricant, but VG 32 mineral oils are rarely the best choice for rolling element bearings in pumps. High-performance synthetic ISO VG 32 oils will often rank as the most suitable selection if different bearing styles must be accommodated in the same bearing housing [4].

In oil mist lubrication systems, it is generally understood that once shaft surface velocities reach in excess of 2000 fpm, windage effects are opposing the flow of oil mist. As this is being observed, uninformed or baffled oil mist users have, in some cases, reverted back to conventional oil lubrication. In sharp contrast, reliability-focused users have, for many decades, installed directed oil mist reclassifiers to overcome windage at >2000 fpm. The mist dispensing opening in these reclassifiers is located ~0.2–0.4 in. from the rolling elements. Thousands of these have been supplied and used with total success. This information is available from dozens of texts and articles [10,13,14].

LIQUID OIL LUBRICATION

As mentioned elsewhere in this text, once the shaft peripheral velocity exceeds 2000 fpm, the oil level should be no higher than a horizontal line tangent to the lowermost bearing periphery. This means there should be no contacting of the oil level with any part of a rolling element, and oil "lifting" is needed.

Going back to earlier chapters in this text, assume that shaft surface speed parameter $DN > 6000$. Therefore, and because initial cost was to be minimized, either oil rings (Subject Category 24) or the shaft-mounted flinger disks shown in previous chapters were chosen. If properly engineered and manufactured, both arrangements are capable of lifting the oil from the bearing housing's sump. Better yet, some such arrangements may even create a randomized oil spray. Shaft-mounted flinger disks are well represented in European-made pumps. If properly designed, their operating shaft peripheral speed range exceeds that of oil rings.

A bearing housing with a cartridge containing the thrust bearing set facilitates retrofitting of a flinger disk. The bearing housing bore that accepts the cartridge is slightly larger than the diameter of the steel flinger disk, making assembly possible. Pay attention to the needed oil return passage at the 6 o'clock bearing positions, and note that no such return passage was provided by a careless manufacturer below the thrust bearing in Fig. 30.5.3.

A small pool of oil will get trapped behind the thrust bearing. Whatever this then leads to is unknown, but plain common sense tells us that trapped oil overheats. Excessive heat causes oil to turn into coke fines—it's not conducive to improving pump reliability.

TWO DIFFERENT *DN* RULES EXPLAINED

When determining oil level settings, either of two empirical rules could be applied. To illustrate rule (1), a 2 in. bore bearing at 3600 rpm, with its *DN* value of 7200, would operate in the risky or ring instability-prone zone > 6000. Equipment with a 3 in. bore bearing operating at 1800 rpm ($DN = 5400$) might use oil rings without undue risk of ring instability. In another example, using rule (2), a 3-in. (76 mm) diameter bearing bore at 3600 rpm would operate with a shaft peripheral velocity of $(\pi D/12) \times (3600) = 2827$ fpm (~14.4 m/s), which would disqualify oil rings from being considered for highly reliable pumps. The fact that a pump manufacturer can point to satisfactory test stand experience at higher peripheral velocities is readily acknowledged, but field situations represent the "real world" where shaft horizontality and oil viscosity, depth of oil ring immersion, bore finish, and out-of-roundness are rarely perfect. We can thus opt for using either the $DN < 6000$ or the surface velocity < 2000 fpm or the lesser of these two "real-world" rules of thumb.

Either way, the vendor's test stand experience is of academic interest at best. Pump manufacturers test under near-ideal conditions of shaft horizontality, oil ring concentricity and immersion, oil level, and lubricant viscosity. As users, we might ask ourselves how often we have seen nonround oil rings or rings that have shaft radius wear marks (from shaft fillet radii) on one side of the ring. If the answer is "never," perhaps another look will be warranted. For the reliability-focused, the wide-ranging field experience that led to these two rules of thumb will govern over all else.

The cartridge approach shown in Chapter 24.2, Fig. 24.2.2, has been in use since about 1960 on many thousands of open-impeller ANSI pumps. These pumps need axial rotor location adjustment; a cartridge allows this type of adjustment. The same cartridge approach may be needed to dimensionally accommodate flinger disks (Fig. 24.2.2) instead of vulnerable oil rings (Fig. 24.2.3). Of course, cartridge-mounted bearings are a cost-adder, and you may hear claims that the benefit-to-cost ratio will not justify upgrading to cartridges. However, with the average API pump repair costing slightly over \$10,200 at a Texas oil refinery and \$11,000 at an oil refinery in Mississippi, we might be surprised at the rapid payback. Even a single avoided failure over the 30-year total life of a pump will probably pay for it many times over.

THE TROUBLE WITH OIL RINGS AND CONSTANT-LEVEL LUBRICATORS

Issues with oil rings are found in many scholarly works [15–18]. On a website post in September 2012, the Malaysian equivalent of the US Occupational Safety and Health Administration (OSHA) alerted users to catastrophic failures brought on by oil rings [19]. All of these sources observed problems with oil rings, although an industry source opined (in 2011) that "ring lubrication is an accepted practice and it would take user consensus to damn it." Of course, history shows us that innovations are rarely driven by consensus or by manufacturers pursuing marketing strategies catering to uninformed buyers. If innovations were driven by consensus, the Wright brothers would have invested in franchising bicycle repair shops instead of developing an engine-powered flying machine.

Meanwhile, keep in mind that virtually every chapter of this text was written for the reliability-focused. Nothing will convince those who accept without questioning dozens of repeat failures of centrifugal pumps at their plants. Many illustrations of failed oil rings are available, Fig. 24.2.1 among them. Studies, observations, and measurements have shown their field reliability in process pumps out of harmony with the quest for higher reliability and availability. Work described in Refs. [12,15] recommends oil ring concentricity within 0.002 in. However, in 2009, shop measurements were performed by the author at a pump user's site in Texas. The oil rings measured in 2009 exceeded the 0.002-in. allowable out-of-roundness tolerances by a factor of 30 [20].

Experience shows that oil rings are rarely the most dependable or least-risk means of lubricant application. They tend to skip around and even abrade (Fig. 24.2.3) unless the shaft system is truly horizontal, unless ring immersion in the lubricant is just right, and unless ring eccentricity, surface finish, and oil viscosity are within tolerance. Taken together, these parameters are not usually found within close limits in actual operating plants.

Reliability-focused purchasers often specify and select pumps with flinger disks. Although sometimes used in slow-speed equipment to merely prevent temperature stratification of the oil (see Fig. 30.5.5), larger diameter flinger disks (Chapter 4.5, Fig. 4.5.2) serve as efficient (nonpressurized) oil distributors at moderate speeds. Of course, the proper flinger disk diameter must be chosen and solid steel flinger disks should be preferred over plastic materials. Insufficient lubricant application results if the diameter is too small to dip into the lubricant; conversely, high operating temperatures can be caused if the disk diameter is much too large or if no thought was given to its overall geometry.

In about 2006, flexible flinger disks were briefly offered because they enabled insertion in some "reduced cost" designs, that is, configurations where the bearing housing bore diameter is smaller than the flinger disk diameter. They were no longer offered for the simple reason that strong high-performance plastics did not deflect into an umbrella shape at the point of insertion. Conversely, "deflectable" plastics allowed insertion but proved too weak in high peripheral speed applications. As was brought out earlier, to accommodate the preferred solid steel flinger disks, bearings must be cartridge-mounted. Using a cartridge design, the effective bearing housing bore (i.e., the cartridge diameter) is made large enough for passage of a steel flinger disk of appropriate diameter.

The patent literature is full of attempts to get around the use of oil rings; roll pins inserted transversely in pump shafts [1, p. 251] and the abovementioned flexible (plastic) flinger disks have brought mixed results and marginal improvement at best. Cheap disks pushed on the shaft became a source of failure and were disallowed by API-610 about 10 years ago. Cheap plastics and disk configurations chosen without the benefit of sound engineering practices have also not been sufficiently reliable. In all, we should never lose sight of the charter and mission of reliability professionals. We believe their goals should be to work in harmony with basic science and to achieve high pump reliability and availability.

Properly engineered steel flinger disks can do the job for decades. We estimate the incremental cost (comprising material, labor, and CNC production machining processes) of an average-size (30 hp) process pump with cartridge-mounted bearings at $300. The value of even a single avoided failure was earlier shown to be over $10,000 and the benefit-to-cost ratio would thus exceed 33 to 1.

The shortcomings of oil rings were known in the 1970s. A then well-known pump manufacturer claimed superior-to-the-competition products. This manufacturer's literature pointed to an "antifriction oil thrower (i.e., a flinger disk), ensuring positive lubrication to eliminate the problems associated with oil rings" [4]. About 2 decades later, in 1999, at least one major pump manufacturer saw fit to examine the situation more closely. In a comprehensive paper and conference presentation, the manufacturer described remedial actions that included grade 46 oil viscosity and oil rings made of high-performance polymers [21]. However, the problem did not go away. Users in Canada soon reported that black oil persisted, and so did repeat failures, even after adopting nonmetallic oil rings.

Black oil can easily be traced to one of two origins. A simple analysis either will point to overheated oil (i.e., carbon) or will detect slivers of elastomeric "dynamic" O-ring material from components that operate too close to sharp-edged O-ring grooves.

CONSTANT-LEVEL LUBRICATORS

The potential malfunction risks of constant-level lubricators are more widely known. A number of makes, models, and brands are in common use, and their "unidirectionality" is described in at least one manufacturer's literature [22]. Perceptive reliability professionals have observed that caulking (where transparent bottles meet die-cast metal bases) will, over time, develop stress cracks (fissures). Rainwater can then reach the oil via capillary action. Accordingly, bottle-type constant-level lubricators are a preventive maintenance item and should be replaced after 4 or 5 years of service [1,4].

If you use constant-level lubricators, carefully read the manufacturers' literature, not that their products are flawed; they observe all the laws of physics. These lubricators must be installed and maintained in harmony with the manufacturers' instructions. Do not leave anything to chance. Suppose the oil level no longer reaches the rolling elements. In that case, be sure you do some troubleshooting and do not immediately blame the operator for not having filled the glass bowl. You may have purchased the least expensive constant-level lubricator, one that lacks pressure balance. This particular lubricator will only work if the bearing housing has a liberally sized vent opening. Without such a vent, any pressure increase in the space above the liquid oil will drive the oil level down. For a while, the top layer of oil will overheat; carbon will form and black oil will (probably) appear in the glass bulb. Increasing temperature in the closed space causes a further pressure increase and the oil level decreases even more. Oil then no longer reaches the rolling elements and another bearing failure is likely to occur.

The lubricator shown earlier in Fig. 4.5.7 is configured for a balance line that ensures that the oil levels in the die-cast lubricator support (or at the edge of a slanted tube in some lubricators) and in the pump bearing housing are always exposed to the same pressure [22]. Undersized balance lines can exist; either a generously dimensioned hard pipe or a suitably sized stainless steel hydraulic balance line is favored. If constant-level lubricators cannot be avoided, a pressure-equalized model or arrangement is recommended.

Again, bearing distress is inevitable if a constant-level lubricator fails to maintain the desired oil level. An incorrect level setting can be caused by a number of factors. It will be clear from basic physics that even small increases in the bearing housing internal pressure can greatly increase the failure risk. Suppose there is heat generation and you have added a (desirable) bearing protector seal. If that no longer allows air to escape *and* there's also a lack of housing internal pressure balance, your bearings will be at risk. Worthington had included housing internal balance holes in Fig. 30.5.5, but the reasons have probably been forgotten…

In any event, without internal balance holes, you are perhaps trapping air between the thrust bearing and the adjacent end cover. The trapped air will be at a slightly higher pressure than the air near the center of the bearing housing. As the housing internal pressures become uneven, it will exceed the ambient pressure to which the oil level at the height adjustment screw in the bulb holder portion of the constant-level lubricator is exposed. According to the most basic laws of physics, a pressure increase in the bearing housing causes the oil level near the bottom of the bearing inner ring shoulder to be pushed down. Lubricant will no longer reach the bearing rolling elements, oil turns black, and the bearing will fail quickly and seemingly randomly.

To restate, at $DN > 6000$ and to satisfy minimum requirements in a reliability-focused plant environment, a stainless steel flinger disk fastened to the shaft will often perform well. Such a disk will be far less prone to cause unforeseen outages than many other presently favored methods. Remember that traditional oil rings will abrade and slow down if they contact a housing internal surface. They are sensitive to oil viscosity and depth of immersion, concentricity, and RMS surface roughness.

If you upgrade to flinger disks, you are accepting the findings of the legacy manufacturer whose advertisement is shown in Ref. [1]. That manufacturer's findings were backed by facts. Still, it must be ascertained that flinger disks are used within their applicable peripheral velocity so as to contact the oil and fling it into the bearing housing [2]. The flinger disk OD must exceed the outside diameter of the thrust bearing, and this dimensional requirement strongly favors placing the outboard (thrust) bearing(s) in a separate cartridge. Providing such a cartridge will add to the cost of a pump, as will the cost of a well-designed flinger disk. However, in most cases, the incremental cost will be considerably less than what it would cost to repair a pump just once.

RANKING THE DIFFERENT LUBE APPLICATION PRACTICES

Although oil ring lubrication is widely used, it is relatively maintenance-intensive and ranks last from the author's experience and risk reduction perspective. Next, flinger disks have been used for many decades and allow operation at higher *DN* values than oil rings. Because they are firmly clamped to the shaft, there is far less sensitivity to installation and maintenance-related deviations. On the other hand, nonclamped flinger disks were tried a few decades ago, and with very disappointing results. API-610 disallows push-on flingers and some other low-cost oil application components [23].

Plant-wide oil mist lubrication systems are ranked ahead of flinger disks. Oil mist has proved superior to conventional lubricant application since the late 1960s. Pump bearing failure reductions ranging from 80% to 90% have been reported by Charles Towne of Shell Oil and many others [24–28]. Charles Towne performed tests on identical process units at Shell Oil and deserves much credit for seminal work on the subject.

The highly beneficial in-plant, real-life results reported by Towne refer to pure oil mist, not purge mist. Pure oil mist is an oil-air mixture with a volumetric ratio of 1:200,000. The oil is atomized to globule form and carried by the air, applied in modern plants as shown in Subject Category 23. Several illustrations, including Fig. 23.3.2, could be used to depict liquid oil spray. Liquid oil spray is sometimes called "jet oil" lubrication [4,8] and differs from oil mist.

These facts were summarized in Ref. [29], which incorporated a number of very important recommendations for the truly reliability-focused:

- It establishes that pump bearing housings need not be symmetrically configured. (Asymmetry is visualized by looking into the pump shaft. The distance to the right edge of the bearing housing is not the same as the distance to the left edge of the bearing housing. The additional volume thus gained will accommodate a small oil pressurization pump; this small pump is to be arranged inside the process pump's bearing housing.)
- A boxlike geometry with a flat cover and ample space to incorporate a wide range of oil pumps is feasible. Boxlike bearing housings for process pumps would open up a host of new and inventive solutions. These might incorporate shaft-driven or other reliable self-contained means or oil application pumps [30]. The oil application pump would possibly take suction from an increased-size oil sump.
- The main process pump shaft need not be in the geometric center of the box.

- Flat surfaces would invite clamp-on, screw-in, or flange-on oil pumps.
- Oil pressurized by the oil application pump would be routed through a filter and hydraulic tubing to spray nozzles incorporated in the end caps. Therefore, the cross-sectional view of a bearing housing with oil spray would be identical to the one shown for oil mist [14].
- Internal pressure equalization and windage issues would never again be a concern.
- The incremental cost of superior bearing housings would be more than matched by the value of avoided failures.

In applications with either oil mist or oil spray, there would be no oil rings, flinger disks, or constant-level lubricators. Because the mist (or spray) application nozzles shown here are relatively close to the bearings, oil mist flow or the stream of liquid oil will overcome windage. While this jet oil or oil spray lube application method seems like a bold idea, the method is extensively documented by MRC and SKF, also in at least seven of our many reference texts, among them Ref. [7]. This lubrication method is very often used in military aircraft and we certainly take no credit for devising it:

- The duty imposed on self-contained oil spray pumps would be quite benign compared with other known, reliable, shaft-driven pumping technologies or services.
- Oil filtration would be easy.
- The elimination of oil rings and constant-level lubricators would be a very positive reliability improvement step.
- Part of the energy requirement of an oil application pump would be regained in the form of reduced bearing frictional losses.

With spray lubrication, much needed oil application innovation would benefit the drive end, and thousands of repeat failures of pumps would no longer occur. However, as of today, little interest has been shown by manufacturers and users to redesign pump bearing housings.

The market drives these developments. So, if buyers and pump owners tolerate repeat failures and the manufacturers benefit from the sale of spare parts, it will be business as usual. Still, and at the risk of stubbornly bucking the trend, as responsible engineers, we should advocate changes in mind-sets. As realists, we are under no illusions as to where some users and manufacturers will be when the dust settles: we will never convince or even reach some of them. All we wanted to do is explain matters to those whose reliability focus extends beyond "business as usual" and who are interested in pushing for lower risk oil application alternatives.

One of the most straightforward ways to drive a housing internal oil pump could be modeled on the right-angle worm drives typically found in small steam turbines. A bevel-geared or similar arrangement that's usually associated with a mechanical governor would be located inside the pump's bearing housing.

It would be one of many highly reliable options that merit consideration for small oil pumps that take suction from the process pump's oil sump and pressurize it. After leaving a spin-on filter, the pressurized oil would enter spray nozzles, which would direct the oil into the process pump bearings.

CONCLUSIONS AND SUMMARY

As of 2016, some process pumps continue to experience costly repeat failures. Motivated reliability professionals and informed users can avoid these failures and will appreciate recommendations on failure risk reduction. For the reliability-focused pump users, a number of conclusions and upgrade recommendations may be of interest:

1. Discontinue using maintenance-intensive oil rings and, if possible, constant-level lubricators.
2. As a matter of routine, the housing or cartridge bore should have a passage at the 6 o'clock position to allow pressure and temperature equalization and oil movement from one side of the bearing to the other. Note that such a passage was shown in Fig. 30.5.3 for the radial bearing, but not for the thrust bearing set.
3. With proper bearing housing protector seals and the right constant-level lubricators, breathers (or vents) are no longer needed on bearing housings. The breathers (or vents) should be removed and the vent openings plugged.
4. If constant-level lubricators are used, a pressure-balanced version should be supplied, and its balance line should be connected to the location from where the breather was removed.
5. Bearings should be mounted in suitably designed cartridges, and loose slinger rings (oil rings) should be either avoided or, in some high *DN* cases, disallowed.
6. Suitably designed flinger disks should be secured to the shaft whenever the oil level is lowered to accommodate the need to maintain acceptable lube oil temperatures (i.e., for pumps operating with *DN* values in excess of 6000).
7. Modern and technically advantageous versions of bearing housing protector seals should be used for both the inboard and outboard bearings. Lip seals are not good enough, and neither are outdated rotating labyrinth seal designs.
8. Understand that the implementation of true reliability thinking must strongly support moves away from traditional bearing housings. These moves should push for exploration of the many alternatives that eliminate oil rings and constant-level lubricators.

Knowledgeable engineers can show that some widely accepted pump components tend to malfunction in the real world. Your own repeat failures attest to the validity of this statement. Moreover, as the industry often moves away from solid training and from taking the time needed to do things right, designing out risk and designing out maintenance become attractive propositions.

DEMAND BETTER PUMPS AND PAY FOR VALUE

In late 2008, the purchasing entity representing a large reliability-focused plant in the United States had thoughtfully and deliberately specified better pumps. The purchaser and user wanted better pumps and they were willing to pay for the improved products. But the buyer's improvement requests were declined by every one of the four pump vendor companies that responded to an invitation to bid. The disappointed owner-user company suggested that we get out the message to users and manufacturers alike: better pumps are possible. Understand why reliability-focused users need them and realize why, for the value-seeking purchaser, certain "standard products" are no longer good enough. Well, it is hoped that the request of this owner-user company was answered in this 2016 recap and update. Progress has been slow; some pump manufacturers were concerned that patents would not be granted on many of these improvement options; others claimed that the market is not demanding these improvements. Meanwhile, we can only hope that someday, somewhere, the message will make inroads.

Anyone can confirm the accuracy of the information by merely looking up the various references. We were careful not to disclose manufacturers' names or failure analysis details that might cause embarrassment. Of course, this chapter of our text also tried not to bore the reader with the customary consultant-conceived generalities. Organizations intent on systematically reducing repeat pump failures will find it easy to move ahead of the ones that are complacent or indifferent. Enlightened managers might even find disincentives to specifying, purchasing, or even tolerating process pumps that fail repeatedly and often catastrophically.

REFERENCES

[1] H.P. Bloch, A. Budris, Pump User's Handbook—Life Extension, fourth ed., Fairmont Press, Lilburn, GA, ISBN: 0-88173-720-8, 2014.

[2] I. Taylor, The most persistent pump-application problems for petroleum and power engineers, in: Energy Technology Conference and Exhibit, Houston, Texas, September 18–22, 1977. ASME Publication 77-Pet-5.

[3] Karassik, et al., Pump Handbook, second ed., McGraw-Hill, New York, NY, ISBN: 0-07-033302-5, 1985.

[4] H.P. Bloch, Pump Wisdom, John Wiley & Sons, New York, NY, ISBN: 9-781118-041239, 2011.

[5] SKF Americas, General Bearing Catalog, 1990. Kulpsville, PA.

[6] Worthington Pump Company, Pump Operation and Maintenance Manual, 1968.

[7] H.P. Bloch, Improving Machinery Reliability, third ed., Gulf Publishing Company, Houston, TX, 1982, 1998.

[8] MRC Bearings General Catalog 60, TRW Engineer's Handbook, second ed., 1982, p. 197.

[9] Eshmann, Hashbargen, Weigand, Ball and Roller Bearings: Theory, Design, and Application, John Wily & Sons, New York, NY, ISBN: 0-471-26283-8, 1985.
[10] H.P. Bloch, Practical Lubrication for Industrial Facilities, second ed., Fairmont Press, Lilburn, GA, ISBN: 088173-579-5, 2009.
[11] SKF USA, Inc., Bearings in Centrifugal Pumps, second ed., 1995. Publication 100-955.
[12] D.F. Wilcock, E. Richard Booser, Bearing Design and Application, McGraw-Hill Publishing Company, New York, NY, 1957.
[13] H.P. Bloch, Oil Mist Lubrication Handbook, first ed., Gulf Publishing Company, Houston, TX, 1987.
[14] H.P. Bloch, A. Shamim, Oil Mist Lubrication—Practical Application, Fairmont Press, Lilburn, GA, ISBN: 088173-256-7, 1998.
[15] R.A. Baudry, L.M. Tichvinsky, Performance of oil rings, Mech. Eng. 59 (1937) 89–92. ASME, J. Basic Eng. 82D (1960) 327–334..
[16] H. Heshmat, O. Pinkus, Experimental study of stable high-speed oil rings, Am. Soc. Mech. Eng. 1984 (Paper). Also J. Tribol. 107 (1) (1985) 14–22.
[17] M.D. Hersey, Discussion of performance of oil rings, Mech. Eng. 59 (1937) 291.
[18] L. Urbiola Soto, Experimental Investigation on Rotating Magnetic Seals, Masters Thesis, Texas A&M University, 2001/2002.
[19] Government of Malaysia, Department of Occupational Safety and Health, http://www.dosh.gov.my/doshv2/index.php?option=com_content&view=article&id=424%3Afire-at-oil-refinery&catid=84%3Asafety-alerts&Itemid=118&lang=en.
[20] H.P. Bloch, Deferred maintenance causes upsurge in BFW pump failures, Hydrocarbon Processing, 2011 (based on consulting work by the author).
[21] S. Bradshaw, Investigations into the contamination of lubricating oils in rolling element pump bearing assemblies, in: Proceedings of the 17th International Pump User's Symposium, Texas A&M University, Houston, TX, 2000.
[22] TRICO Manufacturing Corporation, Pewaukee, Wisconsin, Commercial Literature. www.tricocorp.com, 2008.
[23] American Petroleum Institute, Alexandria, VA, API-610, Centrifugal Pumps, 10th ed., 2009.
[24] C.R. Miannay, Improve bearing life, Hydrocarbon Processing, 1974.
[25] A. Shamim, C.F. Kettleborough, Tribological performance evaluation of oil mist lubrication, J. Energy Res. Technol. 116 (3) (1994) 224–231.
[26] A. Shamim, C.F. Kettleborough, Aerosol aspects of oil mist lubrication generation and penetration in supply lines, in: Presented at the Energy Resources Technology Conference and Exhibition, Houston, Tribology Symposium, PD, vol. 72, ASME, New York, NY, 1995, pp. 133–140.
[27] D. Ehlert, Getting the facts on oil mist lubrication, in: Texas A&M Middle East Turbomachinery Symposium, 2011.
[28] D. Ehlert, Consider closed-loop oil mist lubrication, Hydrocarbon Processing, 2011.
[29] H.P. Bloch, Tutorial at 31st International Pump User's Symposium, Texas A&M University, Houston, TX, 2014.
[30] H.P. Bloch, Inductive pumps solve difficult lubrication problems, Hydrocarbon Processing, 2001.
[31] C.A. Towne, Practical experience with oil mist lubrication, Lubr. Eng. 39 (8) (1983) 496–502.

Chapter 30.6

Periodic Switching: Caution with Parallel Operation of Pumps

INTRODUCTION

Periodically switching (alternately operating) all pumps is recommended for a number of reasons. Also, misunderstandings regarding parallel operation of pumps are clarified. If standby pumps are rarely used, their bearings will degrade and reliability will suffer. Many pumps are exposed to vibration transmitted by adjacent equipment; others in standby mode ingest (and expel) ambient air.

INSTALLED SPARES

Many process plants have "identical" centrifugal pumps installed in a given service. Most of these were probably intended as spare equipment. They would be started up in the event the primary pump had to be serviced or repaired. In best practices plants, the two pumps are switched monthly. Switching is very important because it extends the bearing life of most pumps. For one, the lubricant is thus redistributed and corrosion damage is less likely.

Additionally, the extent of bearing damage due to vibration transmitted from an adjacent operating pump is cut in half. Operating each pump for about a month goes a long way to ensure that the rolling elements in the nonrunning pump do not remain in the same location for too long. More specifically, this reduces the severity and rate of incidents of a failure mode called ball indentation damage or "false brinelling." (The occasional concern that both pumps will wear out at the same time is best refuted by considering twin children. By definition, they were born on the same day, but the probability of their dying of natural causes on the same day is rather remote.)

As plants increase their throughput capacity, they often run both pumps simultaneously. Likewise, when pumping requirements are known to vary greatly, it may be desirable to use several small pumps and stop one or more pumps when the throughput demand drops. The remaining pumps then operate closer to their respective best efficiency points (BEP). In new installations, it is worth keeping in mind that, given a definite flow rate and head, several small pumps operating in parallel may allow increased pump speeds and may lower total initial pump cost.

PUMP CURVES AND THEIR IMPORTANCE

Trouble-free parallel operation is possible only if the head versus flow curves (sometimes called H/Q curves) are relatively steep. A good rule of thumb would call for the head rise from operating point to shutoff of each pump to be 10 or more percent of the total head at BEP. New impellers are often available for existing pump casings and this upgrading may be easy to cost-justify [1].

Numerous references attest to the fact there is no such thing as two truly identical pumps [2]. Each pump has its own H/Q performance curve because of its own unique internal roughness and wear or corrosion-affected clearances. There have been many instances of two pumps operating satisfactorily in parallel for years. Problems can occur suddenly after one pump has been overhauled or a new impeller has been installed. A thorough analysis of alternative solutions is always appropriate. These may include system modifications and the use of a single, new, and more efficient pump instead of continuing to tolerate using two old, fundamentally weak pumps. Please see Chapter 30.12 for more on this topic.

REFERENCES

[1] J.W. Dufour, E. Nelson, Centrifugal Pump Sourcebook, in: McGraw-Hill Publishing Company, New York, NY, ISBN: 0-07-018033-4, 1992.

[2] H.P. Bloch, A. Budris, Pump User's Handbook, in: fourth ed., Fairmont Publishing Company, Lilburn, GA, ISBN: 0-88173-720-8, 2014.

Chapter 30.7

Consider Disk-like Impellers for Viscous Fluids

INTRODUCTION

Finding a suitable pump for plants pumping viscous fluids or fluids that undergo viscosity change need not be a challenge. Disk pumps have been applied in services approaching 300,000 cPs. Flows ranging from 2 to 10,000 gpm and heads up to 1000 ft have been accommodated. High-flow pumps often use several parallel disks; another manufacturer offers eccentric disk pumps. They are collectively called "disk-like impeller pumps," although the disks differ greatly from a traditional impeller.

UNUSUAL OR DIFFICULT SERVICES

Some successful services had up to 65% entrained air and gas; other services were packed with solids approaching 80% by weight. As with everything else, there are drawbacks; these can include lower efficiency, but that seems to be a small price to pay if conditions merit going with this pump type.

Certain foods (think of honey), glues, and plastics can be processed with progressive cavity, gear, and screw pumps, to name a few. These are services where using conventional centrifugal pumps may lead to poor performance and extensive or sometimes costly maintenance. If this describes your problem, you may want to look into the applicability of disk-operated pumps. These have been around for decades. Many have been reconfigured to fit into the ANSI envelope.

Disk pumps embody a somewhat unique pumping concept. They use a noncontact mechanism based solely on the principles of boundary layer and viscous drag. Although the pumpage enters a centrally located suction nozzle just as it would in conventional centrifugal pumps, at least one prominent manufacturer of such pumps (www.discflo.com) explains that pumpage does not impinge on the disk. The boundary layer provides a protective barrier on the moving parts of the pump, minimizing the effects of vapor implosions and cavitation.

Flow through disk pumps is smooth and laminar. This accounts for an NPSHr ranging typically from 35% to 50% of standard centrifugal pumps in the same service condition. Another favorable "idiosyncrasy" of disk pumps is an NPSH curve that is essentially stable right to shutoff. There is no prerotation in the suction end of the pump, which reduces NPSH requirements even further.

Experience shows that nonimpingement disk-operated pumping reduces wear and shear dramatically. It has been shown that the design does not require close tolerances and produces no radial and only insignificant axial loads. This accounts for the demonstrated durability and exceedingly infrequent maintenance requirements of the various types of disk pumps. Consider contacting experienced manufacturers by using the Internet. Become familiar with both simple disk and eccentric disk pumps. Each has its merits.

Chapter 30.8

Why Pump Shaft Failures Happen

INTRODUCTION

Technical publications thrive on questions from reliability engineers. These questions tend to inform editors and readers about prevailing levels of knowledge. Also, the questions often reflect training-related issues confronting industry. In any event, finding answers may require practical knowledge, and editors of trade and professional publications must often match up answer seekers with answer providers. Next, and to the extent possible, editors fold the resulting discourse into presentations or articles of interest to a wider spectrum of readers. Notice how a simple but powerful failure analysis method [force, reactive environment, time, and temperature (FRETT)] is discussed.

WHAT CAN CAUSE SHAFTS TO FAIL

Guidance is often needed when users ponder over the possible causes of catastrophic shaft and impeller failures in process pumps and, occasionally, other rotating shaft machines. "In the course of my work as a Pump and Vibration Specialist," one reader said, "I have encountered a number of such failures. One root cause of such failures that has been suggested is a defective check valve, which allows the process pump to freewheel in reverse due to back flow. However, I have found no documentation that supports or explains this failure mode. In conversations with various pump and field service technicians, it has been suggested that when a pump is started while spinning in reverse, the starting torque exceeds the shaft strength and catastrophic shaft failure occurs. But this explanation is counter to my understanding of electric motor starting torque."

IMPELLERS AND REVERSE ROTATION

Our initial reply highlighted first what our reader, of course, knew: there are many marginal design pumps and a few well-designed, usually more expensive, pumps. When there was still an abundance of common sense (decades ago, perhaps), some wise man wrote that we always get what we pay for. It may take days or decades, but we will ultimately get just that. Of course, that is still true today and it applies to entire machines, their component parts, services, employees, contractors, educators, consultants—whatever.

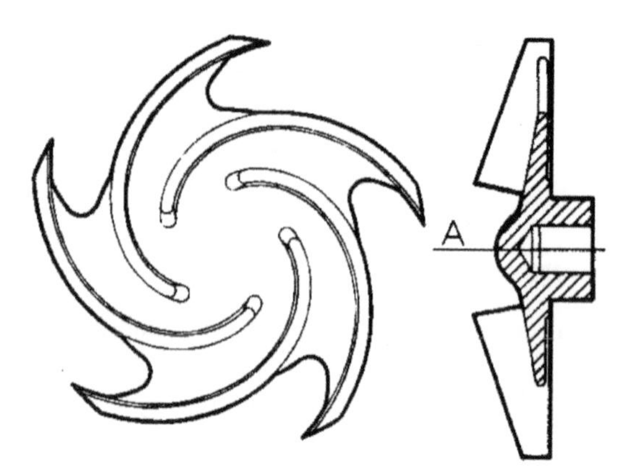

FIG. 30.8.1 Screw-on (dead-end threaded) impeller hub [2].

Back to the point: user experts Ed Nelson and John Dufour [1] noted that nearly all impeller thread arrangements (left hand vs. right hand) for single-stage end-suction pumps are such that the impeller fit gets tighter if the pump shaft rotates in the as-designed direction. Such a screw-on impeller is shown in Fig. 30.8.1. The impeller gets loose (or spins off the shaft) if the pump shaft is rotated in the opposite, or unintended, direction of rotation. Impellers installed with close-fitting keys properly mated to the shaft will not spin off in case of inadvertent reverse rotation. They will cost a bit more, but their properly designed keyed fits make it near-impossible for an impeller to come off in the event of a malfunctioning check valve or incorrect motor polarity causing reverse pump rotation.

REASONS FOR REVERSE ROTATION

But pumps are occasionally running in the wrong direction, and if they do, the reasons are quite easy to find. Pump and steam turbine discharge check valves have been known to leak. Metal distortion, seat erosion, and hinge friction issues can occur over time. Auto-start or standby equipment is more vulnerable because block valves are deliberately left open. In another scenario, an uninformed pump installer may decide to test for motor direction of rotation after the driver is already coupled up to the pump shaft. That's a risky way to verify direction of rotation, which, of course, should have been checked before coupling the driving to the driven shaft. Yet, someone may have decided "to save time" by installing the motor and pump in the as-shipped (generally mounted on a base plate and coupled to an electric motor) condition. In that case, the direction of rotation is left to chance and the probability of an impeller spinning off is 50%. The damage potential is then much greater than if motor direction-of-rotation had been ascertained with the shaft coupling initially removed.

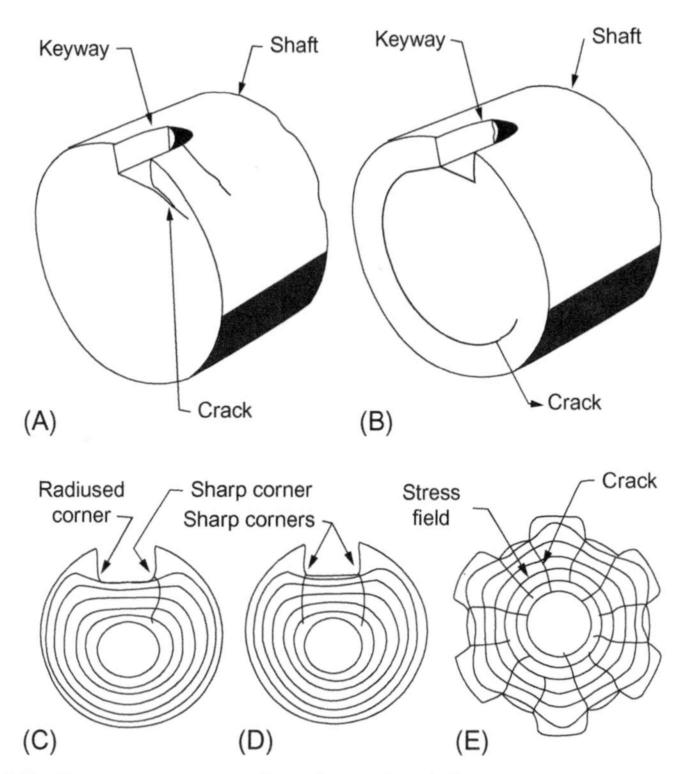

FIG. 30.8.2 Sharp corners are stress risers; they weaken shafts.

KNOW THE IMPELLER FASTENING METHOD

No pump manufacturer has a universal impeller securing method suitable for all pump sizes and service environments. In fact, the fastening method is not usually shown on the manufacturer's standard drawings. Also, relatively few user-purchasers include process pumps in thorough machinery quality assessment (MQA), which stands for up-front MQA [3]. Some impellers are fastened to shafts by standard acorn nuts or similar components that the pump manufacturer buys in bulk quantities from a cost-competitive supplier. Others incorporate half-keys or special key contours (see Fig. 5.1.5, earlier). Many standard pump shafts incorporate elementary key and keyway configurations similar to Fig. 30.8.2, but notice how sharp corners are stress risers that weaken the shaft and promote crack formation (sketches A through E in Fig. 30.8.2). Wanting to save money up front, some users join pump manufacturers by purchasing parts from the lowest-cost subvendor or third-party supplier. That's why a good pump specification usually contains a clause requiring pump cross-sectional views and parts lists that the purchaser reviews during the bid evaluation process. The commercial parts a pump manufacturer obtains from third parties must be identified in exact detail. The purpose is not a secret: years later, an equipment owner-operator may have cause to buy replacement parts (buyout parts) directly from their respective manufacturers or from entities that produce superior, upgraded, components.

Different impellers and bores give clues regarding their attachment methods. Some hub bores have a keyway; many keyways have sharp corners that potentially weaken coupling, shaft, and impeller (see Fig. 30.8.2, sketches A through E). Fasteners can be another source of problems. Hardness and metallurgy must be observed, which brings us back to MQA. Usually, an impeller spins off only if it is not properly secured. But even a keyed impeller fit jeopardizes reliability if the key is loosely fitted. And, whatever their size and speed, pump impellers secured by castellated nuts or tab washers must retain impellers in a manner that does not allow them to come off while operating in any direction. That's why we should examine drawings before we purchase; we should also know how parts or machines work before we buy parts. While this does not mean that one needs 50 years of experience to buy $10,000 pumps, it does mean that, in the interest of reliability and safety, one must ask lots of relevant questions before buying plant assets.

The majority of superior pump designs use keys to secure impellers. A good key fit is a "snug fit," which means that hand fitting is generally advantageous [2]. The more vulnerable sharp-cornered keyways should be avoided. Half-keys are often superior to full keys. Keyways with generous bottom fillet radii and the so-called sled runner geometry have low stress concentration and a more desirable (higher) shaft factor of safety. Well-designed shaft ends also have a generous fillet radius at the shaft shoulder. As strength considerations prompt us to maximize the various radii, their contours must not interfere with the mating radius at the bearing inner ring. Verification takes time and is time well spent.

SHAFT DEFLECTION INDUCES FATIGUE FAILURES

Hydraulic forces act on pump shafts; the magnitude of these forces determines shaft deflection. Shaft diameter, hardness, metallurgy, fillet radius at shaft shoulder, and distance to the nearest bearing also influence how much the shaft will deflect. Shaft deflection is greater when centrifugal pumps operate at throughputs below design point or in excess of design point. Because the hydraulic force action is of unequal magnitude and the shaft rotates, reverse bending will take place and fatigue failures are possible if the design is marginal. Examining the fracture surface and performing a simple stress calculation will give focus to weaknesses and available remedies. Possible risk reduction steps will present themselves, and future failures will be less likely when we thoughtfully select one or more experience-based upgrade option.

MOTOR TORQUE AND SHAFT STRESSES

For the record, the starting torque of many motors is as high as seven times the full "normal" running torque. Agreed, discharge check valves rarely leak to the point of allowing substantial reverse flow. But "lean and mean" plants don't always install these check valves, and if they do, they sometimes forget to include these valves in their preventive maintenance scheduling. In any event, a pump reliability and/or failure analysis review should include the piping and all systems in the pumping loop. Pumps, controls, valves, piping, and other elements mutually interact and all must be considered.

ALL MECHANICAL PARTS FAILURES ATTRIBUTABLE TO "FRETT"

Through diligent training and reading, we learn about "FRETT." We come to accept that any and all mechanical parts can fail only due to one or more of the four cause categories: FRETT. So, because a pump shaft is a mechanical part, it can only fail due to "FRETT." An excessive reverse bending force will exist if the shaft is too slender, if its allowable bending moment or twist-inducing torque input is exceeded, if its metallurgy is not suitable for the fluid environment, if it has been in service for an abnormally long time, or if it was operated at an excessively high temperature. The reader who expressed his opinion about shaft strength and motor torque would have to examine shaft shoulder radii (fillet radius dimension) and the resulting stress intensification factors, obtainable from many handbooks. He would eliminate some of the FRETT causes, perhaps T for time and T for temperature. It might be evident to the reader that no reactive environment (RE) existed in the failed shaft system, and his investigative efforts for finding the root causes of the shaft failure might come back to "F" = force. Examination of motor starting torques may indeed show that starting torques can reach seven times the normal running torque. Applying an inrush current while a shaft system spins in reverse can cause shaft breakage.

In the reader's repeat failure example, it is also possible that several seemingly small deviations had combined. One can easily get away with one or even two deviations, but one rarely gets away with four or five. And what does this tell us as we ponder that, next to an electric motor, a pump is the simplest machine used by modern industry? Say a pump typically has 40 parts and yet fails relatively often. An aircraft jet engine has more than 7000 parts and fails rarely. Why? We believe the jet engine and aircraft manufacturers strive for perfection; they disallow every known deviation. In contrast, pump engineers strive for low cost because that's what the pump purchaser often seems to prefer over pump reliability.

But top manufacturers and good engineers know that normalization of deviance can cause disasters. Their quest to find root causes of failure and not tolerating known deviances requires training and discipline, strict adherence to checklists and procedures, and allocating the time needed to do things right.

In this instance, we were not given enough information to accurately determine (from behind our desk) why the reader's pump shafts failed. We can only vouch for the greatly increased probability that a few seemingly minor deviations from best practice combine. Together, these deviations cause trouble.

Their collective safety factors will vanish, impellers come off, and shafts break. There will never be a good substitute for following procedures and for uncovering what happened in this instance. Replacing parts and restarting the rebuilt machine without addressing and eliminating the true root cause will set owners up for repeat failures. Finding root causes of failure and implementing sound remedial steps are the commonsense course of action. That's exactly what aircraft jet engine manufacturers do, and with great success.

From our response, the reader should have inferred that any and all verbal hints at what may have caused impellers to spin off or shafts to fail must be supported by factual observations and evidence. There will always be a cause-and-effect relationship and it should be our task to uncover this relationship. For the benefit of the reader asking the original question, numerous texts show motor torque curves and explain why starting torques and inrush currents can differ greatly from normal operating torques and currents. Check valve action and condition may indeed be relevant. Impeller attachment is always important. Finally, pump shafts have safety factors that range from none to considerable. Safety factors are affected by shaft stresses, and high stresses can exist in shafts with discontinuities such as steps and keyways. Radiused corners will reduce stress intensification in keyways; these and other upgrades can be found in sturdy machinery; such upgrades can be easily duplicated or replicated by diligent reviewers.

We hoped that our response prompted the original reader to expand his horizons. He, and others, should start by considering better specifications. They should become familiar with keys or half-keys and with keyway ends and how these ends blend back into the shaft with a "sled runner contour." Failure risk can be affected by fillet radii, stress intensification factors, and the many other well-documented contributors to safe pump design and pump failure avoidance.

REFERENCES

[1] J.W. Dufour, W.E. Nelson, Centrifugal Pump Sourcebook, McGraw-Hill, New York, NY, 1992.

[2] H.P. Bloch, Pump Wisdom: Problem Solving for Operators and Specialists, John Wiley & Sons, Hoboken, NJ, 2011.

[3] H.P. Bloch, A.R. Budris, Pump User's Handbook: Life Extension, fourth ed., The Fairmont Press, Lilburn, GA, 2014.

Chapter 30.9

Consider Recessed Impeller Pumps

INTRODUCTION

While a number of manufacturers offer recessed impeller pump configurations that have not markedly advanced from their respective configurational or hydraulic performance constraints for decades, there are others who have been more progressive. Seek them out and become familiar with recessed impeller pumps. Be prepared to pay a little bit more up front and let a good manufacturer assist you in calculating the payback. Today's quality equipment is less maintenance-intensive than machines that were considered acceptable decades ago. Of course, low-maintenance, highly reliable equipment will cost more.

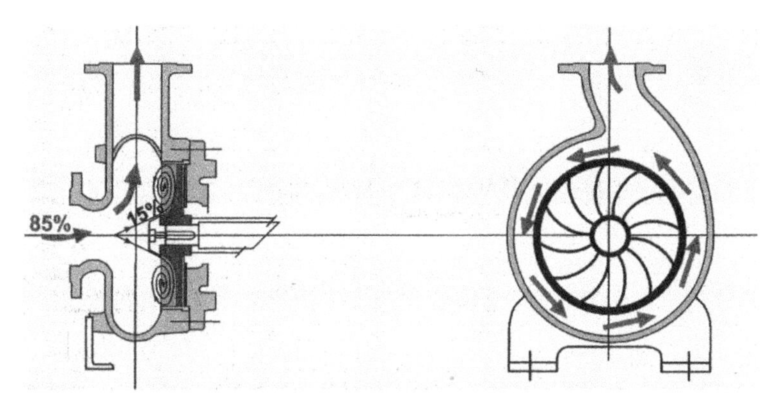

FIG. 30.9.1 Operating principle of a radial discharge fully recessed impeller pump. *(Courtesy Emile Egger, Cressier, Neuchâtel, Switzerland.)*

SUITABILITY EXPLORED

Recessed impeller pumps are ideally suited for a variety of pumping services handling free-flowing slurries, sludge, and fibrous materials. If used in these duties, a standard centrifugal pump may clog, become vapor-bound, or wear excessively. For the applications mentioned here, pumps with fully recessed impellers (Fig. 30.9.1) should be given serious consideration. Recessed "gentle pumping action" impellers incorporate the vortex principle illustrated in Fig. 30.9.1 wherein only an estimated 15% of the total fluid throughput makes contact with the fully recessed impeller. These pumps are typically available in flow capacities approaching 100L/s (1580gpm) and heads ranging to 130m (430ft).

Recessed impeller pumps have been around since the 1930s. Unfortunately for the user, a number of manufacturers offer recessed impeller pump configurations that have not advanced from their respective configurational or hydraulic performance constraints for 40 or more years. It is also fair to point out that some legacy models require a degree of maintenance involvement that was considered acceptable decades ago, but is no longer tolerated by today's best-of-class users.

A number of important characteristics and advancements separate one make or design of recessed impeller pump from another. It is worth understanding and considering how recessed impeller pumps sold worldwide under the Egger or Turo labels have favorably distinguished themselves in this regard. Commercial models became available in the mid-1950s; since then, the original Egger design has seen a number of seemingly small, yet important, upgrades. Successive iterations have consistently advanced relevant efficiency and the ability to handle solids with minimum damage to either the pump or the material being pumped. The overall vortex-type operating principle has remained the same.

HOW RECESSED IMPELLER PUMPS OFTEN DIFFER

Most recessed impeller pumps rotate the liquid and solids inside the casing until the solids reach a speed at which they exit the casing. This recirculation of solids creates wear in the casing and also increases damage to soft solids. Egger has overcome this problem by designing the casing with an "axial spiral" in the casing. Visualize an automobile tire to represent the basic design of a recessed impeller casing. Cutting the tire at the top and then twisting it yields a spiral. In like manner, the spiral contour helps guide solids out of the casing; it prevents solid recirculation. The manufacturer has demonstrated on many occasions that this design substantially improves the true overall hydraulic efficiency of the pump. Additionally, the axial-spiral twist has greatly reduced component wear and damage to solids being pumped. As a further point of interest, the minimum flow capability of a recessed impeller pump is much lower than that of conventional radial-spiral casing design pumps. On the minus side, top centerline discharge implies a measure of vulnerability when pumping large hard solids. Solids such as rocks might, on rare occasions, smash through the casing neck. In some rock feed applications, tangential discharge might be viewed as an advantage.

In many cases, users and engineering design contractors elect to place emphasis on pump efficiency. When asked to define efficiency, they inevitably refer to power draw. That, unfortunately, is seriously wrong. Some pumps achieve seemingly high hydraulic efficiency by simply letting the impeller edge protrude into the casing. Protruding impellers, of course, limit unimpeded passage of solids through the pump. Reliability professionals are urged to rethink what is of true importance here: the efficiency with which both liquids *and* solids are being transported. Many "old-style" recessed impeller designs have simply not progressed much since their initial introduction to the marketplace. Their best operating points (BEPs) are typically in the range of 30–40%. On the other hand, advanced designs incorporating axial-spiral design casing internals and fully recessed impellers will have true and effective BEPs around 50–60%. Less energy goes into the liquid and less power is consumed to forward-feed the solids.

Chapter 30.10

Auditing Vertical Pump Mechanical Reliability

INTRODUCTION

Vertical pumps are important for major process plants. They are available in numerous different configurations and excel in such areas as ease of multistaging, virtually unlimited installation depth below grade, and small footprint, that is, not requiring much real estate. Mechanical reliability assessments typically focus on the components that experience most failures: the thrust bearing, seal circulation system, and column bearings. Not to be overlooked is an examination of the overall design from the point of view of best available configuration, low-risk-of-failure installation, and preventive maintenance practices. On cryogenic services, we prefer to involve materials specialists in our review of material selection.

ASSESSING THE OVERALL DESIGN

Two areas of overall design are of primary interest. For ease of maintenance, individual column sections should be bolted, not screwed together. Also, impeller attachment by means of tapered collets should be disallowed on high-reliability vertical pumps.

Unless we are considering a relatively small size or model, a vertical pump should preferably be equipped with its own oil-lubricated thrust pot instead of depending on the motor thrust bearings to serve the entire assembly. This independent bearing assembly is shown near the top of our illustration.

Additionally, we would typically scrutinize the following for industry experience and definable component life:

- The *DN* value of the bearings should be within the API-acceptable limit of 500,000.
- Column bearings must be suitable for long-term operation in the medium being pumped. Unless the vendor can demonstrate prior experience, we would enlist the assistance of design engineers familiar with nonmetallic sleeve bearings using certain graphite composites (e.g., Graphalloy) or high-performance polymers such as DuPont's (now produced and marketed by Chemours®) Vespel CR-6100 crafted into carefully dimensioned special-contour sleeves.
- On critically important flammable or toxic services, the preliminary, not yet finalized selection of mechanical seals, and their respective support systems would be submitted to the three or four major seal manufacturers represented

at the installation site. These manufacturers would be asked to comment on the reliability aspects of the selection. We would expect seal manufacturers to provide feedback on appropriate material selection, pv values, balance ratios, flush plans, and support system supervisory instrumentation.

- Pump and seal reference locations and the names of contact persons should be disclosed to the purchaser. The purchaser would then verify data accuracy by communicating with some of the referenced user locations and require that mechanical seals be purchased from the most experienced vendor.
- Vibration probes are a plus on large vertical pumps. They should monitor both high-frequency acceleration and low-frequency velocity. Gradually developing bearing defects will show up in the acceleration spectrum long before there are velocity excursions. During shop testing, the manufacturer and the owner's inspector should also verify the absence of resonant structural vibration.
- A competent pump manufacturer will hand-fit keys and provide bottom-radiusing of keyways. He will not use roll pins for key fixation. It can be shown that improved shop practices along these lines will increase the shaft factors of safety.
- Arrange for a shop tour of the vendor's facilities. Observe how workers mount rolling element bearings and point out unacceptable practices whenever necessary.

PAYING ATTENTION TO LUBRICATION

Some lube system designs for low speeds are API-compliant with properly sized water cooling coils and ISO grade 150 mineral oils. However, the vertical pump manufacturer may not be aware of the merits of substituting an ISO grade 100 (equivalent to mineral oil ISO grade 150) diester or polyalphaolefin synthetic lubricant. These synthesized hydrocarbons have greater tenacity (stickiness), improved water contamination tolerance, lower pour points, reduced foaming tendency, and greater oxidation resistance than their mineral oil counterparts.

In his efforts to get a superior product, the user-purchaser may wish to explore the feasibility of immediate or future upgrading of both bearing environment and lube application. There is universal agreement among bearing manufacturers that an oil spray introduced into the bearing cage (ball separator) represents the most advantageous lubricant application method. It greatly reduces the risk of overheating, a primary concern with some of the customary layouts. In particular, triple-bearing thrust pots are best lubricated by an oil spray that douses the uppermost bearing.

In all instances, the oil must be applied via a carefully engineered integral or external oil supply system. A lube oil system originally designed and manufactured for lower viscosity lubricants may not perform as expected if the user selects a higher viscosity lubricant without thorough revalidation.

Chapter 30.11

Upgrading the Mechanical Design of Vertical Multistage Centrifugal Pumps in Low-Temperature Service

INTRODUCTION

The overall mechanical design of most API-style vertical column pumps (Fig. 30.11.1) comes relatively close to meeting the expectations of modern pump users. Such pumps are now often equipped with their own oil-lubricated bearing thrust assembly, whereas, in decades past, they depended on motor thrust bearings only. Of course, some older vertical pumps deserve to be closely reviewed. Based on failure history and criticality of service, older vertical column pumps are candidates for upgrading at the next routine repair or maintenance opportunity.

Determining the adequacy of the mechanical design of large vertical pumps should include a number of items, including the following:

- The *DN* value of the bearings (shaft rpm multiplied by the mean bearing diameter, in mm) should not exceed the experience-based limit of 500,000 [1].
- Disclosure of mechanical seal *pv* values (*pv* = pressure times velocity), also seal component materials and seal balance ratios, is needed. Working with a respected manufacturer, reliability-focused users should verify the *pv* values against prior experience. In case of unusually high *pv* values, the locations and names of contact persons may need to be established.
- The line bearings or column bushings used for shaft stabilization (Fig. 30.11.2) should be made of high-performance polymer materials. Many high-quality bushings nowadays contain carbon-graphite fibers. The typical diametral clearance at maximum bearing operating temperature should be [(0.001) (shaft diameter, in.) + 0.002 in.] inches. As an example, for a nominal shaft diameter of 1.6875 in., the bushing bore should thus be 1.6912″ ± 0.0005″. Three or four axial grooves should be provided in the bushing bore to counteract fretting risk during occasional, but potentially severe, rubbing contact.
- A nominal diametral clearance of 0.010 in. is recommended for the bore of labyrinth bushings not serving as bearings.
- If vibration probes are used, they should monitor both high-frequency acceleration and low-frequency velocity. Gradually developing bearing defects will show up in the acceleration spectrum long before there are velocity excursions. During shop testing, the pump manufacturer should verify the absence of resonant vibration. This is especially important in variable-speed vertical pumps, and resonant vibration must be absent at all anticipated operating speeds.

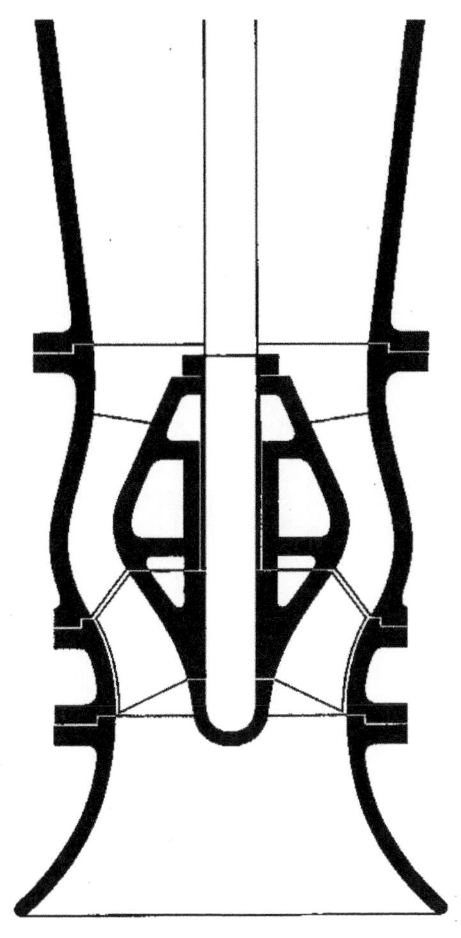

FIG. 30.11.1 Typical vertical pump inlet bowl. A mixed-flow impeller is shown here. Note flanged connections.

- Hand-fitting of keys and bottom-radiusing of keyways should be considered [2], and roll pins should not be used for key fixation. It can be shown that improved shop practices along these lines will increase the shaft factors of safety.
- Pay attention to proper assembly procedures [3]. Bearing manufacturers have long insisted on either supporting the bearing inner ring while pushing on a shaft or, alternatively, while pushing the bearing inner ring on the shaft.
- O-ring selection varies with the fluid being pumped. Teflon wrap over nitrile rubber or Viton cores should be considered for olefin services. The final selection should be approved by an O-ring or mechanical seal manufacturer.
- In cryogenic temperature environments and where dual seals are used, a specially formulated low pour point synthetic lubricant will be advantageous as a barrier fluid.

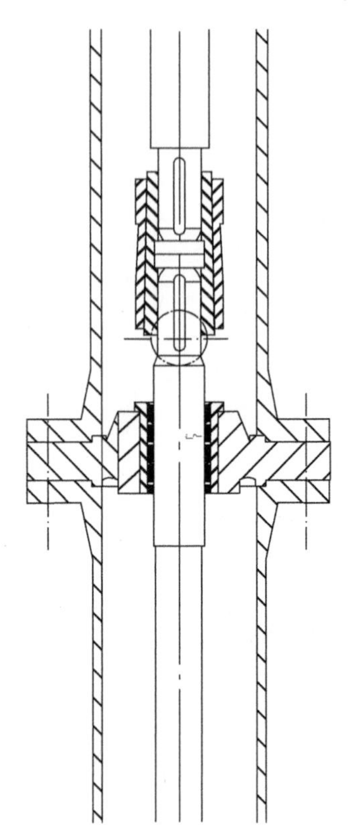

FIG. 30.11.2 Typical column bearing sandwiched between two column flanges. A two-piece split tapered bushing secures two keyed shaft ends in place; a single-piece tapered sleeve fits over the two-piece split tapered component.

- There is universal agreement among bearing manufacturers that an oil spray introduced into the bearing cage (the "ball separator") is the most desirable lubricant application method. An oil spray greatly reduces the risk of overheating—of primary concern in pump geometries where several rolling element bearings are assembled as a stack of two or more bearings.

REFERENCES

[1] H.P. Bloch, Pump Wisdom: Problem Solving for Operators and Specialists, John Wiley & Sons, Hoboken, NJ, ISBN: 9-781118-04123-9, 2011.

[2] H.P. Bloch, F.K. Geitner, Machinery Component Maintenance and Repair, third ed., Gulf Publishing, Houston, TX, ISBN: 0-7506-7726-0, 2004.

[3] H.P. Bloch, F.K. Geitner, Major Process Equipment Maintenance and Repair, second ed., Gulf Publishing Company, Houston, TX, ISBN: 0-88415-663-X, 1997.

Chapter 30.12

Recognizing the Pitfalls of Pumping at Part Load or in Parallel

INTRODUCTION

There is a danger in routinely operating pumps in parallel. Operating process pumps at part load for extended time periods is economically unattractive. Accepting part-load operation as a year-long routine is indicative of normalizing and accepting what, at best, is a deviance from intended best practices. Unfortunately and at worst, always operating critical pumps in parallel can become an extreme breach of safe operating rules that will put people in harm's way.

DEFECT ELIMINATION MORE VALUABLE THAN DEFECT DETECTION

A pump fire at a major refinery (we might call it "X") uncovered the serious potential consequences of operating certain critical hydrocarbon pumps in parallel. The result was a fire that put an entire plant at risk and could easily have led to the loss of human life. Fortunately, it "only" cost the refinery a bit in excess of $5,000,000. Also, an enlightened management group at "X" insisted that the root cause of the fire event had to be determined. Therefore, they carefully listened to the facts and implemented changes.

Before delving into the important lessons, findings, and reminders of any in-depth failure analysis, we should remember the definition of a root cause: *the cause that, if removed, will prevent the primary effect from occurring*. Because of this limited-value definition, staffers at refinery "X" initially considered not identifying high pump vibration before destructive bearing failure to be at fault. They decided, at least initially, that it would be logical to identify, as the primary cause of this pump fire, either the delay in recognizing bearing degradation or perhaps someone's inability to hear changing noise patterns emanating from the pump at issue here.

Well, deeper thinking prevailed in this instance, and the refinery recognized that vibration monitoring does not eliminate the root cause of the problem. Condition monitoring actually implies waiting for a defect to develop and then spotting it. However, the entire concept of reliability engineering is anchored in dedicated involvement in the *elimination* of defects before they ever have an opportunity to announce themselves.

NO TWO PUMPS ARE FULLY IDENTICAL

Pumps, these relatively simple and supposedly well-understood machines, continue to be involved in hazardous failure incidents. There have been many instances of deaths resulting from pump failures. Fortunately, that wasn't the case at "X," although a serious and preventable event did occur.

Knowing that parallel operation requires two pumps to be identical is really not enough. Two seemingly identical pumps are very rarely sufficiently identical in the sense that matters here. Experience shows that even on the standard curves furnished by the manufacturer, many API-style pumps have relatively flat head versus flow characteristics at partial flow, as did the pumps operating at "X." In other words, with flows significantly below BEP (best efficiency point), a minor change in the system characteristic can produce operation at insufficient flow. Changes in the roughness of the piping, or progressive wear on an impeller, can adversely affect the system characteristics. Therefore, one pump might drive the other pump off its curve or into an unacceptable low-flow condition. It is widely known that operation at *zero flow* is not possible for more than a few seconds; similarly, long-term operation at *low flow* may be an undue risk as well. Think of the automobile analogy: just because it may be possible to drive at 55 miles per hour in first gear *without* shifting into second gear doesn't mean that a rational person would go that fast in first gear for more than a few seconds at most.

KNOW ALL ABOUT OPERATING PUMPS IN PARALLEL

However, *that* point should not be the reliability professional's only concern. The real issue is that many refineries operate two pumps in parallel because their present total throughput requirement is 120% of nameplate. So, they reason, operating both pumps in parallel will make the pumps share the total load equally, and each is now assumed to be feeding a throughput of only 60% of nameplate capacity. At 60%, it is thought that a pump is "loafing," but that may really not be the case. A concerned professional would also question the premise that the two pumps share the load equally. The goal of selecting pumps with high hydraulic efficiency often leads to designs with head-flow curve relationships that are much shallower than those illustrated earlier in Fig. 30.1.1. It should be noted that a relatively steep per pump curve is shown here for clarity only. Experience shows that, with shallow curves, it is likely that two seemingly identical pumps delivering a combined flow of 120% operate at 70/50% or even 80/40% flow distribution. In the case of what pump engineers call high suction energy pumps, the uneven split leaves the pump with the lower flow in the unacceptable risk high internal recirculation category.

As is well documented in texts such as Ref. [1] and several Hydraulic Institute research documents, high-energy pumps will suffer from serious run length limiting internal recirculation at certain part loads. Depending on flow, impeller eye diameter, specific gravity, speed, and NPSH characteristics of a given pump, internal recirculation can impose significant additional stresses on impellers, seals, and bearings. Operation at part loads for certain pumps must be restricted to 50, or 60, or 80, or perhaps just 90% of rated normal load. Using Ref. [1], this parameter can be easily calculated and, for the pump that was involved in our refinery fire at location "B," would have limited the permissible flow to the vicinity of 70% of BEP. The bearings in this pump, although properly lubricated with dry sump oil mist and an excellent synthetic lubricant of just the right viscosity, failed frequently and rather consistently after, typically, about 18 months of operation. Refinery "X" had the records to prove it.

EFFICIENCY IS WORTH REAL MONEY

Suppose you were to drive an automobile at a constant 60 mph and were to get 20 miles per gallon. That would mean that you would have covered 60 miles and would have consumed 3 gallons of fuel in 1 h. While at only 6 mph, it would take 10 h to cover the same 60 miles; it would not be logical to expect the fuel consumption in those 10 h (at 6 mph) to amount to only 3 gallons. And there, of course, is the entire point about part-load efficiency—it's not as good as full-load efficiency. Inefficiencies cost money. One single extra kW over a year's time at a power rate of $0.145/kWh will cost a plant $1270.

But that's not all. At least one of our references [1] shows the shaft deflection of a single volute pump at 60% of flow at BEP to be six times that at 100% flow. Suppose this were to load up the bearings to twice their normal load. Bearing life changes as the inverse of the cube of the load; that is, twice the load causes us to achieve only about 12% of normal bearing life.

In our refinery "X" example, this pump and its sister pump operated at roughly 60% of BEP. Also, the engineers at "X" had not paid attention to the relative flatness of the pump performance curves at partial flow conditions. There can be no doubt that one of the two pumps was often operating at flows well below even 60%.

NEGLECTING RISK IS NEGLIGENT IN A RETURN-ON-INVESTMENT CALCULATION

With few exceptions, return-on-investment (ROI) calculations are a key ingredient to decision-making. ROI should determine repair versus replace and API versus non-API equipment and drive a host of other business decisions. When an overseas refinery we will call "Y" was questioned on its ROI calculations, they claimed that a project to replace existing pumps operating in parallel couldn't be justified on maintenance cost avoidance and energy efficiency grounds. When asked about risk, the consequences of an outright failure and secondary damage, a potential fire, and the value of a unit outage, the engineers and managers at "Y" stated that the inclusion of risk was not allowed in their ROI calculations.

A California-based specialist in asset management and risk assessment rightly pointed out that risk is the only reason for doing or not doing things. Risk is the reason why we use winter tires on our cars when the overall weather conditions and past snowfall statistics warrant. Undue risk of accidents is why a 10-year-old cannot obtain a driver's license; avoiding certain health risks prompts prudent and proactive people to exercise several times each week.

Of course, we might agree that at an average (direct) cost of repair of $6000 and a marginal loss of efficiency, the annual savings will perhaps be a mere $12,000, and a considerably more costly pump replacement would not be justified. But that's limited and incorrect reasoning in instances where critical pumps have already given ample warning of the *true* risk by having inappropriately low bearing lives and by being involved in a fire. Having an appropriate fact-based remedial strategy should not be an issue where a reasonable and unbiased assessment would show that the practice of operating certain high-energy pumps in parallel endangers people. If the practice risks losing an entire refinery and can be remedied by such steps as continuous monitoring and *automatic shutdown* upon failure of critical pump components, then *automated means* of monitoring are readily cost-justifiable. Asking a monitoring technician to spot incipient bearing failures and to engage in "timely manual surveillance" is very risky in that instance and assumes that bearings do us the favor of deteriorating gradually when, in fact, they can also fail instantaneously and without warning.

WHAT TO DO TO AVOID UNDUE RISK

Certainly, "business as usual" and similar do-nothing attitudes are out of tune with the professed reliability improvement goals of mature hydrocarbon processing organizations. In many instances where operating personnel have been allowed to habitually operate high-energy pumps in parallel, the practice constitutes an excessive and potentially catastrophic risk. Remedial action starts with an investigation of redesign options. Suppose two pumps had been in continuous parallel operation with 60% of BEP-flow per pump so as to satisfy the total demand of 120% throughput. Perhaps a third pump was used as a standby spare. This may well be the time to realize that a redesigned bowl casting (Fig. 30.12.1) or structurally and geometrically redesigned impellers (Figs. 30.12.2 and 30.12.3) may allow the standby spare pump to achieve a BEF flow of 120%. Operating that single pump at full load may be the best and most cost-effective long-term solution by far.

EXPERIENCE AT A MAJOR UTILITY BUTTRESSES OUR POINT

Here, then, is another experience. A few decades ago, a major utility "C" had three large, multithousand kW boiler feedwater pumps installed. For many years, the utility had been operating two pumps in parallel until, one day, two

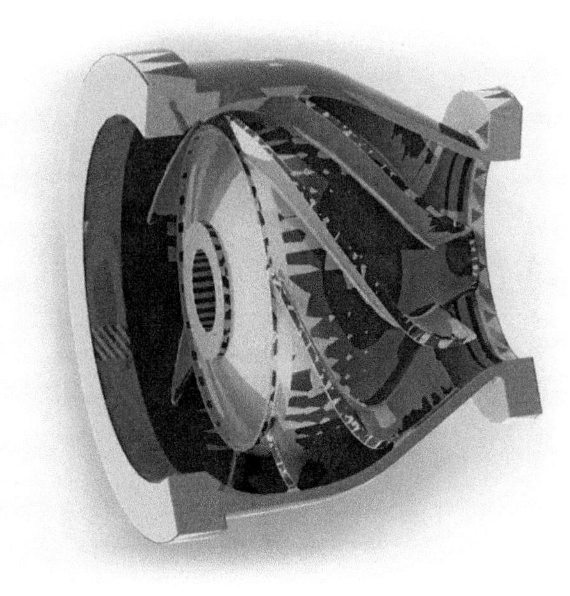

FIG. 30.12.1 Cast parts (bowls) are frequently used in vertical pumps. *(Courtesy Hydro Inc., Chicago, IL, http://www.hydroinc.com.)*

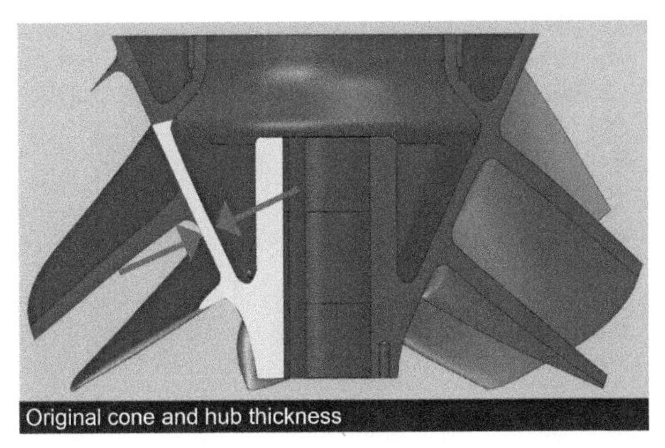

FIG. 30.12.2 Cone and hub thickness comparisons and failure analysis may uncover shrinkage-induced weaknesses in vertical pump parts. *(Courtesy Hydro Inc., Chicago, IL, http://www.hydroinc.com.)*

were no longer sufficient to provide full plant capacity. The decision was made to run all three in parallel, which worked quite nicely, but raised well-justified concerns in the eyes of a perceptive manager. With a regularly scheduled power plant outage not far off, the decision was made to send all three pumps to their original manufacturer for total revision and restoration to as-new condition.

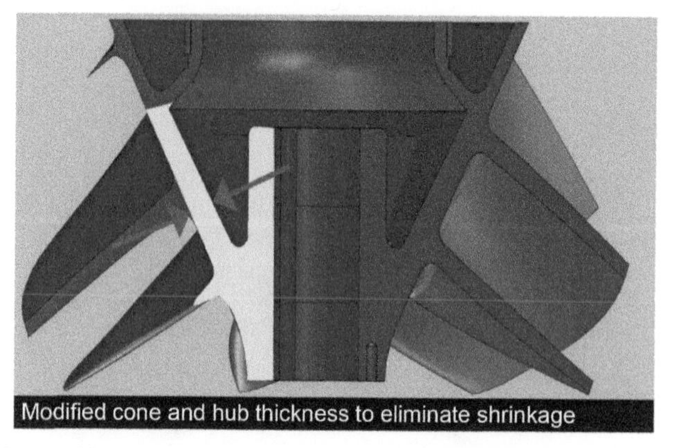

Modified cone and hub thickness to eliminate shrinkage

FIG. 30.12.3 Modifications can address shrinkage-related flaws in cast parts for vertical pumps. *(Courtesy Hydro Inc., Chicago, IL, http://www.hydroinc.com.)*

A few weeks later, the three pumps were back at the power plant and the facility was being restarted. Two of the three rebuilt pumps were put on line and operated in parallel, as they had been for years. However, the two together now produced flow well in excess of what the plant needed for full-capacity operation. One of the two pumps was shut down, and the one single pump staying on line proved sufficient to satisfy the nameplate requirements of the facility. The two others were now back on standby, which had been the original design intent but had been overlooked in the years since the time when, for reasons never fully understood, someone at "C" had decided on parallel operation. When the lost revenue for years of unnecessary operation of at least one huge boiler feedwater pump was calculated, the figure was in the millions of dollars.

CONCLUSION AND ADVICE

After the pump fire event mentioned earlier, the refinery manager at "X" was interested in pursuing the issue. He accepted the premise that reliability engineers must go well beyond arranging for an operating or mechanical workforce member taking frequent data and correctly interpreting the (often minor) differences between "go" and "risk" signals. The manager listened to experience-based opinions that shed light not only on these basics but also on certain issues that touched on the "people side" of failure prevention:

1. First and foremost, a good reliability engineer must advocate features and work processes that reduce failure risk, and these features should include the avoidance of long-term operation of high-energy pumps at part load.
2. Refinery managers must nurture people who are responsible, will understand facts, and divulge the truth at all times. The nurturing includes well-defined and highly relevant training. This training must be pursued without compromise.

3. Politics has no place in the engineering profession and an organization must be structured to shield reliability professionals from corporate or in-plant politics.
4. Passing blame and fruitless counterarguments that are refuted by facts must not be allowed to consume organizational energies.
5. An employee whose name comes up in a failure investigation may strive to save face in many ways, but this person cannot be allowed to disregard or dispute a body of knowledge that is readily accessible. He or she cannot be allowed to delay action by disregarding fundamental facts.
6. We must change our culture from that of guesswork and hearsay to a culture that searches for and applies, in an uncompromising manner, what surely can be called basic and well-documented pump know-how.

So then, going all the way back to the fire on one of two pumps operating in parallel, we believe that the deeper underlying cause of the pump fire event at "X" was rooted in not placing the right people in charge of decision-making, or training, or learning, or the implementation of a fair and balanced routine of being held accountable. It has been said that the fundamental root causes of serious incidents are ultimately linked to risk tolerance. Risk tolerance is a management decision and defining the level of risk tolerance is a key function of leadership.

REFERENCES

[1] H.P. Bloch, A. Budris, Pump User's Handbook: Life Extension, fourth ed., Fairmont Publishing, Lilburn, GA, ISBN: 0-88173-720-8, 2014.

[2] V.S. Lobanoff, R.R. Ross, Centrifugal Pumps: Design and Application, second ed., Gulf Publishing Company, Houston, TX, ISBN: 0-87201-200-X, 1992.

Chapter 30.13

Buying Better Pumps

INTRODUCTION

Experience clearly shows a preponderance of failures that could be avoided, but will continue to persist because of human error. That is also the reason for the randomness with which they occur. There is generally neither the investment in education, personnel, tools, and time nor systematic grooming of a culture of knowledge and loyalty. The latter two would be needed to reduce failure frequencies and improve the predictability of failures.

PREDICTIVE MAINTENANCE TOOLS

Large-scale investments have been made in predictive maintenance devices and tools that are then sitting idle or cannot be used because plant staff members have not been trained, mentored, or taught to interpret reams of data. A good example would be the many computerized maintenance management systems that to this day are being fed useless information such as "bearing replaced," "bearing failed," and "bearing repaired," instead of "bearing failed because loose oil ring abraded and brass chips contaminated the lube oil. Corrections made by upgrading to a clamped-on flinger disk." Of course, the prerequisite is for the facility to practice failure analysis and take remedial action instead of merely replacing parts.

People in authority make frightfully wrong decisions. As an example, they legislate to standardize on just one type of lubricant, or to buy critically important machinery and components from the lowest bidder, or to procure replacement parts without linking them to a well-thought-out specification, or to allow parts to be stocked without first thoroughly inspecting them for dimensional and material-related accuracy or specification compliance. And keep in mind progress made possible through advances in materials technology and fabrication techniques. Fortunately, major components can often be replaced or upgraded for long-term reliable performance.

That said, random failures demand ready availability of good options. Reliability engineers and their associated job functions often try hard to make their responsible supply chain staff or purchasing departments understand that it is impossible for the maintenance department to plan, or accurately predict, spare parts demand. The reason is that a major proportion of equipment and systems fail randomly. A good example would be rolling element bearings in all industries worldwide.

According to a bearing manufacturers' statistics, 91% fail *before* they reach the end of their conservatively estimated design lives, and only 9% reach the end of design life. The millions that prematurely fail every year can all be slotted in one or more of the seven basic failure categories: (1) operator error, (2) design oversights, (3) maintenance mistake, (4) fabrication/processing defect, (5) assembly/installation error, (6) material defects, and (7) operation under unintended conditions.

Because of the randomness of failures in process plants that are not adhering to true best-of-class concepts, it is necessary to have certain parts in stock. We know that some theoreticians extrapolate widely from data that apply to predictable wear-out failures. However, what transpires in hydrocarbon processing facilities is a function of numerous variables, most of which have lots to do with human error. And so, spare parts predictions determined from mathematical models are generally too far off to merit intelligent discourse. Instead, reasonable specifics are a function of equipment type, geographic location, skill levels of workforce members, etc. For decades, it has been understood that relevant answers require auditing or reviewing a particular local situation. Non-OEM repair and upgrade specialists are available to work with users and owners that believe spare parts definition and management in 2016 must not follow the pattern set in 1950. Our earlier Subject Category 22 applies to spare parts options in its entirety.

Subject Category 31

Reciprocating Compressors

Chapter 31.1

Reciprocating Compressors with Floating Pistons

INTRODUCTION TO NONLUBRICATED COMPRESSORS

Process plants have for decades applied nonlubricated reciprocating compressors. Early designs used carbon piston rings and rider bands; these were soon followed by Teflon, filled Teflon composites, and high-performance plastics. In the 1940s, Swiss manufacturer Burckhardt started to market labyrinth piston machines that are not subject to ring and band wear. Today, thousands of these low-maintenance compressors are in successful use worldwide. (Detailed information on many of these options are presented in a comprehensive text "Reciprocating Compressor Operation and Maintenance," [1].)

More recently, Dutch equipment manufacturer Thomassen Compression Systems (TCS) has developed "floating piston technology." This technology presents an elegant solution that is sure to circumvent the wear problems encountered with plastic composites in horizontally opposed piston reciprocating compressors of the nonlabyrinth variety. Using a floating piston configuration, the TCS development promises to place within reach the user industry's requirement for at least three years of uninterrupted run length.

OPERATING PRINCIPLE EXPLAINED

The principle of the floating piston technique is that the weight of the piston is carried by the process gas itself. During the compression stroke, the interior of the piston is filled with process gas at high pressure. Due to the pressure difference, the process gas flows out at the bottom of the piston, ensuring that there is no physical contact between rider bands and cylinder liner. As a result, rider band wear is virtually eliminated, since surface contact exists only at standstill

Petrochemical Machinery Insights. http://dx.doi.org/10.1016/B978-0-12-809272-9.00031-1

and start-up operation. The design of the piston itself has been adapted to the needs embodied in the operating principle:

- To create constant pressure, the rider bands are located between the piston rings.
- Regular compressor valves that react to differences in gas pressure have been installed in the piston faces, enabling gas to enter the interior of the piston.
- Flow nozzles have been integrated in the rider bands and a profile machined on their lower surface to create a gas-bearing layer under the piston.

EXPERIENCE AND KEY ADVANTAGES

In May 1998, a continuously monitored floating piston compressor went into hydrogen service at Dutch State Mines in Geleen, the Netherlands. Successful long-term operation has been reported:

- Practically no rider band wear.
- Increased piston and packing ring life, attributable to the piston maintaining its concentricity location relative to the cylinder.
- Elimination of cylinder lubrication with attendant component failure risk reduction.
- Increased uptime, resulting in higher production rates.
- Maintenance cost avoidance.
- Feasibility of eliminating spare or redundant machines.
- Feasibility of extending range of gases that can be processed in reciprocating compressors.
- Retrofit potential opened up for compressors from many different manufacturers.
- Power reductions (efficiency gains).
- Reduced sound intensity.

Depending on the gas service and other performance parameters, floating pistons in reciprocating compressors are a development worth looking into.

REFERENCE

[1] H.P. Bloch, J.J. Hoefner, Reciprocating Compressors, Gulf Publishing Company, Houston, TX, 1996.

Chapter 31.2

Upgrading Compressor Piston Rods

INTRODUCTION

Since the late 1900s, the high-velocity oxygen fuel (HVOF) coating process has achieved commercial acceptance. After making inroads in gas turbine applications, these pure, dense, hard, and highly adherent coatings are now increasingly found in such compressor components as piston rods (see Fig. 31.2.1).

Thermal spray systems are involved in the process. Both HVOF and its predecessor, the D-Gun system, are thermal spray systems; they have many strengths in common. Both operate at high deposition rates and lay down highly adherent coatings of 0.002–0.020 in. thickness per layer. Either system is quite flexible and will deposit virtually any feedstock that can be melted or softened.

The operating principle behind thermal spray is to produce a high enough temperature to melt virtually anything, from metals to ceramics, and then blast the melted metal out of the chamber of an application gun and onto the metal component to be coated. The faster the melted material leaves the chamber, the less time it spends being contaminated by the ambient oxygen atmosphere it needs to traverse on its way to the target surface and the harder its impact. High-impact speeds increase coating density, hardness, uniformity, cohesion, and adherence. Suitable HVOF systems quite obviously are known to the leading compressor service providers.

A modified and generally advanced HVOF version is a system called HVLF, denoting high-velocity liquid fuel system. It uses liquid kerosene fuel, which is more economical and safer than pressurized explosive gases. When liquid kerosene hits the hot ignition chamber, it vaporizes and raises the chamber pressure. A relatively high pressure boosts particle velocities. Powder feeds are injected into the gun beneath the flame front and the lower pressure "pulls" these feeds into the flame.

Coating compositions must match the application. A number of coating materials are available for piston rods. These include tungsten carbides of different properties. Depending on anticipated compressor piston rod service condition, an experienced coating company will recommend the proper tungsten carbide and bonding matrix best suited for a particular application.

For example, a representative coating combining 83% tungsten carbide with 17% cobalt was applied with an average thickness of 0.037 in. Its bond strength exceeded 8000 psi and its apparent porosity was below 0.25%. At an average coating hardness of 67 Rc, application in the wear-prone packing area of compressor piston rods will no doubt prove beneficial.

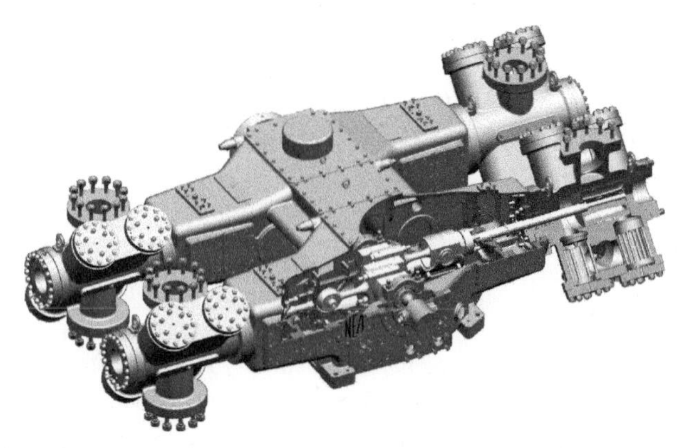

FIG. 31.2.1 A four-throw reciprocating compressor. Cutaway portion highlights its central running elements. Note piston rod, located between crosshead and piston. *(Courtesy Neuman & Esser, www.neuman-esser.com.)*

DOCUMENTING YOUR REPAIR PROCEDURE

You should expect a knowledgeable repair facility to offer you their repair procedures for review and comparison. One of the leading suppliers will normally proceed with an appropriate step-by-step sequence. In general terms, such a sequence incorporates the following:

- Check rod for straightness, amount of wear, thread damage, piston fit size, hardness of packing/wiper section, etc.
- Document the "as received" condition on a sketch.
- Grind packing and wiper ring sections plus at least ½in. at each end of area undersize, so as to remove damage and wear. The edge of the undercut should be configured with a radius to preclude the formation of stress risers.
- Document the before-coating surface hardness.
- If the rod has been hardened or nitrided, grinding must penetrate through this layer until the original hardness substrate is reached. Hardness testing is to be performed and documented.
- Magnetic-particle inspect the rod to verify the absence of cracks throughout.
- Demagnetize the rod to a residual level not to exceed two Gauss.
- Heat-soak the piston rod in an oven at 400°F for 4 hours and allow it to slow-cool so as to remove residual gases.
- Mask and protective-tape all surfaces except those to be grit-blasted.
- Blast with aluminum oxide grit to provide a surface finish of 200–350 RMS before commencing coating operations.
- Coat packing areas with an HVLF (3300–3900ft/s particle velocity) tungsten carbide overlay in all packing areas. Using an infrared thermograph gun aimed directly at the point of impact, verify that the rod temperature does not exceed 350°F during this operation.
- Apply coating so as to achieve a diameter 0.010–0.015in. greater than the specified finished (postgrinding) diameter.

- Attach a coupon of like material substrate to be sprayed with the rod. Submit this coupon to purchaser for follow-up evaluation.
- Finish-grind with diamond wheel and then diamond-hone and superfinish to agreed-upon RMS value. (Note from Ref. [1]: a better than 8 RMS finish may not be desired since oil will not stick to it!)
- Polish all relief grooves and fillet radii.
- Dye penetrant check for indications of blistering, spalling, flaking, cracking, or pitting.
- Seal coated area with a proved PTFE ("Teflon") or proprietary epoxy sealant.
- Identify rod by die-stamping end face and provide report including the following:
  - dimensional data recorded on a sketch,
  - magnaflux results,
  - "as received" and "as corrected/finished" dimensional sketch,
  - coating lengths and thicknesses shown on sketch,
  - type of coating and lot number identifying material and properties, and
  - final RMS finish documented with profilometer tape.

FIG. 31.2.2 A large six-throw reciprocating compressor being assembled at the factory. *(Courtesy Neuman & Esser, www.neuman-esser.com.)*

It should be noted that several competent non-OEM coating companies in the United States are also producing entire new piston rods to OEM suppliers. The base material for these rods is typically AISI 4140, although AISI 4340 is used for sweet gas service and rods larger than 4in. in diameter. One of these very large machines is shown in Fig. 31.2.2. To avoid stress corrosion failure, the rod manufacturer normally pays close attention to the hardness value of base materials in sour gas services.

REFERENCE

[1] H.P. Bloch, J.J. Hoefner, Reciprocating Compressors, Gulf Publishing Company, Houston, TX, 1996.

Chapter 31.3

Gaining Flexibility With Special Liners for Reciprocating Compressor Cylinders

INTRODUCTION

Most reliability engineers with a background and responsibilities in reciprocating compressors are familiar with replaceable cylinder liners for these positive displacement machines. Modern cylinder inserts can often be used to gain compressor efficiency by optimizing compressor throughput. Generally speaking, such inserts are an offshoot of the cylinder liners typically incorporated in API-style process reciprocating compressors. Equipped and upgraded with suitably designed liners, these cylinders can provide a relatively simple solution to one of the most labor-intensive problems of compressor reconfiguration commonly found in gas-gathering applications. With the new cylinders, it may no longer be necessary to remove or replace cylinders, reconfigure piping, or change out pulsation bottles. It is only necessary to change the cylinder's internals to reconfigure a compressor for a new application.

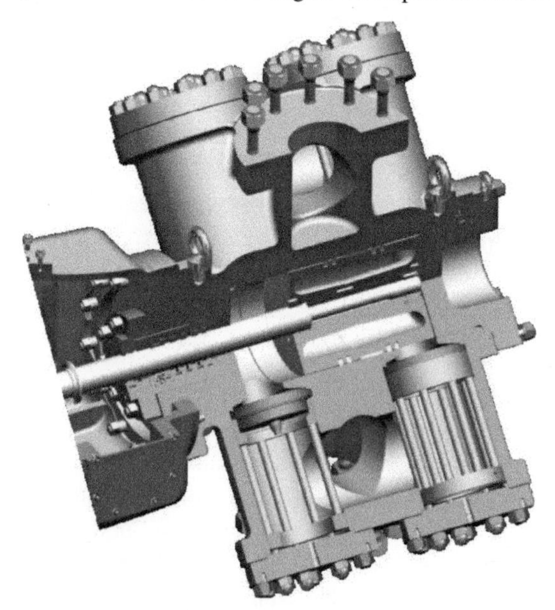

FIG. 31.3.1 Reciprocating compressor piston and cylinder cutaway portions are highlighted. Note valve arrangements. *(Courtesy Neuman & Esser, www.neuman-esser.com.)*

DESIGN OVERVIEW

Special liners with low fixed clearance can provide optimal performance characteristics over a wide range of bore diameters. Special liners often reach this goal with only one cylinder body. Major OEM and non-OEM suppliers accomplished the versatility enhancement task by first designing a cylinder body that is configured for the largest required bore diameter and the highest foreseeable working pressure. They accommodated the needed bore diameter with a liner. Use of a liner to change a cylinder bore diameter is not a new idea; however, traditional lined cylinders tend to have high fixed clearances as the bore diameter is decreased. Using purposefully designed standard conversion components that minimize the fixed clearance is often possible. The bore diameter change allows compressor valve losses within acceptable ranges. These liner components optimize the port geometry relative to the bore size, tailor special port geometries to specific operating conditions, can vary the size and quantity of valves and ports, and usually vary the position of the valves with respect to the bore [1].

Although Fig. 31.3.1 shows a conventional cylinder, we ask you to visualize a purposefully oversized cylinder bore ready to be fitted with liners of different wall thickness. One manufacturer pointed out that one cylinder body, rated at either 1270 or 1500 psig MAWP (maximum allowable working pressure) and weighing about 3500 lbs, can accept bore diameters ranging from 16.0 to 5.5 in. This one cylinder body can accommodate 95–98% of all gas-gathering applications. It replaces the previously required seven or eight different cylinder classes found in gas-gathering applications with comparable fixed clearances and valve losses. By changing out standard internal components available from this manufacturer, the cost of converting to different cylinder diameters would be considerably less than many other options. The full-fledged revamp of certain compressor packages used in upstream (oil field) operations would cost much more.

Cylinders accommodating many different liners are reported to offer unique advantages even if only deployed on the first stage of a gas-gathering compressor. The cylinder can be fitted with a smaller bore diameter when the producing field is new and the suction pressure is high. The liner diameter can later be increased, as the rod load rating of the frame may permit, to maximize the unit's capacity as suction pressure declines. Conversion can be done relatively rapidly and efficiently without removing the cylinder or disturbing the piping. Only the outer head and the valve cap have to be removed to gain access to all the components requiring change to a new bore diameter [1].

For upstream compressors, one such development enables optimal compressor utilization and deployment as operating conditions change. This, of course, translates into more throughput and increased profitability for the owner-operators. By changing only the crank end head and the piston and rod assembly, the same cylinder body can be mounted on any of the many common 5.0–7.0 in. stroke compressor frames that have been manufactured in the last five decades. The entire development may also favor streamlining a facility's spare parts inventory.

REFERENCE

[1] ACI Services, Inc., Cambridge, OH, Commercial bulletins.

Chapter 31.4

Interpreting API Standard 618 for Reciprocating Compressors

INTRODUCTION

A rotating machinery engineer working for an EPC (engineering-procurement-construction) contractor in India is often involved in selecting reciprocating compressors for hydrocarbon processing plants. More than likely, this engineer is also responding to compressor sizing and specification questions. She had a question relating to drive motor rating selection for a large reciprocating compressor and applying the recommendations found in a major API standard, API-618, fifth edition. The matter is of interest to others who are occasionally puzzled by wording or by seeming discrepancies in equipment sizing practices.

ARE THERE DISCREPANCIES?

FIG. 31.4.1 Four-throw reciprocating compressor (the "frame" containing "running gear") before crosshead and cylinders are attached to frame. *(Courtesy Neuman & Esser, www.neuman-esser.com.)*

The possible size of process gas compressors in refineries is hinted at in Fig. 31.4.1. API-618 is a widely used specification covering these machines and Paragraph 7.1.2.2 was at issue here: for motor-driven units, the motor rating, inclusive of service factor, shall be not less than 105% of the power required (including power transmission losses) for the relieving operation specified in 7.1.1.3.

The EPC correspondent next highlighted Paragraph 7.1.1.3: the driver shall be capable of driving the compressor with all stages at full flow and discharging at the relevant relief valve set pressure. She advised that, during her company's interaction with some compressor suppliers (in mid-2015), she and her colleagues came to realize that the manufacturer's internal practice for multistage compressors is to select a motor rating considering the relief valve set pressure of the final stage only, and not that of all the stages relieving together. (A compressor cannot operate with *all* stages relieving each other. The compressor operating with all stages at full flow and operating at the last stage relief valve pressure will be under *maximum* load.)

The EPC's machinery group believed that the stance taken by the various suppliers was not in line with the API-618 recommendations found in Paragraph 7.1.1.3 earlier. Moreover, she noted that not taking into account the relief valve set pressures for each stage would affect not only the motor power selection but also piston rod loads.

The correspondent-engineer's desire to clarify what she perceived as a potential conflict was important. We replied that, on the inside cover of an API standard, we usually find conveyed that API standards represent industry guidelines; compliance is not a regulatory or legal obligation. The EPC's procurement specifications and vendor need to itemize where the vendor's offer differs from applicable API clauses. The parties' written agreement and referenced procurement documents will ultimately govern.

COMMON PRACTICE EXPLAINED

It is common practice for EPCs to obtain several bids and to then do a bid comparison. At that point, the EPC and owner-operator company (the client) must choose between the less conservative and more conservative offers. One of these yields lower reliability than the other. One is more forgiving of operating errors than the other.

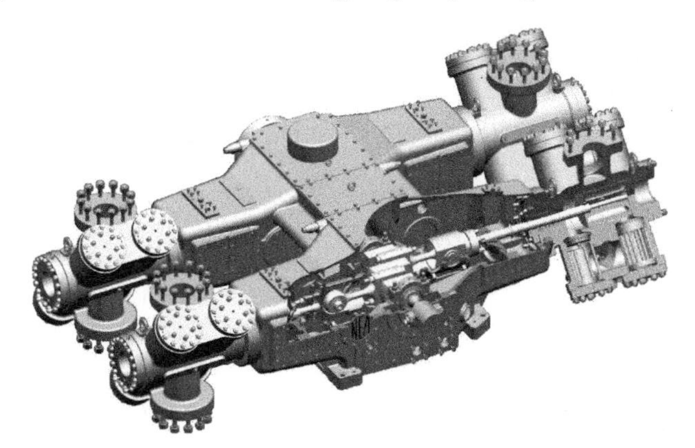

FIG. 31.4.2 A four-throw reciprocating compressor. *(Courtesy Neuman & Esser, www.neuman-esser.com.)*

It is the course of prudence to pick the more conservative offer in view of future uprate flexibility. The machine with the lesser rating will not permit future change-outs to larger cylinders. The user who purchases machines with lower safety factors must anticipate and tolerate one of two events with the passing of time: higher failure risk or more frequent preventive maintenance or both. All of that will cost money in the future. Become familiar with "inner workings" (Fig. 31.4.2), uprate capability, and safety factors; they may differ significantly from vendor to vendor. At least one major manufacturer of process reciprocating compressors offers running gear ("inner workings") designed and configured so as not to create couple action.[1] The manufacturer claims, among other advantages, that this machine has little or no vibration. These compressors have opposing cylinders located on the same common centerline.

Each compressor cylinder has associated with it a relief valve that protects pistons and rods; it relieves at certain cylinder loads. That load, for each cylinder, is at "per stage" *differential pressure* values (per cylinder). Relief valve settings beyond design would cause excessive rod loads. The compressor manufacturer calculates these maximum loads for each cylinder and advises what these relief valve settings should be on a given stage.

The compressor driver horsepower demand at start-up is much less than at full load. Start-up must take place with either gas being recycled through a bypass or suction valves held open by a suitable valve unloader mechanism. No gas is compressed at that time. Once the drive motor is rotating at its design speed and at no load (it can take several seconds before the motor is up to speed), the valve unloading mechanism is usually programmed to progressively revert to its "loaded" operating mode and to thereby increase the load to a demanded throughput.

At full-load throughput and with as-designed differential pressure (in this case referring to final-stage discharge pressure minus first-stage suction pressure), the drive motor rating shall be 1.05 times the maximum demand calculated for the rated operating conditions.

Here is an example: say at a flow of "xyz" cubic feet per minute and with a suction pressure of 14.5 psia, we reach a discharge pressure of 2000 psia. For xyz cfm and a compression ratio of 2000/14.5 = 139, assume we might theoretically need 3000 hp. We add power losses of 7% and thus need 3210 hp. We observe the API clause 7.1.2.2 and purchase a motor with a continuous capability of 3210 times 1.05 = 3370 hp. We have the option of purchasing the motor with different service factors (SF) allowing hot conditions, temporary overload conditions, etc., and we specify, for instance, an SF = 1.25. However, regardless of the SF selected by us, we buy a 3370 hp motor. We do *not* buy a 3370 times 1.25 = 4213 hp motor.

Essentially, the compressor manufacturer looking to do business with the EPC was correct. The API specification is a bit unclear. The service factor covers all reasonable deviations and all piston rods are individually protected by relief valves. The minimum acceptable margin between stage discharge pressure and relief valve set pressure is typically 10%, but will depend on the maximum allowable pressure rating of a particular cylinder and its attendant rod stresses.

[1] A couple is a pair of equal and parallel forces acting in opposite directions and tending to cause rotation about an axis perpendicular to the plane containing them.

Rod stresses are a function of pressure differential across a piston. Sometimes, the maximum allowable rod stress occurs at less than maximum allowable cylinder pressures. All issues seem to relate to robustness of overload capacity, and the guidelines in the API standard are approximate—never absolute.

THE AUTOMOBILE ANALOGY

Machinery engineers can often draw analogies with automobiles. A four-passenger car with a 1600 cm^3 four-cylinder engine will be fine as long it is being operated within the intended limits. Suppose we only paid $15,000, initially. We also looked at another automobile with a 3.2 L six-cylinder engine and considerably more space; it would cost $40,000. The four-passenger four-cylinder vehicle translates into an uncomfortable ride for a six-passenger load. It will be slow on mountain roads, or wherever up-hill driving conditions are usually encountered. The tires will not last as long as anticipated. With six passengers and 1600 cm^3, the fuel consumption may be disappointing. The hydraulic shock absorbers will have to be replaced more often. As the car's owner, we have to look at the advantages and disadvantages.

As regards reciprocating compressor, many machines offered in the marketplace will work, but some will not allow deviating from proper operation. They will be unforgiving. And so, in follow-up correspondence, we were informed that the compressor manufacturer advised relief valve setting less than the minimum value recommended in Paragraph 7.6.5 of API-618 (fifth edition), due to his machine's rod load limitation. Well, that certainly reinforced our views. Reliability engineering is aimed at just that: reliability. We must buy equipment with reasonable safety margins—1.25 in this instance—and should not allow routine operation to nibble away at that margin.

ALWAYS EXAMINE EXPERIENCE

The various API standards are a great help to EPCs and the equipment user community. We carefully examine standards such as API-618 and specify certain available options wherever the standard has clauses requiring us to do so. Then, in the interest of buying low-risk reciprocating compressors, we should go beyond merely invoking the standard. As reliability engineers, we are aware of industry trends such as compressor manufacturers mating a medium-speed frame to a more cautiously designed, strong cylinder and piston assembly. There have been instances where the resulting match was a bit unhappy and where the various parties later met in a court of law. If we want to have the most reliable equipment, we must do detailed checks beyond a bid comparison. Look it up under the index words "machinery quality assessment" (MQA). In any event, choose wisely. True reliability assessment takes time, effort, and knowledge. The cost of MQA is typically 5% of the cost of the machine. This cost is an upfront budget item for best-of-class user companies. Best-of-class user companies invest in better-than-average reciprocating compressors.

Chapter 31.5

Upgrade Kits and Service Bulletins

INTRODUCTION

A top-rated manufacturer of reciprocating compressors keeps its customers informed of relevant updates. Service bulletins are offered to you by compressor manufacturers because they want to promote an ongoing sales and service relationship with a user's organization. However, the operation of your plant involves factors not within the manufacturer's knowledge, and operation of the plant is within your control and responsibility. The responsibility for the plant's reliability, operability, and profitability rests with you. It should not surprise us that an equipment manufacturer will tend to disclaim responsibilities beyond parts, and even that responsibility ends as warranties expire. The manufacturer's disclaimers will always include, but not be limited to, direct, consequential, or special damages. The vendor/manufacturer will use legal language making this position abundantly clear. Moreover, the manufacturer considers his service bulletin confidential and proprietary, places it under copyright and/or other intellectual property protections, and provides these bulletins for authorized use only. So be sure to obtain all kinds of written consent before you involve others.

ABSTRACTS PROVIDE AN OVERVIEW

Suppose the manufacturer wishes to inform its customers that he, the manufacturer, has released a series of kits for uprating certain compressors in one or more frame sizes. He may highlight that a particular upgrade kit allows the owner of a series XYZ compressor to increase its allowable rod load, thereby making possible the revamping of the compressor for process debottlenecking. Or perhaps the manufacturer wants to convey information on increasing the safety margin in order to improve the robustness of the machine in critical services or where there is no backup unit. Bearing loads and component strengths are often a function of running gear dimensions and material selection (Fig. 31.5.1).

To maximize the benefit received from such service bulletins, know your machines and keep track of parts being replaced and invest in upgrade kits that reduce failure risk.

DETAILS DISCLOSE PARTICULARS

Some compressor uprates consist of field-proved redesigned components such as piston rods, crosshead, connecting rod, crankcase bearings, and, in some cases, even the crankshaft material and its heat treatment. Certain applicable changes may be called mandatory due to their safety impact; others are probably optional

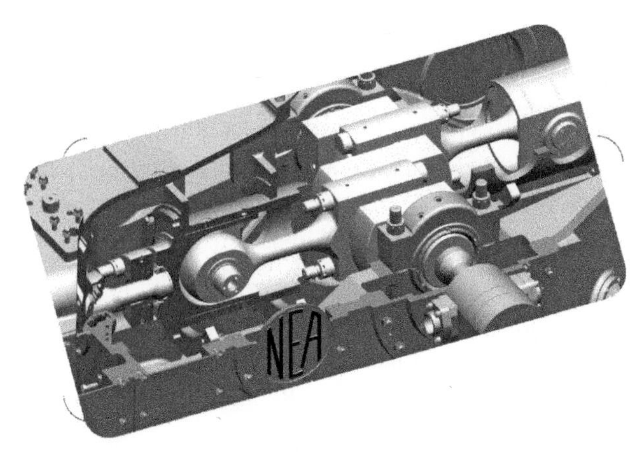

FIG. 31.5.1 Know your machines, keep track of parts being replaced, and invest in upgrade kits that reduce failure risk. *(Courtesy Neuman & Esser, www.neuman-esser.com.)*

or discretionary. There could be new and experience-based maintenance guidelines or best practices for reliable equipment operation.

The manufacturer's upgrade bulletin could specify timing. Typical timing categories are (a) prior to unit start-up or continued operation, (b) at first opportunity/at next shutdown, (c) prior to operation of an affected system, (d) at scheduled component part repair or replacement, and (e) at next scheduled outage.

FIG. 31.5.2 Compressor service organizations with OEM design backup are more effective than "low bid" service entities. *(Courtesy NEAC Compressor Service, www.neacusa.com.)*

SPECIAL SERVICE ORGANIZATIONS

It will be possible in conjunction with repair and service work to communicate your needs to competent service organizations with design know-how that harness the collective experience of both users and manufacturers. Figs. 31.5.2 and 31.5.3 are simple reminders of this fact. Experience shows the overall benefits of entrusting compressor upgrading to service organizations specializing in this type of work. Among other advantages, their highly skilled workforce will be in the best position to call on the manufacturer's design engineers to resolve any unusual issues.

It makes no sense to depend on unskilled labor to dismantle, diagnose, upgrade, and test reciprocating compressors. Compressor service organizations with OEM design backup are more effective than "low bid" service entities. I stand by that statement, knowing full well that it will not be accepted by everyone. Recall, however, that the overriding topic of this book is safety and reliability. Safe plants are reliable plants. Reliable plants are profitable plants. Profitable plants have staying power.

FIG. 31.5.3 Compressor service personnel should have years of solid experience. *(Courtesy NEAC Compressor Service, www.neacusa.com.)*

Subject Category 32

Reliable Plants and Machines Benefit From MQA

Chapter 32.1

Reliable Because of Up-Front Investment in "MQA"

INTRODUCTION

We had tracked down two machinery quality assessment (MQA) engineers who continued doing this type of work long after retirement from major petrochemical companies. We found one of the men working for the old employer as the "machinery lead" at an engineering, procurement, and construction (EPC) firm's office in the UK. The project was for a facility in the Middle East. The second engineer performed the same job function at an EPC contractor's office in Japan; this large project was for a Pacific Rim country. Please note that all of this background information should prove important to the reader. It should prove important because, sad to say, most reliability engineers in all kinds of industries are involved in retroactive fix-ups at plant sites. However, the previous employer of the two retirees was a prominent best-of-class multinational HPI corporation. For a certainty, the former employer, an hydrocarbon processing industry (HPI) company that's usually in the news because of solid profits, had quite obviously been investing some of its money up front. They had done so since the early 1960s and have been both consistent (and consistently successful) with this approach.

UP-FRONT INVESTMENT IN RELIABILITY IS THE BETTER WAY

The term up-front investment implies that thoughtful specifications covering components, procedure, and work processes are being issued [1]. Each of the two engineers mentioned above (and others in related job functions) was trained

Petrochemical Machinery Insights. http://dx.doi.org/10.1016/B978-0-12-809272-9.00032-3

in equipment engineering and had the requisite background and experience to monitor the EPC's work, verifying that none of the purchaser's specification clauses are either overlooked or misunderstood. Both engineers had lots of practical and analytic experience. They are resourceful individuals who know what to question and what guidance to give on matters such as equipment selection based on life-cycle costs, to name just one.

That gets us again to an important question: How do best-of-class companies handle matters dealing with engineering drawings (P&IDs, PFDs, and cross-sectional drawings of rotating or stationary equipment) and other data deemed proprietary by engineering and design contractors and manufacturers or vendors? Well, in the case of plants presently going through the various preoperating phases (plants that are being designed, or for which equipment is yet to be procured, or for which equipment is presently being built), the owner-purchaser assigns MQA engineers. These engineers are subject matter experts who verify specification compliance and organize audits to take place at the vendors' facilities. Acting as owner representatives, they also make periodic follow-up visits to verify order status and, sometimes, manufacturing quality.

Relative to facilities that were operating or being scheduled for expansion or modernization, one owner company answered by giving these specifics:

Our five plants are now 1, 8, 10, 13, and 24 years old. For heat exchanger failures we have done RCA (root cause analysis) and are now recommending specific modifications to the units. We have avoided doing over or re-issuing complete drawings in the past and, instead, simply spell out the specific upgrade in our request for quotation.

Problems arise when the OEMs (Original Equipment Manufacturers) complain; they ostensibly fear having their intellectual property being shared with competitors. While we acknowledge their concerns, they are typically not competitive with after-market suppliers. Given their initial reluctance to assist when the failures originally occurred, we later find it hard to single-source to them after we have done all the work.

GOOD PLANTS AND THE AUDIT AND REVIEW PROCESS

Our introductory paragraphs hold the key to avoiding certain dilemmas expressed in this communication. Placing review engineers (MQA engineers) at a design contractor's office is part of the budgeted cost of a plant [2]. Compared to the incredibly shortsighted "as-cheap-as-possible" approach so often practiced nowadays, the incremental cost for the staffing described earlier has been estimated to require a multiplier of 1.05 or about a 5% add-on. As an example, certain bare bones and, from the future reliability point of view, very risky, seat-of-the-pants plant design may cost $600,000,000. In contrast, the approach whereby reliability is being assessed and imparted at the inception may cost $630,000,000.

Over the decades, we also found that an appropriation request for the *larger* amount would *not* be rejected by sensible top managers or boards of directors. Common sense tells us that it is highly unlikely for only the *lesser* of the two budgets to meet the ultimate owner's return-on-investment criteria. That said, good reliability professionals ascertain that the design engineering firm's cost-estimating manuals do *not* list only the cheapest machines available. Intelligent project design bases or project cost estimates must reflect reliable equipment. Therefore, competent reliability professionals are hard at work conveying these requirements to management.

Having one's engineers at the design contractor's facility ensures that cross-sectional drawings of equipment in your plant will be made available for audit purposes before you buy equipment such as heat exchangers or rotating machinery. The audit will have to take place at the vendor's facilities, and your designated engineers would participate in such audits. As to an existing plant, it might not be possible or allowed to copy the vendor's drawings, but the purchaser is entitled to see what is being offered. The vendor or his designated representative must somehow review drawings and bills of materials with you.

Cross-sectional drawings with critical dimensions *removed* (so as not to give away the vendor's competitive advantage) must be part of the purchase order and are typically listed as a condition of sale [3]. Moreover, a good contract is structured to withhold nominally 15% of the total amount due until the ultimate owner has received all specified documents. What to review is spelled out on 90 pages of appendix material in Ref. [4].

Having full manufacturing drawings stored at a local authorized repair facility is another option that many best-of-class companies have exercised over the years. For instance, in the 1970s, a world-renowned overseas manufacturer of mechanical drive steam turbines agreed to place the proprietary manufacturing drawings on microfiche and store them near the purchaser's plant site in the United States. Today, all of the material would fit on a thumb drive, and one would usually predefine and prenegotiate who the custodian of electronic files is.

As to accurate piping and instrumentation diagrams (P&IDs), these should—of course—reflect the as-built plant. Without them, one cannot safely operate a modern facility. The ultimate installation site must have access to these process-related documents, and we believe this fact is rarely questioned.

On existing heat exchangers and machines, one might consider hiring a competent reverse-engineering specialist and give the OEM a choice: either the OEM works with you (the asset owner) or you entrust the reverse-engineering and modification jobs (which you have apparently engineered) to someone else. It's hard to believe that heat exchangers contain proprietary elements, but marketers may try to claim otherwise.

Machinery reliability audits and reviews can be a tremendously worthwhile investment as long as they are performed by experienced engineers. Of course, this presupposes that a perceptive project manager or owner's representative will see to it that the resulting recommendations are, in fact, implemented.

DON'T TAKE RISKS WITH RELIABILITY

In the 1980s, an early edition of Ref. [3] noted that a petrochemical project in the $500,000,000 range (typical of the mid-1970s) would optimally staff machinery reliability *audits* with four engineers for a 4-month period, and machinery reliability *reviews* with two engineers for a period of 2–3 years. At that time, the total cost of these efforts was found in the league of $650,000–$850,000. While this sounded like a lot of money in 1975, the plant owners contrasted it with the value of a single start-up delay day, say $400,000, and the potential cost of 2 unforeseen days of downtime —perhaps accompanied by the thunder of two tall flare stacks for the better portion of 2 days.

Of course, reliability assurance efforts made before delivery of the machinery are more cost effective than postdelivery or post-start-up endeavors aimed toward the same goals. However, questions may remain how to optimally conduct these efforts, how to man them, and which components or systems to subject to close scrutiny. This is where an analysis of available failure statistics and a close review of available checklists will prove most helpful. A review of the failure statistics of rotating machinery used in modern process plants will also help determine where the company's money should be spent for highest probable returns. Moreover, failure statistics can often be used to determine the value of and justification for intelligent resource allocation.

A thorough review of proposed machinery and comparison of the different proposals is of immense value. Fig. 32.1.1 depicts a multistage hydrocarbon feed pump on the test stand; the image is of interest in more ways than one. Upon closer examination, the MQA engineer will note not only closely-spaced stud bolting at the horizontal split line where the two casing halves meet. There are also a number of bolts farther up on the casing. What purpose do they serve and how does this quality pump differ from a competing offer? What's the pump's incremental cost over a less expensive one and what's that worth to a potential owner-user? That's the essence of MQA.

FIG. 32.1.1 A multistage hydrocarbon feed pump on the test stand. *(Courtesy Dengyosha Machine Works (DMW), Tokyo, Japan.)*

SUCCESSFUL INTERFACING ASSURES UPTIME GOALS ARE MET

In the plant equipment area, owner-contractor interfacing activities typically will occur during the following functional project steps:

- flow sheet and design specification review,
- vendor selection,
- preorder review with vendors,
- vendor drawing review,
- inspection and test at vendor's site,
- review of spare parts and documentation,
- equipment field handling and storage,
- field installation and equipment turnover, and
- precommissioning.

Two interface functions common to the steps just listed are of particular interest in our context. They are *review* and *inspection*. Both have a high potential for assuring project service factor goals are met.

Review functions are sometimes equated with audit activities. Reliability *audits* are defined as any rigorous analysis of a contractor's or vendor's overall design after purchase order issue and before equipment fabrication begins. Reliability *reviews* are defined as a less formal, ongoing assessment of component or subsystem selection, design, execution, or testing. They are aimed at insuring compliance with all applicable specifications. These reviews will also judge the acceptability of certain deviations from applicable specifications. Moreover, an experienced reliability review specialist will provide guidance on a host of items that either could not be or simply had not been specified in writing. He will draw on his field of expertise and start-up experience when making recommendations aimed at ensuring a successful plant start-up and safe, reliable operation of the equipment for years to come. Such a specialist must, by definition, work the interfaces between owner and contractor.

FLOW SHEETS AND SPECIFICATIONS

Flow sheets and specifications are the technical bases for interfaces with contractors' representatives. What are technical specifications? They are quite simply documents that define in writing, together with drawings and flow sheets, plant, and equipment that a purchaser wants a vendor to supply or, conversely, that which a bidder is prepared to offer to a buyer. It defines the technical conditions of a contract; it does not concern itself with the commercial contract conditions that are usually the subject of a separate covering letter or other documents between the two parties. Writing specifications is a skill of technical communication in print. The owner's representative must write a clear, precise statement of what he wants. The contractor or supplier must read, understand, and make an offer in line with the owner's requirements.

The use of a specification requires careful consideration of needs. For example, simply specifying the maximum flow rate of a cooling water pump may be inadequate if the pump is expected to operate over a range of flow rates. Obviously, cooling water system resistance changes, and where a pump may work adequately in laminar flow, it may vibrate in turbulent flow and vice versa.

One thing is certain: building reliability into a plant "up front" is much less expensive than postcommissioning root cause failure analysis (RCFA) after the fact and then arguing over remedial action. Best-of-class companies are fully aware of the value of spending money on intelligent asset selection. They have become best-of-class companies by allocating resources for MQA.

REFERENCES

[1] H.P. Bloch, A Practical Guide to Compressor Technology, second ed., John Wiley & Sons, Hoboken, NJ, 2006. ISBN: 0-471-727930-8.

[2] H.P. Bloch, F. Geitner, Compressors: How to Achieve High Reliability and Availability, McGraw-Hill Companies, New York, NY, ISBN: 978-0-07-177287-7, 2012.

[3] H.P. Bloch, Machinery Reliability Improvement, third ed., Gulf Publishing Company, Houston, TX, ISBN: 0-88415-661-3, 1998.

[4] H.P. Bloch, F. Geitner, Machinery Uptime Improvement, Elsevier-Butterworth-Heinemann, Stoneham, MA, ISBN: 0-7506-7725-2, 2006.

Chapter 32.2

Buy From Knowledgeable and Cooperative Pump Vendors

INTRODUCTION

Using pumps as just one widely used example asset, the reader is alerted to an important fact: with relatively few exceptions, purchasing from the lowest bidder is rarely conducive to getting lowest possible life-cycle performance of an asset. Picking the right bidder requires forethought, specifications, and intelligent follow-up. All of these require effort; all of these are invoked or pursued by best-of-class user-purchaser-owner companies.

EQUIPMENT SELECTION DEMANDS FORETHOUGHT

Selecting from among the many pump manufacturers may be tricky. That said, picking the right bidders may well be an important prerequisite for choosing the best pump. Note that three principal characteristics identify capable, experienced vendors [1]:

- They are in a position to provide extensive experience listings for equipment offered and will submit this information without much hesitation.
- Their centrifugal pumps enjoy a reputation for sound design and infrequent maintenance requirements.
- Their marketing personnel are thoroughly supported by engineering departments. Also, both groups are willing to provide technical data beyond those that are customarily submitted with routine proposals.

FIG. 32.2.1 Barrel-type (vertically split) refinery pump with up-oriented suction and discharge nozzles. *(Courtesy Dengyosha Machine Works, Tokyo, Japan.)*

Vendor competence and willingness to cooperate are shown in a number of ways, but data submittal is the true test. When offering refinery or petrochemical process pumps (Fig. 32.2.1) that are required to comply with the standards of the American Petroleum Institute, that is, the latest edition of API-610, a capable vendor will make diligent efforts to fill in all of the data required in API specification sheets. However, the real depth of technical know-how will show in the way a vendor-manufacturer explains exceptions taken to API-610 or supplementary user's specifications. Most users are willing to waive some specification requirements if the vendor is able to offer sound engineering reasons, but only the best qualified centrifugal pump vendors can state their reasons convincingly.

Pump assembly drawings are another indispensable documentation requirement. Potential design weaknesses can be discovered in the course of reviewing dimensionally accurate cross-sectional drawings. Examination and review of ISBN 0-88173-720-8 or past "HP in Reliability" columns or familiarization with and application of "MQA" (see index words) will disclose dozens of areas to be questioned. There are two compelling reasons to conduct this drawing review during the bid evaluation phase of a project: First, some pump manufacturers may not be able to respond to user requests for accurate drawings after the order is placed; second, the design weakness could be significant enough to require extensive redesign. In the latter case, the purchaser may be better off selecting a different pump model [2].

CHEAP PUMPS ARE RARELY A WISE CHOICE

It is intuitively evident that purchasing the least expensive pump will rarely, if ever, be the wisest choice for users wishing to achieve long run times and low or moderate maintenance outlays. Although a new company may, occasionally, be able to design and manufacture a better pump, it is not likely that such newcomers will suddenly produce a superior product. It would thus be more reasonable to choose from among the most respected existing manufacturers, that is, manufacturers that currently enjoy a proven track record.

The first step should therefore involve selecting and inviting only those bidders that meet a number of predefined criteria. Here's the process by which to determine acceptable vendors for situations demanding high reliability:

- Acceptable vendors must have experience with size, pressure, temperature, flow, and service conditions specified.
- Vendors must have proved capability in manufacturing with the chosen metallurgy and fabrication method, for example, sand casting, weld overlay, etc.
- Vendor's "shop loading" must be able to accommodate your order within the required time frame (time to delivery of product).
- Vendors must have implemented satisfactory quality control and must be able to demonstrate a satisfactory on-time delivery history over the past several (usually 2) years.

- If unionized, vendors must show that there is virtually no risk of labor strife (strikes or work stoppages), while manufacturing of your pumps is in progress.

USE SUPPLEMENTAL SPECIFICATIONS

The second step would be for the owner/purchaser to do the following:

- Specify for low maintenance. As a reliability-focused purchaser, you should realize that selective upgrading of certain components will result in rapid payback. Components that are upgrade candidates have been identified in an HP in Reliability column since 1990. Be sure you specify those. To quote a simple example, review failure statistics for principal failure causes. If bearings are prone to fail, realize that the failure cause may be incorrect lube application or lube contamination. Address these failure causes in your specification.
- Evaluate vendor response. Allow exceptions to the specification if they are both well-explained and valid.
- Future failure analysis and troubleshooting efforts will be greatly impeded unless the equipment design is clearly documented. For a certainty, your plant will require pump cross-sectional views and other documents in future repair and troubleshooting work. Do not allow the vendor to claim that these documents are proprietary and that you, the purchaser, are not entitled to them.
- Therefore, place the vendor under contractual obligation to supply all agreed-upon documents in a predetermined time frame and make it clear that you will withhold 10 or 15% of the total purchase price until all contractual data transmittal requirements have been met.
- On critical orders, contractually arrange for access to a factory contact. Alternatively, insist on the nomination of a "management sponsor." A management sponsor is a vice president or director of manufacturing or a person holding a similar job function at the manufacturer's facility or head offices. You will communicate with this person for redress on issues that could cause impaired quality or delayed delivery.

Following these guidelines will give best assurance of meeting the expectations of reliability-focused owner/purchasers.

REFERENCES

[1] H.P. Bloch, Machinery Reliability Improvement, third ed., Gulf Publishing Company, Houston, TX, 1998.

[2] H.P. Bloch, A. Budris, Pump User's Handbook: Life Extension, fourth ed., Fairmont Publishing Company, Lilburn, GA, 2013.

Chapter 32.3

Selection of Better Equipment Starts With Understanding Available Upgrade Options

INTRODUCTION

The reliability manager of a major multinational petrochemical company asked each of the maintenance departments within the corporation's businesses to submit the top 10 sources of production loss over the past few years. When his staff analyzed the various replies, it was discovered that a majority of the corporation's problems are related to equipment that had been recently commissioned as part of major projects. Having discovered this common thread, the reliability staff asked the maintenance departments how much influence they had in equipment design and selection during the early stages of the project. Due to resource limitations, these maintenance departments had not been heavily involved. The letter asked how others manage this matter during earlier stages of major projects. "Are maintenance/reliability experts included in the project teams or is there any oversight from operations/maintenance on design quality and equipment selection?" he asked.

HOW BEST-OF-CLASS PERFORMERS DO IT

Best-of-class performers manage to avoid the issue mentioned by this reliability manager. They limit their bidders' lists to a few good and consistent vendors. Then they add a corporate specification to their invitation to bid and make it clear to the bidder that the specification clauses must be met and that the purchaser understands that the equipment specified will cost more than the vendor's standard offering.

However, prepare for setbacks, and take a good look at, for example, pumps. In early 2009, a facility in North Dakota had specified better pumps but was turned away by every one of the companies that had responded to its invitation to bid. The disappointed owner-user company then asked us to consider putting the attributes of better drive-ends for pumps in a short article.

There are a number of reasons why well-versed reliability engineers will decline to accept pumps that incorporate the drive-end shown in our earlier Chapter 4.5, Fig. 4.5.8. The short summary of reasons is that he or she takes seriously the obligation to look for *ultimate, not just short-term*, cost of ownership. Serious reliability professionals have learned long ago that price is what one pays and value is what one gets. These professionals are allowed to provide input before a project is finalized, and their input starts at the specification stage. Anyway, without solid input, the pump components of Chapter 4.5, Fig. 4.5.8 might be supplied. The entire drive-end of that previous illustration does not measure up to the expectations of reliability-focused owner-operators. The issue is relevant and may even negatively affect US pump manufacturers unless they pay very close attention.

Here are the details on where and why to pay close attention:

1. Oil rings (Chapter 4.5, Fig. 4.5.8, also subject category 24) are rarely found acceptable by best-of-class plants. Unless perfectly engineered, installed, maintained, and immersed (in the right oil), these rings can skip around and will abrade. In other words, unless the shaft system is truly horizontal, ring immersion in the lubricant is as required, and ring eccentricity, surface finish, and oil viscosity are all within limits, these rings will not perform in accordance with the design intent. Taken together, these necessary parameters or conditions are rarely found within safe limits in real-world situations. Therefore, some users prefer a stainless steel flinger disk fastened to the shaft (Chapter 4.5, Fig. 4.5.2). The disk contacts the oil and flings it into the bearing housing. The disk (OD) must exceed the thrust bearing's outside diameter.
2. In order to allow the steel flinger disk to pass through the housing bore upon assembly, the back-to-back oriented thrust bearing set must be placed in a suitably sized cartridge.
3. Chapter 4.5, Fig. 4.5.8, does not show bearing housing protector seals. An advanced version of such a protector seal must be supplied for both the inboard and outboard bearings. Lip seals are not good enough, and neither are old-style rotating labyrinth seals with "dynamic" O-rings in close proximity to sharp-edged geometries.
4. As a matter of routine, the housing or cartridge bore must have a passage at the 6 o'clock position to allow pressure equalization and oil movement from one side to the other side of the bearing. Note that such a passage is shown in Chapter 4.5, Fig. 4.5.8, for the radial bearing but needs to be added to the thrust bearing set also.
5. Once items (1) through (4) have been implemented, the breathers (or vents) are no longer needed. They should be removed, and one of them should be plugged.
6. A pressure-balanced constant level lubricator should be supplied, and its generously sized balance line should be connected to the second of the two breather ports [1].

Of course, even the failure-prone configuration depicted earlier in Fig. 4.5.8 will generally work "reasonably well," but that characterization is no longer acceptable. Reliability professionals and pump manufacturers must do better and should know how to do better. Yes, we realize pump manufacturers are able to submit test stand data certifying that things work even if the aforementioned upgrade issues are disregarded. Then again, realistic engineers can prove how and why things tend to malfunction in the real world. If such malfunctioning occurs just 10% of the time, considerable repair expenditures will result, and the maintenance budget will be stressed.

Perhaps we might all agree on one premise: as we get further and further away from solid training and from taking the time needed to do things right, we become ever more vulnerable. One way to counteract this vulnerability is by designing-out maintenance. The users have to demand designs that accomplish this goal and have to be willing to pay for these upgrades.

REFERENCE

[1] H.P. Bloch, A. Budris, Pump User's Handbook—Life Extension, fourth ed., Fairmont Press, Lilburn, GA, ISBN: 0-88173-720-8, 2006.

Chapter 32.4

Unreliability, Engineers, and Lawyers

INTRODUCTION

Machines are not always reliable; people make mistakes for quite a number of reasons. While we are not lawyers, we may have knowledge of how machines work; we may have been assigned to a plant or unit where an incident happened. In that case, we may be compelled to answer questions from teams of attorneys who represent different parties in a law suit. Here are a few items or pointers that may benefit us in many ways.

BACKGROUND

On more than one occasion and while employed by a major multinational petrochemical company, I had to answer questions relating to machine failure incidents. Years later, I chose to assist attorneys or law firms in understanding how certain machines function and why certain machines malfunction. On a number of the so-called causes (*sic*), my participation and involvement could be called providing litigation support. To me, litigation support is a term that conveys that we might occasionally choose to work for law firms that pay engineers to explain why parts fail and how humans have something to do with failed equipment. At all times when it was my obligation to do the explaining, I thought it's fair to assume the obligations of opposing law firms were to serve their respective clients and to make (or save) money for these firms. However, just to be clear, as a professional engineer, I consider it my obligation to be nothing but an advocate of the truth and not allow myself to be turned into a blind supporter-advocate of either litigant.

Generally speaking, 50% of the lawyers make one claim; the other 50% make some other claim. That's the basis of the adversarial legal system, and it's not my place to critique it here. Unfortunately, engineering “experts” sometimes become advocates, and that's not good. Don't get me wrong; I like to get paid for my work, but I decline to get involved with someone who doesn't want me to say it as I see it or who argues about the validity of the basic laws of physics.

Let's suppose you, too, agreed not to compromise your integrity. You accepted work from the side that allows you to be truthful and to divulge your expertise in a forthright and candid fashion. Well, you might still have to sit across from the opposing side during a deposition. To be deposed no longer has the same meaning it had in the original English where the term meant that someone was sent into exile. Today, being deposed means that one is asked to give answers under oath. The folks that question you may have an agenda that we can't even allude to without the risk of being sued for slander.

An opposing attorney may want to establish the inadequacies of your work background or performance. Specific questions may relate to the expert's education, expertise, publications, experience, lectures, previous involvement with law firms, compensation, and the like. Even a person's accent was brought in on one occasion.

The questioning attorney will try to freeze the expert into certain positions on all relevant facets of a case and then explore the adequacies or inadequacies of the expert's investigations, the correctness of assumptions, the titles of books and articles consulted, and so forth. Much time is spent on obtaining concessions from the expert, and if he's ever trapped in a contradiction, the questioning attorney may ask to have him impeached—dismissed from the case.

SURVIVING A DEPOSITION

This is not the place to relate any of a number of utterly absurd statements that I've heard while sitting in on engineers giving depositions. Suffice it to say that I've seen more than one engineering "expert" come up with astonishing claims; one even signed an affidavit that simply didn't represent the facts of the case. When his naivety was quickly exposed by opposing counsel, the embarrassed engineer squirmed. All present in the room perceived the ethics of the engineering profession to have moved down another notch. On that day, I decided to jot down my thoughts on how engineers can survive a deposition without compromising their integrity:

1. Stick to the truth. Your attorney knows your findings and your sentiments. If he does not like them, he should not have you sitting there being deposed.
2. Listen to the question. Make sure that you understand it. When in doubt, ask the court reporter to read it back to you.
3. Again, listen to the question! Occasionally, there may be a bit of trickiness or a potential trap in the questioner's phrasing. Ask the attorney to rephrase the question, but do not let him/her use questions to rephrase your answer.
4. Answer the question. If need be, slow down; think it through. Always listen to the question in full before giving your answer.
5. Do not volunteer information and do not be "routinely helpful." By the same token, do not allow inadvertent misconceptions to prevail if these would affect your integrity and professionalism.
6. If you would like to take a break, ask for one. Do not talk with the opposing counsel during coffee breaks. You also cannot confer with your attorney during breaks.
7. It is quite proper to answer "I do not know," assuming, of course, that is the truth. If indeed you do not remember, say so without any reluctance.
8. Listen to your attorney's objections. They could serve one or more of the following purposes: (1) He/she could be making a legal point; (2) he/she could be "sending you a signal" (body language, perhaps); and (3) or he/she may want to break the questioner's stride.

9. Realize that the questioning attorney's statement "nonresponsive" generally means that he or she did not like your answer, no more, no less.
10. Include themes or underlying thoughts in your answer if you have to redirect the deposition toward important points that the prevailing line of questioning might not allow you to make otherwise. Make you answer "unreadable at trial" by inserting your expert knowledge in this fashion.
11. If you are being interrupted, complete your answer. Remember that you have the right to fully answer the question and that you are never restricted to "yes" or "no" replies if doing so would put a spin on the truth.
12. Answer repetitive questions consistently and truthfully.
13. Remember the real purpose of the deposition is to allow you to clearly state your expert opinion within the basic framework of the question and answer format deemed appropriate by our legal system. The purpose is not to let you give this opinion without structure, or to convince lawyers, or to score points, or to vent frustration.
14. Avoid mistakes and inconsistencies that could damage your credibility at trial and your reputation overall.
15. Display proper posture and positive voice.
16. Do not let past questions worry you. Perfection is impossible. Your goal should be excellence.
17. Do not be concerned with upcoming questions.
18. Do not proffer a guess. Only tell what you know, no more, no less.
19. When shown a document, read the entire document.
20. Tell the truth. Stick to the truth.

THE HUMAN FACTOR

Remember that machines fail for a reason. Without a single exception, the reasons are always traceable to decisions, commissions, omissions, or whatever prompted a worker, manager, supervisor, or chief executive officer to exercise judgment or to perhaps decline the use of judgment. For your part, resolve to get involved in litigation support only in causes and subjects that you have mastered. Above all, be convinced that sticking to the truth will never fail you, but becoming an advocate of anything other than the truth may cause serious grief.

Chapter 32.5

How to Trim-Cut Closed Pump Impellers

INTRODUCTION

Reliable plants will have used machinery quality assessment (MQA). Originally conceived for new plant design and construction, MQA now extends beyond what ordinarily meets the eye. We use the trimming of closed impellers as but one of numerous examples of risk-lowering work procedures or component details that are examined by an experienced MQA engineer. Perhaps including it here as a chapter under the overall subject category 32 ("Reliable Plants") also makes the point that a facility should groom, nurture, and reward individuals who can ultimately contribute on this level of detail. But we wish to relate some specifics here.

The hydraulic relationships between pump speed, differential pressure, and power were already well understood in the mid-19th century; they are restated in the formulas given below. What was less well understood are certain refinements that may save power or improve the vibration behavior of centrifugal pumps—quite obviously the most widely used fluid machine type. So, while one might easily conclude that these are maintenance details, we think it's really an important reliable plant detail, one more fitting example out of hundreds. Throughout all of these many examples, organization, motivation, understanding, training, learning, and accountability are interwoven. Impeller trimming, seemingly as mundane a subject as it can possibly get, brings to the reader's attention the interwoven nature of details that make plants reliable.

Experienced reliability professionals will implement least-risk solutions whenever possible. A case in point relates to impeller trimming, meaning reducing the diameter of a fixed-speed centrifugal pump's impeller. Diameter reductions accommodate a lesser head (lower differential pressure) requirement. An approximate relationship is given by the common fan laws, although these laws are only approximate:

$$H' / H = D'^2 / D^2$$

In this expression, H and D are the original values of head and impeller diameter; H' and D' are the reduced (after machining) values of H and D. The pump user or designer is given H'/H and can measure the original impeller diameter, D. With this information, it will be possible to calculate the required impeller diameter D', taking into account the exponential relationship between D and D'.

Good reference texts tend to be more specific, and up-to-date reliability professionals may want to implement all reasonable steps to optimize process pump performance [1]. Widely publicized guidelines for dimensions "A" and "B" are found in Refs. [2,3]. Pioneering fluid machinery expert Dr. Elemer Mackay had made these design modifications after verifying that they improved the quality of manufacturing and repair of high-energy water pumps. His mid-1970s observations on the effects of closely controlling interior pump clearances found their way into the process industries when Amoco's Ed Nelson asked Dr. Makay to solve issues with large process pumps at an oil refinery in Texas City.

Makay's clearance ranges for "A" and "B" (Fig. 32.5.1) were later collected in many other references, among them [3]. But, while modifications to the gaps labeled "A" and "B" in Fig. 32.5.1 are generally of primary interest to pumps in the size range exceeding 300 hp, plain impeller trimming is done in virtually all centrifugal pumps, both large and small. Such trimming of impellers predominates in applications with fixed-speed alternating-current induction motor drivers. However, when trimming, the machine shop or repair provider can choose from at least four trimming options, (A) through (D), in Fig. 32.5.1.

TRIMMING USING "FIELD WISDOM"

Trimming across the entire impeller periphery (Fig. 32.5.1A) is clearly the fastest and least expensive. But, when trimming in accordance with (a), it would be especially important to again realize that the fan laws are approximations. Field-wise reliability professionals want to avoid cutting off too much of the vane tips. So, after using the fan laws, they make the actual cuts only 70% of whatever had been calculated based on these laws. To illustrate, if the fan laws had resulted in a diameter cut from originally $D = 9$ in. to now $D' = 8$ in., a field-wise pump expert would trim to a new diameter of 8.3 in. and then check the pump's new performance. Also, the experienced pump person would readily surmise that the somewhat more time-consuming modifications of Fig. 32.5.1B and C had certain advantages.

Dr. Makay and others explained the advantages of trimming impellers in the vane tip area only. The shop machinist would leave the shroud and backplate intact in order to minimize recirculation-related turbulent flow. This undesirable flow could be exacerbated when making the shroud and impeller backplate diameters identical with the trimmed vanes. The "vane area only" suggestion is shown in Fig. 32.5.1B. Trimming per Fig. 32.5.1B probably also yields a 0.7–1% efficiency gain that, depending on pump size and operating characteristics, could amount to thousands of kW-hours saved in a year's time. Therefore, trimming per Fig. 32.5.1B is recommended for virtually all types of closed ("shrouded") pump impellers. It is one of the many steps taken by those striving for a combination of failure risk reduction and energy conservation.

Many proactive pump professionals also implement trimming per Fig. 32.5.1C. They do so in the interest of lowering vibration risk and vibration severity on impellers with a prior vibration history. Of course, the unsupported shrouds (covers) might have to be "scalloped" in further efforts to separate operating frequencies and their multiples from the resonant frequencies of some shrouds [3]. These resonances could also be related to the number of vanes and an occasional lack of baseplate structural stiffness.

If you have an uncompromising interest in reducing all risks you will opt for trims per Fig. 32.5.1C although, as realists, we know that one would usually get away with (Fig. 32.5.1B). And if there is a (rare) resonance, we often call in a vibration expert and then—probably—weld a few braces into the support. That, then, will cost far more money than what it would have cost to go with (Fig. 32.5.1C) as a matter of routine. After working with the vibration expert, we would pat ourselves on the back for having solved a seemingly elusive vibration problem. A fortunate few might even get promoted for belatedly doing what they should have been doing all along.

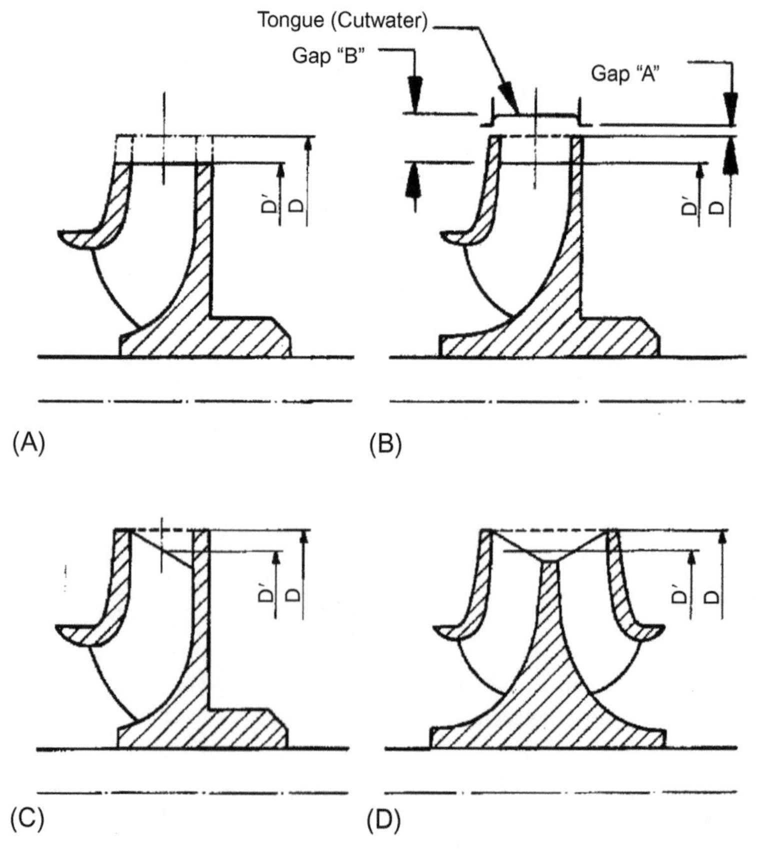

FIG. 32.5.1 Gap designations and impeller trim methods, including vane tips only (left); across the entire outside diameter (center); oblique cut (right), based on Ref. [1].

OK then, no irony: As professionals and value-adders, we should resist implementing Fig. 32.5.1A. If we follow best practices throughout, we will finally get on the road toward fully understanding why pumps fail. Even more important, we will inculcate a new mindset: doing it right is adding more value than doing things cheap or doing them fast. Having the new mindset is the only way to get to virtually zero defects in a few decades.

REFERENCES

[1] H.P. Bloch, Pump Wisdom—Problem Solving for Operators and Specialists, John Wiley & Sons, Hoboken, NJ, 2011.

[2] E. Makay, Problems encountered in boiler feed pump operations, in: Proceedings of 5th Texas A&M Turbomachinery Symposium, 1976.

[3] H.P. Bloch, A. Budris, Pump User's Handbook: Life Extension, fourth ed., Fairmont Press, Liburn, GA, 2014.

Chapter 32.6

Equipment Life Extension Involves Upgrades

INTRODUCTION

Every plant wants to get more life from its machinery. In fact, since about 1990, quite a number of start-up consulting companies have been formed to advise clients on equipment life extension. They use different approaches; some apply large-scale computer-based statistical methods; some are blending traditional estimates with risk-based analysis. All of these have merit, but none of them can provide all the answers with high precision. Still, the key ingredients of any useful endeavor are the asset's past failure history, examining nondestructive testing (NDT) data, and upgrading the weakest link.

EXAMPLES OF WHAT TO UPGRADE

Wherever failure history exists and where the root causes have been analyzed, authoritative answers on remaining life are possible. The same can be said for thoroughly evaluating NDT data. NDT helps giving focus to any attempt to determine remaining life.

On stationary equipment and piping, wall thickness is of great importance. Loss of material decreases the allowable pressure rating. Corrosion and erosion cause factors of safety to be lowered; continued operation then becomes risky. Thickness changes often occur at locations where fluid flow changes its direction; think of elbows. Changes in velocity such as at valves or near restrictions are of interest. Some can be investigated with NDT methods that certainly include X-ray imaging, among others. The extent of fluid-dependent corrosion can be estimated from coupons placed in piping and vessels.

On pumps, failure history and past repair data must be matched with a thorough understanding of upgrade measures that have been taken by successful "best-of-class" (BoC) organizations. BoC organizations, plants, and corporations view every repair event as an opportunity to upgrade. If feasible and cost-justified, they send their pumps to specialty firms that will go beyond repairing the asset. The specialty firm determines if an upgraded impeller would extend the mean time between repairs (MTBR) or increase the power efficiency of the pump. Instead of simply repairing, the owner company is thus aiming to combine maintenance and reliability. Advanced lube application will probably be part of it, as will the extension of oil replacement intervals made possible by synthetic lubricants and advanced bearing housing protection measures.

To what extent superior bearings (ceramic hybrids) are of value must be determined on a pump-by-pump basis. Perhaps a set of angular contact bearings with unequal contact angles should be installed in your problem pumps. The symmetrical sets of angular contact bearings mentioned in the most widely used pump standard may not perform adequately.

Similarly, determining the extent to which superior sealing technology (dual seals, Fig. 32.6.1) is of value must be determined on a service-by-service basis. However, as a general rule, the user industry's old notions about dual seals deserve to be reassessed. Sealing technology has made considerable progress in the past several decades. Virtually, all seals today are cartridge-style configurations, and braided packing is getting displaced by mechanical seals in the hydrocarbon processing (HP) and in the power generation and mining industries.

However, not all manufacturers of mechanical seals use the same acceptance test procedure for their products. A widely used industry standard (API 682) stipulates using air as a test gas for mechanical seal tightness. Of course, these seals are ultimately intended for safe containment of flammable, toxic, or otherwise hazardous liquids. While the standard's expectation is that leakage from those seals does not exceed 5.6 g/h, recent tests showed that merely following this easy testing routine can actually allow orders of magnitude more liquid to escape. It is, therefore, advisable to question seal vendors on the matter and to purchase only products that meet the purchaser's safety and reliability requirements. Surely, we want seals to leak no more than 5.6 g/h when we first install them in our pumps. This is just one more of many compelling reasons for equipment owners to be involved in mechanical seal selection and upgrading. Smart pump owners allow more than one seal manufacturer to be represented in their plant.

Lubricant application and standby bearing preservation are especially important in coastal humid climates and in dust-laden desert climates. Oil mist is the answer. The settling of foundations and pipe supports has to be taken into account. On steam turbines, blade stresses and water quality at the asset must be compared with those in successful long-running installations elsewhere.

In gearboxes, remaining life is largely examined by tooth loading (stresses on tooth face) and temperature measurements. In all instances, synthetic oils from the most experienced oil formulators will greatly extend gear life. (Oil additives are everything. They drive cost, and they drive life; oil cleanliness is equally important.)

Certain warehouse spares (gears, electric motors, etc.) should be upgraded if important and if doing so is likely to speed up recommissioning after an unanticipated future shutdown.

For compressors, engineers should consider the mentioned points. Valve technology and piston velocity are important comparison-worthy parameters on reciprocating compressors. Onstream performance tracking and observation of prior sealing technology are important for centrifugal compressors, and other important machines. They determine seal system upgrade potential. Never overlook couplings and the work procedures used to attach couplings to shafts. They tell a lot about remaining run length.

Whether one ultimately receives life extension guidance from individual consultants or billion-dollar consulting giants with applicable experience is of

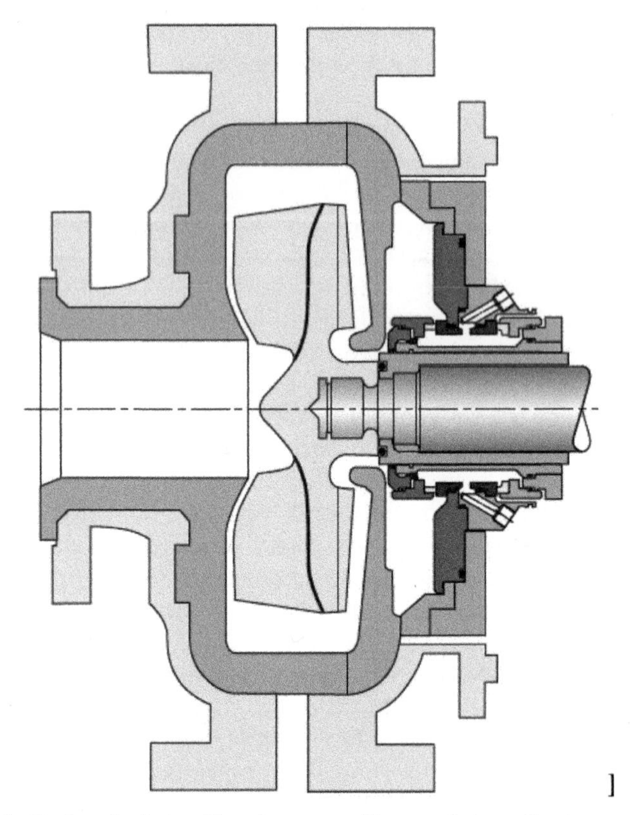

FIG. 32.6.1 Dual mechanical seal in a slurry pump. The space between the sleeve and the inside diameter of the two sets of seal faces is filled with a pressurized barrier fluid—usually clean water. *(Courtesy AESSEAL Inc., Rockford, TN, and Rotherham, UK.)*

no consequence, as long as there is the one common thread: determining where upgrades are possible. Doing so is an important step. It is the real key to imparting longer life to existing equipment and can often be accomplished at relatively low cost. Assessments of remaining life should include detailed advice on how to upgrade weak links. That then implies the following:

- The expert authoritatively spells out recommended upgrade components.
- Recommended upgrade procedures are explained.
- Facilities recommended to do the upgrading are defined.

In short, the entity involved in advising you on equipment life extension must understand the feasibility of component upgrades. Component upgrading is one of the keys to life extension, and deliverables that should be contractually agreed upon with the upgrade provider must be described. Be sure that the consulting company you've asked to give advice on equipment life extension includes these deliverables.

Chapter 32.7

Top Ways to Improve Pump Life

INTRODUCTION

Experienced professionals know that even the premier equipment standards used in the hydrocarbon processing industry (HPI) occasionally fall short of spelling out the best, most reliable, or safest available technology. A widely used industry standard starts out by stating: "the purchaser may desire to modify, delete, or amplify sections of this standard."

The preface to another standard is worded: "standards are not intended to inhibit purchasers or producers from purchasing or producing products made to specifications other than this one." In one more document, a well-known industry standards organization expressly declines any liability or responsibility for loss or damage resulting from the use of its standards or for the violation of any regulation with which the standards publication may conflict.

KNOW ABOUT BETTER BEARINGS AND BEARING PROTECTORS

For reasons of standardization and overall practicality, pump standards allude to thrust bearing sets comprised of two mirror-image mounted 40-degree angular contact bearings oriented back-to-back. In most cases and assuming the bearings receive adequate lubrication, these paired bearings will not be a problem. At high loads, however, one of the two bearings may skid; a skidding bearing will usually overheat and cause premature bearing failure. Perhaps an upgraded set of angular contact bearings, ones with unequal contact angles, should be installed in your problem pumps (see subject category 4). The symmetrical sets

FIG. 32.7.1 Laser alignment in progress. *(Courtesy Easy-Laser and Ludeca.)*

of 40-degree angular contact bearings mentioned in the most widely used pump standard will not always perform adequately. You may need to upgrade in situations where frequent failures are experienced with "standard" parts.

SUPERIOR BEARING HOUSING PROTECTOR SEALS WILL SURPRISE YOU IN SOME PUMPS

Although buying in accordance with proved industry standards is generally recommended, many clauses in these standards deserve clarifying statements. Bearing housing protector seals are a case in point; a good bearing protector seal is a must [1]. But here are a number of precautions and requirements that you should know about:

(1) If the pump manufacturer has provided a bearing housing design *with balance holes that equalize the pressures* on one side of a set of angular contact ("AC") bearings relative to the other side of an AC bearing set, then there will not be a problem when bearing protector seals are being retrofitted.

(2) If the pump manufacturer *has omitted* these internal balance passages, oil may properly reach all bearing elements *as long as the design incorporates an old-style or conventional wide-open labyrinth housing seal*. If one now adds a modern bearing housing protector seal, there will be *air stagnation* on that side. Therefore, air will flow more freely from the side of the angular contact bearing closest to the breather vent (usually near the center of a bearing housing), whereas little or no air flow will escape through a superior bearing housing protector seal. We often call unidirectional air flow "windage." Windage *may* become a factor, and the bearing protector seal often gets blamed, although the problem could have originated with the omission of housing-internal balance holes [1].

(3) Machines and other equipment can often perform with one or two deviations from perfect operation, perfect dimensions, etc. But whenever a few more deviations are added or if the deviations stack up, failures will usually result. Many repeat bearing failures on not-so-well-designed pumps have occurred for the above reasons. The appropriate course of action is to look at all factors, to study pertinent references, and to make sure there are internal pressure balance provisions and other reliability-improving steps. Always remember that improvement steps include using upgraded components, practicing uncompromising workmanship, using well-written procedures, and also ascertaining that these procedures are actually used. Once these pump improvement steps are confirmed, it is highly recommended to use the right bearings, the right lubricant, and lube application method. In line with that thinking, a reliability-focused owner-purchaser would use modern bearing protector seals and steer clear of the risky ones [1].

(4) For decades, best-available work procedures have included sound equipment installation and monitoring routines. Well-installed machinery requires less frequent monitoring than equipment that is installed without proper care. Perhaps the best illustrative example is seen in Fig. 32.7.1, laser alignment. Not only will properly aligned machines vibrate less than improperly aligned ones, but also properly aligned machines will invariably require less frequent monitoring than misaligned machines.

INSIST ON SAFEST POSSIBLE MECHANICAL SEAL TEST PROTOCOLS

You will benefit immensely if you allow the knowledge of a competent MQA engineer to be absorbed by your maintenance technician work force. For ways to allow this knowledge transfer, simply look up the training chapters of this book. Look in the index for "shirt-sleeve seminars."

Ask if in your mechanical seal procurement routine there is a requirement for seal testing by the seal manufacturer or perhaps by the individuals in charge of spare parts at your plant.

Good testing ensures that only installation-ready mechanical seals are ready for storage on your spare parts shelves. Explain to everyone that not all mechanical seals manufacturers utilize the same acceptance test procedure for their products. A widely used industry standard stipulates air as a gas for mechanical seal tightness testing. Of course, these seals are ultimately intended for the safe containment of flammable, toxic, or otherwise hazardous liquids. While the standard's expectation is that leakage from seals in critical services does not exceed 5.6 g/h, rigorous tests showed that merely following a presently allowed and relatively convenient air leakage testing routine can lead to wholly unexpected risks. The test protocol is found in one of the premier industrial standards; yet, using this test can actually allow orders of magnitude—in excess of 1000 g/h—of dangerous liquid to escape from a seal that has just passed the air leakage test. It is, therefore, advisable to question seal vendors on the matter and to purchase only products that meet the purchaser's well-reasoned safety and reliability requirements.

ASCERTAIN SAFETY AND RELIABILITY

Fortunately, many reputable reliability professionals pay more than lip service to equipment safety and reliability. They insist on seals leaking no more than 5.6 g/h of liquid when first installed in their process pumps. They insist on pumps with bearing sets that will not skid. Finally, they will review, specify, and inspect bearing housing internals for avoidable risks. These true professionals will insist on reversing whatever shortcuts may have been taken by persons unknown in times past.

REFERENCE

[1] H.P. Bloch, Pump Wisdom, John Wiley & Sons, Hoboken, NJ, 2011.

Subject Category 33

Repair Quality and Statistics

Chapter 33.1

Reasonable Pump Life Assessed from Published Statistics and Proprietary Consulting Access

EXAMINING PUMP REPAIR RECORDS AND MTBF

Process pump failure statistics are often translated into mean time between failure (MTBF). For what it's worth and so as not to get enmeshed in arguments, many of the best-practice plants in the time period of the early 2000s simply took all their installed pumps, divided this number by the number of repair incidents, and multiplied it by the time period being observed. This is still an appropriate practice in 2016. Taking, as an example, a well-managed and reasonably reliability-focused US refinery with 1200 installed pumps and 156 repair incidents in one year, the MTBF would be (1200/156) = 7.7 years. The refinery would count as a repair incident the replacement of parts—any parts—regardless of cost. In this case, a drain plug worth $3.70 or an impeller costing $7000 would show up the same way on the MTBF statistics. Only the replacement of lube oil would not be counted as a repair.

The best-practice plant's total repair cost for pumps would include all direct labor, materials, indirect labor and overhead, administration cost, the cost of labor to procure parts, and even the prorated cost of pump-related fire incidents. There are references to the stated average cost of pump repairs: $ 10,287 in 1984 and $ 11,000 in 2005. Newer estimates remained in the range of $11,000–$12,000. The normally expected inflation multipliers seem to cancel because earlier detection of flaws now limits the extent of damage. Such earlier detection is thought to be made possible through refinements in instruments and analyzers:

- It has been reasoned that predictive maintenance and similar monitoring strategies have led to corrective maintenance interventions taking place before failures became too severe. Quality control is practiced by best-of-class

Petrochemical Machinery Insights. http://dx.doi.org/10.1016/B978-0-12-809272-9.00033-5

FIG. 33.1.1 Quality control is one of the key contributions of a competent pump repair shop. (*Courtesy Hydro Inc., Chicago, IL, www.hydroinc.com.*)

TABLE 33.1.1 Pump Mean Times Between Failures

ANSI pumps, average, the United States	2.5 years
ANSI/ISO pumps average, Scandinavian P&P plants	3.5 years
API pumps, average, the United States	5.5 years
API pumps, average, Western Europe	6.1 years
API pumps, repair-focused refinery, developing country	1.6 years
API pumps, Caribbean region	3.9 years
API pumps, best-of-class, US refinery, California	9.2 years
All pumps, best-of-class petrochemical plant, the United States (TX)	10.1 years
All pumps, major petrochemical company, the United States (Texas)	7.5 years

companies; these user-owners understand the importance of seeking out and predefining competent pump repair shops (CPRS), Fig. 33.1.1.

- Using the same bare-bones metrics and from published data and observations made in the course of performing maintenance effectiveness studies and reliability audits in the late 1990s and early 2000s, the mean times between failures of Table 33.1.1 have been estimated. As of 2016, they were still considered reasonably accurate.

OTHER STUDIES OF PUMP STATISTICS

We are indebted to, Gordon Buck, the John Crane Company's chief engineer for field operations in Baton Rouge, Louisiana. He had examined the repair records for a number of refinery and chemical plants and obtained meaningful reliability data for centrifugal process pumps. A total of 15 operating plants having nearly 15,000 pumps were included in his survey. The smallest of these plants had about 100 pumps; several plants had over 2000 pumps. All plants were located in the United States. Also, all plants had some sort of pump reliability program in progress; for mechanical seals, they typically aimed for the target MTBFs in Table 33.1.2.

Several of these programs could be considered as "new," others as "renewed" and still others as "established." Many of these plants, but not all, had an alliance contract with John Crane. In some cases, the alliance contract included having a manufacturer's technician or engineer on-site to coordinate various aspects of the program [1].

TABLE 33.1.2 Suggested Refinery Mechanical Seal Target MTBFs

Target for Seal MTBF in Oil Refineries	
Excellent	>90 months
Very good	70/90 months
Average	70 months
Fair	62/70 months
Poor	<62 months

Not all plants are refineries, however, and different results can be expected from different locations. In chemical plants, pumps have traditionally been "throw-away" items as chemical attack limited life. Things have improved in recent years, but the limited space available in DIN and ASME stuffing boxes does limit the type of seal that can be fitted to more compact and simple versions. Lifetimes in chemical installations are generally believed to be around 50–60% of the refinery values.

TARGET PUMP AND COMPONENT LIFETIMES

Based on the lifetime levels being achieved in practice in 2000 and combined with the known "best practice" as outlined in more current consulting updates, the target component lifetimes of Table 33.1.3 are recommended and have often been achieved.

TABLE 33.1.3 Realistic Target Pump and Component Lifetimes

	Refineries	Chemical and Other Plants
Seals		
Excellent	90 months	55 months
Average	70 months	45 months
Couplings	All plants	
	Membrane type	120 months
	Gear type	60 to 80 months
Bearings	All plants	
	Continuous operation	60 months
	Spared operation	120 months
Pumps	All plants	
Based on series system calculation		48 months

It should again be emphasized that many plants are reaching these "target levels." In fact, informal statistics obtained from one of the world's largest multinational oil companies in 2015 easily exceeded these targeted values by approximately 50%. Nevertheless, to ultimately reach the stipulated overall pump lives, the pump components themselves must be operating at the highest levels. An unsuitable seal with a lifetime of 1 month or less will have a catastrophic effect on pump MTBF, as would a badly performing coupling or bearing.

REFERENCE

[1] H.P. Bloch, A.R. Budris, Pump User's Handbook—Life Extension, fourth ed., Fairmont Publishing, Lilburn, GA, ISBN: 0-88173-720-8, 2014.

Chapter 33.2

How to Improve Equipment Repair Quality

INTRODUCTION

Florida-based Philip Crosby Associates, Inc. was a management consultant who enjoyed an excellent reputation for advising industrial managers on how to achieve profits through quality. He lived, defined, and taught the "absolutes of quality management" [1] by emphasizing the following:

1. Quality has to be defined as conformance to requirements, not as goodness.
2. The system for causing quality is prevention, not appraisal.
3. The performance standard must be zero defects, not "that's close enough."
4. The measurement of quality is the price of nonconformance, not indexes.

If we could just imitate Crosby, we would stay well ahead of the game.

SPELLING OUT THE GOALS

Some of the best names in the HPI have Crosby's statements framed, mounted, and displayed in their shops, training rooms, or reception offices. They are sincere in the desire to achieve tangible quality objectives, but may not necessarily understand why lasting progress has, so far, proved elusive. Let me elaborate on the dilemma and make a few suggestions.

On the first item, few companies truly quantify their requirements. Requirements relating to antifriction bearings for pumps, for instance, should be "quantified" to the extent that certain bearing configurations, or cage materials, or internal clearances are disallowed in some centrifugal pumps. In other words, quality can only be achieved with well-defined and technically accurate procurement specifications. Similarly, no quality without stipulating—"quantifying"—and enforcing the required shaft-to-bearing fit tolerances, mounting methods, and use of appropriate assembly tools. And that is precisely where many companies deprive themselves of achieving quality: they do not recognize that appropriate training is an indispensable prerequisite to "conformance to requirements." There is often no understanding that manufacturers may have made mistakes or have delivered products that were incorrectly assembled. These manufacturers may not own up to their mistakes; some other repair facility has to find and rectify the flaw, Fig. 33.2.1. Time needs to be allocated; the root causes of flaws need to be found, and rectifying them may take more than a day. Rectification will need planning. Consider Fig. 33.2.1, and observe what happens in such cases. Three pump stages with impeller exits relative to diffuser entries are depicted in misaligned (above centerline) and properly aligned (below centerline) conditions. A non-OEM will be eager to point out such deficiencies and recommend corrective measures.

Item two tells us quite correctly that the system for causing quality is prevention, not appraisal. Vibration monitoring as most often practiced is appraisal. Granted, you will make some money by finding defective bearings, but you would make orders of magnitude more money by preventing bearings from failing in the first place. Failures are avoided by understanding that bearings (or any other parts) fail for a reason. We proceed to finding the reason and by systematically eliminating the root causes of bearing failures. This last goal can best be achieved by training. Unfortunately, it's usually the training budget that gets cut first in times of economic hardship, or it is the quality of training that suffers when companies hire contract trainers or consultants on the basis of low bid rather than demonstrated competence and experience.

"That's close enough," we often hear from the folks who have been asked to perform, say, machinery alignment or proper foundation grouting. Don't let them get away with it. Explain how with certain types of couplings a misalignment of "x" mils per inch of distance shaft-end-to-shaft-end will multiply the load on adjacent bearings by a quantifiable factor; explain also how bearing and coupling life expectancy will go down as the cube of this load factor. People must be shown why it is reasonable to set the performance standard where we have set it. Moreover, generally marginal extra expenditure of time, effort, and money must be explained and shown to be both feasible and well justified. This demonstration of feasibility and cost justification assessment is best done by someone knowledgeable; it may again require quality training to get everyone to accept our performance targets.

Finally, we have noted a profusion of indexes that attempt to measure quality. Think of the statement: "Only 17% of our pump repairs are unsatisfactory versus 22% last year." Yes, but suppose you've had 406 pump repair events last year and 17% are 69 pumps. Assume, for the sake of illustrations, that you've paid a (very low) average of $6783 for each of these "recycle repairs" and this would therefore amount to $468,000 per year. And think that if you had used cartridge seals instead of conventional seals, you would have avoided 23 of these repairs. Each of these seals would have incrementally cost $203 more; your net savings would have been in the vicinity of $151,340. Better shaft alignment would have avoided a grand total of 8 repairs. The combined extra time spent would have increased your labor expenditures by about $6213, but your net savings would still have amounted to $48,000 and so on. These are just two of many compelling ways of demonstrating the price of nonconformance. If only we would invest more time in well-structured training, failure identification, and other beneficial engineering-driven efforts, we might make more headway toward failure avoidance.

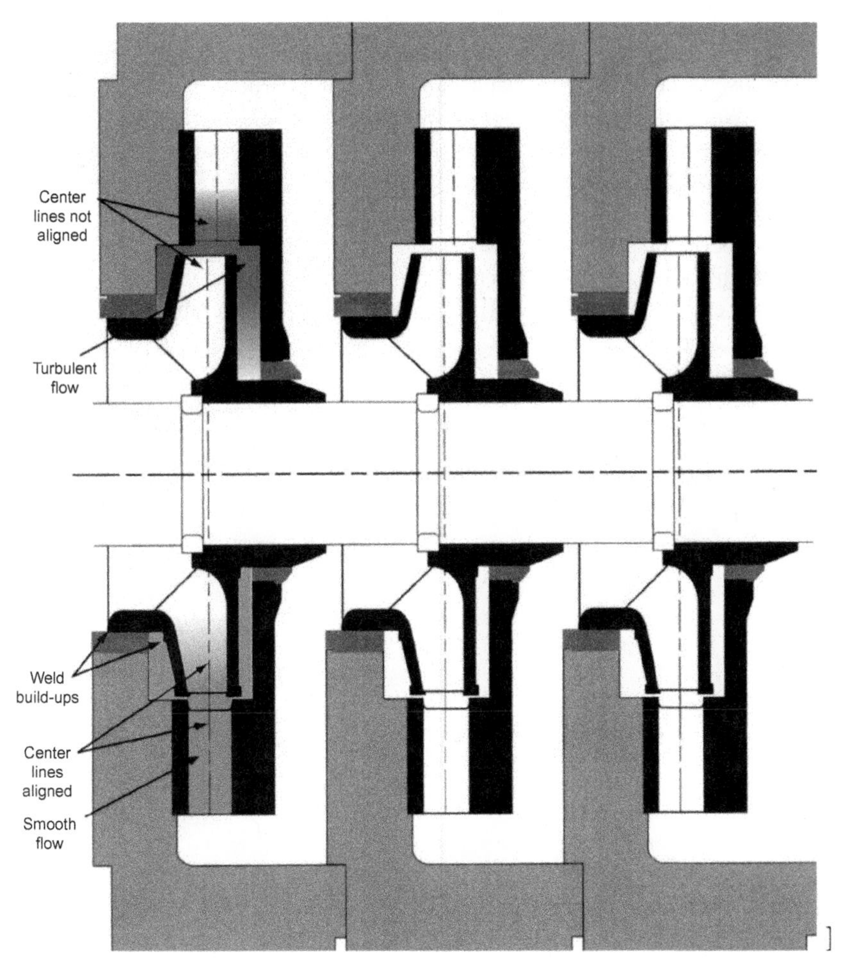

FIG. 33.2.1 Three pump stages with impeller exits relative to diffuser entries shown here misaligned (above centerline) and properly aligned (below centerline). Non-OEMs will point out such deficiencies and propose corrective measures. (*Courtesy Hydro Inc., Chicago, IL, www.hydroinc.com.*)

And that's what it's all about, this reliability improvement effort. We *must* achieve failure avoidance through better training, better specifications, identification and elimination of weak links in the component chain, etc. We *must* find ways of imparting guidance, proactive direction, and motivation to young engineers and to process technicians and equipment maintenance workers regardless of age and experience. Above all, we *must* find the wisdom to groom and reward those who *prevent* failures instead of praising only the overworked fixer whose highly visible restorative efforts are the result of someone not paying heed to the old saying "an ounce of prevention is worth a pound of cure."

HOW WORKING WITH A CPRS MAKES SENSE

We are always careful not to cast aspersions at any particular vendor or manufacturer. We have seen quality change from a given baseline in both positive and negative directions over time. In this instance, we actually mean going back in time from the baseline and going forward with our baseline. Some petrochemical machinery manufacturers have progressed; others have retrogressed. Check them out periodically in a rigorous assessment. Also, consider involving your top reliability engineer ("RE") in the various specification, upgrade implementation and equipment restart activities. The RE's on-site presence (Fig. 33.2.2) frequently represents an excellent refresher training opportunity for both equipment operator and RE.

FIG. 33.2.2 A reliability engineer is present during initial startup and also while re-commissioning an upgraded/improved multistage process pump.

REFERENCE

[1] P. Crosby, Quality Is Free, McGraw-Hill, New York, NY, ISBN: 0-07-014512-1, 1979.

Subject Category 34

Resource Utilization

Chapter 34.1

Notes from a Consulting Assignment

INTRODUCTION

"Machine failed; how do we fix it?" Consulting engineers and reliability professionals often get this type of request. Unfortunately, the client sometimes has neither the time nor the interest to elicit the consulting engineers' input on anything but the obvious. There seems to be a subconscious fear that part of the blame would have to be placed at management inadequacies or that the fault lies with people instead of components and parts.

Still, an experienced consulting engineer may wish to go beyond the obvious and alert the client to issues simmering below the surface. I see it as the engineer's responsibility to make at least an effort to go beyond the obvious. To use an analogy, a medical patient makes an appointment to seek relief for a certain ailment. While the patient may be happy with a few pills and two adhesive bandages, a competent doctor may insist on telling him why he's really sick. Although the patient may ultimately ignore him, the doctor has discharged his responsibility to share his knowledge-based in-depth appraisal.

At the completion of a consulting assignment involving issues with fluid machinery, we asked the client to hear us out on issues that need to be addressed in hundreds of HP plants in the United States and in many places around the globe. We knew these to be universal issues because we had often heard from readers around the world. Via these communications and by attending representative technical conferences, we receive timely updates on "the state of matters."

1. The hydrocarbon processing industry (HPI) struggles with contractor competence issues. If we don't groom the talent at our plants, what makes us think the contractor (with his lower billing rates) will groom these folks for us?
2. Issues of systematic training are not addressed. Much training in the HPI is haphazard and unfocused. Training must add value to the enterprise and

Petrochemical Machinery Insights. http://dx.doi.org/10.1016/B978-0-12-809272-9.00034-7

there has to be a methodology of checks and balances. There also needs to be accountability. Do you have such systems in place?

3. At some companies, involved personnel see resolution of technical issues and the pinpointing of human error as an "I could lose my job" issue. These staffers seem to have little incentive to even understand the merits of carefully considering the outsider's (the consulting engineer's) findings. They quickly make up their minds and decide to "dig in and ride this one out."
4. There is considerable guessing going on in some quarters with regard to field-installed spares. When there is nonredundant ("nonspared") equipment, onstream (continuous) surveillance may be the most intelligent and cost-effective condition monitoring approach. When the installed equipment not only is nonspared but also contains prototype-like elements, condition monitoring and automated operating techniques take on special meaning.
5. There is often an overdependence on the so-called alliance partner. As an example, issues of single-sourcing mechanical seals can soon deprive one of access to other vendors. Other vendors should be counted among your technology resources. Single-sourcing gives one the lowest initial cost, at best. While full-scale changes in one's seal supplier are rarely justified, we have yet to see a true best-of-class company that does not (selectively) work with one or two additional suppliers. We have reason to believe that some salesmen either do not understand what advantages the competition is able to offer or, if they do know, will not hesitate to misrepresent facts.
6. Managers are not always aware of the extent to which "business as usual" attitudes exist at their plants. Perhaps asking some probing questions might give your staff the incentive to do a bit of research and to update their own knowledge base. A manager might ask questions such as my "top ten" in the succeeding text:

   (a) Did the present seal supplier tell you about the vulnerabilities incurred by gas seals (or, for that matter, conventional liquid-lubricated mechanical seals) in certain specific services?
   (b) Does your organization know about certain very cost-effective alternatives to some of your existing sealing devices?
   (c) Are your reliability engineers and technicians aware of, say, hydraulically mounted coupling hubs?
   (d) Do they know about industry experience combining a bidirectional tapered pumping device with API Plan 23? If they do, have they studied its applicability?
   (e) Have your reliability professionals cultivated a relationship that elicits engineering input from alternative suppliers?
   (f) In some instances, supervisory instrumentation must be connected to shutdown logic. Is that done in your plant? Specifically, is that done on the level monitoring instruments for the suction vessels on your most critical nonspared pumps? Did not one of them run dry recently and did this not incur considerable unexpected repair cost?

(g) Are you aware that oil mist *lubrication* does double duty as oil mist *preservation*? What will happen to your little-used installed spare fluid machinery when this "protective blanket" does not exist on a conventionally lubricated standby pump? Will you have to schedule more frequent oil changes?

(h) Are we using the most appropriately configured bearing protector seal at our facility?

(i) Are we favoring "vendor X" because he offers superior technology or because his representative calls on us often and always brings along an assortment of two dozen doughnuts?

(j) What is our annual consumption of general-purpose pump couplings and how does that compare to a best-of-class consumption of only three coupling-related components per hundred pumps per year?

7. Far more effort goes into defensiveness after things have happened than being proactive so things don't happen in the first place. That is a distinguishing issue between average performers and best-of-class performers.
8. Judge where you stand on the following: some organizations allow their staff to raise concerns and walk away from the problem. Others insist that professional staff go beyond just raising a concern. At true best-of-class facilities, a professional is a person who expends effort and adds immeasurably more value by defining if a concern *is valid* or is *not valid*. Here is an arbitrary example: One of your reliability folks proposes using synthetic lubricants in a certain application. Another joins in the conversation and tells you that synthetic lubes will dissolve certain paints. Instruct him to dig deeper. If he does, he will likely discover that the synthetic oil being considered attacks acrylics, which are house paints. Insist he also seeks out the answer to the question of what paint type we have in our gearboxes. Rest assured *that* paint is likely to be fully compatible with your synthetics. If, on the other hand, you decide not to use synthetics because a concern was raised, you may not only adversely affect profits. You may encourage "business as usual" attitudes to propagate and accelerate.

Finally, consider checking out the following. If your organization does not make it easy for top-notch consultants to have access to your top managers, you're in good company. Most organizations are far too busy to allow highly experienced consultants to explain to top managers why their progress does not match expectations. The consultant's access is blocked for reasons the manager does not want to hear. Yet, allowing access to the lowest folks on the totem pole will rarely be productive. These folks are not empowered to implement change through equipment upgrades.

We're left with access to mid-level managers, and that is often unproductive because many mid-level managers find themselves locked in a cycle of turf issues. They also, occasionally, use "brinkmanship," which lets them become a hero by taking chances. They drive the company to the brink by deferring maintenance, allowing trial-and-error solutions, and frequently being just plain indifferent. Please don't be one of them.

Chapter 34.2

Consider the Most Effective Approach to Data Collection

INTRODUCTION

This abbreviated sequence of correspondence is real, although we quite obviously had to hide some names. In particular, I want to apologize to Centralia Refining and its mid-level technical employee Hugo Slowley in the unlikely event they do exist. But the story is quite typical of many discourses with readers or conference participants in the span of a year.

It started with an e-mail from Hugo Slowley; he had been with Centralia Refining Industries (CRI) for decades when we first met years ago. He asked:

> *My question is how often our vibration group should be doing their routes on every piece of equipment. We collect data on all fans, gearboxes, pumps, and motors monthly. I believe this is too frequent for most machines, but thought I would get your opinion before I stuck my head out too far. I have read a lot of articles and see that no one agrees. We seem to be in the "paralysis by analysis" mode of doing business.*

We answered Mr. Slowley by affirming that many reliability managers are unaware of the main reason for gathering vibration data: to get the operators out of the control room. We claimed, somewhat tongue-in-cheek, that operators will not leave their control room if

- ambient temperatures climb above +75°F (they fear the risk of heat stroke),
- the temperature drops below +66°F (there's then real fear of frostbite),
- wind speed exceeds 4 mph (understandably, they fear being blown off their bicycles).

WHY OPERATORS COLLECT DATA

Seriously though, best-of-class (BoC) refineries use a good portable data collector (Fig. 34.1.1) for the primary purpose of ascertaining if and when operators leave their control rooms. For BoCs, vibration monitoring is often of secondary importance.

We explained to Hugo Slowley that best-of-class refineries arrange for experienced vibration analysts to strictly limit their analytic involvement to the interpretation of out-of-limit vibration data. At CRI, the operators do the data collecting for the reasons mentioned above. Assigning both the collecting and analyzing of data to a highly trained vibration technician is a rare practice.

At best-of-class companies, experienced reliability professionals will also examine in-house records of failure frequency, repair cost, and downtime risk. In oil refineries, considerable judgment is required and equipment criticality is important. Data collection frequency is based on this criticality and can vary from monthly to twice yearly.
Mr. Slowley replied:

I forwarded this message to our maintenance manager who, by the way, never had anything to do with maintenance in a refinery until four years ago. He is in his late 30s and most of his staff is fresh out of college. However, the maintenance manager often mentions "Operator Involvement in Reliability," about which he had heard at a conference.

It was easy to see that CRI was not involving its operators in reliability stewardship. To be successful, such an involvement presupposes a well-structured training and technical education program; implementing such a program requires time, competence, monetary resources, and continuity of effort. It represents an investment in the future and cannot ever be a "flavor of the month" thing.

FIG. 34.1.1 In addition to warning of component deterioration, state-of-the-art data collectors serve as "vigilance monitoring tools." They make it possible to ascertain that responsible personnel visit the equipment as scheduled.

To again quote Mr. Slowley:

All or most of our unit supervisors are on too much of a "good buddy" relationship with their operators to make them get out of the control rooms so as to look, feel, and listen to their equipment. They operate by alarms, so to speak. Just recently, we lost four boiler feedwater pumps out of a total of seven. One of them *failed due to the deaerator running dry (bad level indication in freezing weather conditions) and then one of the turbine-driven pumps kicked in due to low flow and pressure and wiped out that pump set. On the following day, we repeated the process on a different set of pumps. Our total cost for expedited repairs was over $200,000. Amazingly, we never lost the boilers; somehow, one pump handled them all. However, the drive motor was severely overloaded and we were fortunate it didn't burn up the windings.*

Not having much more detail, we alerted Hugo to some typically overlooked issues. What happens when too many pumps are online was described [1]. If CRI slow-rolled its turbine-driven pump at below 100 rpm, the slinger rings (oil rings) in the turbine bearing housing will not pick up any oil, and the bearings will fail rapidly. Slinger ring dimensional tolerances are extremely important and rings must be concentric within 0.002 inches. To remain concentric, these rings must have been stress-relieved (annealed). And they work well only on properly aligned and truly horizontal shaft systems [2].

We were certain that CRI had never measured the width and concentricity of its oil rings (slinger rings). CRI had probably purchased them from the lowest bidder, who in all likelihood had skipped the important stress-relief annealing step. Without annealing, these cheap oil rings will distort and skip around on the shaft. In the future, CRI will have to budget the right price for oil rings that don't distort. An incremental cost of $6.67 per ring may avoid two $47,560 pump failures at his refinery.

While engaging in a lively discourse, Mr. Slowley and agreed and confirmed that best-of-class plants get their exceptional (9.4 years) pump MTBF by systematically upgrading and paying attention to every detail. Many put a competent pump repair shop (CPRS) into their planning strategies; that, too, was discussed.

As always, we appreciated the discourse with CRI because it updated us on the state of affairs in small and midsize refineries. Some of these refineries have become progressively less profitable, and the root causes of their problems and issues are often both cumulative and elusive. The communications sequence with Hugo Slowley and Centralia Refining Industries took place in 2013. It filled in the picture, and the picture was not entirely encouraging in 2013.

REFERENCES

[1] H.P. Bloch, Deferred Maintenance Causes Upsurge in BFW Pump Failures, Hydrocarbon Processing, June 2011.

[2] H.P. Bloch, Pump Wisdom: Problem Solving for Operators and Specialists, John Wiley & Sons, Hoboken, NJ, ISBN: 9-781118-041239, 2011.

Chapter 34.3

Why Users Must Update Design Contractors' Understanding

INTRODUCTION

In times past, we have repeatedly pointed out to EPC (engineering/procurement/construction) contractors where, why, or how their equipment selection and installation methods did not reflect best available reliability practice. Their answer inevitably was that the EPC did what the owner-purchaser (the "UP") requested, no more and no less. These experiences simply confirm that the UP must take the lead. The UP must use its human resources to define, understand, specify, and lay out for the EPC what the UP really wants.

The issue resurfaced when a rotating equipment engineer was involved in MQA—machinery quality assessment [1]. He was located far from his home base and halfway around the globe. It was his job to look over the shoulders of the EPC to make sure the multinational owner-operator's long-term reliability interests were understood and carried out. However, as best-of-class owners know, EPCs are primarily geared toward low initial cost and on-schedule plant start-ups. So, reasonable compromises must be found among initial cost, safe long-term equipment performance, life cycle cost (cost of ownership), and project schedules. Best-of-class owner-operators have formalized and institutionalized MQA with the goal of finding the right compromises.

PROTECTING THE CLIENT'S INTEREST

Representing the owner's interest, the engineer was involved in talks with the EPC about acceptable methods of baseplate leveling for centrifugal process pumps. He noted that the pumps at issue did not have adequate access on the mounting pads. Therefore, the machined mounting surfaces could not be used for leveling without removing the pump. The vendor had suggested using the machined surface of the discharge nozzle instead. However, RP-686 ("Recommended Practices for Machinery Installation and Installation Design"—a document issued by the API, the American Petroleum Institute) asks us to never use nozzles for that purpose.

The rotating equipment engineer now consulted a number of relevant books, among them "Pump User's Handbook" and "Machinery Component Maintenance and Repair," both of which have, for years, argued against the vendor's and design contractor's quick, but risky, approaches. These two books and several other experience-based texts strongly recommend that pump and driver be removed from their common baseplate. True, pump and driver had been premounted by the vendor on the baseplate to ascertain bolt locations and fits. Transporting the premounted pump/driver set as a single unit facilitates shipping but should not be considered the best approach to a long-term reliable field installation. In other words, full access to machined surfaces will be needed for proper baseplate leveling. This will require removing both pump and driver until leveling is accomplished.

DOES THE EPC NEED TECHNOLOGY UPDATES?

From this assignment at a distant location, the rotating equipment engineer then restated the main points made in those books. These texts always emphasize equipment reliability and caution against making quick and "expeditious" installation one's primary goal.

In this case, for unexplained reasons, the EPC provider had chosen to specify conventional baseplates for the process pumps on this project. Baseplates prefilled with epoxy would have been viable contenders here [2]. Ideally, the owner's representative involved in MQA might have looked into the matter; he could have asked to examine cost justification as well as long-term reliability issues. All parties may have been surprised by the findings.

But even as we sometimes limit ourselves to the more traditional installation methods, let's be sure to keep in mind the linkage between installation details and ultimate equipment reliability. We must verify there has been no distortion of baseplate mounting surfaces as a result of the shipping and delivery processes.

Unless pumps and drivers are removed from their common baseplates, it will not be possible to confirm that all mounting surfaces are coplanar, parallel, and colinear. Such confirmations would be needed to make the mounting pad portions of baseplates qualify as leveling surfaces. Establishing these as reference surfaces would be important to achieve a precise level. Having achieved level allows one to recheck after completing grouting procedures. At that time, one must ascertain that no distortion of the baseplate's mounting surfaces has occurred due to grout shrinkage (Fig. 34.3.1). Allowing grout shrinkage may lead to potential soft foot alignment issues and, in some instances, resonant vibration.

The reliability engineer's understanding of a clause in API RP-686 stating "never use nozzles" for alignment was correct. Experienced users know that nozzles are not necessarily parallel to fluid machine and driver mounting surfaces. Such out-of-parallelism may make it impossible to achieve precise levels. If lack of parallelism then prompts installers to use jackscrews and apply undue force to a baseplate, another risk will be created: The pump and motor bases may, inadvertently, become distorted or the pump and motor casings may become slightly twisted. Seals and bearings will no longer run absolutely parallel and component life will suffer.

MECHANICAL SEAL OPTIMIZATION

Not all design contractors are sufficiently familiar with desirable features found in mechanical seals for process pumps. Again, the owner-purchaser may have to take the lead in pointing out the desirability of dual seals with superior guide baffle and tapered pumping ring designs (see Fig. 34.3.2; refer also to Subject Category 21, "Mechanical Seals"). Some old-style designs are often less efficient and, ultimately, cost more to maintain.

LESSONS LEARNED

Three lessons are inescapable: (1) Not all EPC companies are knowledgeable about prefilled baseplates for small- and midsize process pumps, (2) reliability focus includes much attention to installation details, and (3) future maintenance cost avoidance requires user input. Do not let a provider or installer tell you that they have "always done it that way." There are instances where the installers have always done it wrong and where the resulting repair frequency has kept an entire plant from ever becoming a best-of-class performer.

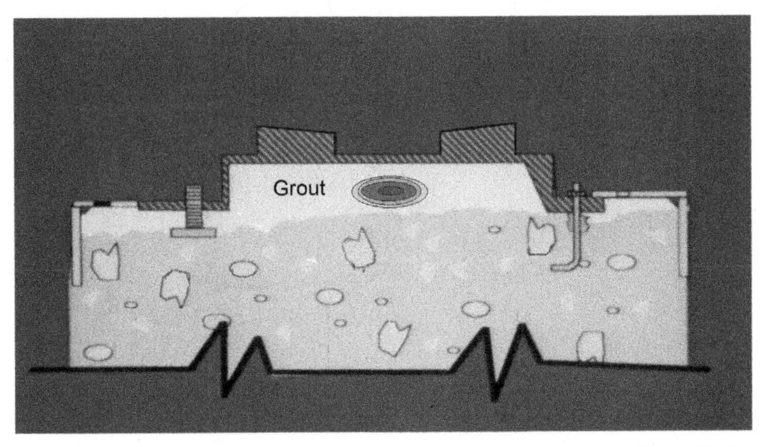

FIG. 34.3.1 Baseplate flaws such as out-of-level mounting pads come to light only if driver and driven equipment are left off and if leveling procedures can be carried out with unobstructed access. Note also that the fully embedded hold-down bolts in this illustration are not acceptable to reliability-focused users.

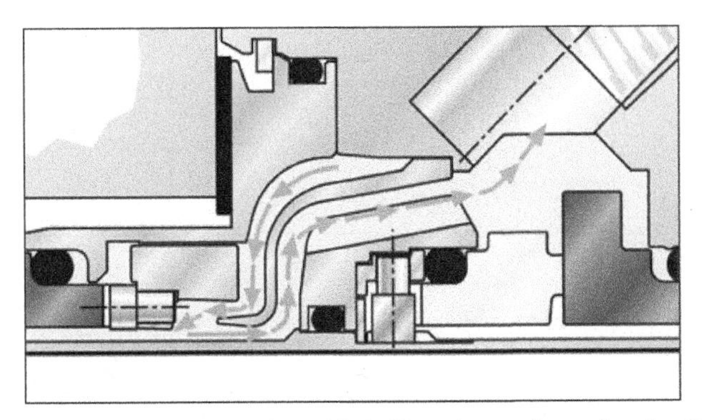

FIG. 34.3.2 Dual mechanical seal with guide baffle and tapered pumping ring for enhanced movement of buffer fluid. *(Courtesy AESSEAL, Rotherham, the United Kingdom, and Rockford, TN)*

REFERENCES

[1] H.P. Bloch, Fred Geitner, Compressors: How to Achieve High Reliability and Availability, McGraw-Hill, New York, NY, ISBN: 978-0-07-177287-7, 2012.

[2] H.P. Bloch, "Pump Wisdom: Problem Solving for Operators and Specialists", John Wiley & Sons, Hoboken, NJ, ISBN: 9-781118-04123-9, 2011.

Chapter 34.4

What to Consider When It's No Longer Safe to Work at Your Workplace

INTRODUCTION

We can rest assured that, in a safety and reliability poll of 1000 industry CEOs, all 1000 would profess the desire to lead and/or head safe and reliable companies. But there is usually a gap between aspiration and achievement. The record will show there is much work to be done in closing the gap, as a maintenance manager's letter showed. He had asked for direction from us and we were ready to give this direction. We think that driving a conscientious maintenance manager away from your business is a matter of poor resource utilization.

A FRUSTRATED MAINTENANCE MANAGER

He wrote about his frustration while working as a factory's maintenance manager and said:

"I see unsafe actions or conditions every day. Whenever I then confront the responsible unit supervisors, they tell me that's how it has always been done here. Or, if I explain to the employees involved that such-and-such is not a safe practice, they threaten me or yell in my face while the production manager stands by and laughs it off. My sense of self-respect wants me to fight back, but my manager training and sense of self-preservation take over and I follow the chain of command.

But nothing happens. I have been here for only 9 months and am already looking for a new job. I love what I do but I cherish working in a safe environment more. I don't want anyone to get hurt or worse. I have brought up some big equipment safety issues with management and even with the company's vice president. The inevitable answer mentions budget constraints and unaffordable cost run-ups. I know that's not a sensible answer.

What advice would you give me?"

We thanked the reader for relating his experience and reaching out to find a solution. We knew quite well that he's not the only one facing this dilemma, and he very likely is concerned with the inevitable cause and effect relationships that pertain in all industrial incidents. These incidents range from the simple and inconsequential to the catastrophic and life-changing. That said, we recommended six tangible steps, followed by a more specific example.

Our example highlights show how this maintenance manager would—or could—move from what is perceived as mere opinion to unassailable hard facts:

1. Start keeping a pocket-size journal (quietly) and make detailed notes when you see something unsafe (what it is, what risk results from it, and what you suggest may solve the issue).
2. Always follow up by informing a supervisor or boss about the situation and your suggestion. Do it in writing and safeguard your own copy of the communication.
3. Write down the recipient's reply and date and time of response.
4. Do not leave the journal laying around anywhere. Get a new pocket-size journal when the old one is full. This will also keep your head clear and make sure you do not lose sensitivity to violations. Committing details to your journal is extremely important. While it may not change anything, it will, nevertheless,
5. protect you in case something bad happens (and it *will* happen, sooner or later),
6. remind you to stay focused on getting a new opportunity,
7. give you new ideas for your next opportunity.

In addition to the aforementioned, this reader and others in similar lower management job functions should ask why intuitively safer and more reliable work processes and procedures are often hard to sell to our superiors. The answer is usually that we tend to convey opinions instead of facts. Opinions can be right or wrong; they can (and will) be disputed. In sharp contrast, facts are facts, regardless of how they are attacked or disputed. That's where our example comes in.

As you examine the two illustrations, you will perhaps express the *opinion* that the oil ring (in Fig. 34.4.1) moves back and forth and will probably abrade. But if you go the additional step and measure a new oil ring before you start the machine (Fig. 34.4.2, left side) and remeasure after operating it for a few

FIG. 34.4.1 An oil ring tending to move around and touching the inside of the bearing housing.

months (Fig. 34.4.2, right side), you will capture the difference between the two measurements. The difference was converted into abrasion product. At that point, you are dealing with *indisputable facts* and should recommend highly tangible remedial steps—actions that prevent this from ever happening again. To make these recommendations, we have to read a few pages in one of many reputable texts that explain details. If we don't have time to read and absorb the pages that give solid explanations, we will revert to being an opinion-giver and will not mature to becoming a fact-submitter. The choice is ours.

Fact-based recommendations will be far more likely to carry weight, especially if we also explain both cost and benefit in quantifiable ways. If our reader gives fact-based and quantitative recommendations and his organization still disregards him, he should take it as a signal to switch jobs. As bright and highly motivated employees, we should endeavor to work only for someone who is delighted to have us. And this may well be the time to send your resume to an organization that supports safety and reliability. We were thinking of reliabilityweb.com and clicking on the jobs menu. The maintenance manager might review the listings and let the website owners know if the disaffected reader wishes to travel or stay local.

But we want to come back to a common thread that weaves through this text: become an informed individual in whatever job function it is you are serving. Be a value-adder, be an example, be consistent. Once you know for a fact that you, as a maintenance manager, are much more than a job-scheduler, you will tactfully fight back and hold your head high. You will calmly explain to the vice president that you know exactly what you are talking about and that you want him to take note of your value system. A sound value system trumps over a perception of self-preservation allowing you to adversely affect lives that are entrusted to you. A person with an uncompromising value system will be a valuable resource. He will yearn for employment by an enterprise that eagerly seeks to employ him. Therefore, he will never be in want of a job.

FIG. 34.4.2 If this is the "before versus after" (left vs. right) oil ring condition, abrasion becomes an indisputable fact. Now, it's no longer just opinion.

ACKNOWLEDGMENT

Terrence O'Hanlon, CEO of Reliabilityweb, an asset management and publishing entity based in Fort Myers (FL), ably participated in compiling this chapter.

Subject Category 35

Root Cause Failure Analysis

Chapter 35.1

When Does a Good Practice Become a Bad Practice

INTRODUCTION

Root cause failure analysis (RCFA) is a skill that must be taught, absorbed, and consistently practiced. We certainly concur with the CEO of a RCFA training company that for several decades has successfully shown professionals to analyze failures and avoid their recurrence. This CEO makes it his business to read all sorts of company newsletters and finds them of interest. He often alerted readers around the world to one that caused him well-justified concern.

METHOD EXPLAINED

A company had trained its six expert failure investigators in the TapRooT® method. As stated in a newsletter from the same company, it was originally decreed that these six corporate staff members would perform all of the company's root cause analyses.

But a follow-up newsletter from the same company soon proclaimed a new "best practice." They had "evolved"—or so they thought—from the established TapRooT® method to an approach that just asked "why" five consecutive times. In essence, they were now teaching all their workers to "ask *why* five times" because "people can then do their own root cause analysis and solve their own problems without help from the corporate staff."

In case you wanted to understand to what the (often highly inadequate) "five-why method" refers, here's how it works its way back toward a perceived solution in a contrived example:

> Why did the Environmental Protection Agency levy a fine? Because the flare went off.
> Why did the flare go off? Because control valve PV-456 opened.

Petrochemical Machinery Insights. http://dx.doi.org/10.1016/B978-0-12-809272-9.00035-9

Why did PV-456 open? Because the unit lost feed.
Why did the unit lose feed? Because pump P-123 had caught on fire.
Why did the pump catch on fire? Because the bearing failed.

There you have it: the bearing failed. But the bearing manufacturer had explained the bearing should have a statistical 90% chance to operate flawlessly at rated load and speed for 40,000 h.

What if the bearing *really* failed because the plant uses oil rings at a DN value (inches of shaft diameter times rpm) of 10,900—well in the range where instabilities or even the slightest amount of shaft out of horizontality causes these oil rings to run downhill and contact the inside of the bearing housing, which causes slivers of brass to flake off and get into the lube oil and cause the destruction of the bearing? What if none of the supervisors or mechanics have been taught to use only flinger disks, oil mist, oil jets, or pump-around liquid oil—*just never to use oil rings*—at these DN values? Why don't they know? The true answer to the question may embarrass several layers of supervision and management.

Surely, this mode of "*five whys*" and "*letting people solve their own problems*" is far from being "best practice." We recall how TapRooT® trainers have explained in talks and articles *why* it is that "asking why five times" and trusting certain other forms of cause-and-effect analysis will not work well and is certainly not used by best-of-class performers. So, there is really no need to dwell on the issue. Instead, we should highlight here the results reported when shop floor people were taught to "ask why" five times as they went about performing investigations.

To be fair, people trained in this "ask-why-five-times" method were usually getting beyond the point of simply placing blame on something (or someone). Placing blame was a rather common conclusion *before* they received the "five-why" training. But they seldom came close to the root causes uncovered in the far more powerful TapRooT® system. And the results of analyzing the same problem varied significantly from one investigator to the next. Why? Because investigators essentially use their past experience to guide them and often stop at symptoms, which they then proceed to address with ineffective corrective actions. After all, this is what they have always seen and have always done.

The company that wrote the newsletter and (mistakenly) thought they had discovered best practice is not alone. In many companies, people think they are performing RCFA when, in fact, they are merely addressing symptoms.

While the above example was devised only to illustrate our point, a few actual examples from facilities applying "simple" RCFA (such as asking "why" five times in succession) are given next.

EXAMPLES OF INEFFECTIVE RCFA RESULTS

Equipment Failure

Defective mechanical seals are seen as the root cause of an equipment failure, and installing new seals is seen as the ultimate corrective action. There was no effort made to determine why the mechanical seals failed or to acknowledge that whenever a mechanical seal fails in a machine, it does so for a reason. Perhaps that's why the company had 23 mechanical seal failures on just two pumps over a 2-year period; you will find it highlighted in Chapter 35.3 of this text.

Inappropriate Action by Operator

Human error ("they just goofed up") is listed as the root cause for a serious mistake made by an operator, and additional training is being prescribed as the appropriate corrective action. In this instance, nobody asks why the previous training has failed or if training is really the most effective way to prevent the error from being made again.

Bad Behavior

Inappropriate behavior is cited as the root cause of an operator not using a procedure. The corrective action for this problem is reemphasizing the need to use procedures. In this instance, nobody asked about the usability of the procedure, about enforcement of procedure usage by management, or if operators were actually being rewarded for *not* using the procedure.

All three of these examples are real. These types of errors and failure to address the root causes harbor the seeds of repetition. Since none of the corrective actions given above cured the source of the problem, all of these failure events repeated themselves at the affected companies. At times, the repetition occurs after years or just months, as evident from the 23 seal failures at one facility. After a series of events such as the 2005 explosion and fire at a major refinery in the United States, this repetition of problems should frighten us. When dealing with flammable, explosive, or toxic substances at hydrocarbon processing plants, we should never allow deviations from the norm to become the new accepted standard. The safety of personnel and profitability of entire plants are at stake when RCFA is not carried out properly—or when some of the best-documented and best-understood pipe corrosion mechanisms are ignored, for whatever reason.

WHEN A "GOOD PRACTICE" IS REALLY A "BAD PRACTICE"

A corporation is at risk when people *think* they are improving performance but are, in fact, wasting effort. This happens when they implement ineffective fixes and lead management to believe that progress is made when, in reality, they continue to misdiagnose underlying causes. Failure to remedy the root causes of problems brings such a company perilously close to major failures. Major failures could

- maim or even kill people,
- lead to major production losses,
- cause significant product quality issues,

- result in significant environmental damage,
- lead to serious and difficult-to-regain loss of goodwill, and/or
- culminate in painful fines from government or regulatory agencies.

The serious accidents that occurred in these instances prove that instead of being a good practice, most—if not all—“quick-and-effortless” analyses are actually *bad practices* that should be shunned like the plague.

Chapter 35.2

Avoiding Failures of Three-Dimensional Compressor Impellers

INTRODUCTION

We want to convey a failure analysis example and chose a closed impeller (one with backward-leaning vanes) as our demonstrative field example for the "FRETT" approach. FRETT stands for force, reactive environment, time, and temperature. Getting back to this being an impeller example, here too, one or two of these four ("FRETT") are always involved in any conceivable component failure, and the impeller example serves us well in this instance. Two versions of impellers are available for compressors, a two-dimensional ("2-D") and a three-dimensional ("3-D") configuration. Fig. 35.2.1 shows a 3-D impeller that was contour milled from one single piece of metal. This type of contour milling is ranked among the favored fabrication processes and is not to be confused with techniques whereby individual blades are being welded to an impeller hub [1].

TWO-DIMENSIONAL VERSUS THREE-DIMENSIONAL IMPELLER BLADING

The backward lean in a 2-D version has the same curvature throughout the blade width. In the case of a single-shaft compressor with multiple stages, the volume at the inlet to the next stage is being reduced, and this lowers the stage efficiency. A compromise comes to mind: reduce the impeller diameter to gain efficiency. However, reducing the diameter would reduce head produced, and more stages would be needed. Now, the shaft would have to be made longer, which would affect the dynamic stability or susceptibility to undesirable vibration behavior of the compressor.

All of these factors must be given consideration for good compressor design practices. The same machine will not incorporate significantly different impeller diameters on the same rotor. In any event, 3-D impellers with contoured ("twisted") blades seem to be better adapted to varying flow conditions. For technical reasons, the axial width of a 3-D stage exceeds that of the equivalent 2-D impeller [2]. Since the axial dimension is wider compared with the 2-D version, the number of impellers ("wheels") that can be installed is restricted due to stability considerations. Moreover, speeds are generally governed by the mechanical tip speed limitations of the impeller material selected by the designer.

The operating range of compressor impellers can also be improved by using a measure of backward lean for the radial portion. Impeller performance and surge-point location are related; the surge point determines the useful operating range and head capability. Comparing 3-D impellers with backward-leaning radial inlet impellers is of interest. The more highly contoured 3-D geometries often feature improved operating and flow range.

CAREFUL DESIGN NEEDED

There are, in certain instances, compelling reasons for using 3-D impellers. Recall, however, that vendors have two manufacturing options: that of producing individual blades and to then weld these on an impeller hub and that of contour machining the entire wheel.

In either case, the designer must realize the importance of industry's decade-old practice of not allowing blade passing frequencies (BPFs) to coincide with impeller natural frequencies. For risk reduction, it will be necessary to validate the vendor's analytic "model" against actual field experience. Aerodynamic excitation could originate in many ways, including in the form of gas returning from downstream impeller(s) to the first-stage impeller inlet, via the balance line. Consideration should also be given to nonuniform and nonrepeatable impeller blade and blade weld geometry and their potential effect on component strength, stiffness, and frequency response.

But what if a failure is experienced? Ref. [3] explains a three-step approach that allows competent reliability professionals to quickly zero in on the most probable root causes of failure. The three steps consist of the so-called seven-root-cause methodology, the "FRETT" examination, and, finally, appropriate model validation procedures.

FAILURE ANALYSIS STEP NO. 1: THE "SEVEN-ROOT-CAUSE CATEGORY" APPROACH

The first of the two failure analysis steps is called the "seven-root-cause category examination." It accepts the premise that all machinery failures fall into one or more of only seven possible cause categories:

- Design errors
- Material defects
- Fabrication and processing errors
- Assembly and installation deficiencies
- Maintenance-related or procedural errors
- Unintended operating conditions
- Operator error

Using logical thought processes, we ask ourselves which of these seven cause categories are influenced by the compressor user and which ones are under the full jurisdiction of the compressor manufacturer. The answer determines the cause categories where failure analysis efforts should be concentrated.

Suppose there was a case where there is no indication in the record that the user/owner of a failed compressor impeller with 17 blades ever operated the machine at conditions other than specified by the purchaser and accepted by the manufacturer. Also, all gas properties had been disclosed to the vendor at the inception of the procurement chain. Suppose that no surge events had occurred during operation, especially since prolonged surge would typically result in thrust bearing damage and no such damage had been experienced.

Assume it had been ascertained that no operator error occurred. Accordingly, categories (6) and (7) are being ruled out. Assume further that during compressor maintenance, there is no logical causal event falling into either the assembly and installation or maintenance/procedural error categories. With (4) and (5) thus being ruled out, one might focus next on item (2), material defects, but again finds no metallurgical evidence of flaws in the base material selected by the manufacturer.

FIG. 35.2.1 Three-dimensional (3-D) compressor impeller. Just as any other mechanical component, impellers can be investigated by using the "FRETT" method of analysis.

The reviewer would thus be left with the two cause categories: (1) design error (2) and fabrication and processing errors.

FAILURE ANALYSIS STEP NO. 2: "FRETT"

It is universally recognized that all machinery component failures or machine distress brought down to the mechanical component level are attributable to one or two of only four failure mode sets. These four possibilities are force, reactive environment, time, and/or temperature. An easily remembered acronym, "*FRETT*," allows us to recall these four possible initiators.

Say, hundreds of compressor impellers in identical gas service had accrued operating hours far in excess of those at issue in this failure event. Hence, the impeller did not fail because the machine ran for too long and we could immediately rule out "time."

One would also rule out "temperature," since (assume you were able to ascertain) the actual compressor operating temperatures had always remained well within the acceptable range. Likewise, one might discount the suspicion that minor corrosion was responsible for the failures if it could be shown that the owner's gas composition did not measurably deviate from that disclosed in the original specification documents.

That clearly would leave the reviewer with "force" as the only remaining logical failure mode set. The issue might be what type of force acted on the impellers or where the force came from. This is where one might rely on data collected by instrumentation and data collectors. Remember the failed impellers had 17 blades and it is well known that failures can occur due to BPFs coinciding with impeller natural frequencies. You would look for amplitude excursions at a frequency of 34xRPM and/or 17xRPM. Moreover, you would search for the presence of acoustic pulsing at BPF in the balance line connecting compressor discharge and first-stage suction inlet.

STEP NO. 3: VALIDATION OR RELATING ANALYTIC MODELS TO FIELD EXPERIENCE

Whenever an impeller design is "modeled" for computer analysis, the analyst will generally make a number of assumptions, and the validity of these assumptions must be ascertained. Common sense tells us that there are four possibilities and corresponding definite conclusions, as follows:

- Analysis says "ok," but blade fails: model or analysis technique is flawed.
- Analysis says "ok," and blade survives: model and analysis technique are ok.
- Analysis says "not ok," and blade also fails: model and analysis technique are ok.
- Analysis says "not ok," but blade survives: model or analysis technique is flawed.

Suppose the manufacturer's analysis claims that the design was acceptable, yet the owner experienced repeat failures. Given the absence of root cause factors other than design and/or fabrication, this leads to the logical conclusion that the manufacturer's model or analysis technique was flawed.

If you have experienced such failures, there might be the appearance of tacit acknowledgment of this conclusion on the part of the vendor. Perhaps, the manufacturer no longer recommends operating these compressors at certain speeds that correspond to the BPF interference mentioned earlier. Or, the manufacturer switched from welded blading to the superior one-piece-machined fabrication method shown in Fig. 35.2.1. That would support the conclusion of a flawed computer model. You would concentrate your research on the success of milled impeller redesigns. Such redesigns could logically be expected to cure the problem.

REFERENCES

[1] H.P. Bloch, A Practical Guide to Compressor Technology, second ed., John Wiley & Sons, Hoboken, NJ, ISBN: 0-471-72793-8, 2006.

[2] H.P. Bloch, Compressors and Process Applications, John Wiley & Sons, Hoboken, NJ, ISBN: 0-471-72792-X, 2006.

[3] H.P. Bloch, F.K. Geitner, Machinery Failure Analysis and Troubleshooting, Gulf Professional Books, Houston, TX, ISBN: 0-88415-662-1, 1998.

Chapter 35.3

Repetitive Pump Seal Failures Can Cause Disasters

INTRODUCTION

Release No. 2004-08-I-NM, issued in October 2005 by the US Chemical Safety and Hazard Investigation Board, addresses an incident at an oil refinery with a history of repeated pump failures. Located in New Mexico, this facility's total of three primary, electric, and steam-driven spare isostripper recirculation pumps had 23 work orders submitted (see Chapter 21, Table 21.6.1) for repair of seal-related problems or pump seizures in the one-year period prior to a fire and explosion. The catastrophic incident occurred during disassembly on April 8, 2004 and caused over $13,000,000 in damage. At least six people were injured, and production at this alkylation unit was shut down for months.

LESSONS FOR THOSE WILLING TO LEARN

The reliability-focused also realize that when a seal failure combines with one or more other deviations from the norm, disasters result. More specifically, they make it their business to know what fit-for-service mechanical seals are available. But, of course, these components have to be properly installed and will usually require a pump-around circuit and dual seals, generically depicted in Fig. 35.3.1, which include a conservatively designed wide-clearance pumping ring.

For a certainty, breakdown maintenance on machinery will lead to complete plant breakdowns, as occurred here. Identifying operating conditions and risky components that could contribute to equipment failure are critical ingredients of a sound mechanical integrity program.

Finally, consider instituting a new organizational regime by valuing, developing, and promoting competent leaders instead of favoring those who can fabricate a profit margin in the next quarter. Understand that house-of-cards based or built-on-sand next quarter profits are crumbling at random, whereas house-of-brick organizations have few surprises and will keep standing.

In this particular instance, the handwriting was on the wall. A fundamental lack of proper reliability engineering can be traced back to an unheard-of tolerance for repeat failures. Management intervention and an insistence on root cause failure analysis were long overdue when this catastrophic event finally happened. Notice that we decided to place this event in the subject category Root Failure Analysis because that's the first order of business here. Attention

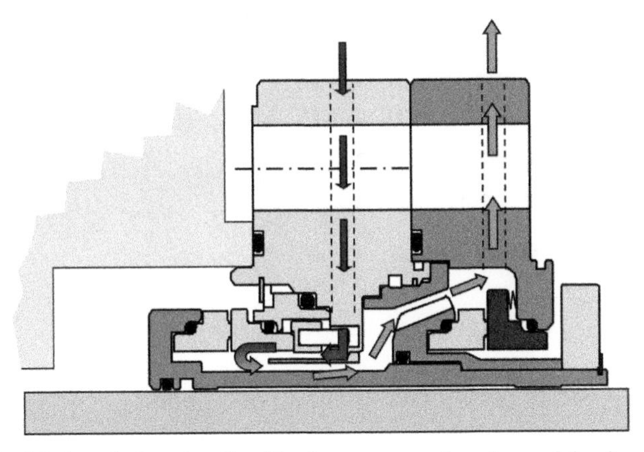

FIG. 35.3.1 Modern dual seal with wide-clearance pumping ring and barrier fluid pump-around circuit. A dual seal and a plan 53 system is the appropriate specification for this particular application.

to seal type selection is next in the order of importance, and the lack of well-focused training probably runs a close third.

A reliability professional must be able to ask the right questions. After the second of 23 failures, a professional would insist on questioning the seal vendor on where these seals had been used in the exact same service. The answer to this question will allow a professional to determine if seal type and flush plan are identical to what's being used elsewhere. The answer allows one to determine if the issue is a local one, that is, an issue that has to do with local skills or work practices, or with installation and assembly issues existing at this facility. Make the vendor your technology resource instead of merely the parts provider. Deal with more than just one manufacturer of mechanical seals. No refinery has ever become "best of class" by dealing with a single source of mechanical seals or by treating mechanical seals as a commodity no different from paper clips.

Subject Category 36

Run-Length Determination

Chapter 36.1

How Can Remaining Run-Length Potential be Determined?

INTRODUCTION

With so much money at stake, how should one approach the issue and how might a reasonably precise numerical answer be obtained? Well, let's agree beforehand that obtaining a very precise numerical answer will forever elude us since there are far too many variables involved. Let's look at a rather straightforward steam turbine, for example.

The life expectancy of steam turbine components is prone to be influenced by steam purity and steam quality, also known as moisture content. Steam generation facilities experience occasional upsets, and acceptable control of water treatment is not always obtained. Operation at varying speeds may subject the blading to different steady state and alternating stresses. Lube-oil quality and purity can easily affect bearings, valve positioning hydraulics, and overspeed trip-actuating mechanisms. Electronic governing systems may be influenced by ambient conditions and component drift; all might be influenced by maintenance oversights and/or operator error.

Making a reasonably accurate assessment of prudent operating time to the next scheduled turnaround inspection is difficult. The responsible engineer will have to review all of the above and many more factors. A thorough investigation of the experience and service condition of similar machines, comparison of stress levels in identical blading operating elsewhere with the stress levels experienced in the machine at your plant, evaluation of "their" maintenance procedures against "our" maintenance procedures, etc. would be appropriate.

We have seen over the years that much of this information is easiest to obtain during preprocurement review of the particular design offered by a turbomachinery manufacturer. We know of relatively few procurement situations where capable vendors would not have shown willingness to explain in detail their

Petrochemical Machinery Insights. http://dx.doi.org/10.1016/B978-0-12-809272-9.00036-0

prior experience for the service conditions imposed on the proposed machinery. In other words, the capability and the ability to demonstrate satisfactory experience go hand in hand.

Service conditions include parameters such as start-up (heat-up) time, output rating versus actual power delivery, speed, steam temperature, and pressure. Similarly, capable vendors can usually demonstrate mechanical design experience for applicable critical parameters that could include

- bearing span;
- bearing design, loading, size, and clearance;
- blade design—structural and thermodynamic;
- casing size and design and exhaust orientation;
- casing-joint design configuration, gasketing, and bolting;
- coupling design and arrangement;
- extraction control, single and double extraction;
- material selection;
- nozzle design;
- number of stages and staging arrangement;
- power/transmission components (design and arrangement);
- rotor dynamics;
- sealing system.

To then lay the groundwork for answering future questions on mean-time-between-turnaround inspections, the responsible engineer would do well to expand his review of a vendor's mechanical design experience. Follow up by requesting operating and maintenance feedback from other installations. We consider this the important statistics link with the manufacturer's sales group. As an aspect of networking and making the equipment manufacturer one of your information resources, close communications are valuable.

Subject Category 37

Screw Compressors and MQA

Chapter 37.1

Machinery Quality Assessment (MQA) Needed on Twin-Screw Compressors

INTRODUCTION

As is so often the case, an urgent request for consultation was passed on by a colleague, but very little information accompanied his initial request. It was known that the two relatively large horsepower-input twin-screw compressors were driven through a common gearbox perhaps similar to the well-engineered general layout in Fig. 37.1.1. Things got fuzzy when a follow-up note mentioned that the machines were

> *"running at speeds between 2000 and 3500 rpm and failed suddenly and without warning. The original equipment manufacturer (OEM) seems to ignore our request for reasonable information, let alone solid explanations. The owner company assigned a young engineer to this project; his basic competency is in controls engineering. Do you have any data regarding acceptable versus suspect vibration spectra or amplitudes? If possible, please alert me to relevant case studies."*

Well, we can always learn from the past; it's rare that any of us are the first ones to encounter a particular problem with machinery in the hydrocarbon processing industry, the HPI. We could look up a very similar failure case described in [1], where resonant torsional vibration wrecked a screw compressor. Speaking in general, issues can be resolved by sitting down with the vendor's design engineers at the factory location—typically an up-front preplanned machinery quality assessment (MQA) activity. One could then, belatedly, determine if torsional vibration testing was done covering all operating speeds. Whenever variable capacity is required by the process, speed variations may not be the one's best choice.

Petrochemical Machinery Insights. http://dx.doi.org/10.1016/B978-0-12-809272-9.00037-2

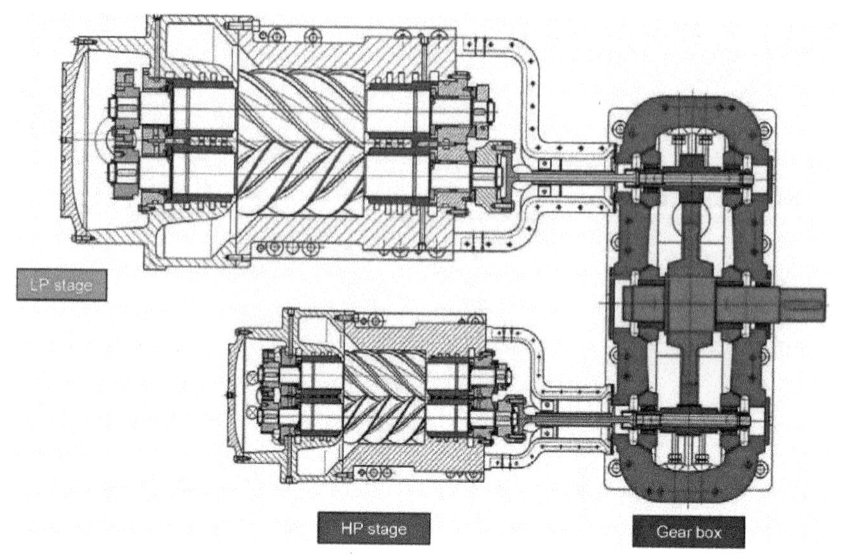

FIG. 37.1.1 Well-engineered twin-screw compressor set with low-pressure and high-pressure casings connected to a common drive gear box.

Explanations Slow in Coming

Expecting to obtain a written explanation or admission of negligence via e-mail or telephone is a utopian hope. Nobody in any chain of command of any of the many parties involved will admit anything for reasons of litigation risk. However, discussing and reviewing test data (if these exist) might disclose certain prototype features or operating modes. User references could then be examined, and plans could be made to cooperatively rectify the overall problem. Just to reemphasize, such issues are usually uncovered and addressed as part of budgeted and preplanned "MQA" activities [2].

But with the failure at issue having occurred yesterday or today, the owner company's management team must ask themselves other key questions: why are we attempting to solve a problem that costs us a million dollars each year by (presently) assigning these tasks to an I&E person without machinery background? Where is the engineer's mentor? Are we really saving anything by not involving an experienced consulting company? It should be of interest that Ref. [1] points to torsional critical speed problems that manifested themselves in the form of lateral vibration once the crests of the male and female rotor lobes came into contact. Two years later, a very similar problem was experienced elsewhere. This time, another vendor-manufacturer was involved and the young I&E engineer was expected to know (or quickly learn) the details of vendor liaison, prior publications, and what we collectively call MQA.

At this point in time (i.e., after the fact), the owner-operators may come to realize that they purchased a failure-prone machine. They may have (unwittingly) become the manufacturer's proving ground. Yet, even if that is indeed the case, the manufacturer will never admit it. Also, nothing will (usually) be solved by engaging in long-range e-mail correspondence. Instead, the owner-purchaser's reliability professionals must sit down with the vendor at the vendor's factory. During this meeting, each side should go over data and a list of questions submitted a few days in advance of the meeting. Reference lists will have to be examined, and unusual or newly designed components must be discussed with the designers.

It would have been far better for the equipment owner to have performed MQA when buying major capital equipment [3]. Long-term reliable operation will result and is worth the extra expenditure. Good MQA adds 5% to the cost of the machinery budget of a plant—obviously much less when considered as a percentage of the value of all plant assets. Best-of-class performers are best-of-class because they include MQA in their budgets before asking for the budget to be approved. A company without MQA is leaving itself exposed to many avoidable risks and is very unlikely to reach its full profitability potential.

FIG. 37.1.2 Comparing a very large with a very small (arrow "B") screw compressor rotor. *(Courtesy Aerzen USA, Kulpsville, PA; also Aerzener Maschinenfabrik, Aerzen, Germany.)*

And in case you were wondering why we used screw compressors as one of the machines worthy of MQA, their rotors range from sizes you can hold in the palm of your hand—see arrow "B" in Fig. 37.1.2—to the massive size also depicted in Fig. 37.1.2. Best-of-class users favor best-of-class manufacturers; the former make the latter their highly cooperative tutors and technology providers. Both sides benefit.

REFERENCES

[1] H.P. Bloch, F.K. Geitner, Machinery Failure Analysis and Troubleshooting, fourth ed., Elsevier Publishing, Kidlington/Oxford, UK and Waltham/MA, USA, 2012, pp. 630–634.

[2] H.P. Bloch, F.K. Geitner, Machinery Failure Analysis and Troubleshooting, fourth ed., Elsevier Publishing, Kidlington/Oxford, UK and Waltham/MA, USA, 2012, p. 696. 708.

[3] H.P. Bloch, F.K. Geitner, Compressors: How to Achieve High Reliability and Availability, McGraw-Hill, New York, NY, 2012.

Subject Category 38

Specifications and Standards

Chapter 38.1

Standards and Disclaimers

INTRODUCTION

Reliability specialists are encouraged to make good use of industry standards. Well-written standards improve equipment uniformity and quality. Take, for example, the NEMA (National Electrical Manufacturers Association) standards. Their foreword explains scope and purpose and could serve as a model for what standards are all about.

TYPICAL SCOPE OF STANDARDS

The clauses I found in an old NEMA publication (MG 1-2003) are now superseded by more recent revisions. All have been developed by NEMA's Motor Generator Section and approved as NEMA standards. These standards, and the standards issued by API, AGMA (American Gear Manufacturers Association) and others are intended to assist equipment users in the proper selection and application of machinery. These standards are revised periodically to provide for changes in user needs, advances in technology, and changing economic trends. All persons having experience in the selection, use, or manufacture of electric motors and generators are encouraged to submit recommendations that will improve the usefulness of these standards. Inquiries, comments, and proposed or recommended revisions should be submitted to the Motor and Generator Section by contacting the vice president of engineering at the National Electrical Manufacturers Association.

The best judgment of the Motor and Generator Section on the performance and construction of motors and generators is represented in these standards. They are based upon sound engineering principles, research, and records of test and field experience. Also involved is an appreciation of the problems of manufacture, installation, and use derived from consultation with and information

Petrochemical Machinery Insights. http://dx.doi.org/10.1016/B978-0-12-809272-9.00038-4

obtained from manufacturers, users, inspection authorities, and others having specialized experience. For machines intended for general applications, information as to user needs was determined by the individual companies through normal commercial contact with users. For some motors intended for definite applications, the organizations that participated in the development of the standards are listed at the beginning of those definite-purpose motor standards.

Returning to MG 1-2003, practical information concerning performance, safety, test, construction, and manufacture of alternating-current and direct-current motors and generators within the product scopes defined in the applicable section or sections of this publication. Although some definite-purpose motors and generators are included, the standards do not apply to machines such as generators and traction motors for railroads, motors for mining locomotives, arc-welding generators, automotive accessory and toy motors and generators, machines mounted on airborne craft, etc.

In the preparation and revision of the NEMA standards, consideration has been given to the work of other organizations whose standards are in any way related to motors and generators. Credit is given to all those whose standards may have been helpful in the preparation of a particular volume.

NEMA and similar standards publications undergo revisions and/or updates in 5–7 year intervals. Prior to publication, the NEMA Standards and Authorized Engineering Information that reappear in a particular publication unchanged since the preceding edition, are reaffirmed. The same process is generally true for other prominent standards.

The standards or guidelines presented in a NEMA standards publication are considered technically sound at the time they are approved for publication. They are not a substitute for a product seller's or user's own judgment with respect to the particular product referenced in the standard or guideline, and NEMA does not undertake to guaranty the performance of any individual manufacturer's products by virtue of this standard or guide. Thus, NEMA expressly disclaims any responsibility for damages arising from the use, application, or reliance by others on the information contained in these standards or guidelines.

DISCLAIMERS IN STANDARDS

This now logically leads to a "notice and disclaimer." As is to be expected, the information in a reputable standards publication was considered technically sound by the consensus of persons engaged in the development and approval of the document at the time it was developed. Consensus does not necessarily mean that there is unanimous agreement among every person participating in the development of a document.

All NEMA, and the great majority of similar prominent standards and/or guidelines, are developed through a voluntary consensus standards development and periodic revision process. This process brings together volunteers and/or seeks out the views of persons who have an interest in the topic covered by this publication. While NEMA administers the process and establishes rules to promote fairness in the development of consensus, it does not write the document, and it does not independently test, evaluate, or verify the accuracy or completeness of any information or the soundness of any judgments contained in its standards and guideline publications.

NEMA disclaims liability for any personal injury, property, or other damages of any nature whatsoever, whether special, indirect, consequential, or compensatory, directly or indirectly resulting from the publication of, use of, application of, or reliance on this document. NEMA disclaims and makes no guaranty or warranty, expressed or implied, as to the accuracy or completeness of any information published therein, and disclaims and makes no warranty that the information in this document will fulfill any of your particular purposes or needs. NEMA does not undertake to guarantee the performance of any individual manufacturer or seller's products or services by virtue of this standard or guide. Reliability engineers should meditate on this wording. It means, to this author, "do not hide behind a standard clause when expert advice points in a better direction and you have experienced an incident that should cause you to reassess a particular situation."

In publishing and making standards and related documents available, NEMA is not undertaking to render professional or other services for or on behalf of any person or entity, nor is NEMA undertaking to perform any duty owed by any person or entity to someone else. Anyone using NEMA standards/documents should rely on his or her own independent judgment or, as appropriate, seek the advice of a competent professional in determining the exercise of reasonable care in any given circumstances. As an experienced reliability professional, I fully endorse what I see here. And NEMA certainly parallels what we read in the introduction to numerous API standards. Information and other standards on the topic covered by a particular publication may be available from other sources, which the user may wish to consult for additional views or information not covered by whatever publication may have been initially available.

NEMA explains that it has no power, nor does it undertake to police or enforce compliance with the contents of its standard documents. NEMA does not certify, test, or inspect products, designs, or installations for safety or health purposes. Any certification or other statement of compliance with any health- or safety-related information in a NEMA document shall not be attributable to NEMA; it is solely the responsibility of the certifier or maker of the statement.

WHERE DOES THIS LEAVE YOU?

Being reasonably well acquainted with this and other standards and, especially, having an awareness of third-party certification agencies and their practices, we want to assure our readers that there is no substitute for user experience. This issue was recently brought to the fore when a previously cooperative and industry-leading service provider refused to endorse the time-tested, well-thought-out practices of its previous customers. The service provider fully realized that the customers, very knowledgeable and safety-conscious equipment users, employed these practices. However, the practices in question seemingly contradicted the rather illogical stipulations of a third-party certification provider, and so, fear of confrontation and litigation prompted the service provider to side with the "listing agency."

Insightful Best-of-Class companies have drawn an important conclusion with respect to industry standards: if you know that your practices (meaning practices that deviate from the standard) are correct, time-tested, and technically sound, put your findings in writing, and for good measure, subject them to peer review. If applicable, send a letter with your observations to the appropriate standards organization. Just as NEMA points out, seek the advice of a competent professional. Also, consider finding and hiring another service provider, and never hesitate to disregard agencies or entities that clearly impede progress. It may be difficult to see every hidden reason, but ignorance or misguided priorities can be part of the agenda.

Chapter 38.2

Expanding the Scope of Industry Specifications: Include Installation Specifications in Addenda

INTRODUCTION

In so many ways, it is neither the purpose nor the intent of this text to give full coverage to every conceivable topic. That said, the narrative in this chapter will not convey detailed field erection and installation specifications for the many rotating machines found in modern process plants. However, field erection and installation specifications are definitely needed by reliability-focused plants. Moreover, they must be reviewed, understood, and approved by a competent machinery engineer who directly represents the owner plant and has a personal stake in its long-term reliable operation.

SPECIFICATIONS AND ADDENDA

We found that all of these specifications have a few things in common:

- The scope of a standard is always explained first. For instance, a field erection and installation standard would cover mandatory requirements governing installation and erection for compressors and drivers mounted on baseplates or soleplates.
- Additional information is almost always superimposed on existing industry standards. An asterisk might be used to indicate that additional information is required. Here, the contractor may have to *specify*, whereas the owner's machinery engineer may have to *approve* relevant information.
- Many industry standards are not intended to be invoked without the purchaser's addenda. Therefore, a summary of additional requirements is usually provided. Owner-purchasers often provide a separate tabulation of applicable cross-references; this tabulation usually lists documents that have to be used in conjunction with the particular industry standard.
- Design requirements are clearly explained. Here is an example:
  Concrete foundations must be properly sized and proportioned for adequate machinery support and prevailing piping forces. The complete compressor train (compressor, gear, and motor or other drivers) must have a common foundation. Foundations must rest on natural rock or entirely on solid earth or good, well-compacted, and stabilized soil. They must be supported on pilings that have a rigid continuous cap or slab cover.
  Foundations must be isolated from all other structures such as walls, other foundations, or operating platforms. They have to be designed to avoid

resonant vibration frequencies at operating speeds, 40–50% of operating speeds, rotor-critical speeds, gear-meshing frequencies, two-times operating speeds, and known, specified background vibration frequencies.

The temperature surrounding a foundation must be analyzed to verify uniformity to prevent any distortion and misalignment. Concrete foundations must also be properly cured (approximately 28 days) before being "loaded" with a machine and driver.

Foundation arrangements are described in considerable detail in a good specification and its customary addenda. Anchor bolts must be designed by specialty firms and must be sleeved. In most instances, a civil engineer will provide and/or certify a foundation drawing or separate foundation specification.

Around the perimeter, a W-8 or larger I-beam must be properly anchored to the foundation for supporting small piping, conduit, and instruments. Pay special attention to auxiliary structures and include all piping details in your review. Experience shows that least-cost designs tend to be implemented without realizing that safety and accessibility are of primary concern here. The owner's engineer ascertains that proper attention is given here.

Typical compressor, gear, and motor foundation arrangements and baseplates must be completely filled with epoxy grout. Soleplates must be completely supported with epoxy grout.

Reinforcing rods, ties, or any steel members must be a minimum of 2 in. (~50 mm) below the concrete surface to permit chipping away 1 in. of concrete without interference.

A minimum space of 1 in. (~25 mm) must be provided between the foundation and chock block for proper grout flow. The maximum distance between the foundation and baseplate should not exceed 4 in. (~100 mm). The minimum distance between the foundation and baseplate should not be less than 2 1/4 in. (~55 mm).

For epoxy chock applications, the distance between the baseplate or soleplate and top of grout should be 1 in. (~25 mm), unless otherwise approved by the owner's machinery engineer.

Chock block arrangement and installation are described. Chock blocks must be properly sized to distribute anchor bolt and machine loads to not exceed 10% of the weakest compressive strength material in the foundation structure (the customary design is 300 psi for concrete).

Instructions and appropriate illustrations of field erection and assembly tools must be provided. For instance, a special hydraulic coupling hub-to-output-shaft installation tool will probably be used in virtually any modern plant.

In this day and age, much lip service is paid to reliability concepts and uptime optimization. These two goals often clash with the desire to award plant construction to engineering and construction contractors who, in turn, are almost entirely driven by cost and schedule concerns. In those instances, an owner-operator firm would be foolish not to give their utmost attention to the existence of specifications that reflect reliability focus and uptime extension goals. Reliability professionals are then assigned to ascertain compliance with these specifications. Look it up under the index identifier machinery quality assessment (MQA).

Subject Category 39

Steam Turbines and Steam Traps

Chapter 39.1

Selecting Steam Turbines in a "Lean" Environment

INTRODUCTION

Lean, in our opinion, is a slogan that has come and gone. Whatever its original meaning might have been, it has been misapplied to the detriment of many. It has led to mind-sets of buying cheap; it may even have driven good manufacturers into extinction.

An engineer in South America encountered the situation described in our Main Focus, above. He explained that he worked as a mechanical engineer on power plant designs at a major corporation and was aiming to clear up some doubts related to steam turbine technical specifications. More specifically, his corporation was developing a combined-cycle power plant project that included an 86 MW condensing-type steam turbine with one reheat entry. The HP (high-pressure) inlet steam is at 110 bar/540 °C and the reheat is being designed for 24 bar.

He was in correspondence with several respected steam turbine manufacturers, and some of them were proposing a "standard-type" machine; in other words, they offered a turbine with a single casing and a single rotor direct-coupled to the generator. But there are also some manufacturers that proposed a "cross-compound-type" machine. Such machines consisted of a turbine with two casings and two rotors. In one offer, the HP rotor was coupled to the generator by gearbox and the IP/LP casing was direct-coupled to the generator.

The engineer-correspondent asked for an opinion on the "cross-compound" machine, its general technically feasibility, and potential operating and maintenance (O&M) problems.

Petrochemical Machinery Insights. http://dx.doi.org/10.1016/B978-0-12-809272-9.00039-6

DETAILED INVESTIGATION NEEDED

We drafted an answer to give general guidance and to map out the approach one would recommend. In essence, the best way to arrive at a definitive judgment is to

(a) look at the guaranteed efficiencies of the two different offers and keep in mind the overall steam balance of the facility,
(b) make a decision as to how well trained the operators will be, and
(c) closely examine the respective field and service experience histories of the two different steam turbine offers.

Complying with the basic requirements of (a), (b), and (c) requires considerable diligence, time, and effort. The reviewer should include a thorough check of the gearbox design and should accept the fact that time is needed to draw up a comprehensive comparison between the two offers. It would even be appropriate to ask if the original inquiry went to the right bidders. It is always prudent to solicit bids from manufacturers that have ample experience with both direct-drive generator turbines and the more complex compound/reheat multicasing machine.

Time permitting, consider adding a few bidders willing to comment on the very advisability of double-shell machines. A double-shell construction machine prevents inlet steam coming into direct contact with the outer casing joint. Double-shell machines are (probably) less in need of careful operator attention but, during the maintenance cycle, might require rather competent maintenance skills.

"Cross-compound" machines are probably found on shipboard, but predominantly at inlet pressures slightly lower than 110bar. Again, there is much inquiring to be done before a decision can be made on a land-based installation. Ships have downtime while in a harbor. In contrast, land-based utilities desire to reach long uninterrupted run lengths. For them, downtime is very expensive. As regards items to be reviewed, one might investigate, among other things, the lubrication system. In a cross-compound machine, the input and output shafts are at different levels, and the lubrication system serves not only the turbine and generator bearings but also the gearbox. Who makes the gearbox and how are the gears lubricated?

Initial cost, operating cost (efficiency), and long-term reliability are of interest and must be considered as life cycle cost factors. All are of equal concern, and, without making a final judgment one way or the other, many different options should be explored before reaching a conclusion. Although one should make good use of vendor input and defer to their demonstrated experience, expect double-shell machines to cost more money and cross-compound machines to require more than the average maintenance commitment. And the "simple" machine would also stay in the running until all the relevant data have been obtained and ready for comparison.

DON'T GET CAUGHT BY THE "LEAN AND MEAN" CRAZE

A perceptive reader may have seen how our answer alludes to the subject of suitability analyses or prepurchase selection work that needs to be done. We were reminded of the pitfalls of "lean and mean" when another facility experienced several extreme failures on smaller two-stage back-pressure mechanical drive steam turbines. For a number of years, these turbines had been driving refrigeration compressors without incidents. Then, about two years ago, the refrigeration gas composition was changed to accommodate new (and well-justified) environmental concerns. The new gas conditions mandated a speed change for the steam turbine drivers, and multiple catastrophic blade failures have occurred since then.

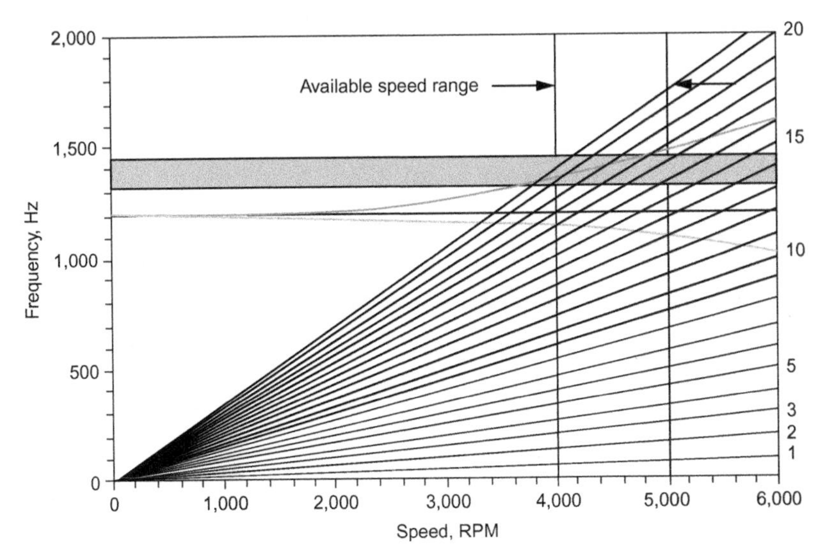

FIG. 39.1.1 A typical Campbell diagram shows which rotor speeds must be avoided so as not to cause resonating vibration of the machine's rotor components.

It seems the equipment owner was unaware of the need to look at the vibration modes of the blades in these steam turbines. A Campbell diagram, or interference diagram (Fig. 39.1.1), is used to indicate what speeds to avoid so as to safeguard blade life in a particular stage. Because almost all blade failures are caused by vibratory stresses, many reliability-conscious purchasers are requesting Campbell diagrams with turbine quotes or orders. A Campbell diagram is a graph with turbine speed (r/min) plotted on the horizontal axis and frequency, in cycles per second, plotted on the vertical axis. Also drawn in are the blade frequencies and the stage-exciting frequencies. When a blade frequency and an exciting frequency coincide or intersect, it is called resonance. Stress magnitudes are greatly amplified at resonance.

CONSIDER SWEEPING UPGRADES TO GENERAL PURPOSE STEAM TURBINES [API-611]

A mechanical engineering consultant at a Middle Eastern Corporate Engineering ("MECE") location took tangible steps to infuse modern reliability thinking in old general purpose steam turbines. Fig. 39.1.2 shows the impressive result.

There are now pneumatic actuators at the trip valve and speed indicators on the OB (outboard) bearing, with a separate low cost PLC-based overspeed electronic trip system. Many improvements were applied by encouraging manufacturers such as Elliott and Dresser Rand to follow MECE's year 2009 GP (general purpose) standard which required many design changes to the vendors' old-style established lines. The consulting engineer worked with Elliott to achieve a heavy duty design which eliminates the conventional spring-loaded trip valve and applies, instead, a pneumatic trip actuator. Its fail-safe spring return also eliminates all mechanical trip devices normally found on steam turbine shafts. The first of the resulting "hybrid" GP machines was commissioned in 2015 at a major refinery on the Red Sea; it drives a deaerator pump. Note again that there no longer are mechanical trip linkages, no unsafe over-speed events, all done at a relatively low price and with features which have taken such turbine designs from their World War II-vintage up to this latest design. The proven hybrids have been available on the market since 2010.

FIG. 39.1.2 One of several Elliott Steam Turbines with state-of-art controls installed in a Red Sea refinery. *(Courtesy Abdulrahman Al-Khowaiter.)*

To conclude, over the past few years, the often mindless interpretations of "lean and mean" have impeded reliability. As just one example, neither time nor budgets were allocated to understand what happens when steam turbine speeds are reset for operation away from the original governor adjustment range. The result has been a much higher probability of steam turbine blade failures. Consider this comment a plea to know if and when it is proper to be lean, or green, or whatever. Evaluating interference diagrams and steam turbine blade stresses is a mandatory task that can never be overlooked in a modern plant. Include the upgrade steps described by MECE.

Chapter 39.2

Steam Turbine Overspeed Protection Retrofits

INTRODUCTION

Not too long ago, the overwhelming majority of small-turbine overspeed trip devices were mechanical; they incorporated a spring-activated trip bolt that moves out and contacts a trip lever whenever a preset overspeed is reached. A number of factors make it necessary to check and possibly recalibrate these built-in devices on a yearly basis. Needless to say, process plants are not always performing this preventive maintenance activity in a timely and procedurally correct fashion. Unscheduled downtime events and serious injury, even deaths, have resulted from such oversights.

With electronics and even robotics having made huge gains in the recent past, well-engineered and highly reliable electronic overspeed prevention (OSP) systems are now available for new and retrofit applications. These systems offer end users the security and dependability of a two-out-of-three voting system with the flexibility and ease of do-it-yourself retrofit installation and configuration.

One relatively recent OSP system incorporates a two-out-of-three voting feature that monitors turbine speed and will initiate a trip command to prevent overspeed events. Designed and manufactured by Des Moines, Iowa-based Compressor Controls Corporation (www.cccglobal.com), the system includes three identical speed modules that individually measure a frequency input signal from a passive or active magnetic pickup sensor. A supervisory module continually monitors the three speed modules for proper operation, which helps to eliminate unnecessary downtime and increases system availability.

MULTILEVEL PASSWORD ENSURES SECURITY

The OSP system provides users with a multilevel password function for added security. Each level provides access to higher system functions, including peak speed reset, overspeed test mode, configuration mode, and set new passwords.

The three speed modules provide a greater level of dependability. Open-pickup detection on "passive" sensors is provided as well as "dynamic" sensor failure detection via the supervisory module, which knows when a speed module should be receiving a valid frequency input. Active sensor power is provided as an output as well. Each speed module can display its current speed and current set point. Additionally, each speed module includes four dedicated LEDs for indicating overspeed test, trip, MPU failure, and fault conditions.

SELF-DIAGNOSTICS: SUPERVISORY MODULE

CCC's overspeed protection system also features complete continuous self-diagnostics. The supervisory module monitors proper operation of the speed modules and power supplies. A four-line by 20-character alphanumeric display is integrated into the operator interface. A test mode provides testing of each speed module including the verification of the voting relays' operation. This module also includes a dedicated LED for indicating an alarm condition.

FLEXIBILITY

Two power supply modules provide redundant power to the relay board, speed modules, and supervisor. A failed power supply module can be replaced without interrupting the operation of the OSP system. With CCC's OSP system, users no longer have to order separate systems for specific power requirements or required trip logic. De-energize-to-trip or energize-to-trip logic is jumper selectable. By meeting a broad range of installation requirements, this OSP system can be installed with any turbomachinery train for virtually any application. What's more, installation and configuration can be completed by your own in-house personnel.

API/ISO COMPLIANCE

The "Guardian" OSP system is API-670- and ISO-compliant, making integration with an existing system easier. Even though some competitive systems are available with simplex, redundant, and triplex speed switches, not all of them comply with the latest API and ISO requirements. Many OSP systems lack self-diagnostic testing and data communication capability. These are standard features with CCC's OSP system. Together with the many other advanced features, the resulting package opens up safety, reliability, and maintenance enhancement possibilities that merit serious consideration.

Chapter 39.3

Consider Bearing Protection for Small Steam Turbines

INTRODUCTION

In most small machines, there is a need to limit both contaminant ingress and oil leakage. Inexpensive lip seals are sometimes used for sealing at the bearing housing, but lip seals typically last only about 2000 operating hours—three months. When lip seals are too tight, they cause shaft wear and, in some cases, lubricant discoloration known as "black oil." Once lip seals have worn and no longer seal tightly, oil is lost through leakage. This fact is recognized by the API-610 standard for process pumps, which disallows lip seals and calls for either rotating labyrinth-style or contacting face seals. Modern bearing protector applications accommodate small steam turbines and are a subject of interest here.

LEAKAGE EVENTS

Small steam turbines often suffer from steam leakage at both drive- and governor-end sealing glands. Each bearing housing (Fig. 39.3.1) is located adjacent to one of these two glands, which contain carbon rings. It is a well-known

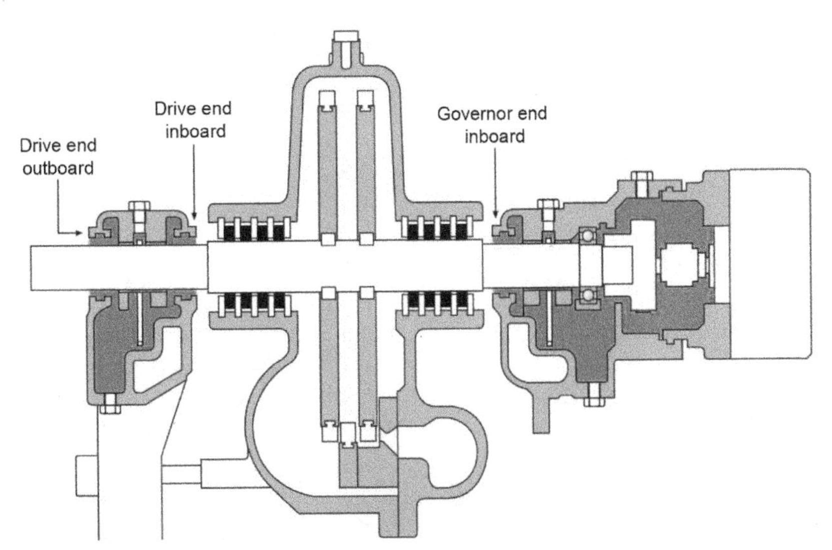

FIG. 39.3.1 Generic sketch of a small steam turbine. The drive end is on the left. A mechanical or electronic governor can be accommodated on the thrust end at the right.

FIG. 39.3.2 Steam leakage will often contaminate the adjacent bearing housing.

fact that, as soon as the internally split carbon rings start to wear, high-pressure and high-velocity leakage steam (Fig. 39.3.2) finds its way into the bearing housings. Traditional labyrinth seals have proven ineffective in many such cases and only solidly engineered bearing protector seals now manage to block leakage steam passage.

The bearing housing protector seal in Fig. 39.3.3 was designed for steam turbines. It incorporates a small- and a large-diameter dynamic O-ring. Recognize the similarity with the bearing protector seals highlighted in Subject Category 14. This bearing protector seal, too, is highly stable and not likely to wobble on the shaft; it is also field-repairable. With sufficient shaft rotational speed, one of the rotating ("dynamic") O-rings is flung outward and away from the larger O-ring. The larger cross-sectional O-ring is then free to move axially and a microgap opens up.

When the turbine is stopped, the outer of the two dynamic O-rings will move back to its standstill position. At standstill, the outer O-ring contracts and touches the larger cross-sectional O-ring. In this highly purposeful design, the larger cross-sectional O-ring touches a relatively large contoured area. Because contact pressure = force/area, a good design aims for low pressure. Good designs differ greatly from technologically outdated configurations wherein contact with the sharp edges of an O-ring groove will cause O-ring damage.

Fortunately, concerns as to the time it might take to upgrade to advanced bearing protector seals have been alleviated. In June 2009, a refinery in the Netherlands requested the installation of the bearing protector seal shown in Fig. 39.3.3 in one of its 350 kW/3000 rpm steam turbines. No modifications were allowed on the existing equipment, and installation of three modern bearing housing protector seals on the first machine had to take place during a scheduled plant shutdown in June 2009.

With no detailed drawings of the bearing housings available, the exact installation geometry could only be finalized after dismantling the small steam turbine. One of the main problems was the short outboard length: less than 0.25 in. (6.35 mm) was available due to the presence of steam deflectors and oil flingers. But the manufacturer's engineers were able to modify the advanced design in Fig. 39.3.3 to fit into the existing groove of the OEM labyrinth seals. Delivery was made within one week of taking measurements of the steam turbine and bearing housings, and the turbine has been running flawlessly for years.

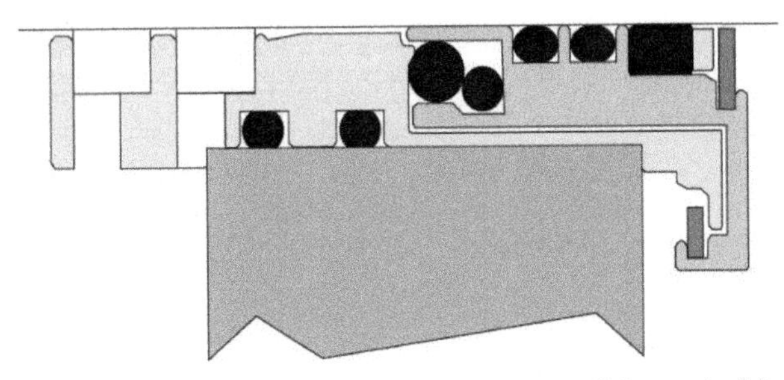

FIG. 39.3.3 Modern steam turbine bearing protector seal made to fit into each of the three bearing housing end caps of most small steam turbines. *(Courtesy AESSEAL Inc., Rotherham, the United Kingdom, and Rockford, TN.)*

Our point is that highly cost-effective equipment upgrades are possible at hundreds of refineries and petrochemical plants. Nevertheless, superior bearing protector products for use in steam turbines must be purposefully developed. The type described here has important advantages compared with standard products typically used in pumps:

- It is suitable for high temperatures.
- It incorporates Aflas® O-rings as the standard elastomer.
- Extra axial clearance is provided to accommodate thermal expansion.
- High-temperature graphite gaskets are incorporated in this design.

There should no longer be any reason for water intrusion into the bearing housings of small steam turbines at reliability-focused industrial facilities.

Chapter 39.4

Consider "Water Washing" for Steam Turbines

INTRODUCTION

Occasions may arise when deposits form on the internal parts of steam turbines. The accumulation of these deposits may be indicated by a gradual increase in stage pressures over a period of time with no evidence of vibration, rubbing, or other distress. Such deposits have a marked detrimental effect on turbine efficiency and capacity. When deposits cause extensive plugging, thrust bearing failure, wheel rubbing, and other serious problems can result. Fortunately, the deposits can be washed off with the steam turbine staying in operation.

CLASSIFICATION OF DEPOSITS

Deposits are classified as water-insoluble and water-soluble. The characteristics of the deposits should be determined by analysis of samples, and corrective measures should be taken to eliminate these deposits during future operation. When it has been determined that deposits have formed on the internal parts of the turbine, three methods may be employed to remove the deposits:

1. Turbine shut down, casing opened, and deposits removed manually
2. Turbine shut down and allowed to cool, the deposits cracking off because of temperature changes
3. Online water washing (while running); essentially removal of deposits when deposits are water-soluble

Since deposits tend to accumulate to a greater extent during steady high-load conditions (e.g., base load generator drive and ethylene process drive), the application and plant operating conditions will dictate which method will best serve to restore the turbine to optimum performance. In a generator-drive service, shutting down the unit or water washing at low speed and reduced load may create minimum plant upset. The size of process drivers and plant operating conditions present different circumstances.

METHODS OF WATER WASHING

Turbine washing at full speed (onstream cleaning) can be and has been successfully accomplished on many mechanical drive steam turbines. Considerable hazard attends any method of water washing, and full-speed washing is more hazardous than washing at reduced speed. But this can be accomplished provided great care and judgment are exercised. While we know of no steam turbine manufacturer who would guarantee the safety of turbines for any washing cycle, capable manufacturers recognize that deposits do occur. They will, therefore, help operators as much as possible in dealing with the problems until effective prevention is established.

Saturated steam washing by water injection is the conventional and well-tried method of removing water-soluble deposits from turbines. The amount and rate of superheat to be removed and the amount of steam flow required for operation determine the water injection rate. It is the injection of large quantities of liquid (such as may be required on process drivers) that creates potential problems.

The nature of a typical impulse turbine lends itself to full-speed water washing. Axial clearances between first-stage buckets and nozzles and between moving buckets and diaphragms will range from 0.050 to 0.090 in. The typical Ni-resist labyrinth packing radial clearance when the unit is cold will be approximately 0.007 in. The labyrinth will seal on the shaft only; the moving blades will not require seals. With impulse turbines, these liberal clearances help minimize the hazards associated with water washing. Nevertheless, numerous reaction turbines have also been successfully water-washed.

Water injection is accomplished by a piping arrangement for the atomization and injection of water into the steam supply to ensure a gradual and uniform reduction in the temperature of the turbine inlet steam until it reaches 10–15 °F superheat. It is probably a safe rule that the temperature should not be reduced faster than 25 °F in 15 min or 100 °F/h. Figs. 39.4.1 and 39.4.2 show suggested piping arrangements for the admission of water and steam, and a simple assembly of fabricated pipes to form a desuperheater is found in the text referenced below. Failure of water injection pumps presents a great hazard, especially at maximum injection rates. To guard against pump failure, untreated boiler feedwater is used since these pumps are usually the most reliable in a plant.

If plant operating conditions allow, the vacuum on a condensing turbine should be reduced to 5–10 mmHg; for noncondensing turbines, the exhaust pressure should be reduced to atmospheric pressure. Note that on any noncondensing unit requiring full-speed washing, the manufacturer should be consulted about minimum allowable exhaust pressures. Extraction turbines should be run with the extraction line shut off.

A steam gauge and thermometer should be installed between the trip-throttle valve and the governor-controlled valves. The thermometer should preferably be a recording type and should be very responsive to small changes in temperature. Low-speed wash, as illustrated in Fig. 39.4.1, represents a well-understood method of deposit removal.

To start the washing procedure, it is normally recommended to operate the turbine on trip-throttle valve control at one-fifth to one-fourth normal speed with no load. The live-steam valve to the mixer would now be opened and the boiler stop valve closed, after which the trip-throttle valve and the governor valves may be opened wide and the speed controlled by the small live-steam valve to the mixer. Water is then supplied to the mixing chamber in quantities sufficient to reduce the steam temperature at the recommended rate until 10–15 °F (6–9 °C) superheat at turbine inlet is reached. During the washing cycle, the exhaust steam should be discharged to the sewer.

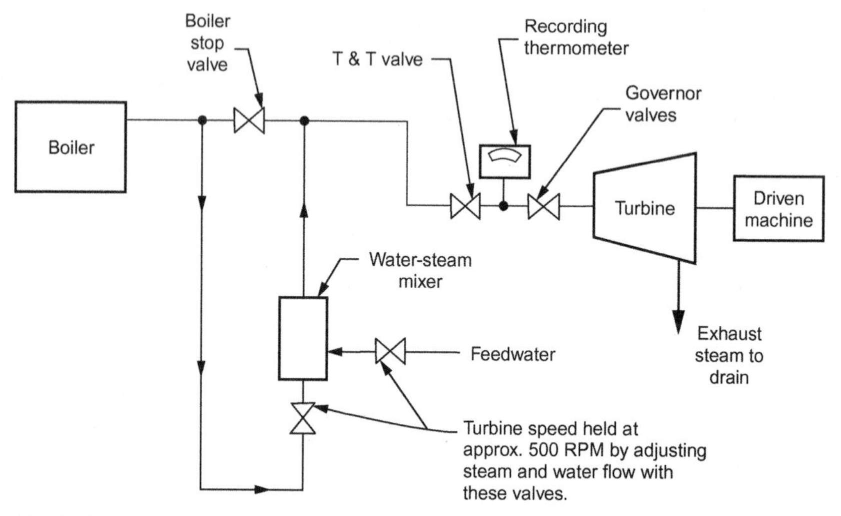

FIG. 39.4.1 Schematic of low-speed water wash system.

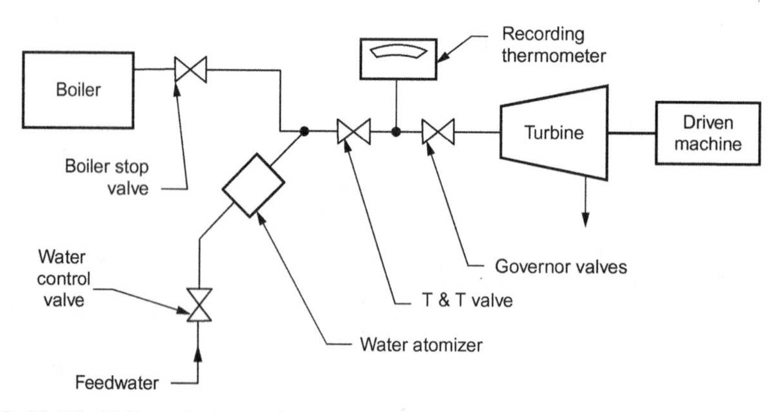

FIG. 39.4.2 Full-speed water wash arrangement.

Full-speed water washing (Fig. 39.4.2) uses a water atomizer instead of the water-steam mixer arrangement. Both Figs. 39.4.1 and 39.4.2 illustrate operating principles and the required minimum instrumentation. Advanced controls usually go beyond the instruments shown here, but will adhere to the principles described here.

Chapter 39.5

Steam Trap Technology Worth Understanding

INTRODUCTION

Steam is water vapor. At elevated pressure and temperature, steam contains much energy. As it travels to its destination in pipes, some of the steam may recondense and again become water. Water must be removed in steam traps. There are many types of steam traps—some old and some new technology. The implications are obvious: reliability, high versus low, and maintenance outlays, costly versus less costly.

It is becoming more understood that the steam system asset plays a major role in the determination of plant profitability. Consider that it is the major source of mass process heat that, when optimized, provides the best chance for high-quality production. Over time, energy prices have moved dramatically with resultant changes in generated steam cost. It seems that the cost of producing steam has correlated to a plant's focus toward long-term proactive elimination of avoidable steam loss; when costs are high, there is an increased focus relative to when costs are low. However, this type of thought process overlooks the main reason to manage a steam trap population—to help maintain an optimized steam system that can operate with minimized hammer to avoid catastrophic damage or steam leaks, corrosion, and erosion. It also helps minimize back pressure and hammering in the condensate return lines, which improves system reliability as described below.

PROFITABILITY IS A FACTOR

The steam system asset is playing an increasingly major role in the determination of plant profitability. As a result of steadily increasing energy prices, steam cost has more than doubled from just a few years ago. These higher energy costs appear to represent a new standard as opposed to just a temporary spike and have caused a new focus toward long-term proactive elimination of avoidable steam loss. Modern steam trap technology is needed for thoughtful energy conservation (Figs. 39.5.1 and 39.5.2).

Open steam leaks from piping or equipment are usually easy to identify and schedule for their repair. The mystifying activity over the years has been how to consistently maintain a program that will accurately identify *internal* system leaks such as those from steam traps. Plants generally want to correct leaking traps, but have not always been able to assign trained personnel to the task

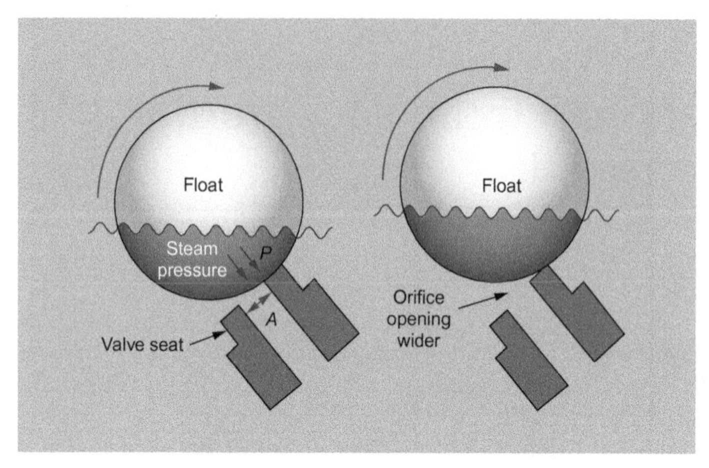

FIG. 39.5.1 Basic operating principle of one type of many advanced steam trap models. *(Courtesy TLV Corporation, Kakogawa, Japan, and Charlotte, TN.)*

FIG. 39.5.2 Steam traps are available in many shapes, sizes, and configurations. *(Courtesy TLV Corporation, Kakogawa, Japan, and Charlotte, TN.)*

or choose an accurate diagnostic instrument to identify each trap's condition. The facts show that if personnel are not well trained and experienced, it can be extremely difficult to accurately determine the condition of certain types of steam traps. In addition, maintaining the records in a custom database takes substantial time and resources that may not be available in some plants. This is where it may be useful to consider the diagnostic data collection, data logging, and database management systems shown in Fig. 39.5.3.

FIG. 39.5.3 TrapMan® advanced steam leakage condition monitoring device. *(Courtesy TLV Corporation, Kakogawa, Japan, and Charlotte, TN.)*

Traps can fail either "hot" (leaking and blowing) or "cold" (blocked and subcooled). Hot failures represent considerable energy loss value, but cold failures impact directly on the reliability of the downstream equipment (e.g., turbines and nozzles) and personnel safety. Identifying both types of failures is important to provide for the most efficient steam system.

Temperature is the first measurement made by the device shown in Fig. 39.5.3 to determine that a trap is not fully or partially blocked. Temperature is used to identify whether a trap is blocked or subcooled, as opposed to "out of service." Normally, "out of service" traps are those that have failed hot and represent energy loss. Repairing cold traps will improve the overall system reliability by eliminating consequential damage. TrapMan® also simultaneously measures the ultrasonic level of the steam trap and compares it to empirical data to make a flow judgment, including either just condensate only (condition "good") or steam loss (condition "failed"). The measurement is centered at high frequency, well above the normal auditory level. This high ultrasonic centering is important because a more clear differentiation between condensate and steam occurs at such an ultrasonic level.

MONITORING CONCEPT

The basic concept is that as the condensate flow increases, the associated ultrasonic noise level also increases. This continues to a certain ultrasonic threshold whereupon higher ultrasonic readings indicate steam leakage through a steam trap. As the leak rate increases, a higher level of ultrasound is generated. The higher readings are correlated to the empirical test base to determine the level of steam leakage. A report is then generated by the included TrapManager® database software that can be used for powerful analysis of the steam system. Among its many functional capabilities is the evaluation of steam loss of each failed trap. These data can be exported and uploaded into a site's maintenance software to compare return on investment (ROI) and enter an appropriate maintenance response on the activity log. As of 2016, TrapManager® software could list the type of failures for over 4400 specific models for various manufacturers of steam traps to help decisions about the recommended plant trap standard. Additionally, it can be used to assign value to the steam loss of each failed trap.

Subject Category 40

Suction Strainers for Fluid Machinery

Chapter 40.1

Pump Suction Strainers

INTRODUCTION

Whenever strainers are used because the upstream equipment is flawed, be sure to understand the important requirements imposed on strainers by reliability-focused engineers. These engineers recognize, first and foremost, that a distinction is to be made between temporary and permanent strainers.

Temporary strainers are generally installed with the tip pointed in the upstream direction, which places at least some of the material in compression instead of tension. These temporary strainers must be removed about one week after commissioning the piping loop. They are often fabricated on-site, using the general configuration shown in Fig. 40.1.1. However, variants do exist and are often constructed with support rods instead of the guard screen described below.

As a general rule, permanent strainers are designed to be left in place and must be cleanable without shutting down the pump or other fluid machine. Permanent strainers are typically available from a variety of commercial sources, must be made of high-grade corrosion-resistant materials, and, in the larger sizes, can be quite expensive.

Using the strainer guidelines found in Ref. [1], we observe the following:

1. Strainers (both temporary and permanent designs) may be cone- or basket-shaped; they should be installed between the suction flange and the suction block valve. The preferred orientation is shown in Fig. 40.1.1; the small diameter of the cone is upstream; the large opening is downstream.
2. The mesh size of strainers should be selected to stop all objects too large to pass through the pump main flow passage.

Petrochemical Machinery Insights. http://dx.doi.org/10.1016/B978-0-12-809272-9.00040-2

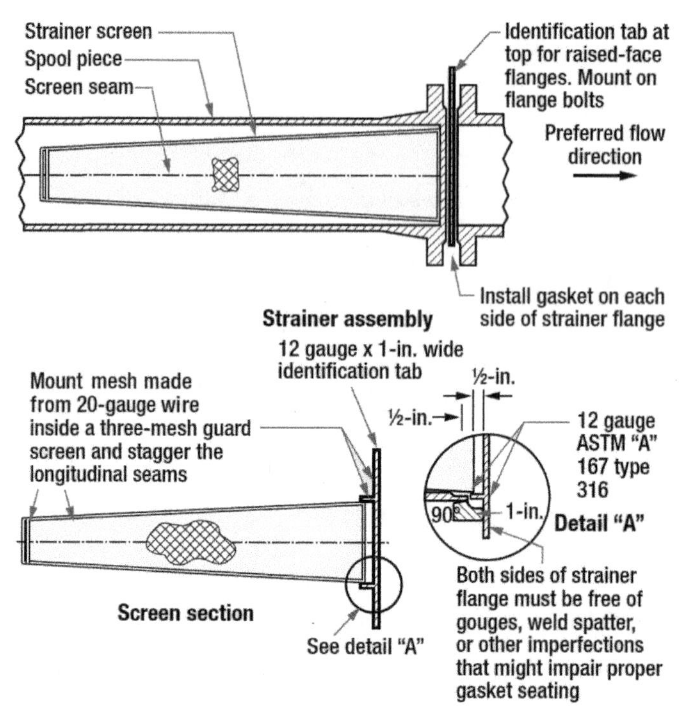

FIG. 40.1.1 Typical temporary suction strainer for pumps.

3. Temporary strainers must be used during pre-start-up flushing and initial (one-week) operating periods, *unless permanent strainers* are specified.
4. Piping layout should permit removal of strainers without disturbing pump alignment; spool pieces are typically used.
5. If *permanent* strainers are selected, the design and location of these strainers must permit cleaning without removing the strainer body.
6. Arrangement of permanent strainers should permit cleaning without interrupting the pumping service.
7. For installations with permanent strainers and equipped with a spare pump, a permanent strainer should be installed in the suction line of each pump.
8. Twin strainers or self-cleaning strainers may be used for pumps without spares.
9. Y-type strainers should be restricted to 2 in. maximum size.
10. Suction lines for proportioning pumps should be chemically or mechanically cleaned to permit operation without strainers.

There are some very important points we wish to reemphasize:

First, best practices (BPC) or best-of-class (BoC) companies distinguish between *start-up* strainers and *permanent* strainers. They insist on the removal of *start-up* strainers long before they will have become a serious disintegration risk. Also, BoCs have established that strainers are *not* needed upstream of most conventional process pumps after the initial start-up period. Second, it follows that once BoCs have determined strainers should be left in place for some reason, they allocate the resources needed to upgrade entire systems in order to reduce failure risk and maximize equipment uptime. Third, because BoCs are serious about maximizing pump uptime, they insist on best practices being implemented at all times. At those facilities, deviations from best practice have to be justified in writing, and a manager is asked to accept responsibility in those instances.

SIZING OF STRAINERS

Referring back to Fig. 40.1.1, even an experienced reliability professional may find some customary terminology on *wire gauge* and *wire mesh* confusing.

Wire gauge (alternatively spelled "wire gage") is indicative of electric conductor wire diameter, with primary implications for allowable current flow (amperes). According to working tables published for standard annealed solid copper wire (Ref. [2]), 20 gauge wire has a diameter of 0.032 in. and, at a temperature of about 40°F, has a resistance of approximately 11 Ω per 1000 ft of length. While suction strainers have nothing to do with electric current, we assume that the same wire diameter (i.e., roughly 1/32th of an inch) also applies to the stainless steel wire used in wire mesh. For the pump suction strainer in Fig. 40.1.1, an as-yet-undefined size of "mesh" is to be placed inside the three-mesh guard screen.

As regards terminology, wire mesh and wire cloth are terms used interchangeably in North America. Either term denotes a metal wire weave, or wire cloth, with its various implications regarding available flow-through area as a percentage of the total square inches of wire cloth area. The notes on the illustration (Fig. 40.1.1) recommend placing a *wire mesh made from 20 gauge wire inside a three-mesh guard screen.* So, to reemphasize, while wire mesh made from 1/32 in. wire (20 gauge) is to be placed *inside a three-mesh* guard screen, nothing is said about 20-mesh wire cloth.

Next, an explanation of "mesh," with particular reference to the three-mesh guard screen, likely recommended to strengthen or to back up an as-yet-unspecified finer mesh:

- But suppose seeing "three-mesh guard screen" puzzles you in Fig. 40.1.1. That simply indicates to the constructing welder that a re-enforcing screen is needed. This re-enforcing screen has three openings per inch in each direction. The notation "three-mesh guard screen" does not refer to the wire diameter or wire gauge. That would be 3 wires per inch in the x-direction (horizontal) and 3 wires per inch in the y-direction (vertical). Similarly, 8-mesh would be 8 wires in the x-direction and 8 wires in the y-direction, 200-mesh would be 200 wires in the x-direction and 200 wires in the y-direction, etc.

- In a three-mesh screen, there are, therefore, $3 \times 3 = 9$ openings. Moreover, in a three-mesh screen, the distance from the center of one wire to the center of an adjacent wire is 1/3 of an inch, or 0.333 in.
- One of many manufacturers of wire cloth offers three-mesh product made with wires ranging in diameter from 0.031–0.162". Irrespective of wire size, there would be 9 openings, and the distance from the center of a wire to the center of an adjacent wire would always be 0.333 in.
- If one were to pick the 0.162" diameter wire, one would have greater strength than if one chose the 0.031" wire.
- Likewise, if one picked the 0.162" wire diameter, the resulting open area would only be 26%, whereas if we had chosen the 0.031" wire diameter, we would have 82% open area (Ref. [3]).

A reasonable choice would be a three-mesh guard screen with a wire diameter of, say, 0.135". To reiterate, the wire-to-wire center distance would be 0.333", and we would have nine openings. For this wire diameter, the tables in Ref. [3] give an opening width of 0.198" (remember that opening width plus wire diameter equals 0.333"). Also, the tables tell us that we have ~35% of the area open for liquid flow and ~65% of the area would be taken up by the 0.135" diameter guard screen wires.

Next, one would determine the wire mesh of the cone within a cone, which was to utilize the 20 gauge wires. That particular wire cloth, to be placed inside the three-mesh guard screen, should use 20 gauge—0.032"— wire. Ref. [2] shows this wire diameter about midway in its six-mesh table. In six-mesh wire cloth with 20 gauge wires, the openings are 0.1347". Adding the two numbers (0.032" + 0.1347") = 0.1667"—six squares per inch.

Using the above explanations and Fig. 40.1.1 will equip us to ask a draftsperson to design a pointed cone—the three-mesh guard screen—with (based on experience) an open surface area in both guard screen and inserted finer mesh that's approximately three times the cross-sectional area of the pipe or spool piece. The six-mesh metal cloth would go on the inside of the cone, at which time the device finally qualifies being called a strainer.

While it would be preferable to install the strainer such that the pointed tip of the cone encounters the flowing liquid first, it would still work with flow the other way around. But one would endeavor to ascertain the welder does the job properly and the materials of construction are picked in accordance with any corrosion concerns.

Regarding suction strainers it's worth recalling our basic premises: first, strainers are primarily installed to catch hard hats, welding rods, and beer bottles left in the piping. Second, strainers are intended to catch tower packing that dislodges from upstream equipment. Note that these different contingencies exist regardless of whether one is dealing with clean or dirty services. Experience shows the entire issue being more relevant during equipment commissioning. If the premise is different from what we have described here, there's probably a need for filtration equipment. In other words, strainers are not filters, and filters are not strainers.

Thus, permanent strainers would have the same mesh as temporary strainers. However, permanent strainers (a) would be made from corrosion-resistant materials; (b) if small enough, would be inserted in a "pipe-Y" with valved blowdown provision; (c) or would be part of a redundant twin (parallel) installation.

On any permanent strainer installation, it would be wise to monitor the delta p (the pressure drop) between the upstream and downstream sides of the mesh. Clogged strainers can cause serious machine malfunction and costly damage.

ECCENTRIC REDUCERS AND STRAIGHT RUNS OF PIPE AT PUMP SUCTION

Whenever fluid flows through pipes, it has to overcome pipe friction. Its pressure is thus reduced. Pressure losses are lower if flow velocity is slowed down. From an overall point of view and taking into account that fluid quantity equals fluid velocity multiplied by cross-sectional area of flow, pipe sizing criteria favor larger pipe diameters on the low-pressure (suction) side and smaller pipe diameter on the higher-pressure (discharge) side of pumps. Reducers are usually required to adapt suction pipe diameters to pump suction nozzle sizes. The orientation and configuration of reducers require special care; refer to Chapter 28.1 for details.

REFERENCES

[1] Mark's Standard Handbook for Mechanical Engineers, seventh ed., McGraw-Hill, New York, NY, 1969.

[2] www.mcnichols.com/product/wiremesh.

[3] www.darbywiremesh.com/wire-mesh-glossary.html.

Subject Category 41

Tooling and Scaffolding

Chapter 41.1

Consider Suspended Access Scaffolding

INTRODUCTION

You may always have left scaffolding and access-to-work questions to an outside contractor to decide. But plant safety and timely work completion are everybody's business, and as a reliability professional, safe and on-schedule work execution undoubtedly ranks high among your concerns. Indeed, properly selected access scaffolding can be critical to the success of emergency repairs, planned equipment turnarounds, and field modifications. This fact should prompt us to make sure that owners and outside contractors are aware of suspended access systems. There are many cases where the use of these systems will optimize or facilitate how equipment overhauls and/or a number of overlapping tasks are carried out.

Figs. 41.1.1–41.1.3 are among the hundreds of applications available to users in all types of industries. The different variations are limited only by the imagination and resourcefulness of a perceptive user. Working in conjunction with established contractor-suppliers, suspended access scaffolding has greatly contributed to users' bottom lines.

VERSATILITY AND STRENGTH

Each of the modular platform designs shown in Figs. 41.1.1–41.1.3 is of high quality. These suspended access systems can be assembled from just a few components and quickly reconfigured to fit almost any shape or size. Assembly work is not labor-intensive, and each component is easily handled by just one person. No special tools or skills are needed, although the developer of QuikDeck suspended access systems can provide its own experts on short notice.

Petrochemical Machinery Insights. http://dx.doi.org/10.1016/B978-0-12-809272-9.00041-4

The load capacity of the systems we looked at ranges from 25 psf (122 kg/m^2) to over 75 psf (366 kg/m^2); the system gets its strength from structural steel members, not wire rope. All structural parts are hot-dipped galvanized to retain long service life and high safety factors.

FIG. 41.1.1 Suspended access scaffolding under compressor decks is no different from its use here on an offshore installation. *(Courtesy Safway Group Holding LLC, Waukesha, WI.)*

We noted that the workspace is flat and wide open (Fig. 41.1.3). This can be a distinct advantage when working on process equipment on offshore platforms and under pipe racks where painting and repair will thus be facilitated. Many boiler maintenance tasks fall into the restricted access category, and it helps that the QuikDeck configurations are almost limitless. They include a wide range of specialized applications where new construction, structural rehabilitation, painting maintenance, and all kinds of repairs often overlap.

FIG. 41.1.2 Safway's QuikDeck suspended scaffolding, here used on an offshore platform. *(Courtesy Safway Group Holding LLC, Waukesha, WI.)*

In essence, even seemingly routine scaffolding tasks merit the "second look" of conscientious staffers at reliability-focused and cost-conscious facilities. It should be reassuring that the service provider offers training and engineering services to either a purchasing company or a "use-if-needed" (renting) entity. Turnkey packages include engineering drawings and a variety of delivery options; all of these are approved by a registered professional engineer (PE).

FIG. 41.1.3 Suspended scaffolding is available in virtually limitless configurations. *(Courtesy Safway Group Holding LLC, Waukesha, WI.)*

Subject Category 42

Training Strategies for Success

Chapter 42.1

Finding Answers to Library and Training Dilemmas

INTRODUCTION

Before being updated and reworked to form the core of this book, hundreds of the *HP in Reliability* columns generated feedback, discourse, and questions. The questions, in particular, allowed us to assess the state of knowledge in the industry. Sometimes, though, these questions gave clear indication that the questioners had not availed themselves of all the tools at their disposal.

AVOID HAPHAZARD PURSUITS

For reliability professionals to not insist on the conscientious pursuit of doing the right things and doing things right is a negative reflection on their diligence and competence. Their tolerance for repeat failures can boggle the mind. We have even seen (false) claims that a particular equipment problem is unique and has never been experienced elsewhere.

The statistical probability of a petrochemical plant or refinery experiencing repeat failures due to phenomena never experienced elsewhere is close to zero. It follows that troubleshooters or reliability professionals will find the solution to their problems in the literature to which they all have access. But access to a book does not mean that the book is being read. This prompted one of many of Mark Twain's sayings about the man who has a book but does not read. To again paraphrase Mark Twain, that man is no different from the man who cannot read. Think about it: a late 19th-century observer, Mark Twain, places *some* of us on the level of illiterates.

Petrochemical Machinery Insights. http://dx.doi.org/10.1016/B978-0-12-809272-9.00042-6

LITERATURE DETAILS

The principal engineer of a major EPC (engineering-procurement-construction) firm at a US West Coast location commented on literature issues in a letter. He regretted the lack of people's effort to find answers by exerting themselves. When engineers have questions, they should understand and research an issue before depending totally on input from others. This principal engineer observed that, when we leave an engineering school, we have—hopefully—a grasp of the basics of certain engineering fundamentals. Initially, none of us have either the experience or equipment and industry-specific knowledge to do much useful work. Being able to teach ourselves as needed is the key element for doing useful work. But he pointed out that teaching ourselves, that is, learning without access to a mentor, requires an unusually elevated attitude and motivation. There also needs to be a willingness to exert ourselves. You cannot exert yourself if more than half of your time is spent on recreational activities.

Study is work. Employees who study and exert themselves to gain knowledge also qualify for receiving support—mentoring—in the workplace. Moreover, study and the expending of training effort encompass giving support and acting as mentors for others. But remember that we can only teach what we ourselves know.

Current social and management trends cut away at both ends of the spectrum. Social trends make sustained work on a single subject less likely. If a subject can't be broken down into 30-second segments, it's often viewed as too hard. It's difficult to get such attitudes reset and, with some people, one never succeeds in resetting or adjusting attitudes. But the employer has to make a conscious effort to nurture changes in attitudes and must be willing to subsidize such small steps as attending local ASME meetings. Another approach would be to make good use of an essential reliability library, which, by the way, is tabulated in Appendix II to this text.

COMMENTS ON LIBRARY TRENDS

While social trends are disturbing to some readers, management trends are distressing to others. One observer noted that it had been more than 20 years since he had seen a plant site or major engineering office with more than a mediocre collection of reference material that covered basic engineering needs. In fact, the majority of sites he had visited no longer had a reference library. The observer contended that if it could not be found on the Internet, information was treated as if it didn't exist.

Some companies avail themselves of a "pay-for-use" service. A pay-for-use resource can heighten an employee's job satisfaction. Because it makes retrieving data less tedious, engineers get to the problem-solving stage (the fun part) more quickly. A few of the more enlightened companies purchase subscriptions to specialized engineering libraries. But while an online library and handheld wireless gadgets filled with technical texts are both legitimate cost-saving methods, companies using these methods are in the minority.

Properly used, online resources can often help a team of engineers and scientists to rapidly find the information they need. Some online resources claim to deliver reliable data with speed and ease, thanks to the complete searchability of their collections and round-the-clock accessibility from any Internet connection. It can be reasoned that rapid access to the right technical information improves a company's bottom line by reducing data retrieval time and compressing the overall product development or problem-solving schedule.

Of course, we might also give full credit to individuals in different industries who managed to formalize access to mentors. Mentors may be in-house resources or nonresident SMEs (subject matter experts) willing to converse with the individual seeking advice. There are also employees who collected excellent personal libraries or loaded their electronic gadgets with technical books at their own expense. Indeed, having one's own collection of reference texts in paper or electronic form shows dedication to technical excellence. Assembling such personal reference libraries sometimes becomes the most reasonable way we work because there are no better choices. Yet, the most productive reliability engineers grew up with reference libraries and access to a network of competent mentors. We should conclude that we can follow in their footsteps, if we make the effort.

In essence, we are of the opinion that employers would do well to create alternatives to pure self-teaching and insistence on personal reference libraries. Reasonable alternatives involve access to other resources and development through targeted training. In another chapter of this text, we mention four distinct phases for such training.

MERGING TRAINING AND NETWORKING

Regrettably, training is an easy budget item to cut in hard times, and the refining and chemical processing industries are, more often than not, claiming to experience hard times. That claim was proffered when a barrel of oil costs $20 (because a year earlier it had been $15/bbl) and decades later when it was at $100/bbl, whereas two years earlier, it had been at $50. It was always possible to make these types of arguments. With both training seminars and conference participation being cut across the industry, a few dollars are being saved—or so it seems.

But there still are zero-cost, highly effective training methods and these should never be abandoned. Creative and determined individuals can do a lot to help themselves. All that's required is the right attitude and finding small, local opportunities such as attending local ASME, Vibration Institute, or AIChE chapter meetings, which does not cost much. Look to networking with other people in similar situations, organize in-house shirt-sleeve seminars, mandate the quick perusal of certain no-cost trade journals, and work on keeping up to date. And never overlook the basics. There will never be a substitute for diligence and craftsmanship. Insist on it; teach it; reward it.

Lack of self-help because an individual persists in harboring the wrong attitude is relatively rare. When it does occur, it is—hopefully—confined to an individual and not widespread within an entire plant. That is not the case with the occasional management failure; such failures are more insidious and, ultimately, much more serious. Management failure will inevitably include lack of effort in improving attitudes across the board. Starving the organization of the few resources it would take to support those with good attitudes puts an entire organization on the path to failure.

Chapter 42.2

Attend User Symposia for Continuous Improvement

INTRODUCTION

Since the early 1970s, Texas A&M University's International Turbomachinery Symposium has guided the most profitable user companies toward achieving equipment reliability optimization. Similarly, a few years later, Texas A&M's International Pump Users Symposium started to provide the same professional networking opportunity and value-adding guidance. The money spent for attending these (usually) productive combinations of lectures, discussion group sessions, tutorials, case history presentations, and floor exhibits is likely to represent the best value your training dollar can possibly buy today. When your employees return from attending one of these conferences or symposia, make them write down, in ten sentences or less, what they learned and how they propose to follow up on what they have learned. What will they do so as to inform and assist the entire organization? Insist on this type of networking and dissemination of relevant information.

WHO MEETS AT CONFERENCES AND SYMPOSIA AND WHY THEY MEET

Each year, hundreds of reliability professionals attended the two symposia in Houston, Texas; the attendees and/or course participants see and hear about emerging technology. During networking or discussion group sessions, attendees are given access to user feedback and experience that allow them to better understand the root causes of elusive and repeated rotating equipment problems. By seeing and listening, attendees and participants position themselves to successfully implement cost-saving failure avoidance measures of the type we would wish to share in this chapter of our text.

WHAT PUMP FAILURES REALLY COST YOUR PLANT

When pumps fail, the cost can be staggering. In 1996, a South American refinery repaired 700 of their 1000 installed pumps. In that year, these pumps consumed well over 1300 rolling element bearings. For reasons of energy balance, the 500 operating pumps were driven by small steam turbines, while the 500 standby pumps were electric motor-driven. A major contributing cause to this inordinate number of bearing failures was lube oil contamination. Contamination problems could have been avoided by plugging the vents and sealing the bearing housings with face-type spring or magnet-activated seals described in Subject Category 14.

The South American refinery's pump maintenance expenditures were not unlike those of a US oil refinery that, in 1984, performed 1198 repairs on their 2754 installed pumps. Of these, 40% were repaired in the shop and 60% in the field. Based on work order tracking, the direct dollar expenditures per repair amounted to $5380. Direct costs for pump repairs thus amounted to over $6,400,000, but the actual charges were $10,287 per pump repair after incremental burden consisting of employee benefits, refinery administration, mechanical-technical overhead, and materials procurement costs was added. It therefore cost the US refinery in excess of $12,000,000 to repair 1198 pumps in 1984. Interestingly, the per pump repair cost estimate for 2016 is only marginally higher. It appears that progressively fine-tuning predictive maintenance tools (condition monitoring technology) allows failure trends to be detected before failures become excessively expensive or catastrophic.

PUMP MEAN TIMES BETWEEN FAILURE (MTBF) VARY WIDELY

We find it interesting to note that the abovementioned US refinery had a pump MTBF of 2754/1198 or 2.3 years. We know of refineries that today (in 2016) have process pump MTBFs as high as 9.4 years. Much of the data and information made available at International Pump Users Symposia point out what the low MTBF performers overlooked or disregarded. There is still the occasional user facility that applies oil mist in the "wet sump" mode to bearing housings equipped with constant-level lubricators that had their respective oil levels exposed to ambient air; see Subject Categories 4 and 17. The slight pressure increase in the bearing housing lowered the oil level in the housing and caused it to increase in the surge chamber of the lubricator, just as Bernoulli's law first stated in about 1750. Application of the pressure-balanced constant-level lubricator in one of these earlier subject categories could have saved this user a few hundred thousand dollars.

EMERGING TECHNOLOGY WORTH TRACKING

In the exhibit area, symposium attendees are often made aware of new and important developments. One of these highlighted the merits of laser surface texturing that treats one of the seal mating surfaces with thousands of micropores. Another one touted the merits of diamond-infused seal faces. There are faces wherein it is thought that each pore acts as miniature hydrodynamic bearing creating a small separating force that reduces friction and lowers face temperatures by as much as 40°F (~16°C). It stands to reason that reduced seal wear and extended seal lives have been experienced at several installations that now use superior seal surfacing approaches. More data on superior surface treatment options can be found on the websites of four major mechanical seal manufacturers AESSEAL, John Crane, Eagle Burgmann, and Flowserve. Be sure to check out the scientific facts before the expert marketers overwhelm you and sell you something of little value. Diamond coated seal faces are a case in point; see Subject Category 21 before making decisions on these expensive surface coatings.

Still, there seems to be universal agreement that

- user-oriented symposia are of immense value to reliability-focused plants,
- repair-oriented process plants spend large amounts of money on repeat repairs and are becoming less and less profitable,
- TAMU's International Pump and Turbomachinery Symposia have provided serious professionals with information that is readily transferable and desperately needed in an environment where, regrettably and all too often, layoffs, force reductions, and cancelation of training funds take place without first thinking through the consequences.

Chapter 42.3

Fact-Checking after Attending Reliability Conferences

INTRODUCTION

Attending technical conferences should have, as one of its primary goals to observe industry practices, to pick up facts and to spot trends. The time one spends at these conferences can be put to outstanding use by networking and checking one's own way of doing things against the methods, work processes, and procedures used by others. The "others" we want to concentrate on are almost always the best of class (BoC), our BoC competitors. As we listen to presentations or attend workshops and discussion group sessions, we try to capture items deemed of interest to our facility or company. Our managers then expect us to brief them, and also our peers, on important findings. We maximize the value of attending conferences by disseminating the information and submitting recommendations regarding better ways of doing our reliability-focused tasks to our own management. This is an important feedback loop. You will find 12 examples, below, in the abbreviated format that space constraints allow.

POPULAR BELIEFS ARE NOT ALWAYS CORRECT

At these conference venues, we occasionally listened to popular beliefs that are not always correct. Just as in a single year seemingly reasonable people in the United States reported thousands of unidentified flying objects (UFOs) and aliens briefly treating them to a view of Pluto, there can be reports of mystifying events in the hydrocarbon processing industry (HPI). Two people might have access to the same background information but give their stories a completely different spin. Spins can turn observations into mere anecdotes of questionable veracity. So, a perceptive conference attendee must come to grips with snippets of information. Should the conference report relate the snippets as amusing anecdotes or should the report withhold the snippets because they add no value? Most of us use reasonable judgment and try not to allow erroneous information to go unchallenged. But, again, we are at the conference to absorb useful facts.

TRUE VERSUS FALSE, AS VALIDATED AT USER'S CONFERENCES

Well, here are a few recollections, claims, counterclaims, and issues we picked up at three reliability conferences in a single year:

1. In BoC plants, the use of nonpolluting closed oil mist systems is on the increase. These systems include electric motor drivers and all standby equipment at BoCs.
   Finding: Fact
2. Oil mist can cause motor insulation to degrade.
   Finding: True for motors made in the 1940s and early 1950s. *Not correct* for motors with epoxy insulation, typically made from 1965 to the present (2016)
3. Pump vendors always use the best available lube application; therefore, oil rings and constant-level lubricators are not to be questioned.
   Finding: *Not factual*
4. Some reciprocating compressor purchasers insist on metal disk pack couplings.
   Finding: *Some do*, but do so at their own peril. Insistence on metal disk couplings is rarely serving the purchaser's best interests. Elastomeric elements are usually (although not always) safer and more reliable in these machines.
5. Some coupling manufacturers know little about coupling stiffness under actual operating conditions.
   Finding: *True*, unfortunately. Repeat failures have resulted.
6. Large electric motors are again made in the United States by at least one competent legacy manufacturer.
   Finding: *True*, and we were pleased to find out.
7. Plants that spend more money on maintenance are typically those that don't spend reasonable amounts of resources on reliability improvements.
   Finding: True. The issue was neatly summarized by Alan Poling (Solomon Associates) at a global reliability forum in 2012; parts of it are woven into Table 42.3.1.
8. Plants that have replaced traditional repair thinking and now use reliability thinking invest 5% of actual equipment cost in upfront machinery quality assessment (MQA).
   Finding: *True*. Effective MQA has been successfully practiced since the early 1960s [1,2].
9. Recently developed formulations of PEEK (polyetheretherketone) and carbon fiber-reinforced Vespel® bearings have joined proprietary formulations of polyimide resins as an advantageous pump wear part material.
   Finding: *True*. But be mindful of possible increases in axial load on the pump's thrust bearing that can result from close-clearance wear rings. And use the correct interference fit. Tight interference fits may cause a closing in of the bore. The ring can seize and initiate considerable damage. Different high-pressure polymers may have similar high-temperature exposure capabilities but vastly different moduli of expansion. Be especially careful if you use some of these materials as sleeve bearings in fluid machines. Be certain the relevant and/or desired clearances exist after assembly and installation are complete.

TABLE 42.3.1 The Relationship Between Reliability, Maintenance, and Industry Standards

- Reliability and maintenance are inextricably linked
- One cannot cost-cut one's path to improved reliability
- Maintenance costs are driven by reliability or the lack thereof
- Best performers achieve high reliability at low cost!
- Poor performers have high cost with low reliability!
- Each 1% increase in mechanical availability can translate into a 10% maintenance cost reduction!
- One cannot legislate prohibition of stupid actions, but one can certainly attempt to limit them with innovation

10. San Antonio/Texas-based Southwest Research Institute (SwRI) has refined and validated methods to better predict turbomachinery vibration. Rotor behavior modeling, often expressed as the log decrement in critical damping values, is now more accurate.
    Finding: *True*
11. There is a narrowing of the gap in reliability performance and profitability of top-quartile companies (upper 25%) compared with fourth-quartile companies (lower 25%).
    Finding: *Not true*. Regrettably, the gap is widening.

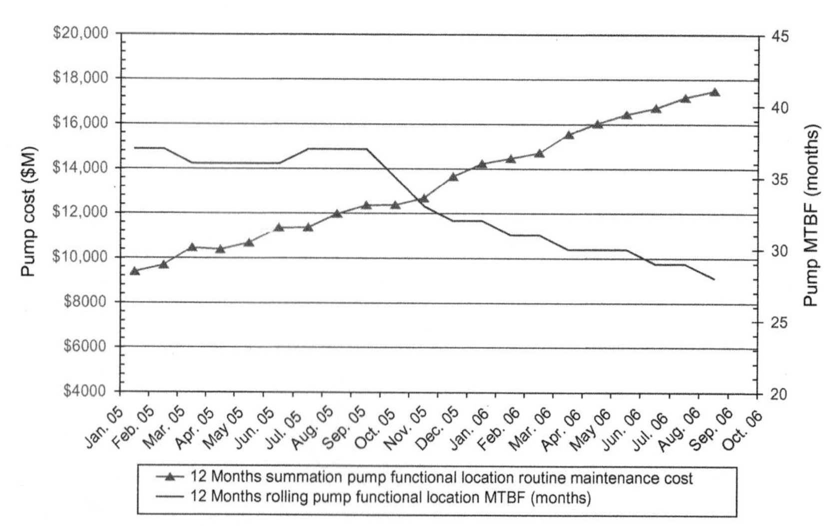

FIG. 42.3.1 As MTBF goes down, maintenance costs increase. Although this illustration represents a particular process unit, it is representative of general correlations.

12. As more maintenance money is spent, pump MTBF increases.
Finding: The exact *opposite* is true; see Fig. 42.3.1.

13. Diamond-infused mechanical seal faces excel over all others and will be worth the extra money outlay.
Finding: That claim has been thoroughly refuted by rigorous testing done in 2015 and 2016.

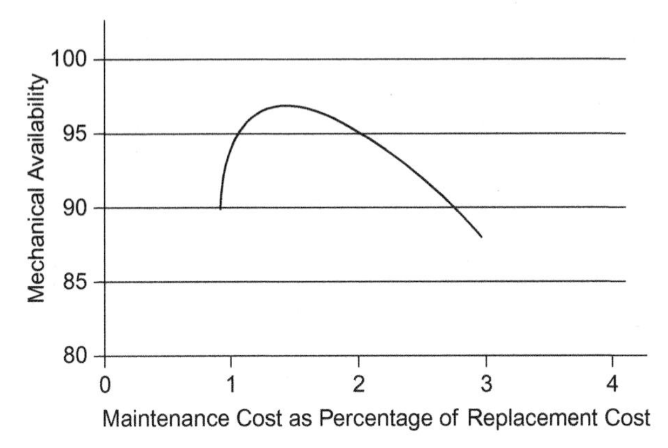

FIG. 42.3.2 Mechanical availability versus maintenance cost.

14. Plotting mechanical availability versus maintenance cost shows a straight-line relationship.
Finding: False. Per Fig. 42.3.2, it's a curve that peaks a little below 2%.

We must do fact-checking by being resourceful in more ways than one.

REFERENCES

[1] H.P. Bloch, Improving Machinery Reliability, third ed., Gulf Publishing Company, Houston, TX, 1998, ISBN: 88415-661-3.

[2] H.P. Bloch, Pump Wisdom: Problem Solving for Operators and Specialists, John Wiley and Sons, Hoboken, NJ, 2011, ISBN: 9-781118-041239.

Chapter 42.4

Why Continuing Education Budgets are Dropped First

INTRODUCTION

It's a fact that an oil refinery in the Western United States achieves a pump MTBF of over 9 years. The facility has 1200 pumps and, in a given year, repairs 130 of these. It's also a fact that an oil refinery "elsewhere" has about the same number of pumps installed and does 260 repairs each year; their MTBF is roughly 4.5 years. It is known that, including overhead expenses, the average pump repair costs a typical US refinery $13,000. Each year, the refinery "elsewhere" spends $1.7 million more for pump repairs than the refinery on the West Coast.

CONSIDER THE SAVINGS POTENTIAL

If one would teach the "elsewhere" oil refinery to safely extend run time and increase pump MTBF to 5.5 years, this facility would save $600,000 per year. If this achievement required sending 20 employees to a local college campus where they would attend a two-day continuing education course costing $12,000, the "elsewhere" refinery would recover its investment 50 times over, in just one single year. Well thought-out continuing education for adults must continue irrespective of the state of the economy.

Why, then, is continuing education often the first budget item to be deferred or dropped in an economic downturn? Here's the short answer: management will not support what it deems of little value. But why would continuing education in equipment reliability improvement or failure and maintenance cost avoidance be deemed of little value? The answer involves us, the reliability professionals. I will try to spell out the answer, knowing full well that it risks alienating a few readers.

The below-average performance of one particular process unit was plotted in Chapter 42.3, Fig. 42.3.1. Note how its MTBF decays, even as the plant's repair expenditures for pumps increase. Best practices professionals, the ones we wish to commend, will use these statistics to make a powerful case for continuing education and will do so as a matter of routine. It is in their role statements to groom successors and to understand where the reliability performance of their plant is ranked relative to comparable industry competition. These are the professionals that will obtain funding for continuing education because they have proof that their plant would suffer avoidable repeat failures without such education. As responsible professionals, they know failure statistics and will have informed their managers if, for instance, pump mean times between failures did not measure up to best of class.

ADULT TRAINING AT LOCAL COLLEGES

Many key institutions of learning are cultivating working relationships with the local industry. A liaison person or local industry training coordinator ("LITCO") facilitates and coordinates the scheduling and related needs of a mutually beneficial involvement. All sides to the agreement will consider it profitable for both industry and local colleges. All must consider it more than an ordinary business deal because dedication to excellence is essential.

Dedication to excellence is far more than a catchy phrase. It adds value when we're moved or motivated to act. Having seen good and not-so-good LITCOs, it seems evident that they should be highly qualified individuals. Whether you work for the industry or an institution of industrial learning, make it your goal to support good LITCOs. Assist these LITCOs in making a strong case as they—hopefully—attempt to nudge below-average industry performers toward better uptime and improved reliability achievements.

The same goes for institutions of learning. States such as Louisiana, Oklahoma, and Texas do not let LITCOs get away with taking the position that there are no funds to send people to be trained. Targeted (which is the opposite of aimless) training is essential for operating plants safely and reliably. The payback for targeted training can be enormous. An institution of industrial learning should never accept the bogus argument that the local industry would be bankrupted if its employees were to attend a solid continuing education course.

That said, we commend the thoughtful reliability professionals whose objective is to help people to continuously improve. These are professionals who quantify value and both report and support management in fully grasping the value of targeted and tangible improvement. Whichever job functions or individuals do not share these objectives are seriously encouraged to make changes. Above all, we urge the occasional misguided faculty member of an institution of industrial learning to consider their role. Some might really add value by spreading the word about the effectiveness of true learning. In some refineries and on a per-attendee basis, the cost of targeted training would be a pittance. For some, the benefit-to-cost ratio of implementing lessons learned from experts would be huge. Whatever good continuing education manages to put into the gray matter between the ears of a willing worker can never be taken away. At no time would it ever hurt the economy to nurture, mentor, and train a well-informed workforce of professionals and craftsmen.

Chapter 42.5

Early Data Collection Facilitates Future Analysis

INTRODUCTION

Years have gone by since a disaster occurred in a nuclear facility in Fukushima, Japan, on 11 March 2011. Much additional data on the nuclear disaster on the east coast of Japan's main island will have become available; a tsunami and earthquake were involved. In the first days after the disaster, the public only knew the wave of seawater had reached a height of 15 m and that all emergency generators got flooded. The question was: How does one tackle the failure analysis in this instance? How does one collect data and maximize the contributions of mentors? The data collection process is the same as it would be in a far more mundane event.

EXAMINING FACTS FIRST

As the author of this text, I will never claim special insights in some disasters. Also, I would not profess to have the answers to the hundreds of questions that have already been raised with regard to some high-profile events. Nevertheless, and for what it's worth, I would like to highlight a few of the detailed questions one would ask as soon as even the first factual data become available. While this has nothing to do with nuclear engineering and design, it has to do with elusive failure causes of rotating equipment such as emergency power generators and centrifugal pumps.

On the tsunami disaster, one would look at the issue of 13 emergency generators reportedly failing in a very short span of time and would, for the sake of this text, assume that this report is verified as accurate. One would separate the listing of potential reasons into four cause categories:

(1) Issues with the diesel engines driving emergency generators
(2) Cooling water pump hydraulics unsuitable for parallel operation
(3) Issues with bearing lubrication after water intrusion
(4) Serious concerns with generator location

Excessive water or solid contamination of diesel fuel (cause Category 1) would have caused engines to cease operating; ascertaining (or ruling out) such a cause is not too difficult. While it's on our list of questions, we might not be certain that it also shows up on the lists prepared by other investigators.

BOOKS AS TUTORS

Early data collection may give guidance relating to "where have I read about this before?" As an example and regarding cause Category 2, recall one of the chapters of this text deals with operating pumps in parallel. Therefore, one would examine the head/flow (H/Q) curves of the cooling water pumps and ask if a counterproductive attempt was made to operate too many of these pumps simultaneously. Depending on prevailing H/Q relationships, operation at the resulting low flow—possibly even zero flow—could have caused one or more pumps to fail. Although this is a well-known issue, it is also an issue that is often overlooked [1].

As to the third cause category, in or about 2001, a nuclear facility on the US East Coast experienced frequent bearing failures on emergency power-generating equipment. The most probable cause was a fanlike blower wheel that had been fitted to the shaft at a location very close to the bearing housing—you will find a report in the literature. While the fan may have promoted increased airflow to the bearing housing, it very probably caused a pressure difference between the inboard and outboard sides of the bearing. We recall that oil starvation was found on one side of the bearing. No such starvation problems were reported on equipment that did not come equipped with the "novel" fanlike blower. Innovation can have unforeseen consequences.

Still on item/Category 3, one would certainly want to know the viscosity grade of the lubricant used in the generator bearings. At times, one's misguided attempts to "standardize" on certain lubricants have proved costly, but we seem to be very slow learners. Whenever oil needs to be lifted into the bearings with slinger rings, oils that are too high in viscosity will cause slinger rings to malfunction. Slinger rings that are out of round will also malfunction, as will slinger rings that are too deeply immersed in an oil sump. And just think what happens when some of these deviations occur simultaneously, or when violent shaking (back to the earthquake) causes the slinger rings to dislodge, or cold ambient (seawater) temperatures will turn the oil into a more viscous fluid.

Finally, in Category 4, there will be questions as to where one best locates emergency equipment. High and dry? Recall from early data gathering that the tidal wave was 15 m high. Again, where should the emergency generators be located? Sheltered or in the open? On springs or on flexible wobble plates? We don't have all the questions, let alone all the answers. Nevertheless, we should examine issues common to certain maintenance practices, certain "standardization" practices, and become familiar with the many elusive problems that befall pumps and other rotating equipment assets. When a number of these deviations combine, we are risking disaster.

When we use 100-year floods as our yardstick, are we underestimating data reflecting the past two decades? Are we weighting things in a manner favoring certain biases?

MANUFACTURERS AND SUPPLIERS AS TUTORS

Experienced manufacturers and suppliers can be your tutors. These tutors should be the best qualified vendors, and purchasers should put into their budgets the cost of the best technology equipment they provide. The relevance of this plea was reemphasized when a manufacturer presumably proposed plastic internal parts for a line of high-temperature (due to high compression ratio) compressors. Working with a highly experienced equipment rebuilder, perhaps one that upgrades the widest possible range of process pumps, might have allowed the purchaser to understand what is, and what is not, technologically feasible. The tutor would surely have pointed out the pitfalls of certain components.

Finally, we believe that a reliability-focused purchaser must examine how systems work and how parts function. If, as an additional example, we had to choose from the large variety of steam traps available today (see index word "steam traps"), we would make it our first endeavor to understand how different trap types function and how they each perform long term. We would obtain that information from a manufacturer whose product slate encompasses the widest possible range and the best possible quality. We should purchase from him and gladly pay him a premium for becoming our tutor.

REFERENCE

[1] H.P. Bloch, Pump Wisdom: Problem Solving for Operators and Specialists, John Wiley and Sons, Hoboken, NJ, ISBN: 9-781118-041239, 2011.

Chapter 42.6

Studying the Best-of-Class Training and Organizing Lineup

INTRODUCTION

Professed commitment to reliability is not unlike professed commitment to safety. Just as not even a small grocery store will admit to unsafe practices, no industrial corporation will ever confess to have little interest in asset reliability. Still, best practices plants (BPPs) are much more effective at identifying and implementing the best, or most appropriate, reliability organization. Many BBPs map out and make available dual paths of advancement for administrative versus technical employees. We will highlight the subtopic later in this brief (yet comprehensive) narrative.

Some organizations opt to divide their staff along traditional lines into technical services, operations, and maintenance divisions or departments or just plain work functions. These companies often place their reliability personnel under the maintenance management umbrella, but then (hopefully) reconsider when reliability professionals end up immersed in fighting the "crisis of the day," as it were.

SOUND ORGANIZATIONAL SETUP NEEDED

While it has been our experience that organizational alignments are less important than the technical expertise, resourcefulness, motivation, drive, and job satisfaction of individual employees, there are obvious advantages to an intelligent lineup. What, then, is an "intelligent lineup" or sound organizational setup? And what kind of grooming and nurturing and training is received by reliability professionals at thoughtful best-of-class performers (BoCs)? We will explore the details and give precise guidance based on many decades of actual experience.

To begin with, two straightforward, short definitions of "maintenance" and "reliability" are in order. Without buying into these definitions and then sticking to them, reliability engineering and maintenance activities tend to become a messy mix. Moreover, training is, at best, an afterthought at companies that don't make a clear distinction between reliability and maintenance:

- The function of a maintenance department is to routinely maintain equipment in operable condition. It is thus implied that this department is tasked with restoring equipment to as-designed or as-bought condition.

- Reliability groups are involved in structured evaluations of upgrade opportunities. They perform life cycle cost studies and develop implementation strategies whenever component upgrading makes economic sense.

For a reliability improvement group to function most effectively, its members have to be shielded from the day-to-day preventive and routine equipment repair and restoration involvement. BPPs and BoC facilities often issue guidelines or predefine a metric that triggers involvement by the reliability group instead of leaving things to the maintenance workforce. Examples might include equipment that fails for the third time in a running 12-month period, equipment distress that has or could have caused injury to personnel, and failures that caused an aggregate loss in excess of $50,000.

At industrial entities too small to have a separate reliability group, reliability section, department, or division, one or more designated reliability professionals are administratively assigned to the maintenance department on a one- to 2-year rotational basis. However, in harmony with the two bulleted items given above, their work scope and tasks are kept distinctly separate from each other.

By far, the most important organizational assist in accomplishing the long-term reliability objectives of an industrial enterprise is totally focused employee training. While this requirement may be understood to cover *all* employees regardless of job function, we are here confining our discussion to a plant's reliability workforce. A good organization will map out a training plan that is the equivalent of a binding contract between employer and employee. There has to be mutual accountability in terms of proficiency achieved through this targeted training.

OPERATOR INVOLVEMENT IN RELIABILITY EFFORTS

Modern process plants also train their operating technicians to have a general understanding of relevant manufacturing processes, process safety, basic asset preservation, and even interpersonal skills.

Operator involvement in reliability efforts ensures the preservation of a plant's assets. It can be labeled operator-driven reliability, or ODR. With ODR, operator activities include the electronic collection of vibration and temperature data and spotting deviations from the norm. Operator activities do not, however, encompass data analysis; data analysis is the reliability technician's or reliability professional's task. Additional operator activities at BoCs include routine mechanical tasks such as the replacement of gauges and sight glasses and assisting craftspeople engaged in the verification of critical shutdown features and instruments. Also, operating personnel participate in electric motor testing and electric motor connecting/disconnecting routines.

The value of entire functional departments tasked with both data capture and analysis merits a critical look. Departments that combine data capture and analysis may not be efficient; they risk involving expert analyst personnel in mundane data collection routines—looking at, or closely examining, 1000 points only to find them all in order. It should be noted that operators are the first line of defense, the first ones to spot deviations from normal operation. For optimum effectiveness, they should be used in that capacity, that is, data collection should be assigned to operators.

There are—at any process plant—three job functions that influence reliability and safety: operations, maintenance/mechanical, and process/technical. The analogy to automobiles is similar. To be reliable, a car will have to be properly driven, maintained, and designed. In a BoC process plant, reliability professionals from new hires to subject matter experts (SMEs) nearing retirement provide linkage among these three job functions. Linkage takes place in the form of (and during) "shirt-sleeve seminars." Although the presentation style is rather informal, the activity itself is mandatory and disregarding it will inevitably produce costly gaps. A plant without shirt-sleeve seminars lacks linkage. Shirt-sleeve seminars are in-house activities that cost the company nothing. We will elaborate on shirt-sleeve seminars later in this narrative.

KEYS TO A PRODUCTIVE RELIABILITY WORKFORCE

It has been said that an underappreciated workforce is an unmotivated, unhappy, and inefficient workforce. Such a workforce will rarely, if ever, perform well in areas of safety and reliability. How, then, will a company's highly interdependent safety, reliability, and profitability goals be achieved? Training is an incredibly important part of it, and exposure to a "steering committee," as will be discussed later, lends great visibility and fosters recognition; it makes for a happier workforce.

In the early 1950s, world-renowned efficiency expert W. Edwards Deming provided the answer to creating a productive reliability workforce. He stipulated 14 "points of quality" that fully met the objectives of both employer and employee and are as true and relevant today as they have ever been. W.E. Deming had aimed his experience-based recommendations at the industrialized world's manufacturing industries; his contributions are mentioned several times in this text.

We had earlier adapted his 14 points to pump technology topics. Here, we have attempted to transcribe his 14 points into wording that applies more generally to the entire process plant reliability environment [1]. Here's our expanded recap:

- View every maintenance event as an opportunity to upgrade. Investigate its feasibility beforehand; be proactive.
- Ask some serious questions when there are costly repeat failures. There needs to be a measure of accountability. Recognize, though, that people benefit from coaching, not intimidation.

- Ask the responsible worker to certify that his or her work product meets the quality and accuracy standards stipulated in your work procedures and checklists. That presupposes that procedures and checklists exist.
- Understand and redefine the function of your purchasing department. Support this department with component specifications for critical parts and then insist on specification compliance. "Substitutes" or noncompliant offers require review and approval by the specifying reliability professional.
- Define and then insist on daily interaction between process (operations), mechanical (maintenance), and reliability (technical and project) workforces.
- Teach and apply root cause failure analysis from the lowest to the highest organizational levels.
- Define, practice, teach, and encourage employee resourcefulness while ascertaining that Management of Change (MOC) is in place and fully adhered to. Maximize input from knowledgeable vendors, and be prepared to pay vendors with application engineering service for their effort and assistance. Don't "reinvent the wheel."
- Show personal ethics and evenhandedness that are valued and respected by your workforce.
- Never tolerate the type of competition among staff groups that causes them to withhold critical information from each other or from affiliates.
- Eliminate "flavor of the month" routines and meaningless slogans.
- Reward productivity and relevant contributions; let it be known that time spent at the office is in itself not a meaningful indicator of employee effectiveness.
- Encourage pride in workmanship, timeliness, dependability, and providing good service. Employer and employee honor their mutual commitments.
- Map out a program of personal and company-sponsored mandatory training.
- Exercise leadership and provide direction and feedback.

DEMING'S METHOD STREAMLINED AND ADAPTED TO OUR TIME

In early 2000, Canadian consulting company Systems Approach Strategies ("SAS" [1,2]) developed a training course that brings Deming's method into even sharper focus. SAS concluded that companies can be energized with *empathy* and, using the acronym CARE to indicate the first letters of four bulleted words, conveyed an important observation: companies excel when management gives consistent evidence of

- clear direction and support,
- adequate and appropriate training,
- recognition and reward,
- empathy.

The last item, empathy, is by far the most important and also the most neglected. Yet, it represents the foundation of the CARE concept. Without empathy, without the ability to put oneself into the shoes of the people one manages, a manager will never know them, certainly won't understand them, and will never bring them to their full potential as employees and people.

ROLE STATEMENTS MANDATORY FOR PROGRESS

The four CARE items represent rather fundamental principles of management. While empathy forms the foundation, it alone will not deliver full results for any given organization. The drive toward assured success starts with clear direction and support. Clear direction must be expressed in writing. For example, reliability professionals must receive this clear direction in the form of a role statement [3].

How well the employee fulfills his or her role will have to be discussed during periodic performance appraisals. The outcome of a performance appraisal session is a contributing factor in salary and promotion-related decisions by the employer.

A typical role statement may comprise 10 or more points. It can be compiled by the employer or the employee, but the other party must buy into it. It may be a negotiated document when first developed, but it will become a binding contract soon after the parties first see it. The style and detail of a 10-point role statement used in modern industry are shown next. Note that it includes, but will probably not be limited to, the following:

1. Assistance role.
   - Establishment of equipment failure records and stewardship of accurate data logging by others. Know where we are in comparison with best-of-class performers.
   - Review of preventive maintenance procedures that will have been compiled by maintenance personnel.
   - Review of maintenance intervals. Understand when, where, and why we deviate from best practices.
2. Evaluation of new materials and recommendation of changes, as warranted by life cycle cost (LCC) studies.
3. Investigation of special, or recurring, equipment problems. Examples are as follows:
   - Ownership of failures that occur for the third time in any 12-month period
   - Coaching others in root cause failure analysis
   - Definition of upgrade and failure avoidance options
4. Serving as contact person for original equipment manufacturers.
   - Understanding how existing equipment differs from models that are being manufactured today
   - Being able and prepared to explain if upgrading existing equipment to state-of-the-art status is feasible and/or cost-justified

5. Serving as contact person for other plant groups.
   - Communicate with counterparts in operations and maintenance departments.
   - Participate in (management's) service factor committee meetings.
6. Develop priority lists and keep them current.
   - Understand the basic economics of downtime. Request extension of outage duration where end results would yield rapid payback.
   - Activate resources in case of unexpected outage opportunities.
7. Identify critical spare parts.
   - Arrange for incoming inspection of critical spare parts prior to placement in storage locations.
   - Arrange for inspection of large parts at vendor/manufacturer's facilities prior to authorizing shipment to plant site.
   - Define conditions allowing procurement from non-OEMs.
8. Review maintenance costs and service factors.
   - Compare against best-of-class performance.
   - Recommend organizational adjustments.
   - Compare cost of replacing versus repairing; recommend best value.
9. Periodically communicate important findings to local and affiliate management.
   - Fulfill a networking and information-sharing function.
   - Arrange for key contributors to make brief oral presentations to mid-level managers (share the credit and give visibility to others).
10. Develop training plans for self and other reliability team contributors.

Again, the above 10-item listing represents a role statement for equipment reliability engineers. While it represents a summary that can be expanded or modified to address specific needs, it is representative of the written "clear direction" that is being taught in the CARE program.

The "support" element in the first of the four CARE points is reinforced by items 9 and 10 in the reliability professional's role statement. In one highly successful and profitable company, an astute plant manager organized a mid-level management "steering committee" that every week invited a different lower-level employee to make a 10-min presentation on how he or she performed their work. The vibration technician explained how early detection of flaws saved the company time and money. An instrument technician demonstrated the key ingredients of an online instrument testing program. Each reliability issue or program had a mid-level management sponsor or "champion" who saw to it that a program stayed on track and that organizational and other obstacles were removed.

Training plans were to be initiated by the employee, meaning he or she had to give considerable thought to long-term professional growth. The initial training proposal by the employee was reviewed, supplemented, modified, often amplified, but always given serious consideration and constructive encouragement by managers.

ADEQUATE AND APPROPRIATE TRAINING

Note that our earlier statements on "clear direction and support" introduced the training issue. Let's face it, we are losing the ability to apply basic mathematics and physics to equipment issues in our workplace situations. As an example, hundreds of millions of dollars are lost each year due to erroneous lubrication techniques alone. The subject is not dealt with in a pragmatic sense in the engineering colleges of industrialized nations. The connection between Bernoulli's law taught in high school physics classes and the proper operation of constant-level lubricators is lost on a new generation of computer-literate engineers.

Managers chase after the "magic bullet"—salvation must be in "high tech," some think. Others are enamored by metrics and play strange games with failure statistics in order to shine in industry comparisons. It seems we have truly neglected to understand the importance of the nonglamorous basics. We are no longer interested in time-consuming details. We have encouraged our senior contributors to retire early. All too often, no thought is given to the consequences. Assumptions are made that one could hire contractors to do the thinking for us, and not many decision-makers see the fallacy in this reasoning. It should be obvious that contract personnel are often less qualified or have little incentive to determine the life cycle cost of different alternatives. Therefore, our reliability professionals must learn to find the root causes of repeat failures of machinery. We have become "big picture" men, from the maintenance technician all the way up to the company CEO. We can't be bothered by details, have no time for details, and are not usually rewarded for dealing with details. Most of us are lionized for quickly stitching stuff together and berated for going through the more tedious steps of preventing failures from occurring in the first place. It's time to adjust our thinking and a good training road map will help to get us there.

But, as some outstanding performers have clearly shown, attention to detail is perhaps the most important step they took to become BoCs. These BoCs have developed and continue to insist on adequate and appropriate training. This training deals with not only concepts and principles but also hundreds of details.

Employees at best-of-class companies develop their own short- and long-range training plans. Time and money are budgeted and the training plan is signed off by the employee and his or her manager. A training plan must have the status of a contract. It can only be altered by mutual consent or in case of dire emergency.

The training plan for a machinery-technical employee at a major petrochemical company was published in [4]. It, and similar plans, typically consists of four columns, as replicated in the general training road map below:

THE TRAINING ROADMAP

Career years	"Knowledge of"	"Work capability in"	"Leading expertise in"
1	Company organization	Interpretation of flow sheets, piping, and instrument diagrams	
	Rotating equipment types	Elementary technical support tasks, for example, alignment and vibration monitoring	
	Company's communication routines	Essential computer calculations	
	Relevant R&D studies, vendor capabilities, in-house technical files		
2	Pump and compressor design	Design specification consulting and support	
	Machinery reliability appraisal techniques	Machinery performance testing	
	Gear design	Start-up assistance, all fluid machines	
	Major refining processes		
3	Machinery design audits	Company standard updates	
	Machinery piping	General technical service tasks and elementary troubleshooting	
	Major chemical processes	Machine-electronic interfaces	
4	Materials handling equation	General troubleshooting	"Shirt-sleeve seminars" (conduct informal training in reliability topics)
	Hyper compressors	Machinery quality assessment and verification	

(Continued)

Career years	"Knowledge of"	"Work capability in"	"Leading expertise in"
5	Thin-film evaporators	Start-up advisory tasks	Machinery optimization
		Appraisal documentation update tasks	Machinery maintenance
	Plastics extruders	Hyper compressor specifics	
6	Fiber processing, general approach	Machinery design audits	Machinery selection
7	Patent and publication matters	Technical publications	Machinery failure analysis

A career development training plan was developed along the same lines [3].

Here's the format we have seen for imparting knowledge to new, intermediate, and advanced machinery engineers.

I. NEW ENGINEER (plant mechanical engineer hire)
Career years 1 and 2, possibly years 1–5

A. *On-the-job training*

Rotational assignments within the plant in various groups to be exposed to different job functions for familiarization. Areas to be covered should include machinery, mechanical, inspection, electrical, instrumentation, operations, and maintenance.

B. *In-house training* (applicable to headquarters/central engineering locations)
Plant and/or corporate standard development/revisions and updates

- Courses in the above
- Courses dealing with industry standards (API, NEMA, etc.)
- Machinery (compressors, pumps, steam and gas turbines, gears, turboexpanders, etc., per Refs. [5–11])
- Failure analysis and troubleshooting (seven root cause method and "FRETT," per Ref. [8])
- Practical lubrication technology for machinery (per Ref. [9])
- Machinery vibration monitoring and optimized analysis (per Ref. [10])
- Predictive monitoring (lube oil analysis, valve temperature monitoring, etc.)

C. *Outside training pursuits* (suggested minimum once/yr, preferred frequency twice/yr)

1. General vendor-type information courses. Examples are as follows:
   - GE and/or Siemens gas turbine maintenance seminar
   - Major mechanical seal manufacturers' training courses
   - Elliott or D-R compressor technology, selection, application, and maintenance seminar

   - Compressor Controls (CCC), Triconex, and Woodward Governor company courses
   - Major turbomachinery manufacturer's lube and seal oil systems maintenance course
   - Coupling manufacturer's training course
2. Texas A&M University Turbomachinery Symposium
3. Texas A&M University International Pump Users Symposium
4. Approved advancement-related public courses
   - Machinery failure analysis and prevention
   - Machinery maintenance cost-saving opportunities
   - Compressor and steam turbine technology
   - Machinery for process plants
   - Reciprocating compressor operation and maintenance
   - Piping technology
   - Practical mechanical engineering calculation methods
5. AFPM and IMC Refinery and Process Plant Reliability Conferences and Exhibitions
6. TapRooT and/or similar RCFA courses

D. *Personal training* (mandatory review of tables of contents of applicable trade journals, books, conference proceedings, etc. Mandatory collection and cataloging/copying articles of potential future value). Here are some examples of trade journals although it should be noted that consolidations occur and names occasionally change:

Hydrocarbon Processing	Mechanical Engineering
Maintenance Technology	Diesel Progress
Oil and Gas Journal	Diesel and Gas Turbine Worldwide
Chemical Engineering	Turbomachinery International
Control Design	Modern Pumping Today
Gas Turbine World	Lubes and Greases
Chemical Processing	Sulzer and Mitsubishi Technical Reviews
Mitsubishi Technical Review	Hydraulics and Pneumatics
Machinery Lubrication	Power
Plant Engineering	World Pumps
Pumps and Systems	Compressed Air
Evolution (SKF Bearing Publication)	Practicing Oil Analysis
Uptime Magazine	NASA Tech Briefs
Compressor Tech Two	Plant Services
Pump Engineer	

Books to be reviewed should include texts on machinery reliability assessment (which include checklists and procedures [11,12] and popular texts on pumps [13], Weibull analysis, reciprocating and metering pumps, and electric motor texts and books dealing with gear technology).

II. INTERMEDIATE ENGINEER (plant mechanical/machinery engineer), years 3–5, possibly 3–8

Rotational assignment—2-year assignment at affiliate location, possibly at central engineering or headquarters

- Involvement in field troubleshooting and upgrading issues
- Familiarization with equipment, work procedures, data logging practices, etc.
- Spare parts procurement practices (probability studies)
- Life cycle costing involvement
- Maintainability and surveillability input
- Structured networking involvement (provide feedback to other groups)

Outside training pursuits

- Extension of earlier exposure
- Attendance at relevant trade shows and exhibitions (provide feedback to others)
- Attendance at ASME, AFPM, STLE, and related conferences (also provide feedback)
- Speaker at local ASME/STLE/Vibration Institute meetings

Personal training and continuing education

- Developing short articles for trade journals and/or similar publications
- Developing short courses (initial aim: in-plant presentations and intra-affiliate presentations)
- Advanced self-study of material on probability, statistics, automation, and mgmt. of change
- Studies in applicable economics

III. ADVANCED ENGINEER (corporate specialist, core engineering specialist, and SME), years 9 and later, depending on exposure and achievements under IIA/B/C, above

- International conferences (speaker/participant)
- Peer group interfaces (e.g., on discussion panels and industry standard committees)
- Development and presentation of technical papers at national/international engineering conferences
- Pursuing book publishing opportunities (case histories, teaching tools, and work procedures)
- Regular contributions to trade journals
- Development of consultant skills

FOCUS ON SHARED RESPONSIBILITY AND VERIFIED SELF-MOTIVATION

(1) *Trade Journals.* Reviewing trade journals is the first in a sequence of training obligations recognized by the employer and employee. As was discussed in (D), above, there are many trade journals that are being mailed by their respective publishers to an interested readership. However, recall that training and information sharing are mandatory endeavors at BoCs. The different trade journals arriving at a BoC facility are routed to designated professionals for mandatory screening within 3 days of their arriving at the professional's desk. Such screening could take 5 min; actually reading several articles of interest can take 2 h. The professional is expected to skip over topics or articles judged of no consequence to his plant, in which case the entire job takes 5 min. He or she then discards the publication. But suppose he or she takes note of a keyword of interest in one of the headings. This would prompt a closer review and a decision to forward a copy of the article to a colleague for additional reading or for filing. During a future performance appraisal, it will be easy to ascertain the degree of seriousness and diligence with which a professional handled the implied technology update. Promotions and salary increases belong to those who use trade journals to educate themselves on relevant technology advancements, vendor capabilities, and so forth.

(2) *Shirt-Sleeve Seminars.* A second and equally important obligation of both parties, that is, professional and manager/supervisor, is the conscientious scheduling and presentation of "shirt-sleeve seminars." At the end of a safety meeting, a reliability professional rolls up his sleeves, so to speak, and makes a 4–7 min presentation on a subject related to asset reliability at that facility. There are many hundreds of topics that could be presented at these sessions. Preparation for making such presentations is a marvelous opportunity for the presenter to learn from older employees, or from texts, or from vendors and manufacturers. The permanent part of the presentation should be confined to a single, double-sided sheet of paper, laminated in plastic, and three-hole-punched for inclusion in a suitable three-ring binder. As an example, the topic might be how to remove a coupling from a pump shaft (side 1) and how to remove such couplings from a shaft without causing damage (side 2).

Hundreds of topics can be derived from the publication review indicated in (1), above. Many hundreds can be found in reference books of interest to reliability professionals; some of these are listed in our many references. Much of the material suitable for single-sheet two-sided shirt-sleeve seminar presentations is easily found in current technical texts. A forward-looking and reliability-focused company facilitates acquisition of a technical library and encourages reading in more than one way. Perhaps it all starts with a supervisor or manager being aware of the existence of relevant texts. You will find them listed in Appendix II.

It has been said that about 100 pages of a 300-page technical book are taken up by illustrations. The remaining 200 pages of narrative can be read and digested in a year. Expecting this kind of reading is not an imposition on the person seeking knowledge, nor is paying for such a book out of company funds a drain on company profits. Assembling a company technical library is one of the hallmarks of a modern facility, and so is reading for true professionals.

(3) *Local Meetings of Engineering Societies*. Next, understand the third mutually agreed-to training opportunity. The reliability professional is our future SME. One of his or her obligations is to attend or participate at local technical society meetings. Participating or attending local meetings of an ASME, Vibration Institute, or other professional society's local subsection is a networking opportunity that has no match. Such local meetings may take place twice a year late in the afternoon. Attending may take 2–3 h and cost less than $50. If the knowledge transfer that takes place at such meetings avoids the failure of even one single API-style pump per year, payback exceeds cost by 100:1 or more.

At best-of-class (BoC) companies, the young or mid-level professional donates his 6 h per year. His or her employer cheerfully pays the yearly membership fee for two engineering societies per professional. The employer also reimburses its professional employees for petty cash expenses (travel and meals) they incur while attending the local meetings described here. Of course, senior-level professionals attend as well. As true SMEs, they accept their implicit obligation to teach and give back to the profession as mentors and tutors.

(4) *Lunch-and-Learn Sessions*. These are one- to one-and-a-half-hour meetings arranged by vendors. The meetings take place at the reliability professional's plant location or at a convenient nearby meeting facility. A vendor representative explains his company's products and invites clients and potential customers to these brief sessions. A professional in training serves as the liaison and recommends others who should attend (e.g., mechanics, machinists, technicians, and engineers). His or her liaison activities include ascertaining that the content of the vendor representative's presentation adds value and is not perceived as mere "sales talk." We consider these our fourth training opportunity and, again, a no-cost activity.

(5) *Steering Committees*. Presentations to the BoCs in plant steering committee were briefly mentioned above; these presentations represent the fifth training opportunity. The reliability professional (technician or engineer) is making a semiformal appearance so as to brief and inform managers. Mid-level managers are members of this steering committee, and the presenter explains and documents how he or she carry out their roles, have spearheaded or accomplished improvements, etc. These presentations are giving due visibility to a professional employee's contributions and also serve as an educational update for the steering committee members.

(6) *Seminars and Conferences*. Perhaps you have noticed that the first five training opportunities do not take the future SMEs away from their home base. Also, the cost of involvement in any of the first five training-related activities is either zero or, at most, pocket change. It's only after exposure to each of the first five training activities that best-of-class companies advocate and support away-from-home training involvements. These best-of-class companies typically allocate 10 training days per year for each professional who is being groomed and nurtured for a value-adding professional career. He or she must thoughtfully select and plan attendance at the courses or conferences of value.

Attending training seminars away from the plant is never an unmonitored activity. The particular professional's manager or supervisor must keep track of how the professional takes care of his/her training commitment. When handing in an expense statement, the professional must also submit a brief write-up of course content or lessons learned. This brief write-up must be shared with other professionals; it represents a tangible networking opportunity that should be considered mandatory, not optional.

RECOGNITION AND REWARD

One of the most important (yet little-known) facts is that the majority of professionals in the active job market seek different employments for reasons other than better pay. This situation is analogous to divorces. Few marriages break up because of the intense desire to find a new partner whose income exceeds that of the previous one. Most marriages break up because of lack of respect; untruthfulness; immoral, selfish, or insensitive conduct; or just plain incompatibility. Most employer-employee relationships are wrecked for the same reasons.

Recognition is implicit whenever a professional is involved in presentations to a steering committee. Also, recognition and reward often come in the form of sincere expressions of appreciation for whatever good qualities, or commendable performance, are displayed by the employee. Still, a few well-chosen words given privately are usually better than public praise. All too often, public praise generates envy in others and may make life more difficult for the recipient of praise. Rewards in the form of certificates of recognition to be hung on the office wall come perilously close to being meaningless, and employers would be wise to consider how negatively these pieces of paper are often perceived. If you want to do something positive for an employee, hand them a certificate for $300 worth of technical books, or a $200 gift certificate for dinner at an upscale restaurant, or a new floor covering, or whatever reaffirms that the employee's contributions are properly valued.

Several major petrochemical companies frequently reward top technical performers with $5000–$40,000 year-end bonuses for exceptional resourcefulness, or for the implementation of cost-saving measures, or for being "doers" instead of "talkers." In 2015, a Houston-based company paid every one of its 1100 employees—from janitor to top manager—a bonus of $100,000. While that might be unusual, it is nevertheless just as factual as some minimum-wage workers not being able to make ends meet on their dismal take-home pay.

In any event, there is nothing a company likes more than having its professional employees go on record with a firm, well-documented recommendation for specific action, rather than compiling lists of open-ended options for managers to consider. Top technical performers make solidly researched recommendations, showing their effect on risk reduction and downtime avoidance or demonstrating their production and quality improvement impact. Good professionals act on facts; average professionals act on opinions. Purveyor of opinions can deprive their employer companies of valuable solutions. They can lock their employers in expensive approaches that really belong in the dark ages where anecdotes were repeated and embellished. Whenever these anecdotes become the standard operating mode of a company, this entity will quickly lose its footing.

DUAL CAREER PATHS

Top performing companies have created two career paths, also called a "dual ladder of advancement," for their personnel. Upward mobility and rewards or recognition by promotion are possible in either the administrative or technical ladder of advancement. This is perhaps the only sound and proved way to retain key personnel in such industries as hydrocarbon processing. There is full income and recognition parity between the administrative and technical job functions given in the table below:

[Admin. Side]	[Technical Side]
Group leader	Project engineer
Section supervisor	Staff engineer
Senior section supervisor	Senior staff engineer
Department head	Engineering associate
Division manager	Senior engineering associate
Plant manager	Scientific advisor
Vice president	Senior scientific advisor

It has been shown that recognition and reward approaches have much to do with management style. There are many gradations and cultural differences that make one approach preferred over the next one. It is not possible to either know or judge them all. Suffice it to say that a thoughtless reward and recognition system is a serious impediment to employee satisfaction. Conversely, a thoughtful program is always a highly positive step.

EMPATHY, THE OVERLOOKED CONTRIBUTOR TO ASSET PRESERVATION

Empathy is an intimate understanding or fellow-feeling. Empathy is a quality that causes one to readily comprehend the feelings, thoughts, and motives of a fellow human being. You may think that this "intimate understanding" has no place at the office or on the factory floor. We ask you to think again.

Say, an employee is late coming to work and the manager rebukes her before, or instead of, tactfully inquiring as to the reason for the tardiness. Assume that this employee has a sick child at home. Does the rebuke make her a more efficient or happier worker? We all know the answer to that question.

In contrast, let's say the manager would understand how empathy works or would remember how he would like to be treated if it were his child that is sick. Let's say the manager would, therefore, offer the employee such options as doing the work at or from her home. The most likely result of his showing empathy and compassion would be that instead of getting 80% efficiency out of the unhappy worker at the office, he gets 120 % efficiency from the appreciative worker at home. All parties would benefit from empathy and compassion in the workplace.

We are fully aware of the standard objection to reacting with empathy: "The workers will take advantage of me. I would look like a pushover and not like the firm leader that I want to project." Let's just end the discussion by stating unequivocally that the vast majority of professional employees respond better to kindness than to harshness. Using such traits as compassion, cooperation, communication, and consideration will result in a more productive, satisfied, motivated, and loyal workforce than many managers could ever imagine.

Yes, empathy is doing more to retain this most valuable asset, that is, your professional employees—than money, slogans, exhortations, and threats. *Empathy*, indeed, is the foundation of the ingredients of *CARE* [1,2] and is the hallmark of a long-term best-of-class company.

Finally, let's return to our theme: targeted and well-thought-out training. No company has ever reached its true potential without a trained, well-motivated workforce. The rough outlines of successfully training reliability professionals are available from hundreds of articles and books. But success requires consistently dealing with details; success is in the grasp of employers and employees having a common stake in the matter. Success is achieved when both sides view intelligent training of mutual benefit. It's truly multifaceted and requires the thoughtful implementation of many details. You will find them in this write-up.

REFERENCES

[1] Systems Approach Strategies, (SAS, Tel. 905–430-8744, also website www.systemsapproach.com).
[2] M. Larkin, D. Shea-Larkin, CARE-Energize your Company with Empathy, Stoddard, Toronto, ON, 2000.
[3] H.P. Bloch, Improving Machinery Reliability, third ed., Butterworth-Heinemann, Woburn, MA, 1998.
[4] H.P. Bloch, F. Geitner, Introduction to Machinery Reliability Assessment, second ed., Van Nostrand-Reinhold, New York, 1993.
[5] H.P. Bloch, Practical Guide to Compressor Technology, second ed., Wiley & Sons, Hoboken, NJ, 2006.
[6] H.P. Bloch, Practical Guide to Steam Turbine Technology, McGraw-Hill, New York, 1996.

[7] H.P. Bloch, J.J. Hoefner, Reciprocating Compressor Operation and Maintenance, Butterworth-Heinemann, Woburn, MA, 1996.
[8] H.P. Bloch, C. Soares, Turboexpander Technology and Applications, Butterworth-Heinemann, Woburn, MA, 2001.
[9] H.P. Bloch, F. Geitner, Machinery Failure Analysis and Troubleshooting, fourth ed., Elsevier, Oxford/Woburn, MA, 2013.
[10] H.P. Bloch, Practical Lubrication for Industrial Facilities, second ed., Fairmont Press, Lilburn, GA, 2009.
[11] R.C. Eisenmann Sr., R.C. Eisenmann Jr., Machinery Malfunction Diagnosis and Correction, Prentice-Hall, Upper Saddle River, NJ, 1998.
[12] J.W. Dufour, W. Ed Nelson, Centrifugal Pump Sourcebook, McGraw-Hill, New York, 1992.
[13] H.P. Bloch, Pump Wisdom: Problem Solving for Operators and Specialists, John Wiley and Sons, Hoboken, NJ, 2011, ISBN: 9-781118-041239.

Chapter 42.7

Membership in Technical Societies

INTRODUCTION

Candid discussions with reliability managers are somewhat rare. That's why a conversation with the reliability manager of a prominent multinational oil company was intriguing. One of the questions of interest to this manager alluded to a regrettable fact: many companies that continue to experience repeat failures of rotating equipment (especially process pumps) have employees that are members of certain technical societies. He had questions on the merits of encouraging membership.

WHAT MEMBERSHIPS ACCOMPLISH

We might indeed ask ourselves just how effective these members of technical or professional societies have been over the years. Have they made a significant dent in the frequency of repeat failure incidents? Have they been motivated to understand, advocate, describe, and promote targeted and truly result-oriented training? Have they been successful (or even interested) in putting a stop to buying from the lowest bidder? Have they accepted the professional's role of exposing products that add little or no value to an enterprise?

As we all know, blindly purchasing from the lowest bidder and cutting out *real training* and mentoring led to equipment that is maintenance-intensive or unreliable. In the process machinery sector, the emphasis should have been on maintenance avoidance decades ago. Note that such an emphasis would require intelligent procurement, active support of the best vendors, and issuing knowledge-based (and then fully enforced) specifications. Professional knowledge comes from pertinent education. For true professionals, acquiring a pertinent education implies conscientious reading.

In the instance of existing equipment, adding value would require that every maintenance event be viewed as an opportunity to upgrade. The feasibility of upgrading must be determined well ahead of the maintenance event. Whenever justified, the maintenance/reliability professional must seek an active involvement in calculating, and making known, the monetary value of these endeavors.

The ever-present *pseudo*-training sometimes offered by technical societies must be identified for what it is: a waste of time. *Real* training is a two-way street. It is well defined, supported in equal measure by the employer and employee, requires motivation and effort by all concerned parties, and represents an investment in personal time. The intrinsic present and future value of such training is intuitively evident; such training is certainly pursued in the education of children and adolescents. Linking it to the economy and discontinuing training in hard economic times are patently absurd. The future belongs to those with knowledge and wisdom. The future does not belong to the uninformed. It is no different with adults that need to be informed and educated at their workplace or in subjects relevant to the survival of their workplace.

To make a solid case in favor of membership becoming informed is an opportunity that many technical societies have not understood. Some are outsourcing speaker selection and other related decisions to third-party organizers. A number of technical societies expect competent retirees to self-fund their appearance at such gatherings; many organizations and organizers do not want to go beyond adding members and collecting membership fees. They quite obviously would not wish to get involved in separating the value-adders from the purveyors of consultant-conceived generalities.

CONVERSATIONS REGARDING VALUE-ADDING

Back to the conversation with the reliability manager: As to one of his employees desiring to belong to technical societies, consider it a sign of an employee trying to better themselves, but don't assume that membership makes it more likely for them to add more value to your enterprise. There are other, far more accurate, indicators that can be spelled out in the employees' role statements. The employee's achievements (and shortfalls) can be discussed during the next performance appraisal meeting with that particular employee.

Chapter 42.8

Leaders Allocate Time for Technology Updates

INTRODUCTION

Just as a good medical doctor is expected to know much about modern treatment options, we would expect reliability professionals to be informed on technology advances. There are numerous training and technology update opportunities and the reliability professional must seek these out. Conscientious reliability professionals take advantage of books, articles, editorial columns, conference proceedings, vendor-organized training sessions, courses, and plain "opportune times."

OPPORTUNE TIMES

Opportune times present themselves in many different ways. Two hours between job assignments and a half-day postponement of a scheduled meeting are opportunities for the reliability professional to update his or her understanding of numerous topics and issues. If you have checklists in your computer, why not use the time for "toolmaking?" We have often used the term toolmaking to describe updating or revising checklists, work processes, and procedures based on an article that we might have filed away a few months ago or a conference paper that we read just last week.

Here are three examples we consider relevant and pertinent because each can be found in recent publications. The very fact that you seem fit to read this text greatly increases the probability that these examples apply to your plant. Anyway, remember that you're reaching for opportunities to look into these and similar matters.

Start with recent illustrations and documentation referring to mechanical seals with bidirectional tapered pumping impellers that effectively pump the flush liquid through a small heat exchanger (Fig. 42.8.1). Which of your hundreds of process pumps would qualify for, and benefit from, this arrangement? Have you asked your seal supplier? Could you learn from the supplier's answers? Should you communicate with this innovative, experienced manufacturer?

Next, consider using opportunistic time to review recent illustrations and documentation dealing with oil rings, also known as "slinger rings." These components pick up oil from a bearing housing sump and direct the lubricant into rolling element bearings. Elsewhere in this text, you will find a recommendation asking you to measure the as-installed width of an oil ring and to again measure the as-found ring at the next repair event. The difference in measured

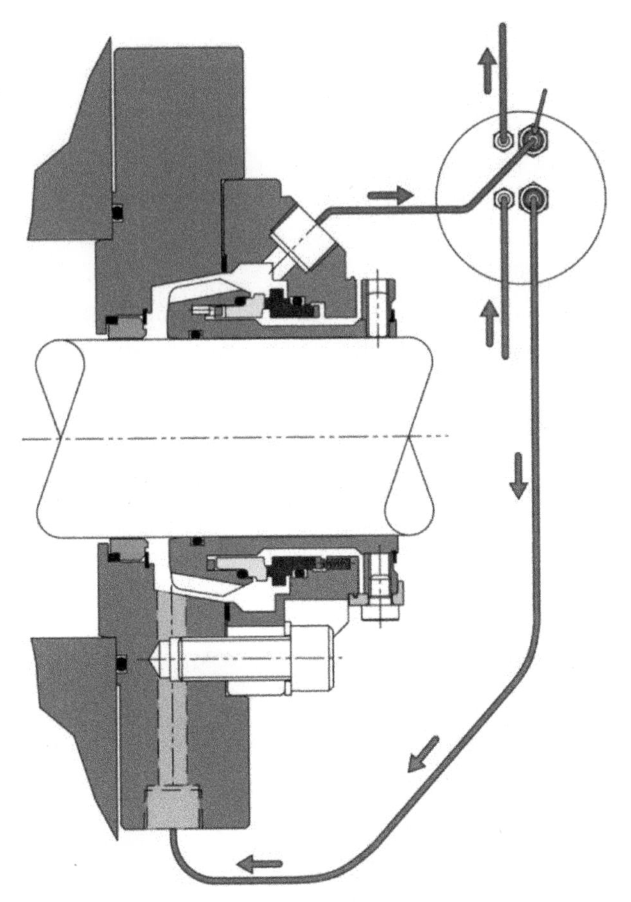

FIG. 42.8.1 Mechanical seal with bidirectional tapered pumping device [1].

widths relates to a volume of abrasive dust that, with virtual certainty, contaminated the oil and curtailed the bearing's life. So then, might you benefit from allocating time to study the shaft drive principles of Fig. 42.8.2? Could a similar drive arrangement power a small oil pump, and could pressurized oil be routed to a spray nozzle mounted a few millimeters from the rolling elements? Would that not eliminate vulnerable oil rings and constant-level lubricators?

Finally, there is Fig. 42.8.3, one of numerous feasible antiswirl labyrinth configurations. These are needed to improve the rotordynamic stability of certain process gas compressors. Could it be that choosing a particular configuration would extend the safe operating speed range of one of your compressors?

All three are mere examples and suggestions on how to make good use of "opportune times" at work. Self-motivated professionals are already educating themselves and need none of our reminders. But others, and especially

FIG. 42.8.2 Journal region of a small turbine. Note the worm gear arrangement on the left—a similar layout could be used to power a spray-producing lube oil pump for reliable lubrication of process pumps.

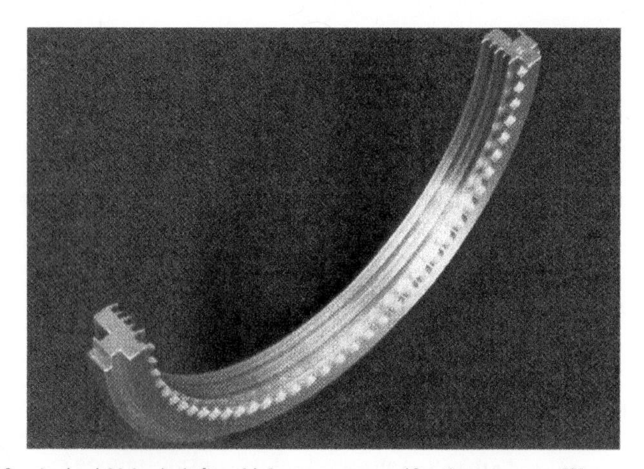

FIG. 42.8.3 Antiswirl labyrinth for a high-pressure centrifugal compressor [2].

managers endeavoring to give guidance to future managers, might wish to do a bit of nudging in this regard. Good managers know that checklists, sound work processes, and well-written procedures are of great value to technical support staff. Checklists merit periodic updating and should be reviewed by personnel representing the three job functions that come into almost daily contact with process equipment: the operators, mechanical/maintenance workforce members, and project/technical personnel.

All job functions must periodically update their knowledge and reliability professionals should pay much attention to detail. They might start with an insistence on recording important measurements. To the extent appropriate, they should be on the lookout for applicable new technology. Concentrating on opportunities to stay informed on bearing lubrication and sealing issues is of great importance and pays rapid returns on time invested.

REFERENCES

[1] H.P. Bloch, Pump Wisdom, John Wiley and Sons, Hoboken, NJ, ISBN: 978-1-118-04123-9, 2011.

[2] H.P. Bloch, F.K. Geitner, Compressors: How to Achieve High Reliability and Availability, McGraw-Hill, New York, NY, ISBN: 978-0-07-177287-7, 2012.

Chapter 42.9

Acquire a Marketable Skill

INTRODUCTION

For the foreseeable time, the "job of the future" will exist in a world that wrestles with issues of outsourcing, offshore design and manufacturing, and generally uncertain employment prospects. Even the aspiring engineers among the job seekers need to develop effective ways of finding and keeping employment in a murky environment.

Guidance is helpful because there is the question of where you should be headed in this sea of instability. We quite obviously live at a time when professed experts expound diametrically opposite views and often have the audacity of serving up their views with an air of infallibility. The answer as to where you should head is multifaceted, but being balanced and learning a marketable skill instead of going after an ill-defined "education" is certainly a good start.

READING SKILLS AND QUOTING FROM MARK TWAIN

The relatively few retired engineers willing to pass down their experience to other generations are unanimous in at least one recommendation: the learning process of an engineer starts with acquiring reading skills. Today, people with what passes for "education" will often have to accept jobs that require little or no education. In contrast, and with far fewer exceptions, people with *marketable skills* are likely to find gainful employment. By inference, there will always be a demand for marketable skills but there may not be a demand for "education." As a corollary, there are—and will be—people with engineering degrees that are in control of their lives. There will also exist other engineers who will have little or no such control.

Over the past four decades, we have met with, and spoken to, many engineers in all age groups who claimed that matters are out of their hands. They told us they are unable to influence decisions that have significant future impact on either them or their respective employers. Why should that be the case? Can we not choose to do (or not do) all kinds of things that affect our future? For instance, we can choose to watch worthless TV programs aimed at the dismally uninformed least common viewer denominator. Violent video games and over 95% of the available TV programs will not benefit us at all. As of 2015, the average American watches over 8h of TV each day. Suppose you were free to choose between "Seconds from Disaster" and "The Three Stooges." One of these programs would at least tend to groom you to be an informed person; the other is simply a big waste of your time. And notice that we have decided to not even mention programs that pollute the mind, programs not unlike contaminated food that can wreak havoc on the body—or even kill.

KNOW WHAT A GOOD BOOK CAN DO FOR YOU

Alternatively (and *not* alternately), we can decide to read a good book or relevant technical text so as to upgrade our language skills and increase our knowledge base. Reading is essential to the development of language skills. To communicate their ideas and add value to an enterprise, engineers must have a better vocabulary than many other crafts or professions. A widely read publication thus stated, quite correctly, that "many people lose their jobs not because of lack of technical skill, but because they lack the ability to communicate effectively." This observation is of great importance, since one obviously cannot communicate or express oneself with flawed grammar.

A person with true marketable skills will not treat language with contempt and risk being refused a potential job offer or jeopardizing a job presently held. When you write, avoid spelling mistakes. Make sound and proper communication one of your most important learning objectives. Good communicators will always enjoy a far better probability of finding and maintaining employment than the noncaring rest of the bunch. So, pay close attention to Mark Twain, who said: "A man who chooses not to read is just as ignorant as a man who cannot read."

KNOW THE DEFINITION OF "ENGINEER"

If you are young and wish to acquire engineering skills, start by knowing the definition of "engineer." In some countries of continental Europe and Latin America, the mere title "engineer" is limited by law to people with an engineering degree, and the use of the title by others (even persons with much work experience) is illegal. In Italy, the title is limited to people who, besides holding an engineering degree, have passed a professional proficiency exam.

Laws exist in all US states and in Canada that limit the use of certain *derivative or explanatory* engineering titles. In particular, the title "professional engineer" is limited, as are often titles indicating a specific, regulated branch of engineering (such as "civil engineer" or "mechanical engineer"). Nevertheless, most US states do not restrict unlicensed persons from calling themselves an "engineer" or indicating branches or specialties not covered by the licensing acts, though the legal situation regarding the title of "engineer" in Canada is unsettled. Again, the situation is entirely different for engineers wishing to use the designation P.E., the letter "P" indicating "professional." Use of the P.E. title is restricted to licensure by individual states. The same is true in Canada, where the designation "P. Eng." is being used. The status of maintenance and reliability engineers is unclear and has been the subject of debate for decades.

In general, engineers in the United States are not held in the same esteem as, say, medical doctors. Some of this is cultural and may never change. If, in the United States, Hubert Google is a medical doctor, he's likely addressed as Dr. Google. If he's an engineer, he's more likely addressed as Hubie. We will leave it to the reader to determine which of the two ways of addressing professionals conveys more respect. Nevertheless, it might be worth noting the specific experience-based remedies that had allowed in the past and will allow in the future perceptive engineers to climb out of the rut in which others apparently find themselves. But first, some background information.

SUCCESS AS AN ENGINEER: THERE ARE CHOICES TO BE MADE

In 2016, guidance from more experienced coworkers is often unavailable. The novice engineer may end up reporting to someone who never expressed his, or the company's, expectations in plain English. Moreover, expectations have often been misdirected by a superior whose own background has nothing to do with the various engineering disciplines reporting to him.

The situation was different in the 1950s and 1960. A mechanical engineer's career was then largely influenced by supervisors and managers who had moved through the same, or at least similar, knowledge-related career steps. Guidance and direction given in the mid-twentieth century were thus far more focused than offered today by early twenty-first-century generalists and money-management types. The worldview of today's boss is often shaped by motives and forces that differ substantially from those a few decades ago. As one of the consequences, far fewer engineers today are being enabled and empowered to act as decision-makers.

Remember that we make many choices every day. While time and unforeseen occurrences befall anyone, our lives are largely influenced by the choices we make. A young engineer can choose to get virtually all of his or her post-college training in the form of on-the-job learning. Although there is nothing wrong with absorbing and thinking on-the-job, engineers that want to be in control of their lives must buttress and supplement this learning with mature reading habits. Mature reading habits will unquestionably accelerate the acquisition of thoroughly marketable skills in more structured ways than traditional on-the-job learning.

THE ENGINEER AND ON-THE-JOB TRAINING

A measure of on-the-job training is always appropriate. It acknowledges that we can always learn from others. However, we must guard against accepting and absorbing as "fact" whatever others tell us; it certainly will not always be of true benefit. Conversely, neither will the act of discarding everything that others have done before us be a very smart approach. In essence, either extreme must be avoided, and science must always trump gullibility and sales pitches. Testing and understanding "the mechanics of things" and even thoroughly examining underlying thought processes are always sensible choices.

This again implies that you should seek a balanced view and that finding and consistently practicing this balance require a conscious effort. While it is certainly never too late to cultivate a balanced view, it is obviously best to do so early in one's life. This cultivating requires an investment in time; it certainly implies reading and thinking not only on one's employer's time but also on one's own time. On the other end of the spectrum, we should not "study things to death" since there are many endeavors that simply do not merit investigation beyond a certain point. Again, this is a matter involving common sense and balance.

SHARED LEARNING AND SPECIALIZATION ARE IMPORTANT

When a person learns or adds experience in a field that is logically related to his or her job function, both employee and employer stand to benefit far beyond their original expectations. The employee gains a sense of self-worth that will allow him or her to confidently look ahead to an otherwise hazy employment future. By nurturing in an employee the desire to learn, an employer stands to gain a value-adding contributor. This contributor's future ability to make go-no-go decisions that are based on fully understanding risks and consequences can be worth a fortune. A smart employer, therefore, makes training a shared responsibility. Both employer and employee will consistently and conscientiously carry out their respective obligations.

If you now accept that there's merit in having control over your life and future, it should be your desire to increase your marketable knowledge. You would take steps to systematically acquire a definable specialty and strive to know, ultimately, how you measure up against the real-world competition. Having accepted and acted on these premises allows you to go into a job interview with greater confidence.

THE JOB INTERVIEW AND BEYOND

So, let's just assume you are a novice mechanical engineer with the goal of specialization in rotating machinery for oil refineries and petrochemical plants or *reliability improvement of fluid machinery* (pumps, turbines, and compressors). Note that this arbitrarily chosen specialization goal is not as *narrow* as, say, "small metering pumps." An overly *narrow* area might not serve you in the long term if, for instance, small metering pumps were suddenly being replaced by "miniature nano-electronic stroking pistons" or whatever. Likewise, an overly *broad* area of specialization (such as "machinery and equipment") might be presumed to include bookbinding and packaging, shoe manufacturing, and ten thousand other types of machines. Expressing the desire to cover such a wide area during a job interview will likely be perceived as shallow or unrealistic.

Don't come to your job interview in beach clothing and flip-flop shoes. The interviewer may not admit that he or she was stunned by the multicolored depiction on your neck of Saint George slaying the dragon, but, rest assured, the job will be offered to someone else. However, suppose you can count yourself among the fortunate ones not handicapped by this potential problem. In that case, show initiative during the job interview. If you're an engineer about to graduate, ask about the training opportunities made available at, or endorsed by, the prospective employer's facility. A serious interviewee must have a goal in mind and this goal must involve professional growth and learning. Learning is obviously a two-component process. While one party offers it and the other absorbs it, the ultimate benefits are shared by both. That being the case, each has a commitment to make, and serious forethought and mutual cooperation are needed to achieve optimized professional training.

Already during a job interview, a graduating engineer would be wise to explore his or her projected role. Certainly soon after starting work, the engineer should be strongly interested in receiving a written role statement from his or her superior. If no such statement is forthcoming, the engineer may put his or her understanding on paper and ask the responsible manager for review, input, or concurrence. Unless there is agreement on the engineer's role, a rating such as "performance exceeding expectation" is as elusive as the same person simultaneously dancing at two separate weddings fifty miles apart.

As an example of systematically adding value to employer and employee, a company could identify a self-motivated employee and ask this person if he or she would be willing to be the custodian of an electronically stored and searchable engineering library dealing with turbomachinery, pumps, gears, shaft couplings, etc. He or she would then be asked to locate useful conference proceedings, published articles, and related information on the chosen topic. The material needs to be indexed and, in one form or another, made accessible to one's peers and other individuals that would be helped by the reference material.

During performance appraisals, the employee and the reviewer/appraiser would make an objective assessment of accomplishments by way of comparison with the previously agreed-upon role statement. Such an assessment would comprise all pertinent training issues and would obviously include measuring the employee's performance with regard to reading and disseminating technical material.

SPECIFIC STEPS IN THE TRAINING AND LEARNING PROCESS

Just to reemphasize: the first and perhaps most important step in an engineer's training is accepting that the most important learning process *begins* at graduation. Yet, even an engineer with a commendable history of prior employment should concede that profession-related training starts in earnest after you have accepted a job offer. Our premise is very simple: to be gainfully employed with continuity, you need to set yourself apart from the indifferent crowd. While the specific training plans differ for various roles and job functions, the *principles* remain the same. Therefore, as we next list components of a training plan for the role of "reliability improvement of fluid machinery," remember that the general principles derived from this example apply to every other aspect of engineering.

READING TRADE JOURNALS

In the interest of continually obtaining work-specific technology updates and related training, the developing engineer must peruse applicable trade journals. He or she should scan and—by either human eye vision or electronic scanner—retain articles on topics of potential interest. You can interpret "scanning" as tearing out pages or viewing and making copies of or reading, filing away, and cataloging articles.

Companies with well-defined training plans arrange for applicable *Trade Journal 1* to be given to employee "A." As he notices an article dealing with shaft couplings, he sends copies to colleagues or coworkers "B," "C," "D," etc. Applicable *Trade Journal 2* starts its route at the desk of employee "B" who notices articles of pivoted shoe bearings and wear-resistant V-belts. "B" makes copies of these and sends these copies to "A," "C," "D," etc.; likewise, "C" sends articles to "A," "B," "D," "E," and so on.

This once-per-month review task typically takes less than 10 min (per month!) and allows each participant to acquire a data bank of relevant cross-references. I have personally had an experience decades ago when I looked for a reference article and then contacted its author, asking for—and cheerfully receiving—priceless guidance on a subject matter related to his article.

TECHNICAL BOOKS: ONE PAGE A DAY, OR 200 PAGES PER YEAR

Few engineers purchase or thoroughly read technical texts after completing their formal education. Fortunately, however, there are some concerned employers who recommend that their staff read and absorb relevant technical texts. In 2003, one such employer encouraged his responsible professional employees to purchase as many books as they could reasonably assimilate or digest in a year's time. During performance appraisals, the effectiveness of this policy is continually being tested, validated, ascertained, and reaffirmed. Another company purchased pertinent technical texts and required each technical employee to read a page per day. To the extent feasible and reasonable, these professionals are then asked to jot down what they discern as differences between *their* work processes, hardware details, failure frequencies, maintenance intervals, work processes, etc., versus what *others* (competitors) are doing in these fields of endeavor. The training value is immense. Certainly, the return on the investment of the time it takes to read a page a day and to make a two-sentence notation each week is huge. There can be no doubt that this well-focused training is priceless and benefits all parties for years to come.

TRAINING THROUGH "SHIRT-SLEEVE SEMINARS"

In the 1970s, one highly profitable company arranged for its equipment reliability technicians and engineers to share the responsibility of making 7–10 min presentations at the end of each routinely scheduled and mandatory safety meeting. The presenters had to first educate themselves on such topics as "how to properly install a centrifugal pump" or "why steam turbines must be preheated before operation." Following the safety meeting and after making their 7–10 min add-on presentation, the equipment reliability technician or engineer would distribute laminated-in-plastic copies of these equipment-related single-sheet guidelines. Each employee had a three-ring binder in which they would place their copies. Plant management made sure that these guidelines were being used and adhered to by mechanical workforce and operating personnel.

In this manner, the "shirt-sleeve seminar" presenters taught themselves and passed on their findings to the entire plant. At this location, equipment failures due to human error and other causes were minimized and everyone profited from this approach. There should be no reason for not adopting it elsewhere with equal success.

By accepting help and by being willing to help others succeed, engineers will make important contributions. Moreover, they will gain a sense of self-worth if they truly pursue targeted training. Engineers who succeed in their quest for acquiring a marketable skill will likely prosper. If they acquire the requisite skills both during formal studies and after graduating from engineering school, they can face the future with considerable confidence.

Once you've been accepted by an employer and are on your way to work every day, resolve to *add value*. Think ahead; perhaps on your way to work in the morning hours, dwell on the specifics of *adding value* on that day. Then, on the way home from work, ask how successful you've been in adding value to the enterprise. On the next day, make the needed adjustments.

Finally, remember that in your job, you may occasionally encounter leaders that either cannot or will not lead. When this happens, don't give up. Only dead fish swim always with the stream.

FAVORABLE RESULTS ANTICIPATED

Self-motivated engineers or technicians who implement and stick to the approaches briefly described here are very likely becoming employees who offer solutions to problems. Instead of becoming folks expressing "concern" over potential problems, they will delineate the discrete steps needed to avoid problems.

There are, then, a few reminders for future maintenance and reliability professionals to ponder. First and foremost, not all that is labeled *education* is beneficial. Some education can be so academic as to lack substance; it would not pass as a marketable skill. Stick to marketable skills.

It's the same with training. Take charge and make it relevant. Instead of waiting for skill-enhancing training opportunities to present themselves, take a lead role in creating some of these opportunities. Recognize that virtually every marketable skill is acquired by tangible, "targeted," and well-thought-out training steps. And it's the *marketable skills* that will get us through life far better than a mere *education*.

Subject Category 43

Turboexpanders

Chapter 43.1

A Turboexpander Overview

INTRODUCTION

The primary objective of turboexpanders is to recover power, which then translates into conserving energy. Contemporary turboexpanders do this either by recovering energy from cold gas (cryogenic type) or from hot process gases at temperatures of over 1000°F (538°C). Current commercial models exist in the power range of 75 kW to 100 or more MW, making many applications feasible. Fig. 43.1.1 shows a well-insulated power recovery expander on the manufacturer's test stand in Jeannette, Pennsylvania.

FIG. 43.1.1 Power recovery expander (hot gas expander) for refinery service on manufacturer's test stand. *(Courtesy Elliott Group, Jeannette, PA.)*

The stator blading or a hot gas power recovery expander is shown in Fig. 43.1.2. Once installed at the user's factory, power recovery expanders can

Petrochemical Machinery Insights. http://dx.doi.org/10.1016/B978-0-12-809272-9.00043-8

drive electric generators, main air blowers, and other machines. The cost justification for blowers is not difficult to derive, and competent manufacturers have decades of experience with these machines in catalytic cracking units in oil refineries.

FIG. 43.1.2 Stator blading in a power recovery expander (hot gas expander) for refinery service. This machine has an output in the 36–38 MW range. *(Courtesy Elliott Group, Jeannette, PA.)*

TURBOEXPANDERS FOR ENERGY CONVERSION

Changing market conditions, accentuated by growing environmental awareness on a global scale, are improving market receptivity for the cryogenic turboexpander. Machinery manufacturers, quick to sense this market potential, have developed design features within their turboexpander ranges that offer user-friendly features, promoting ease of maintenance and operation, and aid design optimization. New nozzle technology deals with multiphase flow in the nozzles upstream of the rotor. They have been called flashing liquid turbines; the industrial demand for flashing liquid turbines is not new. Actually, it has already existed in the 1960s. However, since that time, many lessons have been learned on "how to" and "how not to" design flashing liquid turbines. Be sure you investigate this technology before making your choices.

Substantial energy can be recovered using low-grade waste heat, process gas, or waste gas pressure letdown. Centrifugal (radial inflow) turboexpanders (Figs. 43.1.1–43.1.4) are well adapted to such energy conservation schemes and, with recent developments that have increased their reliability, are suitable for unattended service on a 24-hour, 7-day/week operational basis. Some of the recent developments include better shaft seals, magnetic bearings, thrust bearing monitoring, and superior control devices.

FIG. 43.1.3 Two-stage turboexpander-compressor with active magnetic bearings for ethylene processing, front view. *(Photo courtesy of L.A. Turbine and Matt Vaugh Photography, Valencia, CA.)*

In the past, the use of the turboexpander as an energy recovery device was limited for a number of reasons:

- The return on capital investment did not justify a power recovery system unless more than several thousand horsepower was recovered.
- Finding a market for recovered power was difficult when there appeared no immediate use for it within the plant.
- Continuity and reliability of this energy source were required if it was to be used as "base load," which required standby equipment, spares, and appropriate operator attention.
- Lack of confidence in new power recovery schemes that were not yet proven made both government and private industries reluctant to invest in these systems.

In the late 1990s and as we entered into the 21st century, there has been a substantial shift in conditions and user attitudes. With the increasing cost of power, the return on capital investment has vastly improved. It's true that more energy became rapidly available since 2010, but as you read this text perhaps a full decade later, we don't really know where it went within another decade.

For a while, a more favorable regulatory climate and changes in attitude of utility companies toward returning electricity to their grid have made novel power producing schemes practical and attractive.

FIG. 43.1.4 Two-stage turboexpander-compressor with active magnetic bearings for ethylene processing, back view. *(Photo courtesy of L.A. Turbine and Matt Vaugh Photography, Valencia, CA.)*

High-efficiency expanders and their relatively short payback period made even smaller units economically attractive. These machines have demonstrated a high degree of reliability. Hundreds of units have been in continuous uninterrupted service for many years; this has removed the need for backup equipment and has demonstrated that unattended operation is entirely feasible.

Subject Category 44

V-Belt Drives

Chapter 44.1

Factors Often Overlooked in V-Belt Applications

INTRODUCTION

V-belt applications excel over most other available drives on the basis of low cost and mature design. They are uncomplicated in every way. Accordingly, V-belt drives abound and are assumed to require no further additional consideration. That, perhaps, is the problem.

MULTIPLE BELTS

Because a single belt might not have sufficient torque transmission capability, multiple belt sheaves are selected to allow higher overall torque ratings. However, these belts must be purchased with reasonable care. As an example, they

- must consist of matched sets,
- must not be allowed to have a "flopping" (longer than the rest) belt, and
- must not be assumed the only wear part needing replacement.

On the last point, it is generally assumed that belts are the sacrificial part. In other words, belts would be replaced as part of routine maintenance or when they begin to slip.

Slippage is easy to prove. Say, we had four belts on a particular drive. They are a "matched" set and each belt has a peripheral length of 100 in. With the equipment in the nonrunning condition, we simply draw a line across all four belts and then run the arrangement for 5 min. We then stop the motor and examine if any of the lines drawn across the four belts are now more than 10 in. (10%) away from the line farthest away on one of the other three belts. The difference is the slippage experienced during 5 min of operation.

Petrochemical Machinery Insights. http://dx.doi.org/10.1016/B978-0-12-809272-9.00044-X

FIG. 44.1.1 Laser alignment in progress on a belt-driven machine. *(Courtesy Ludeca, Inc., Miami, FL, www.ludeca.com.)*

Laser optic alignment tools are mandatory for proper alignment, and service companies can be of great help in demonstrating the value of proper belt alignment. Moreover, an experienced service company can teach and train others in the use of these modern tools (Fig. 44.1.1).

While it would now not be uncommon to expect belt wear and to schedule belt replacement, it would be prudent to use a sheave gauge and to look for sheave wear. If deemed excessive on even just one of the eight sheave grooves (recall that you will have four grooves each in the driving and driven sheaves), then sheave replacement is indicated.

Subject Category 45

Vibration Technology

Chapter 45.1

Machine Condition Inspection Systems

INTRODUCTION

Best-of-class (BoC) companies, corporations, and plants are grooming and rewarding a skilled and dedicated group of reliability professionals. In many plants, the members of this group are involved in both the collection and analysis of machinery vibration and other equipment condition data. You undoubtedly agree that an experienced and well-trained team can spot early indications of problems; they can also outline timely remedial action plans that can reduce maintenance and downtime costs.

WHY DATA COLLECTION SHOULD BE AN OPERATOR FUNCTION

Using your reliability professionals to collect data is certainly no longer considered "state of the art." For years, BoC companies realized maximum returns on their investment in monetary and human resources by letting the plant's reliability professionals spend the bulk of their time on data analysis, life cycle cost calculations, and the development of cost-justified upgrade and life extension measures. Data collection and reporting deviations are the operator's job.

Look at this analogy: we would not consider it cost-effective to let an automobile service technician check out 100 vulnerable components or subassemblies on our personal car, expecting perhaps two of these to require repair. We would, instead, prefer to analyze and pay only for items where we, the car's owner-operators, noticed deviations from normal response, sound, temperature, fluid level, etc. In other words, we would do the condition monitoring (data collecting) and leave the detailed analysis to the specialists or experts.

Petrochemical Machinery Insights. http://dx.doi.org/10.1016/B978-0-12-809272-9.00045-1

FIG. 45.1.1 Portable data collector portion of a computerized Machinery Reliability Inspection System. *(Courtesy SKF Condition Monitoring, San Diego, CA.)*

Likewise, today's best practices or BoC companies consider it the operator's job to collect data indicative of equipment performance. Operators do this with a data collector that merits the designation machine condition or Machinery Reliability Inspection System (Fig. 45.1.1). Deviations found with this "go-no-go" or "acceptable-marginal-unacceptable" instrument are referred to the vibration technician or reliability professional for analysis and follow-up. The analyst's valuable person's time can now be fully devoted to the relatively few equipment items that really deserve attention.

Perhaps the most important side benefit of letting operators collect relevant equipment data is getting them to know their equipment more intimately. While collecting these data in a truly effortless and accurate fashion, they may see, hear, and smell what's going on in their area. Steam leaks are observed, developing coupling distress is heard, and oil having soaked into steam turbine insulation reaches the operator's nose. He or she reports the deviation in the data collector and a specialist follows through with remedial action planning.

HOW MACHINE CONDITION INSPECTION SYSTEMS WORK

Until recently, machinery data collection instruments were often heavy and complex. They seemed to have been designed by engineers for use by engineers, whereas the more modern data collectors are designed to give operators the ability to rapidly and accurately monitor how closely a given parameter approaches the predefined limit value. When this limit is exceeded, the deviation is flagged on a screen and also in the computer memory.

Years ago, equipment locations such as bearing housings were often fitted with a measuring stud containing a microchip (see Chapter 45.2; Fig. 45.2.2). The operator would place the collector head over the stud and establish a connection by simply rotating it a quarter of a turn. The collector head recognized the stud location and accurately captured data on such parameters as vibration and temperature. In the 1990s, such instrument stored thousands of data points and instantly alerted the operator to relevant deviations. All of the data were later downloaded into a computer for trending, storage, plotting, or the automatic generation of exception reports. These exception reports were then acted upon by the reliability group or the plant's mechanical workforce.

There is ample proof that operator involvement in the data collection effort has been greatly facilitated by modern machine reliability inspection instruments. Virtually, all data collectors today are operating wirelessly. The resulting profitability improvement makes this the right approach for the overwhelming majority of process plants.

Chapter 45.2

Field-Installed Vibration Amplitudes for Centrifugal Pumps

INTRODUCTION

Vibration monitoring is a very important subject, but vibration avoidance should always be our main goal. Accept the fact that detrimental vibration can be avoided by proper design, proper maintenance, and proper installation. If you use vibration monitoring for the sole purpose of knowing when it's time to replace components, you might be in the business of changing parts. Productive technical employees don't want to be parts changers; they strive to be reliability improvers. That said, it would be wise to use vibration monitoring only for the purpose of finding the root cause. Then we follow up by putting effort into upgrading so as to prevent vibration from occurring in the first place. As a minimum, we might resolve to extend the time interval before the next vibration event occurs. In this manner, we use vibration monitoring and analysis as important tools to achieve what we're really hoping to accomplish: increased reliability asset preservation, plant profitability, and safety.

In 2000 and thereafter, the New Jersey-based Hydraulic Institute issued Standards (and amendments to) ANSI/HI9.6.2. This standard deals with *Allowable Vibration of Centrifugal Pumps*. Both horizontal and vertical pumps are covered [1], and "Hydraulic Institute allowable" all-pass (or "RMS") vibration is given in Table 45.2.1 as a function of pump type and style.

Just as the standards of the American Petroleum Institute, the standards of the Hydraulic Institute reflect decades of combined user and manufacturer experience in pump design and application.

CAUSES OF EXCESSIVE VIBRATION

The potential causes of excessive pump vibration are many [2] and the most prevalent ones include the following:

1. Rotor unbalance (new residual impeller/rotor unbalance or unbalance caused by impeller metal removal or wear).
2. Shaft (coupling) misalignment.
3. Liquid turbulence due to operation too far below the pump best efficiency flow rate (BEP).
4. Cavitation due to insufficient NPSH margin.
5. Pressure pulsations from impeller vane-casing tongue (cutwater) interaction in high discharge energy pumps.

TABLE 45.2.1 Allowable Field-Installed Pump Vibration Values

Pump Type	Less Than Power (HP)	Vibration RMS in./s	Greater Than Power (HP)	Vibration RMS in./s
End suction ANSI B73	20	0.125	100	0.18
Vertical in-line, sep. coupled ANSI B73.2	20	0.125	100	0.18
End suction and vertical in-line close- coupled	20	0.14	100	0.21
End suction, frame- mounted	20	0.14	100	0.21
End suction, API-610, preferred operation region (POR)	All	0.12	All	0.12
End suction, API-610, allowable operation region (AOR)	All	0.16	All	0.16
End suction, paper stock	10	0.14	200	0.21
End suction, solids handling—horizontal	10	0.22	400	0.31
End suction, solids handling—vertical	10	0.26	400	0.34
End suction, hard metal/rubber-lined, horizontal and vertical	10	0.3	100	0.4
Between bearings, single and multistage	20	0.12	200	0.22
Vertical turbine pump (VTP)	100	0.24	1000	0.28
VTP, mixed flow, propeller, short set	100	0.2	3000	0.28

Once a pump has been determined to have a high "total/all-pass" vibration level, the next step is to identify the cause. This would be the time to obtain a filtered vibration analysis (see Table 45.2.2). The first step in the analysis should be to capture, and then evaluate, the multiples ("harmonics") of pump running speed.

TABLE 45.2.2 Sources of Specific Vibration Excitations

Frequency	Source
0.1 × running speed	Diffuser stall
0.8 × running speed	Impeller stall (recirculation)
1 × running speed	Unbalance or bent shaft
2 × running speed	Misalignment
Number of vanes × running speed	Vane/volute gap and cavitation

While this is the first step in the analysis process, other possible causes of vibration may be more complex to analyze.

A wide variety of portable data collectors and devices for both monitoring and analyzing vibration events are available; the device shown in Fig. 45.2.1 is but one of many. Devices interacting with a measuring stud attached to the bearing to be monitored (see Chapter 45.1; Fig. 45.1.1) were common in the 1980s. Today, in 2016, wireless readings are the norm. As the decades progress, we would expect many additional advancements because technology is in a constant state of flux. Therefore, the selection is usually facilitated by engaging manufacturers that can also provide a teaching and training role. User feedback is often obtained at reliability conferences and from companies with decade-long experience hosting and/or exhibiting at these venues.

BEARING LIFE VERSUS VIBRATION

It should be intuitively evident that high vibration is not conducive to long equipment life. Vibratory forces are superimposed on normal bearing loads; their magnitude is relatively easy to calculate. Of course, users often ask to what extent there is linkage between vibration severity and bearing life reduction. An experience-based plot (see earlier Subject Category 2; Fig. 2.1.2) might be helpful. The plot reflects the author's percentage-of-life reduction estimates for general purpose equipment and is based on years of experience. You will find it reasonably accurate for process pumps.

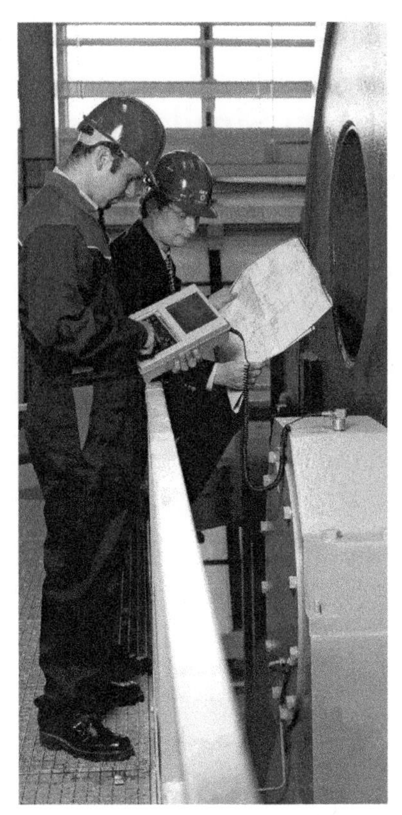

FIG. 45.2.1 Monitoring approach including measuring stud containing microchip for location identification. *(Courtesy SKF Condition Monitoring, San Diego, CA.)*

REFERENCES

[1] Centrifugal/Vertical Pump Vibration, ANSI/HI 9.6.4, 2000, Hydraulic Institute, Parsippany, New Jersey.

[2] H.P. Bloch, A. Budris, Pump User's Handbook: Life Extension, fourth ed., Fairmont Publishing Company, Lilburn, GA, 2013.

Chapter 45.3

Vibration Limits: Similar but no Consensus

MAIN FOCUS

Over the many decades since the late 1940s, industry standards and guidelines have been produced in many industrialized countries. Although attempts were made to reach agreement on the subject of vibration limits, there is no consensus on the matter. There are many explanations; among them the fact that machines purchased with higher-than-usual design conservatism or machines placed on a well-configured foundation are more vibration-tolerant than others.

An experienced reliability manager had similar questions on his mind when he asked our opinion on the subject of turbomachinery vibration. He and his staff had been having discussions about the vibratory behavior and applicable settings on one of their refinery's older (1940s vintage) turbine generators. At the time of our communication, these values were set at peak-to-peak shaft displacements of 3.0 mils (alarm), 4.5 mils (danger), and 6.0 mils (trip). A mil, of course, equals 0.001 in. or 25 μm; these values are stated with reference to the machine casing.

The staffers were familiar with ISO 7919 (Table 45.3.1) and a certain divergence between this international standard and experience-based practices in the United States. The reliability-focused group of engineers at this refinery wanted to understand and reconcile the ISO standard's corresponding acceptable limits of 4.7 (alarm), 6.9 (danger), and 9.0 mils (trip). These engineers also realized that a graph in "Machinery Failure Analysis and Troubleshooting" ISBN 978-0-12-386045-3, fourth ed., pp. 465 and reproduced here as Fig. 45.3.1, is obviously more limiting than ISO 7919.

In any event, the textbook and its various references were mainly intended for application to centrifugal compressors ("CCs"). The upper values of operability in Fig. 45.3.1 (5.5 mils in the case of 3600 rpm turbocompressors) are based on decades of experience with a large sampling of machines. Over the many years of collecting data, none of the close to 100 compressors in the sample deviated from the rules and guidelines. Essentially then, major US refineries have often used these values and, operating at speeds not exceeding 5000 rpm, <2.0 mils was considered "good" for new CCs during acceptance tests. Values up to and not exceeding 3.5 mils were "acceptable" for actual field-installed and operating CCs. A vibration range from 3.5 to 6.0 mils would have been considered marginal and shaft displacements above 6.0 mils unsafe and perhaps inoperable.

DIFFERENCES BETWEEN ISO GUIDELINES AND US PRACTICES

The reliability manager was taking note of and summarizing ISO 7919, which contains the numbers listed in Table 45.3.1; it shows four zones of interest and labels them A through D.

For the refinery's turbine generators running at 3600 rpm, the following values and descriptors would apply:

TABLE 45.3.1 Generalized Vibration Guidelines for Turbine Generators Operating at 3600 rpm

	ISO Max. (μm)	ISO Max. (mils)	US Typical Range (mils)
Zone A	80	3.2	<2
Zone B	150	6.0	2–3.5
Zone C	225	9.0	3.5–5.5
Zone D	>225	>9.0	>5.5

	ISO Max. (μm)	ISO Max. (mils)	US Typical Range
Zone A	80	3.2	<2 mils
The vibration of *newly commissioned* machines would normally fall within this zone (however, the limit is not to be invoked here because the refinery is obviously dealing with an "old" machine).			
Zone B	150	6.0	2–3.5 mils
Machines with vibration within this zone (<6 mils in this instance) are normally considered acceptable for unrestricted long-term operation.			
Zone C	225	9.0	3.5–5.5 mils
Machines with vibration within this zone are normally considered unsatisfactory for long-term continuous operation. Generally, the machine may be operated for a limited period in this condition until a suitable opportunity arises for remedial action. (Per ISO, this scheduling for early repair would allow operation of the refinery's turbine generators with shaft vibratory displacements below 9.0 mils.)			
Zone D	>225	>9.0	>5.5 mils
Vibration values within this zone are normally considered to be of sufficient severity to cause damage to the machine. Thus, operation above 9 mils would not be allowed by the ISO standard.			

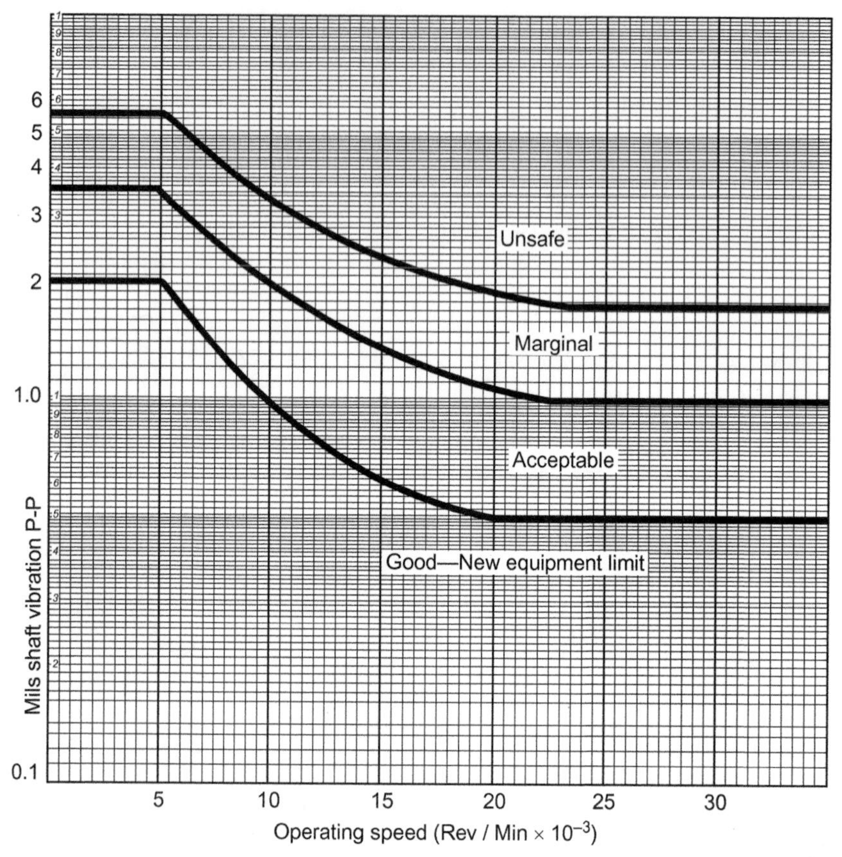

FIG. 45.3.1 Vibration guidelines for dynamic compressors in oil refineries and petrochemical plants. *(Courtesy Sig Zierau.)*

In a subparagraph, ISO 7919 next explains recommended alarm and trip settings. It does, however, allow some discretion by the user.

Alarm value (the first alarm) should be set higher than the baseline by an amount equal to 25% of the upper limit of zone B, that is, 3.2 mils baseline + 1.5 mils (25% of 6 mils) = 4.7 mils.

Danger setting using a midpoint between the alarm setting of 4.7 mils and the trip setting of 9 mils, that is, 6.9 mils.

Trip setting using the upper limits of zone C, that is, 9.0 mils.

The refinery asked where this would leave them and what the recommended course of action would be. The request also asked us to attempt to reconcile US practices and the ISO 7919 recommendations.

In fact, one should bear in mind that the recommended set points should be considered when installing a vibration monitoring system on machines for which there is little or no operating experience. If a machine consistently operates at low vibration levels (e.g., 1.5 mils peak-to-peak), it would be advisable to set the alarm level for that machine about 50% above the normal vibration level (thus far below the 3.2 or 4.7 mils recommendation). A vibration analyst should investigate the vibration signals if that "acceptable" level is exceeded because a change might indicate a growing problem.

The analyst should, of course, investigate any vibration level that is out of the ordinary for that particular machine. It could be caused by a scratch on the shaft that was introduced during the last repair or inspection. Such a scratch is not of concern and can readily be detected when displaying the vibration signal on an oscilloscope or in the "time domain" mode of a vibration analyzer.

In some reported cases, turbocompressors have survived operation at 7 mils. However, under no circumstances should a machine operate under these conditions without a thorough investigation that evaluates the vibration levels along the entire shaft and compares them with bearing and labyrinth seal clearances.

RECOMMENDED VALUES AND SETTINGS

Shaft vibration imposes additional load on bearings and coupling components. For turbine generators in the 4–10 MW size range, a test stand limit of 2 mils is readily achievable for manufacturers that use up-to-date rotor balancing and equipment design methods. There is thus no reason to compromise at this stage and a prudent user would usually insist on full compliance.

Once the turbine generator is operating and has perhaps done so for years, stable vibratory performance is more important than absolute values. In other words, operation at a constant 4 mils over the past four years should be of less concern than operation at 2.9 mils for four years and a jump to 3.9 mils that occurred yesterday. An alarm value of 4.7 mils seems quite reasonable. It would alert the operators that the limit for unrestricted long-term operation is being approached in the ISO case and has been exceeded in the case of typical US practice. A "danger" setting of 6.9 mils, indicating advice to shut down at the earliest opportunity, will incur at least some risk of failure before a safe shutdown is triggered. In contrast, shutting down when 5.5 mils are exceeded would virtually eliminate the risk of catastrophic failure. In an obvious compromise, we would use a shutdown value of 6.5 mils for these 3600 constant-speed turbine generators.

Remembering our first trigonometry lesson in school, we might even realize that 6.5 mils observed by a shaft displacement monitoring probe mounted at a 45-degree angle may, occasionally, relate to a true and actual shaft displacement that will have to be multiplied by the square root of 2. That would then turn 6.5 mils into 9.1 mils. How's that for reconciling the various discrepancies?

Perhaps of even greater importance is the difference in risk tolerance. A repair-focused user will consider himself shielded by compliance with the ISO standard. Conversely, a reliability-focused, risk-averse user will stay closer to the limits indicated in our US practices column.

Chapter 45.4

Vibration Limits for Gas Turbines

INTRODUCTION

Vibration matters take up entire libraries. For some, vibration monitoring and analysis is a lifetime endeavor. Others spend their entire careers on vibration avoidance. Because of the volumes written on the subject, we simply want to reaffirm that most issues would best be addressed in the context of data collected and operational experience with a particular fluid machine or model. Typical guidelines are still sought by many readers—in this instance for gas turbines. So as not to disappoint these readers, here are some general observations worthy of note.

DO TURBINE SIZE AND DUTY MATTER?

It has been argued that gas turbines should be segregated by size, or by duty cycles such as base load or peaking, or even by such lineage as industrial or aeroderivative. Indeed, the best course of action might be to communicate with the manufacturer and look closely at his overall experience.

That is not to suggest that we want to always do what the manufacturer recommends. His agenda may be to err on the safest possible side or on the side that generates more revenue for the manufacturer's spare parts business or whatever. Nevertheless, it is always beneficial to communicate with the manufacturer and to then compare his proposed vibration limits with those typically accepted by a wide group of users. The logical follow-up questions would then explore the ramifications of using his data versus industry-typical data.

While ISO and gas turbine manufacturers' standards are also widely available, reliability professionals refer specifically to API-616, covering gas turbines for industry, and API-670, covering protection systems. Opinions may also be offered by General Electric's Bently Nevada affiliate or by SKF, and other experienced manufacturers of condition surveillance systems. Understandably, these manufacturers will recommend the types of transducers and cable configurations they find cost-effective or are used to providing.

COMPARING STANDARDS AGAINST EXPERIENCE

Next, we generally advocate that reliability professionals communicate with the manufacturer and ask the manufacturer to reconcile deviations from ISO 7919 guidelines. This "MQA" process will define the vendor's conservatism. In consensus plots similar to Fig. 45.4.1 [1], the predominant frequencies detected by the monitoring transducers and related analyzers or instruments are commonly

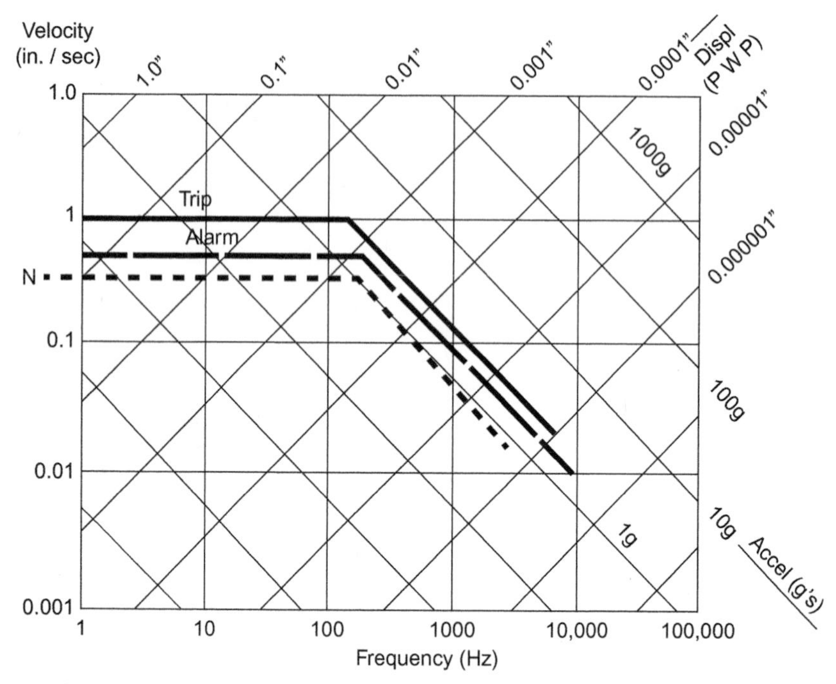

FIG. 45.4.1 Consensus plot of gas turbine vibration limits. Normal operation is within the "N" range.

shown on the abscissa (the x-axis). The units displayed on the ordinate are displacement (a), velocity (b), and acceleration (C). The four possible operating regions [1] are labeled N (normal), P (problems), A (abnormal), and D (dangerous).

As a general rule, we would want to be in region N when the machine is new, will periodically monitor after the measured parameter moves to region P, plan for an orderly shutdown and repair in region A, and refuse to operate in region D. When asked to consider deviations from these rules of thumb, prudent engineers will do lots of homework and document why they agree or disagree with such deviations. Safety-conscious reliability professionals will never allow operation of gas turbines in a trip zone. Human lives are at stake; they are far more important than profits.

REFERENCE

[1] M. Boyce, Gas Turbine Engineering Handbook, Gulf Professional Books, Houston, TX, ISBN: 0-87201-878-4, 1982.

Chapter 45.5

Asset Condition Monitoring With Ultrasound*

INTRODUCTION

Ultrasound testing technology is prevalent across diverse fields. It is useful in medicine for soft tissue repair and imaging. As a nondestructive technology, it measures material thickness and detects flaws in steel structures. And it contributes to the reliability of the manufacturing sector where it is used for monitoring the condition of assets and revealing sources of energy waste.

As a condition monitoring technology, ultrasound has hundreds of useful applications in almost any manufacturing sector. It is most often utilized for bearing condition monitoring; determination of time to bearing regreasing; analysis of low-speed rotating assets; steam trap testing; valve bypass events (leakage); electrical discharge detection such as corona, arcing, and tracking; compressed air leak management; and discovering leaks in shell and tube heat exchangers. It is a powerful equipment condition monitoring technology that maintenance professionals find extremely useful.

WHAT IS "ULTRASOUND"

Ultrasound detectors enable us to hear frequencies above the range heard by the human ear. A piezoelectric crystal mounted in either an airborne or structure-borne sensor is energized by high-frequency sound pressure waves. The crystal gives off a very low voltage that is amplified and then mixed with a lower frequency. This process, called heterodyning, transforms ultrasound from its original frequency to one that humans can hear. All the while, the qualities and characteristics of the source sound are preserved.

For asset maintenance purposes, there are three phenomena that excite an ultrasound sensor. They are categorized as friction, impacting, and turbulence (FIT). Friction is produced by two surface molecules rubbing together. Think about bearings over- or underlubricated, gears, chain drives, couplings, and belts. Impacting is produced by two surface molecules coming together with force. Think about early-stage bearing defects, grease contamination, and worn or broken gear teeth. Turbulence is produced by pressure or vacuum leaks, electrical partial discharge, and internal bypass in valves and steam traps to name but a few. Electrical partial discharge is described as sparks that occur within the insulation of medium and high voltage electrical assets.

---

* Contributed by Allan Rienstra, SDT International.

DEVELOPMENTS SINCE 2010

Two important innovations were added to the technology since 2010. They are incorporated in modern handheld instruments (Fig. 45.5.1) that can also be downloaded and displayed on computer screens. Both of these advances have contributed to ultrasound joining other frontline technologies for determining early-stage defects and diagnosing faults in slow-speed rotating machinery.

The first important advance is dynamic signal analysis, a means by which continuous data are acquired for a user-defined time frame. One important application is to assess condition on slow-speed rotating assets. Condition monitoring of variable frequency drives (VFDs) is also made possible. Dynamic data are recorded and then analyzed using time and spectrum plots similar to vibration monitoring.

A second breakthrough is the advent of four static condition indicators. In the past, static ultrasound data were reported as a single-decibel reading. That didn't provide a lot of significance due to the random, nonsinusoidal nature of a bearing defect. The four condition indicators made static ultrasound data meaningful and are thoroughly explained on a major provider's Internet site (SDT International).

Ultrasound has carved a significant role for itself in the world of condition monitoring. It works well in difficult environments, while modern advancements return accurate and repeatable data. The two advancements dynamic signal analysis and four condition indicators add versatility and allow machine diagnosis to be done at an earlier stage, possibly by less skilled technicians.

However, ultrasound is not a replacement for other condition monitoring technologies. Used in conjunction with conventional condition monitoring methods, ultrasound measurements and trending can contribute to the overall failure reduction and reliability improvement goals of industrial organizations.

FIG. 45.5.1 Ultrasound monitoring in progress. *(Courtesy Allan Rienstra, SDT International.)*

Chapter 45.6

Justification for Machine Condition Monitoring

INTRODUCTION

Continuous operational condition monitoring implies that machine condition data are collected and evaluated in real time. Of special interest here are data that could rapidly change and are telltale signs of unexpected degradation of the condition of the machine. Vibration amplitude excursions indicate rapid changes and will give early warning of disaster. A timely shutdown will minimize possible consequential damage and the resulting extended downtime.

Both frequency and duration of shutdowns can be reduced with carefully targeted planning of maintenance work. Such detailed planning allows streamlined manpower and spare parts allocation. But, intelligent allocation requires information on machine condition to project wear progression and probable time to failure.

In condition-based maintenance, machines are only shut down if their condition demands it. Parts are only changed if a damage criterion is reached, and the total anticipated survivability of parts in terms of remaining life and wear reserves is being exploited. This, then, while achieving a reduction in material costs, may only be possible if one has reasonably accurate data on the machine's condition. One obviously gauges the existing wear potential in "traditional" wear parts, for example, piston rings or packing, and thereby optimizes machine operating life.

Efficiency is another indicator of the condition of a machine. The primary purpose of efficiency monitoring in fluid machinery is to record changes in process or machine parameters that influence energy transfer to the medium being moved or compressed. Low efficiency causes fluid temperature rise and increases power consumption.

The efficiency of a reciprocating gas compressor may be determined, for example, by finding how indicated compression power relates to the driver power consumption.

WHAT TO MONITOR AND WHY

Recall that the objectives described here are general and are not specific to reciprocating compressors; they apply to numerous industrial machines. However, designing an effective strategy for condition monitoring parallels the strategy used for reciprocating compressors. In these positive displacement machines, their mechanical features—and perhaps their most vulnerable parts—must be analyzed in terms of maintenance frequency.

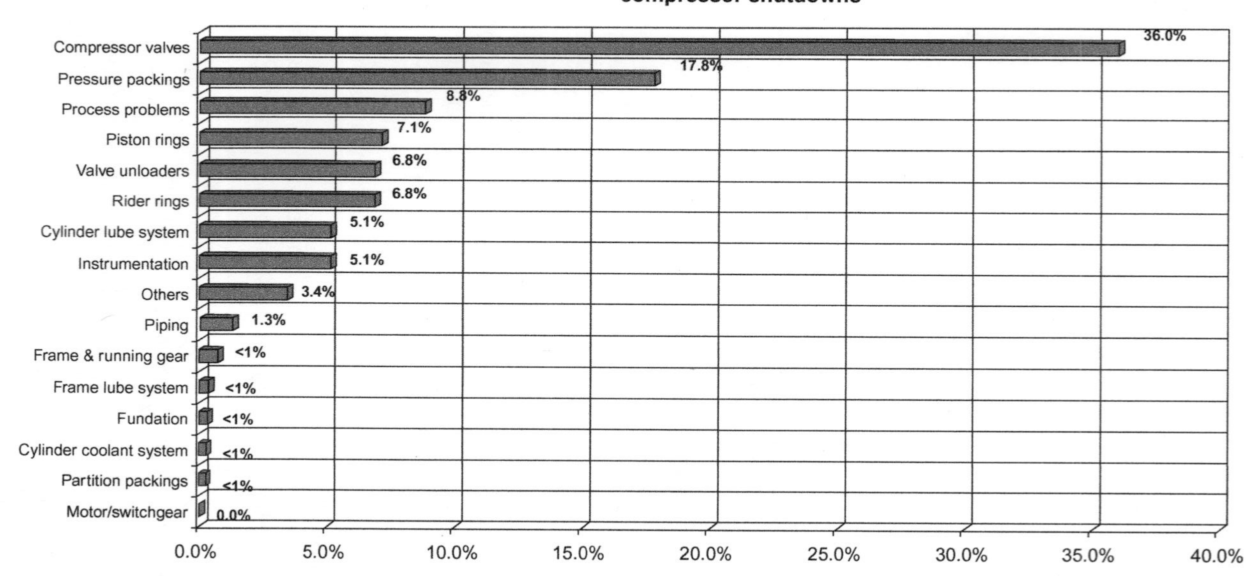

FIG. 45.6.1 Primary causes of unscheduled reciprocating compressor shutdowns. *(Courtesy Dresser-Rand Company, Olean, New York.)*

This strategizing was done by Dresser-Rand who, in 1997, made a survey of 200 operators and designers on the causes of unscheduled shutdowns of reciprocating compressors. The results (Fig. 45.6.1) clearly indicate the relative maintenance intensity of certain component groups.

These statistics showed, in 1997, that eight component groups were responsible for 94% of unscheduled shutdowns. Valve defects are obviously responsible for most of the unscheduled maintenance events. Excepting valves, general experience since 2009 indicates similar results. Valves have benefited from new high-performance plastics and composite materials.

As of 2016, there are monitoring systems being marketed for every component listed in the statistics for rotating machinery. Nevertheless, the specific requirements of a machine must be considered when choosing a system. As just one example, analyses of entire subassemblies or operating point-specific threshold checks is appropriate to detect valve damage on reciprocating compressors.

To date, the most important methods are employed and marketed by a leading provider including measurement of the valve pocket temperature, vibration analysis, and p-V diagram analysis. Reciprocating compressor monitoring has matured to the point where automatic shutdown logic is favored over operator intervention and human decision-making. Case histories attest to the wisdom of this shutdown strategy [1].

VALVE POCKET TEMPERATURES

Measuring the gas temperature in the valve pocket is the simplest and most cost-effective method of valve condition monitoring. The gas temperature in the valve chamber can be measured with a temperature probe observing the upflow of the suction valve and downflow of the discharge valve. If there is an obvious increase in temperature at one valve, one can assume that there is damage, e.g., a leak, at this location.

This method has the advantage of being inexpensive and simple to install in virtually all types of valves. Generally speaking, the measuring points are integrated directly into the process control system and can therefore be fed to larger monitoring systems. The occasionally difficult interpretation of plain measurement data in multistage reciprocating machines or where there are varying pressures on the intake and discharge sides is noted. However, these challenges can be met by adopting operating condition-dependent threshold monitoring. With moderate investment in technical effort and cost, an examination of valve pocket temperatures does, in many instances, provide important information on the pressure-retaining capability or "sealing condition" of the valves. Accurate records must be kept so as to troubleshoot the underlying causes of, say, the same valve location always being the first to show an excessive temperature. In that case, an elementary troubleshooting chart would point to the possibility of a defective (eroded) valve seat.

REFERENCE

[1] Prognost Systems Inc., Houston, TX and Prognost Systems GmbH, Rheine, Germany.

Subject Category 46

Warranties

Chapter 46.1

Consider Amended Warranties

INTRODUCTION

If you have ever encountered situations where an equipment manufacturer declines to offer warranty coverage, allow us to share three case histories. In each of these, competent reliability engineers requested and obtained amended warranties from equipment manufacturers. The affected parties advanced from there. No project manager was getting upset; in no case was there any negative impact on project costs and schedules.

THE CASE OF THE INTEGRALLY GEARED COMPRESSOR

We relate this case history as a tribute to a highly competent and experienced RP (reliability professional) who was deeply involved in applying machinery quality assessment ("MQA"; [1]) for a facility in one of the US Great Plains states. MQA, in this instance, was related to the prepurchase determination of the future reliability of a very large integrally gear-driven multistage process gas compressor in carbon dioxide service. The machine was based on the modular construction featured in Fig. 46.1.1 and had eight stages. Its motor driver was in the 20,000 kW-size category.

Our RP realized that the compressor, although designed to incorporate proven mechanical (oil) seals, would, some day in the future, benefit from dry gas seals. He convinced the manufacturer that all relevant seal cavities must have dimensions capable of accommodating dry gas seals at some time in the future. He had a letter of compliance in his carry-on case before boarding a plane for his flight from Frankfurt back to Minneapolis and a connecting flight to his home destination. That's all the credit I can still give this outstanding engineer and able colleague. I remember him fondly and often met with him in the late 1990s and the time until 2010. Before he left us in 2015, he lived in a city named after the man who was dismissed from his post as chancellor by Kaiser Wilhelm II in 1890.

Petrochemical Machinery Insights. http://dx.doi.org/10.1016/B978-0-12-809272-9.00046-3

FIG. 46.1.1 Multistage integrally geared compressor for CO2 service. *(Courtesy MAN Turbo, Oberhausen, Germany.)*

THE CASE OF THE RECIPROCATING PUMP

In the late 1980s, a major manufacturer was asked to provide two 4500 kW vertical-piston high-pressure pumps to a plant making decaffeinated powdered coffee. After the pumps were installed at the user's site, the user's reliability engineer explained that his company was thoroughly familiar with the performance of dibasic ester synthetic lubricants and wished to use these superior fluids. When the pump manufacturer then threatened to rescind all warranties, the parties negotiated a compromise agreement for using dibasic ester in one of the two pumps. Warranties would be valid for everything except crankshaft and bearings of the pump with dibasic ester lubricant. The pump's liquid end would receive whatever usual warranties were in effect at the time. The second pump would be run with the manufacturer's recommended mineral oil. After a few weeks of operation, the pump with the traditional mineral oil failed massively, whereas the pump with the relatively unknown synthetic dibasic ester lubricant ran with no distress whatsoever. The moral of the story: a competent reliability professional is often ahead of an uninformed manufacturer. Today, few (if any) vertical-piston high-pressure pumps use mineral oils.

THE CASE OF NO WARRANTIES ON CENTRIFUGAL PUMPS

Starting in 1965, a major US refinery placed orders for process pumps and specified that they, the user, would connect these pumps to one of the refinery's many unit-wide oil mist systems. Several pump vendors submitted proposals with exclusionary clauses, noting that warranty coverage for their process pumps excluded pumps with oil mist lubrication. The purchaser-owner's engineers expressed willingness to accept warranty clauses removing bearings from coverage. However, in return for making this concession, the manufacturer had to expressly reconfirm that full warranties stayed in effect for the hydraulic performance of their pumps and that power draw (overall efficiency) would be calculated with mechanical seals installed and with couplings absorbing the extra power resulting from a reasonably anticipated angle of shaft misalignment. The pump vendors were asked to choose wisely; they knew it was the buyer's prerogative to drop unreasonable vendors from the preapproved vendors' list [2].

We can conclude that a smart customer will hold the pump manufacturer's feet to the fire by insisting that the manufacturer give his usual warranty on pump hydraulic performance. That's a given on which we should never compromise—simply because what goes on in the bearing lubricant application region has nothing to do with head, flow, cavitation, NPSH, etc. If the pump manufacturer is uninformed on advantageous lubrication matters, pump owner-purchasers will sometimes make concessions on bearing life warranties, but will never yield on the issue of hydraulic warranties. The two are separate subjects.

Other informed purchasers later joined the then leading oil mist users Shell and Exxon (then Esso) who, in 1966, had taken the position that pump manufacturers must honor hydraulic warranties, or else innovators will no longer buy their pumps. With regard to oil mist addition, conversions, retrofits, and upgrades, one user company put pump manufacturers on notice that they had to make one of two possible choices: (1) a pump manufacturer participates and cooperates in the oil mist-related upgrading activities. In that case, the vendor was entitled to use the information in any of his existing and/or future pump designs. (2) A user-innovator will make appropriate conversions (oil mist upgrades) in its refinery repair shop and withhold all of the important upgrade technology from an uncooperative pump manufacturer.

REFERENCES

[1] H.P. Bloch, F.K. Geitner, Compressors: How to Achieve High Reliability and Availability, McGraw-Hill Publishing, New York, NY, 2012.

[2] H.P. Bloch, Pump Wisdom: Problem Solving for Operators and Specialists, John Wiley & Sons, Hoboken, NJ, 2011.

Subject Category 47

Water Hammer

Chapter 47.1

Consider the Destructive Force of Water Hammer

INTRODUCTION

Envision a piping system within which the flow of liquid is suddenly stopped; the liquid tries to continue in the same direction. In the area where the velocity change occurs, the liquid pressure increases dramatically, due to the momentum force. As it rebounds, it increases the pressure in the region near it and forms an acoustic pressure wave. This pressure wave travels down the pipe at the speed of sound in the liquid. If we assume the liquid is water flowing in rigid pipe at ambient temperature, the wave velocity is 4720 ft/s (1438.7 m/s). The acoustic wave will be reflected when it encounters an obstruction, such as a pump, fitting, or valve.

The fundamental water hammer equations include a number of assumptions as to the magnitude and duration of hydraulic transients in pumping systems. For accurate predictions, the validity of some of these simplifications must be tested on a case-by-case basis. However, it is generally assumed that

- the fluid in the pipe system is elastic, of homogeneous density, and always in the liquid state;
- the pipe wall material or conduit is homogeneous and elastic in all directions;
- the velocities and pressures in the pipeline that are always flowing fully are uniformly distributed over any cross section of the pipe;
- the velocity head in the pipeline is negligible relative to the pressure changes;
- at any time during the pump transient and when operation is in the normal operating zone, there is agreement of the pump speed and torque characteristics that correspond to the transient head and flow that exist at that moment at the pump;

Petrochemical Machinery Insights. http://dx.doi.org/10.1016/B978-0-12-809272-9.00047-5

- the length between the inlet and outlet of the pump is so short that water hammer waves instantly propagate between these two points;
- windage effects of the rotating elements of the pump and motor are negligible during the transients;
- water levels at the intake and discharge receivers do not change during the transients.

FACTORS AFFECTING WATER HAMMER

High- and low-head pumping systems [1]: It is interesting to note that water hammer is of greater significance in low-head pumping systems than in high-head systems. For a given rate of velocity change, the pressure changes in high- and low-head pumping systems are approximately the same order of magnitude. Therefore, a given head rise would be a larger proportion of the pumping head in a low-head pumping system than in a high-head system.

Discharge line profile: While this is rarely a problem in the HPI, there are instances where the pump discharge line profile is based on economic, topographic, and land right-of-way considerations. In those cases, surge tanks may have to be placed so that their height above the natural ground line will be much less than what would be necessary if there were no high ground near the plant.

Rigid water column theory: Although we are demonstrating it in the simplified equations below, the rigid water column theory may not be sufficiently accurate for the computation of water hammer in pump discharge lines. This theory assumes that the liquid is incompressible and the pipe walls are rigid.

Additional factors that either influence the severity or perhaps even cause water hammer include check valve action, product wave velocity, pipe size, number of pumps operating, and even the flywheel effect of pump impellers. From a practical viewpoint, reliability professionals suspecting water hammer should consider reviewing the numerous websites of engineering and consulting firms specializing in the subject. Abatement actions can range from inexpensive standpipes to expensive control devices. There are even occasions where an increase in pipe size may be justified to avoid the use of more expensive controls. Evidently then, for a rigorous analysis, one would need complete pump characteristics.

SIMPLIFIED SCREENING CALCULATIONS

Simplified screening calculations can be used to indicate if water hammer is severe. The potential magnitude is dependent on the speed of valve closure and the liquid velocity in the pipe prior to the start of valve closure. If the speed of the valve closure, in seconds, is greater than the total length of pipe (L), in feet, divided by 1000, then momentum theory (Newton's second law of motion) applies:

$$F_{\mathrm{m}} = (p/g)Q(V_1 - V_2)$$

where

F_m = momentum force (lbs),
p = fluid density (lbs/ft^3),
g = acceleration due to gravity (32.2 ft/s^2),
Q = rate of flow (ft^3/s),
V_1 = initial velocity (ft/s),
V_2 = final velocity (ft/s).

On the other hand, if the time to close the valve is less than $L/1000$, then the acoustic shock wave/elastic column theory applies. In that case,

$$P = p \times A \times V / \left(g \times 144\,\text{in.}^2 / \text{ft}^2\right)$$

where

P = pressure rise (psi),
p = liquid density (lbf/ft^3),
A = velocity of sound in water (ft/s),
V = velocity of the liquid in the pipe (ft/s),
g = acceleration due to gravity (32.2 ft/s^2).

As an example, if a pump is working at 200 psig (1379 kpa), and the liquid is traveling at the maximum recommended discharge velocity of 15 ft/s (4.6 m/s), the instantaneous pressure inside the casing would jump to 1158 psig (7984 kpa). This assumes the liquid is ambient water and the valve closure time is short enough for the elastic column theory to apply. Pump casings are not usually designed for this magnitude of pressure, especially if the casing is made of a brittle material, such as cast iron. Even if the casing is constructed of a more ductile material, the shock wave may still cause some components to permanently deform and, ultimately, fail [1].

It should be noted that the pump is not the only component affected by the water hammer phenomenon. Valves, sprinkler heads, and pipe fittings are also at risk of catastrophic damage. Pipe hangers and pump foundations have been known to fail because of water hammer. Especially PVC pipe and fittings are very susceptible to damage from water hammer.

In summary, water hammer can be controlled through proper valve closure rates (with slow closing valves), the addition of diaphragm tanks or similar accumulators to absorb the pressure surge, and relief valves to release the pressure. An engineered solution is obviously preferred over the trial-and-error approach, but screening studies are still appropriate.

REFERENCE

[1] H.P. Bloch, A.R. Budris, Pump User's Handbook Life Extension, fourth ed., The Fairmont Press, Lilburn, GA, 2013.

Postscript/Epilogue

Despite Gaining "Insights", Industry Will Always Face Challenges

Our title "Petrochem Machinery Insights" is inadequate and cannot do justice to the vast and, inevitably, interwoven subjects of equipment design, manufacturing, installation, training, managing, and so forth. But whatever the insights that were in fact conveyed, they can still impart knowledge. By inference and because the application of knowledge is called wisdom, there will be no wisdom without knowledge. Be prepared, though: once we are ready to act on true wisdom, we will encounter challenges. At that point in the journey, we may have to become far more than mere bystanders or observers. We will have to act in the face of resistance to action. Change may be painful for many people. In fact, change causes some to be distressed.

When asked about challenges facing the hydrocarbon processing industry (HPI) or, for that matter, any other industry sector, my typical answer has always been the same: the challenges faced today are bundled up—are comingled or can be found—in the lessons taught many decades ago. But lessons learned and explained decades ago were often disregarded by managers whose focus was short-range. The focus *today*—whenever it is that you're reading this text—is probably even more short-range than it was in the late 20th or early 21st centuries.

Suppose you read this text in 2016 or even a full two decades later: chances are what the preceding paragraph conveyed will still be accurate. Why is that? We can examine some of the reasons while keeping in mind that all of the people with whom we come in contact can be divided into three groups:

1. The ones who already know the subject and understand what course of action to take in the best interest of stakeholders.
2. The ones who are not teachable. Trying to communicate with them will be both frustrating and a waste of time.

3. The ones who are presently uninformed and would see (a) merit in you showing them the benefit derived from listening to you and (b) would then act on your advice.

While written with much of the hindsight that covers my working life in the period from 1950 to 2015, the text quite obviously targets the third group and is directed to overlapping generations alive before and after the decades of interest to a particular reader. If you are selling a product, we want you to concentrate on group 3. If you work for the companies that have diligently and thoughtfully decided to employ you as a professional or as an expert craftsman or value-adding manager-organizer, consider yourself a stakeholder. You really *do* have a stake in your company; you want the company to have staying power and you want it to enjoy prosperity.

Now think of yourself as a user-purchaser. In a way, the user-purchasers are stakeholders also. Their stake is in seeing the provider company prosper so they, the users, can get service in the future. Needless to say, if both seller-provider and buyer-user prosper, all parties will thrive in what we then rightly call a win-win situation. In a win-win relationship, both sides will conduct themselves within a framework of "CCC," meaning communication, cooperation, and consideration. I heard about CCC many decades ago at a meeting of friends who considered each other part of an extended family. It took little time to convince me that practicing CCC is good for business. It's good for developing and marketing products. CCC is also good for developing people and for keeping entities focused on common goals. Without question, CCC is good for a marriage.

But again, you will face challenges and you will not be unique in having to face some of them head-on, others more insidious and covert. Regardless of your job function—be it employer or employee, manager or nonmanager, or a person focusing his eyes on today or on the future—you will have to face issues and problems. Chances are you will become keenly interested in understanding tangible solutions. So, please pay close attention to this summary. It is offered in the full knowledge that these pages cannot even begin to cover every conceivable issue of interest. But this summary or postscript is also a reminder that we, individually, can make a very big difference.

PROBLEM, CAUSES, AND SOLUTIONS

Today, we deal with a largely uninformed workforce. This adjective is not to be misunderstood; even a genius can be uninformed.

In the mid-1950s, we understood that the United States ranked first worldwide in terms of mathematics skills and science education. A few decades later, in 2015, the country ranked 35th out of 64. In Feb of 2015, Pew Research called this ranking "unimpressive." This should be of interest to us because the individuals who are now in school or have just started employment are the country's present or future technicians, craftsmen, maintenance-reliability professionals, and equipment operators. The needs and wants of many of these individuals are probably not in mesh; some are mere consumers who misperceive the value they actually add. Their actions and inactions may not be well aligned with the notion of adding value. There could be a significant gap between their expectations and the facts or realities of life. Turning an uninformed or unrealistic person—a consumer—into a value-adder will require considerable patience and forethought.

Root causes. The root causes of finding ourselves surrounded by an uninformed workforce go very far back and can be traced to generally shallow leadership. Shallow leaders cannot (or will not) give guidance, or mentoring, or provide a nurturing environment. In-depth guidance, appropriate mentorship, and nurturing take time, effort, and dedication. Today, all of these attributes are in short supply.

Why are many uniformed? Not all workers are motivated to learn and the fault lies not on one side alone. Both sides are responsible for the low esteem in which we hold learning. Learning is not always rewarded. We allow ourselves to be distracted. Industry leaders are not rewarding the one who brings them the facts; all too often, not enough time is allocated to capture and convey these facts. Worse yet, managers often allow facts and opinions to be comingled. Strong opinions often reap rewards, even promotions. Majority *opinions* seem to vastly outnumber factual *findings*. Be this as it may: In the final analysis, acting on *facts* brings rewards; acting on *opinions* very often wastes time and money. I saw it again in late 2015 when I visited a major refinery that had once more acted on someone's opinion. The results were not pleasant. Telling them the facts caused discomfort and it became clear that I had stepped on many toes, figuratively speaking.

FACTS VERSUS OPINIONS

For now, I'm asking you to recall the fact that a severe hurricane struck the US Gulf Coast in Aug 2005. Opinion has it that it was the storm that hit and devastated New Orleans. But a set of facts point to the levies that kept back Lake Pontchartrain giving way. To some, that's a real dilemma because it means that the federal government and the original construction decisions of the government's corps of engineers were responsible for the widespread flooding. What's the difference? Well, if we correctly identify flaws in the levies or in the design factors of safety of floodgates, we stand a chance to take meaningful corrective steps and avoid future disasters. If, on the other hand, we base our thinking on the opinion that the storm that hit New Orleans was too monstrously severe or that "stuff" happens, we will just sit and wait until "stuff" will happen again. Mitigation of future natural disasters would be the planting and nurturing of dune grasses and mangrove swamps as barriers against severe wave action on seashores. Levies could be strengthened; one could study the actions the Netherlands took to ward off intrusions of the sea into low-lying land. One would have to spend some money.

Let's be sure we get the meaning of all this: long-range thinking has merit far beyond what meets the eye. Long-range thinking is the opposite of instant gratification. Instant gratification is not leadership. After accepting this as a statement of fact, we must vigorously apply it to our future equipment reliability pursuits. Note how some industrialized countries explain the origins of their prosperous machine design and equipment manufacturing industries. The strengths, prestige, and achievements are rooted in 120 years of apprenticeship (crafts) training. Sound engineering decisions are based on well-timed internships for future engineers. The list is endless. You get what you deserve; if not right away, then certainly ultimately.

Consider the above a pointed or highly focused message; please don't confuse it with negative thinking, which it is not. Let the message establish that first we have to (1) identify a problem using facts in evidence. Facts are backed by science. From there, we proceed to the next step and (2) outline our available options. Finally, (3) we must recommend a solid and well-researched solution. Indeed, there are experience-based general remedies and sensible steps; we will try to recap them on the remaining pages of this text.

RISK-AND-REWARD SYSTEMS

As we follow the focused message prescription and examine industry trends, we often see an unhealthy risk-and-reward system. Accountability (or the lack of it) becomes a huge challenge. Intellectual dishonesty exists on a widening scale. When I originally set out to compile this book, we all assumed that a prominent automobile manufacturer made clean diesels, but now we know better. They were clean diesels during testing, but not while in operation on streets, roads, and highways. Also, we may recall how years earlier certain advertisements had something to do with clean coal, but then: What happened? I'll stop here and will leave it to others to answer the questions.

Today, we very often see engineers trying to imitate the conduct and behavior of lawyers. The (understandable) aim and job of lawyers is to explain a client's limited responsibility or to make a compelling argument in favor of the nonculpability of their clients. In sharp contrast, it should be a reliability engineer's aim and job to clearly define and outline safe and sustainable asset management. Substantive asset management is a detail task that, regrettably, either is unappreciated or remains unrewarded. It should be no surprise that such detail tasks are, therefore, widely shunned. In some companies, an incompetent manager receives far better pay than a highly principled and competent well-rounded engineer.

PROBLEM AND SOLUTION

All too often, budgets are defined by the lowest possible outlay. The cost estimating manuals at EPC (engineering/procurement/construction) firms often only show least cost equipment. If these cost estimating manuals showed operating/

maintenance/catastrophic failure-optimized equipment, the budget would need a multiplier of, say, 1.07. Offering to build plants at 1.07 times someone else's offer, the EPC would lose out to the competition. Why? Because EPCs are all too often selected on the basis of low billing rate, lowest total as-bid price, or artificially contrived (and ultimately found misleading) yardstick or indicator. Many such selection criteria have little to do with how reliable your plant will be five years after it starts making a product.

Problems and solutions are still our subtopic. Industry-wide, much lip service is paid to asset reliability. However, asset reliability and lowest initial cost of assets are almost always opposites. They can only be reconciled/optimized by experienced and well-informed professionals. Assuming these experts are still around, they must be given early and rather unobstructed access to management. Few, if any, of these professionals are granted the necessary access. Access is one of the solutions. Too many layers in between represent hidden agendas and "turf issues"; they are part of the problem.

Because true experts are no longer groomed and nurtured, or because they are brought in far too late in a project definition-and-implementation sequence, we are now stuck in an endless cycle of reinventing "new" initiatives. Again, grooming, training, nurturing, and enabling are among the interacting solutions. Long-range solutions are not spelled out in consultant-conceived generalities. Consultant-conceived generalities do nothing to define actionable implementation steps.

So, we've made some observations in the various subject categories of this book. These observations can be broken down into hundreds of pages of detail. An entire branch of publications, trade journals, and conferences has sprung up to engage in debates on the matter. The debates rarely lead anywhere because, as we just read in the pages of this book, actionable implementation strategies require lots and lots of detail. Details will need to be learned, conveyed, and supervised. You don't get what you expect; you get what you inspect. Hence, inspection is one of the solutions. Inspection must be against objective and quantifiable targets. While many KPIs (key performance indicators) are valuable, some are decidedly not. We have to learn how to separate the wheat from the chaff. Well-trained, properly rewarded, carefully groomed, and nurtured professionals will be better able to make this wheat versus chaff separation than an opinion-voicing uninformed individual.

We have to learn that not all wisdom is instantly downloadable from the Internet. The Internet is full of dots, but connecting the dots takes time and learning. Without learning, we sink deeper and deeper into what some have called "digital dementia." Let's talk about the reward systems we see in place. A supervisor with precise detailed experience might be discouraged or dissuaded in his professional and ethics-based pursuits. At fault is an often inappropriate and often *unjust* risk-and-reward system. Someone will send him the signal: "We don't need you. We don't have enough failures here to justify paying you more than we pay so-and-so." This supervisor quickly learns that only the "let-it-break and then fix-it-fast with duct tape" practitioners are well rewarded. Those who *prevent* things from breaking are often viewed as laggards or sluggish performers. A more equitable reward system is one of the solutions.

These are all seemingly harsh and unpopular judgments. They cause great annoyance and discomfort because they allude to the need to drastically change course. But discomfort is part of change. In other words, the very mention of change tends to make some people uncomfortable. Verbalizing an unpopular judgment is like telling a mother her baby is ugly—usually not a well-received message, regardless of "facts in evidence." But that's an entirely inappropriate comparison because you will gain nothing by telling a mother your opinion. She cannot change the appearance of her child.

In sharp contrast, a plant asset can be valued on the basis of facts, carefully leaving off and not even mentioning unsupported opinions. You, your employer, your clients, and society as a whole may gain immensely if they're consistently told the truth about your company and the condition of its assets. Oh yes, changes would be required, but change will be possible. Making changes would have to start with conceding the futility of continuing on a course that years ago was already recognized (by unbiased observers) as leading to complications down the road. We are now "down the road," so to speak.

OUR BIAS

All imperfect humans are biased in some ways. Forty years ago, I was biased in favor of keeping my good job. Today, I'm biased in favor of quickly coming to the point because there's little time left and I don't want to waste anyone's time—least of all my own. I've also come to realize that one can only teach those who want to be taught. They would have to be good listeners, and good listeners are now clearly in the minority. But time is of the essence. Don't waste your time on someone who talks about his recent trip to planet Neptune where a group of aliens had taken him last week in a pancake-shaped spaceship. Chances are that he will be on his next trip before you can figure out what makes him the only one who takes these trips in the first place.

Seriously though, we need remedies for the serious problems we experience. While no sweeping initiatives may be needed to become best-of-class, many of the prevailing mind-sets do have to change. Accountabilities must be defined and adhered to and many of the present reward systems must be adjusted. And as I wind up this book in an attempt to stay close to the stipulated page count, I do not want to leave you with mere generalities. Instead, I want to leave you with substantive, experience-based recommendations and reemphasize solutions that have worked flawlessly. These solutions are among many that nudged certain companies toward becoming best-of-class performers. It has always been clear to me that best-of-class companies embraced seven very distinctive performance steps. You might say these are steps that have been successfully implemented by smart companies. They are certainly steps that an upper manager can copy and duplicate and implement without any hesitation:

1. Share educational responsibilities; take seriously the grooming of successors. Pick people with potential and treat them well. They will no longer be uninformed.
2. Lead by example: Be intellectually curious and be very resourceful. The day still only has 24 hours—delegate! Again, the staffers will become informed!
3. Proclaim the end of reliability and maintenance being subservient to operations. Strongly advocate "switching hats" whereby the head of the maintenance department switches jobs with the head of operations and vice versa. The hat-switching will take place at random intervals; one or two years later, the two heads switch hats again. Note how both sides will be eager to be fully informed at all times! Observe how they will from then on communicate, cooperate with each other, and make decisions that benefit all parties.
4. Insist on specifications being developed for reliability, for demonstrably low life cycle cost. Ascertain that entire budgets are governed by reliability thinking.
5. Responsibility to look ahead and to analyze rests with reliability professionals. Reliability professionals are top technical employees who can explain why they *are*, or perhaps *are not*, doing certain things in certain ways. Do not allow them to be hidden somewhere in the general maintenance workforce.
6. Disallow operation that eats away at the safety margin of machines. Better and safer machines will be purchased if you change entire cultures.
7. Nurture absolute accountability by assigning project managers to live with their decisions. Let them operate the plant or process unit for three years and observe their performance. The results will be truly astonishing!

As you consider and follow through with these seven mandatory steps, make sure you always combine them with practicing the three Cs alluded to in step 3, above: communication, coöperation, and consideration. Whenever the three qualities CCC coexist on every level, an enterprise is headed for success. If one of the three CCC legs is short, the enterprise will be among the severely handicapped and could even altogether falter.

Glossary[1]

Accelerated life testing Testing to verify design reliability of machinery/equipment much sooner than if operating typically. This is intended especially for new technology, design changes, and ongoing development.

Acceptance test (qualification test) A test to determine machinery/equipment conformance to the qualification requirements in its equipment specifications.

Accessibility The amount of working space available around a component sufficient to diagnose, troubleshoot, and complete maintenance activities safely and effectively. Provision must be made for movement of necessary tools and equipment with consideration for human ergonomic limitations.

Allocation The process by which a top-level quantitative requirement is assigned to lower hardware items/subsystems in relation to system-level reliability and maintainability goals.

Assets The physical resources of a business, such as a plant facility, fleets, or their parts or components.

Asset management The systematic planning and control of a physical resource throughout its economic life.

Availability The probability that a system or piece of equipment will, when used under specified conditions, operate satisfactorily and effectively. Also, the percentage of time or number of occurrences for which a product will operate properly when called upon.

CBM See "Condition-based maintenance."

Changeout Remove a component or part and replace it with a new or rebuilt one.

CMMS Computerized maintenance management system.

Component A constituent part of an asset, usual modular or replaceable, that is serialized and interchangeable.

Concept Basic idea or generalization.

Condition-based maintenance Maintenance based on the measured condition of an asset.

Confidence limit An indication of the degree of confidence one can place in an estimate based on statistical data. Confidence limits are set by confidence coefficients. A confidence coefficient of 0.95, for instance, means that a given statement derived from statistical data will be right 95% of the time on the average.

Configuration The arrangement and contour of the physical and functional characteristics of systems, equipment, and related items of hardware or software; the shape of a thing at a given time. The specific parts used to construct a machine.

1. Note: Some of the definitions in this glossary were selected from MIL-STD-721C and the SAE publication Reliability and Maintenance Guidelines for Manufacturing Machinery and Equipment.

Corrective maintenance Unscheduled maintenance or repair actions, performed as a result of failures or deficiencies, to restore items to a specific condition. See also "Unscheduled maintenance" and "Repair."

Cost-effectiveness A measure of system effectiveness versus life cycle cost.

Critical Describes items especially important to product performance and more vital to operation than noncritical items.

Defect A condition that causes deviation from design or expected performance.

Dependability A measure of the degree to which an item is operable and capable of performing its required function at any (random) time during a specified mission profile given item availability at the start of the mission.

Discounted cash flow analysis A method of making investment decisions using the time value of money.

Distributions See "Probability distribution."

Downtime That portion of calendar time during which an item or piece of equipment is not able to perform its intended function fully.

Durability life (expected life) A measure of useful life, defining the number of operating hours (or cycles) until overhaul is expected or required.

Emergency maintenance Corrective, unscheduled repairs.

Engineering The profession in which knowledge of the mathematical and natural sciences is applied with judgment to develop ways to utilize economically the materials and forces of nature.

Environment The aggregate of all conditions influencing a product or service, or nearby equipment, actions of people, conditions of temperature, humidity, salt spray, acceleration, shock, vibration, radiation, and contaminants in the surrounding area.

Equipment All items of a durable nature capable of continuing or repetitive utilization by an individual or organization.

Exponential distribution A statistical distribution in logarithmic form that often describes the pattern of events over time.

Failure Inability to perform the basic function or to perform it within specified limits; malfunction.

Failure analysis The logical, systematic examination of an item or its design, to identify and analyze the probability, causes, and consequences of real or potential malfunction.

Failure effect The consequence of failure.

Failure mode The manner by which a failure is observed. Generally, a failure mode describes the way the failure occurs and its impact on equipment operation.

Failure mode effect analysis (FMEA) Identification and evaluation of what items are expected to fail and the resulting consequences of failure.

Failure rate The number of failures per unit measure of life (cycles, time, miles, events, and the like) as applicable for the item.

Fault tree analysis (FTA) A top-down approach to failure analysis starting with an undesirable event and determining all the ways it can happen.

FMEA See "Failure mode effect analysis."

FMECA Failure mode, effect, and criticality analysis—a logical progressive method used to understand the causes of failures and their subsequent effect on production, safety, cost, quality, etc.—see also "Failure mode effect analysis."

Function A separate and distinct action required to achieve a given objective, to be accomplished by the use of hardware, computer programs, personnel, facilities, procedural data, or a combination thereof or an operation a system must perform to fulfill its mission or reach its objective.

Hardware A physical object or physical objects, as distinguished from capability or function. A generic term dealing with physical items of equipment—tools, instruments, components, and parts—as opposed to funds, personnel, services, programs, and plans, which are termed "software."

Hazard function The instantaneous failure rate at time, *t*.

Infant mortality Early failures that exist until debugging eliminates faulty components, improper assemblies, and other user and manufacturer learning problems and until the failure rate lowers.

Item A generic term used to identify a specific entity under consideration. Items may be parts, components, assemblies, subassemblies, accessories, groups, equipment, or attachments.

Life cycle The series of phases or events that constitute the total existence of anything. The entire "cradle to grave" scenario of a product from the time concept planning is started until the product is finally discarded.

Life cycle cost All costs associated with the system life cycle, including research and development, production, operation, support, and termination.

Life units A measure of use duration applicable to the item (e.g., operating hours, cycles, distance, lots, coils, and pieces).

Maintainability The inherent characteristics of a design or installation that determine the ease, economy, safety, and accuracy with which maintenance actions can be performed. Also, the ability to restore a product to service or to perform preventive maintenance within required limits.

Maintainability testing Maintainability testing is used to demonstrate MTTR (mean time to repair). Once MTTR of a critical component is defined and the appropriate personnel trained in the proper procedure, we can test to investigate if the function can be performed in the stated MTTR. It should be stressed that this is not a test of the person's skills, but rather a test of the procedure and design of the equipment.

Maintenance Work performed to maintain machinery and equipment in its original operating condition to the extent possible; includes scheduled and unscheduled maintenance, but does not include minor construction or change work.

Management The effective, efficient, economical leadership of people and use of money, materials, time, and space to achieve predetermined objectives. It is a process of establishing and attaining objectives and carrying out responsibilities that include planning, organizing, directing, staffing, controlling, and evaluating.

Material All items used or needed in any business, industry, or operation as distinguished from personnel.

Mean time between failure (MTBF) The average time/distance/events a product delivers between breakdowns.

Mean time between maintenance (MTBM) The average time between both corrective and preventive actions.

Mean time between replacement (MTBR) Average use of an item between replacements due to malfunction or any other reason.

Mean time to repair (MTTR) The average time it takes to fix a failed item.

Median The quantity or value of an item in a series of quantities or values, so positioned in the series, that when arranged in order of numerical quantity or value, there are an equal number of values of greater magnitude and of lesser magnitude.

Mission profile A time-phased description of the events and environments an item experiences from initiation to completion of a specified mission, to include the criteria of mission success or critical failure.

Model Simulation of an event, process, or product physically, verbally, or mathematically.

Modification Change in configuration.

Normal Statistical distribution commonly described as a "bell curve." Mean, mode, and median are the same in the normal distribution.

MTBR See "Mean time between repair."

MTTR See "Mean time to repair."

On-condition maintenance Inspection of characteristics that will warn of pending failure and performance of preventive maintenance after the warning threshold but before total failure.

Operating time Time during which equipment is performing in a manner acceptable to the operator.

Overhaul A comprehensive inspection and restoration of machinery/equipment, or one of its major parts, to an acceptable condition at a durability time or usage limit.

Predictive and preventive (scheduled and planned) maintenance All actions performed in an attempt to retain a machine in specified condition by providing systematic inspection, detection, and prevention of incipient failures.

Predictive maintenance Predictive maintenance is a method that involves a minimum of intervention. In its simplest form, it is based on the old adage "don't touch, just look." In the context of process machinery, predictive maintenance is practiced through machinery health monitoring methods such as vibration and performance analysis.

Preventive maintenance (PM) Actions performed in an attempt to keep an item in a specified operating condition by means of systematic periodic inspection, detection, and prevention of incipient failure. See also "Scheduled maintenance."

Proactive A style of initiative that is anticipatory and planned for.

Probability distribution Whenever there is an event E that may have outcomes E_1, E_2, ..., E_n, whose probabilities of occurrence are p_1, p_2, ..., p_n, one speaks of the set of probability numbers as the pd (probability density) associated with the various ways in which the event may occur. The word "probability distribution" refers therefore to the way in which the available supply of probability, that is, unity, is "distributed" over the various things that may happen.

Production A term used to designate manufacturing or fabrication in an organized enterprise.

Random Any change whose occurrence is not predictable with respect to time or events.

Re-rating Alteration of a machine, a system, or a function by redesign or review for change in performance, mostly, but not always, for increased capacity.

RCFA See "Root cause failure analysis."

RCM See "Reliability-centered maintenance."

Rebuild/recondition Total teardown and reassembly of a product, usually to the latest configuration. See also "Revamp."

Redundance (redundancy) Two or more parts, components, or systems joined functionally so that if one fails, some or all of the remaining components are capable of continuing with function accomplishment; fail-safe; backup.

Refurbish Clean and replace worn parts on a selective basis to make the product usable to a customer. Less involved than rebuild.

Reliability (R) The probability that an item will perform its intended function without failure for a specified time period under specified conditions.

Reliability-centered maintenance Optimizing maintenance intervention and tactics to meet predetermined reliability goals.

Reliability growth Machine reliability improvement as a result of identifying and eliminating machinery or equipment failure causes during machine testing and operation.

Reliability modeling A model that uses individual component reliabilities to define reliability of a subsystem. Allows for analysis of parallel versus series systems and defines low-reliability components of a subsystem.

Reliability testing Reliability testing is used to demonstrate MTBF (mean time between failure). Once the MTBF of a critical component is defined, a test can be performed (with a measure of confidence) to demonstrate this MTBF. A measure of confidence is built into statistically designed test plans, guaranteeing that if the MTBF requirement has not been achieved, there is a low probability that the test will be passed.

Repair The restoration or replacement of components of facilities or equipment as necessitated by wear, tear, damage, or failure to return the facility or equipment to efficient operating condition.

Repair parts Individual parts or assemblies required for the maintenance or repair of equipment, systems, or spares. Such repair parts may be repairable or nonrepairable assemblies or one-piece items. Consumable supplies used in maintenance, such as wiping rags, solvent, and lubricants, are not considered repair parts.

Repairable item Durable item determined by application of engineering, economic, and other factors to be restorable to serviceable condition through regular repair procedures.

Replaceable item Hardware that is functionally interchangeable with another item but differs physically from the original part to the extent that installation of the replacement requires such operations as drilling, reaming, cutting, filing, and shimming in addition to normal attachment or installation operations.

Return on capital employed ROCE.

Return on net assets RONA.

Revamp Change as to upgrade or modernize.

ROCE Return on capital employed.

RONA Return on net assets.

Root cause failure analysis A formalized systematic approach effort to determine the underlying cause of a failure. This effort is generally separate from repair activities but should be part of the repair cycle. It usually entails a detailed technical analysis of the failure mode by a team of experts.

Safety Elimination of hazardous conditions that could cause injury. Protection against failure, breakage, and accident.

Scheduled maintenance Preplanned actions performed to keep an item in specified operating condition by means of systematic inspection, detection, and prevention of incipient failure. Sometimes called preventive maintenance, but actually a subset of PM.

Scheduled (planned) downtime The elapsed time that the machine is down for scheduled maintenance or turned off for other reasons.

Spares Components, assemblies, and equipment that are completely interchangeable with like items and can be used to replace items removed during maintenance.

Specifications Documents or verbal communication that clearly and accurately describe the essential technical requirements for materials, items, equipment, systems, or services, including the procedures by which it will be determined that the requirements have been met. Documents may include performance, support, preservation, packaging, packing, and marking requirements.

Standards Established or accepted rules, models, or criteria by which the degree of user satisfaction of a product or an act is determined or against which comparisons are made.

Standard deviation A measure of average dispersion or departure from the mean of numbers, computed as the square root of the average of the squares of the differences between the numbers and their arithmetic mean. It is also a measure of uncertainty when applied to probability density distribution.

Standard item An item for common use described accurately by a standard document or drawing.

Standby Assets installed or available but not in use.

Surveillability A qualitative factor influencing reliability. It contains such considerations as accessibility for surveillance and monitoring of a machine or its function(s).

System Assembly of correlated hardware, software, methods, procedures, and people, or any combination of these, all arranged or ordered toward a common objective.

Time The universal measure of duration.

Time to repair (TTR) Total clock time from the occurrence of failure of a component or system to the time when the component or system is restored to service (i.e., capable of producing good parts or performing operations within acceptable limits). Typical elements of repair time are diagnostic time, troubleshooting time, waiting time for spare parts, replacement/fixing of broken parts, testing time, and restoring.

Total downtime The elapsed time during which a machine is not capable of operating to specifications.Total downtime = scheduled downtime + unscheduled downtime.

Total productive maintenance Company-wide equipment management program emphasizing operator involvement in equipment maintenance and continuous improvement in equipment effectiveness.

TPM See "Total productive maintenance."

Training The pragmatic approach to supplementing education with particular knowledge and assistance in developing special skills, helping people to learn to practice an art, science, trade, profession, or related activity. Basically more specialized than education and involves learning what to do rather than why it is done.

Troubleshooting Locating or isolating and identifying discrepancies or malfunctions of equipment and determining the corrective action required.

Unscheduled (unplanned) downtime The elapsed time that the machine is incapable of operating to specifications because of unanticipated breakdowns.

Unscheduled maintenance (UM) Emergency maintenance (EM) or corrective maintenance (CM) to restore a failed item to usable condition. Often referred to as breakdown maintenance.

Useful life The number of life units from manufacture to when the item has an unrepairable failure or unacceptable failure rate.

Up In a condition suitable for use.

Uptime The capacity to produce and provide goods and services.

Utilization factor Use or availability.

Warranty Guarantee that an item will perform as specified for at least a specified time.

Wear-out The process that results in an increase of the failure rate or probability of failure with increasing number of life units.

Appendix I

CPRS Assessment Scoring Matrix

Once an assessment is made, it is important that each surveyed category is measured. Follow-up is needed to ensure that conformance criteria are met. Any categories found to be unacceptable need to be revisited, and changes made to bring them up to either full conformance or a defined level of acceptability. A scoring matrix will help.

Needless to say, the scoring matrix can be expanded to include other items of interest. The "comments" segment lends itself to cataloging highly detailed information.

Category	Does Not Meet Requirements	Meets Requirements	Exceeds Requirements	Comments
Human assets and experience				
Customer base and satisfaction				
Safety, environment, capability				
Scheduling system, visual flow				
Quality and quality systems				
Documentation management				
Outside/procured services				

Appendix II

The Essential Reliability Library

H.P. Bloch	Improving Machinery Reliability, third ed., Gulf Publishing Company, Houston, TX, 1998. ISBN 0-88415-661-3
H.P. Bloch, F. Geitner	Machinery Failure Analysis and Troubleshooting, fourth ed., Elsevier Publishing, Oxford, UK and Waltham, MA, 2012, ISBN 978-0-12-386045-3
H.P. Bloch, F. Geitner	Machinery Component Maintenance and Repair, third ed., Gulf Publishing Company, Houston, TX, 2004. ISBN 0-87201-781-8
H.P. Bloch, F. Geitner:	Major Process Equipment Maintenance and Repair, second ed., Gulf Publishing Company, Houston, TX, 1997. ISBN 0-88415-663-X
H.P. Bloch, A. Shamim	Oil Mist Handbook: Practical Applications, Fairmont Press, Lilburn, GA, 1998. ISBN 0-88173-256-7
H.P. Bloch, C. Soares	Process Plant Machinery, second ed., Elsevier Publishing, New York, NY, 1998. ISBN 0-7506-7081-9
H.P. Bloch, J.J. Hoefner	Reciprocating Compressors: Operation and Maintenance, Gulf Publishing Company, Houston, TX, 1996. ISBN 0-88415-525-0
H.P. Bloch	Practical Guide to Compressor Technology, second ed., John Wiley & Sons, Hoboken, NJ, 2006. ISBN 0-471-72793-8 (first ed. available also in Spanish language from McGraw-Hill)
H.P. Bloch, M. Singh	Practical Guide to Steam Turbine Technology, second ed., New York, NY, 1995, 2009, ISBN 978-0-07-150821-6 (first ed. available also in Spanish language from McGraw-Hill)
H.P. Bloch	Practical Lubrication for Industrial Facilities, second ed., Fairmont Press, Lilburn, GA, 2000, 2009. ISBN 0-88173-579-5

(Continued)

H.P. Bloch, C. Soares	Turboexpanders and Process Applications, Elsevier Publishing, New York, NY, 2001. ISBN 0-88415-509-9
H.P. Bloch	Pump Wisdom: Problem Solving for Operators and Specialists, John Wiley & Sons, Hoboken, NJ, 2011. ISBN 9-781118-041239
H.P. Bloch, A. Budris	Pump User's Handbook: Life Extension, fourth ed., Fairmont Press, Lilburn, GA, 2004, 2006, 2013. ISBN 0-88173-720-8
H.P. Bloch, F. Geitner	Maximizing Machinery Uptime, Gulf Publishing Company, Houston, TX, 2006. ISBN 0-7506-7725-2
H.P. Bloch, A. Godse	Compressors and Modern Process Applications, John Wiley & Sons, Hoboken, NJ, 2006. ISBN 0-471-72792-X
H P. Bloch, F. Geitner	Compressors: How to Achieve High Reliability and Availability, McGraw-Hill Companies, New York, NY, 2012. ISBN 978-0-07-177287-7
K.P. Bloch	Rethinking Bhopal: A Definitive Guide to Investigating, Preventing, and Learning from Industrial Disasters, Elsevier Publishing, New York, NY, 2016, ISBN 9780128037782
M.P. Boyce	Gas Turbine Engineering Handbook, Gulf Publishing Company, Houston, TX, 1982. ISBN 0-87201-878-4
R.N. Brown	Compressors, Selection and Sizing, second ed., Gulf Publishing Company, Houston, TX, 1997. ISBN 0-88415-164-6
J.W. Dufour, J.E. Nelson	Centrifugal Pump Sourcebook, McGraw-Hill, New York, NY, 1993. ISBN 0-07-018033-4
R.C. Eisenmann Sr., R.C. Eisenmann Jr.	Machinery Malfunction Diagnosis and Correction, 1997. ISBN 0-13-240946-1[a]
W.E. Forsthoffer	Forsthoffer's Rotating Equipment Handbooks (set of 5), Elsevier Publishing, New York, NY, 2006. ISBN 1-856-17472-7
T.L. Henshaw	Reciprocating Pumps, Van Nostrand Reinhold Publishing Company, New York, NY, 1987. ISBN 0-442-23251-9
I.J. Karassik, et al.	Pump Handbook, second ed., McGraw-Hill, New York, NY, 1986. ISBN 0-07-033302-5
V.S. Lobanoff, R.R. Ross	Centrifugal Pumps: Design and Application, second ed., Gulf Publishing Company, Houston, TX, 1992. ISBN 0-87201-200-X

(Continued)

J.S. Mitchell	Physical Asset Management Handbook, fourth ed., Reliabilityweb.com, Fort Myers, FL, 2012, ISBN 9780985361938
R. Neumaier	Hermetic Pumps, second ed., Fairmont Press, Lilburn, GA, 2000. ISBN 3-929682-26-5
M. Paradies, L. Unger	TapRooT®: The System for Root Cause Analysis, Problem Investigation, and Proactive Improvement, System Improvements, Knoxville, TN. ISBN 1-893130-02-9 (Copyright 2000, also later editions)
A. Sofronas	Analytical Troubleshooting of Process Machinery and Pressure Vessels: Including Real-World Case Studies, John Wiley & Sons, Hoboken, NJ, 2006. ISBN 0-471-73211-7
A. Sofronas	Case Histories in Vibration Analysis and Metal Fatigue for the Practicing Engineer, John Wiley & Sons, Hoboken, NJ, 2012. ISBN 978-1-118-16946-9

[a]Out-of-print, but generally available from Amazon.com.

Index

Note: Page numbers followed by *f* indicate figures and *t* indicate tables.

D

N

O

P

S

U

V

W